Applied Animal Ethics

Applied Animal Ethics

Leland S. Shapiro

Africa • Australia • Canada • Denmark • Japan • Mexico • New Zealand • Philippines
Puerto Rico • Singapore • Spain • United Kingdom • United States

Delmar Staff:
Publisher: Susan Simpfenderfer
Executive Marketing Manager: Donna Lewis
Executive Editor: Marlene McHugh Pratt
Executive Production Manager: Wendy Troeger
Developmental Editor: Andrea Edwards Myers
Production Editor: Carolyn Miller

Printed in the United States of America
2 3 4 5 6 7 8 9 10 XXX 04 03 02 01 00 99

For more information, contact Delmar, 3 Columbia Circle, PO Box 15015, Albany, NY 12212-0515; or find us on the World Wide Web at http://www.delmar.com

Library of Congress Cataloging-in-Publication Data
Shapiro, Leland S.
Applied animal ethics / by Leland S. Shapiro.
p. cm.
Includes index.
ISBN 0-8273-8494-7
1. Animal rights—Moral and ethical aspects. 2. Animal welfare—Moral and ethical aspects. 3. Environmental ethics. 4.
-animal relationships—Moral and ethical aspects. I. Titl
HV4708.S52 1999
179'.3—dc21

Table of Contents

Preface

The purpose of this book is not to change beliefs, but to induce tolerance of different and often emotionally charged opposing views. As a hopeful dividend, the reading and teaching of this text may also result in a more humane treatment of both animals and humans.

Applied Animal Ethics will instruct a serious student in the technique of gathering data, reading and creating statistical diagrams and charts, researching not only one's preconceived notions, but most importantly, the calm dissection of what will probably be disturbing alternative beliefs and interpretations of data. Statements commonly held to be sacrosanct will be critically analyzed for scientific proof. Even pictures and other illustrations will be examined as to relevance and authority (e.g., Was a picture "staged"? When and where was it taken? By whom and for what purpose?)

Readers of this text may be disappointed by not finding a definitive statement by the author on "ethics" (e.g., whether animals should or should not be used in research; what animals should be eaten by man, if any; how such animals should be slaughtered and under what conditions, etc.) This is for the reader to decide. *Applied Animal Ethics* does not pontificate for Hinduism or Jewish Orthodoxy, for the American Medical Association or for the American Fund for Alternatives to Animal Research. It is for the reader to form his or her own ethical standards. Probably very few students will change their views after reading this book and its suggested avenues for further research. This is all well and good. What will be derived from this text is an understanding and acceptance of the fact that all groups dealing with the various phases of the treatment of animals (and humans) are driven by humane interests even if their end goals may be entirely opposite of one another.

This text combines a basic background in theoretical ethics with the applications of controversial philosophies. It is in the application of beliefs that we can really understand the full implications of what some might call "dogma", but which to others might be considered "revealed truths." *Applied Animal Ethics* presupposes that we are and will continue to use animals in research, to slaughter, and to eat. However a full discussion of what is humane, what is abuse, what is extreme behavior will be encouraged, the vegetarian and the meat-eater, the anti-vivisectionist and the researcher—all will have their full say. Perhaps only in America could such a text be published, and a course like this be taught, and passions calmed for an intelligent and productive airing of ideas and philosophies. Indeed, it is in the market place of ideas that truths (not truth) can be presented. And only in America, and perhaps only now can we begin to understand and appreciate that all viewpoints have validity and are worthy of presentation and attention.

One man's faith allows him to eat everything, but another man,....eats only vegetables. The man who eats everything must not look down on him who does not. And the man who does not eat everything, must not condemn the man who does. (Romans 14:2-3)

Acknowledgments

I want to especially thank the animal caretakers and scientists at Amgen, Inc. for allowing me to intern through their research facility. The experience gave me first-hand knowledge of "typical" modern day animal research. Dr. Barbara Orlans at Georgetown University, was particularly helpful in gathering information, films, and contacts following a course at the Kennedy Institute of Ethics. Dr. Orlans text, *In the Name of Science,* is a recommended text for the area of animal ethics and will be referred to throughout this text. Dr. Temple Grandin, an internationally recognized expert on the humane handling and slaughter of food animals, provided me with reprint articles and designs of her own for our section on farm animals. Most of the section on the basic legal system affecting animal welfare was researched by Robyn Abrams, attorney at law. Dr. Jeff Chan, Reverend Ito, and Chief Phil Stevens, provided insight as to the Buddist and Native American cultural beliefs on animal welfare. Dr. David M. Coffey, Agricultural Education and Rural Sociology specialist at the Western Kentucky University, Dr. Sally Walshaw at Michigan State University, Dr. Gary Comstock at Iowa State University, and Drs. Hugh LaFollette and Tony DeLucia at East Tennessee State University provided copies of their course syllabi, from which I borrowed liberally for my first course in Animal Ethics at Pierce College. Many of their ideas have been incorporated into this text.

Most of all, I want to thank my friend the dairy cow. This may seem a bit strange for a professor to do. However, the close relationship I had (for 19 years) as a dairy farmer gave me a deeper appreciation for the feelings, thoughts, and needs of animals. The dairy cow responds to kindness by producing more milk, better conception rates, and healthier udders. The kinder, gentler and cleaner one is, the more profitable the dairy business becomes. Too bad all businesses can't be run in such a fashion. Or can they??

Several individuals at Delmar provided invaluable assistance and encouragement in guiding me through the writing and editing of this book. I would like to thank Carolyn Miller, Andrea Myers, Zina Lawrence, Wendy Troeger, and Cathy Esperti. In addition, my children, Ilana and Aaron have been very patient while allowing me the time away from them to research and write *Applied Animal Ethics,* and I am truly most appreciative.

The author and Delmar would like to thank those individuals who reviewed the manuscript and offered suggestions, feedback, and assistance. Their work is greatly appreciated.

Dan Brown, Cornell University, Ithaca, New York; David S. Buchanan, Oklahoma State University, Stillwater, Oklahoma; Janet L. Romich, DVM, MS, Madison Area Technical College, Madison, Wisconsin; Ronnie Silcox, University of Georgia, Athens, Georgia.

About the Author

Dr. Leland S. Shapiro is the director of the pre-veterinary science program at Los Angeles Pierce College. He has been a professor of animal sciences for nearly 24 years and is a member of the American Dairy Science Association; Dairy Shrine Club; Gamma, Sigma, Delta Honor Society of Agriculture; and Association of Veterinary Technician Educators, Inc. Professor Shapiro is a member of the college's ethics committee and has completed two post-doctoral studies in Bio-Ethics. Dr. Shapiro is a UC Davis "Mentor of Veterinary Medicine," and the recipient of several local college teaching awards as well as the prestigious NISOD Excellence in Teaching Award, in Austin Texas.

SECTION 1

Applied Animal Ethics

CHAPTER 1

Why Study Animal Ethics

A course in applied animal ethics can have many objectives, and the objectives for applied versus theoretical courses will vary considerably. *Applied Animal Ethics* considers the following five major objectives as basic to such a course.

Objectives

The student should:

1. Analyze, observe, and interpret philosophical differences to acquire a better understanding of self and others, their viewpoints, and the actions of others.
2. Use critical-thinking skills to determine philosophical thoughts of self and others.
3. Create and modify understanding of a concept by organizing information from differing viewpoints.
4. Construct meaning from published materials through reading, observing, discussing, and writing.
5. Demonstrate an open mind to alternative perspectives.

Those who do not know their opponents' arguments do not completely understand their own.

(Bender and Leone, *Religion in America: Opposing Viewpoints*, 1989)

Good laboratory animal care and good science are inseparable. Scientists cannot depend on the results of their experiments if research animals are not healthy. Stressed animals experience changes in their internal environment, which possibly changes the true value of many experimental results. The biomedical research community has and must continue to condemn cruelty to animals in any form and for any purpose.

Laboratory animals are entitled to the best possible care that can be provided. Humane ethics demand it, the Animal Welfare Act decrees it, and good science defines it.

(Dr. F. Barbara Orlans, Kennedy Institute of Ethics, summer 1994)

Dr. Orlans' description of ethical treatment of animals is what most Americans perceive to be "animal ethics." Animal ethics, however, is not limited to research facilities. A course in applied animal ethics is extremely important for people entering the veterinary field. At times, veterinary technicians and the veterinarian must justify procedures necessary for the improvement of animal health and well-being.

In addition, such a course can be beneficial to the horse, dog, and cat owner, to the rodeo athlete, to those who eat meat, and to those who do not eat meat. Too many individuals in our society either do not care about the pain and suffering of others, do not care about what their neighbor thinks or feels, or simply have decided that their viewpoint is the only one that matters. Some are willing to do almost anything to get their viewpoint across (including arson, murder, and other criminal activity). The latter actions, of course, should never be tolerated and must be condemned by those who truly believe in a democratic society. Methods of disagreeing are important, but hopefully most individuals will eventually come to some agreement on ethical issues that society finds acceptable.

For close to 400 years, animals have been utilized in research laboratories to help scientists acquire new knowledge about the function and malfunction of the human and animal bodies. For almost as long, there has been a movement to prevent such use of animals. Some insist that animals have an equal moral standing with humans, while others claim that the animal models do not adequately and/or accurately measure human responses to medicaments, toxins, or other biologic hazards. Even if it is true that animal research saves human lives, animal rights activists argue that this fact is irrelevant. We could also save lives by conducting forcible experiments on humans, but we do not because these acts would be immoral. The same idea, they claim, applies to animals: Using animals for research is just as immoral as using humans for research. Michael Fox, Humane Society of the United States (HSUS) vice-president, wrote in his book *The Inhumane Society* that "The life of an ant and that of my child should be granted equal consideration" (see Figure 1–1).

In addition, many animal rights organizations espouse vegetarianism—not for health reasons, but for moralistic beliefs. Because of pressures put on legislative bodies by the various animal welfare/rights groups and extremist movements, ethics courses are being suggested as requirements for any individual or institution that uses animals in education or research. Unless the research, educational, and agricultural communities create their own ethics

"Well, I guess I just have to accept that they both deserve equal consideration."

Figure 1–1

course program, one will be provided for them by others who might not completely understand the necessities of their work.

Animal scientists, and those involved with animal agriculture, need to be able to justify to the non-scientific, non-agricultural community the need to use animals in their respective areas of employment. These needs will have to be defended with facts, because many individuals might only see the emotional aspect of the situation. Individual ethics or morals are biased by our religious, cultural, and professional surroundings. Courses taught in animal ethics are therefore likely to be biased one way or the other, as well.

Most of the current animal ethics courses being taught were created and are run by philosophers. Philosophy is simply the critical study of fundamental beliefs. No "correct" philosophy exists. In the primary and secondary schools of California, however, there seems to be a growing one-sided presentation of animal ethics.

Although quintessentially California schools have taught traditional courses in the life sciences that included a limited dissection of cadavers, a changing bias has been occurring. According to biology faculty and various animal research organizations, millions of dollars have been funneled into our public schools—indirectly and directly—by animal rights organizations. These organizations generally present what many describe as "only one side of this volatile equation."

In California, some examples of these one-sided presentations include the following situations:[1] At Hosler Junior High School in Lynwood, California, Darrell Jacobson's seventh-grade class collects money for People for the Ethical Treatment of Animals (PETA). The class is taught that *all* animal testing is cruel and inhumane. Mr. Jacobson has his students write letters to companies on PETA's list of companies performing animal testing. Boston-based Gillette has received many such letters from Jacobson's students (and others). Jacobson shows his seventh graders videos of animal vivisection, de-beaking chickens, and laboratory chimpanzees infected with the HIV virus. At Wilbur Avenue School in Tarzana, California, Rhea Damon's third-grade students are encouraged to write letters to various agencies protesting underwater acoustical testing that could affect whales and dolphins. Ervis Tsakirides' history and Latin students at the private Crossroads High School in Santa Monica are taught that those who empathize with animals will be more sensitive to issues such as world hunger, gay rights, and women's rights (see Chapter 3 on Nazi Germany). Mr. Tsakirides weaves animal rights issues into all his classes, including Latin. Pro-animal teaching tools are provided by such organizations as PETA and Last Chance for Animals (a non-profit group in the San Fernando Valley).

Even at the college level, professors outside the life sciences departments are giving only one side of the story in their courses. At El Camino College in the South Bay (south of Los Angeles), Ms. Diana Crossman, a speech instructor, encourages her students to give speeches *against* animal testing and *promoting vegetarianism*. The other side of this controversial matter is never presented. A debate or discussion with varying viewpoints rarely occurs; rather, there is simply an espousal of one side.

Animal rights organizations counter that there was never debate or discussion in previous years, when biology instructors required that all students participate in animal dissection. And, in animal science classes, they say that vegetarian arguments were rarely discussed. Most schools today offer alternative assignments for those students not wishing to participate in these activities due to moral or religious objections. Most students going through America's public school systems, however, do not take animal science or animal anatomy courses involving dissection. Thus, their only exposure to this argument might be from the popular press—or from the "one-sided presentation" claims animal scientists.

The opposition to animal research is nothing new. In 1910, Dr. William Welch warned the American Medical Association (AMA) House of Delegates about individuals and groups seeking to stop biomedical research:

> ***"Agitation for the prohibition of experiments on animals conducted under the guise of a humane purpose is fundamentally inhumane; for if it were to succeed, the best hopes of humanity for further escape from physical suffering and disease would be destroyed."***

As of 1997, 20 million animals are used annually in research. This number is down 40 percent since 1968. Rodents (rats and mice) specifically bred for research make up 90–95 percent of research animals in the United States. From 1–1.5 percent of research animals are dogs and cats, and approximately 0.5 percent are non-human primates. The aforementioned animals and their use in research will be the primary focus of discussion in this text.

In addition, discussion of animal rights and welfare will include animals used for food production, for companion animals, for sport, and/or for exhibition and work. Approximately 1.3 billion cattle, 1.2 billion sheep, 526 million goats, 140 million buffalo, 19 million camels, 13 million yaks, 13 million llamas, 11.3 billion chickens, 60 million horses, 527 million ducks, 43 million donkeys, 15 million mules, and 864 million swine in the world are used for either food production, work, or sport.

Domestically (within the United States), farm and other domestic animals number as follows:[2]

Cattle . . . . 100 million (including 10 million dairy cows)
Chickens . . . . 6.4 billion broilers (281 million laying hens)
Turkeys. . . . . 285 million
Swine . . . . 57 million
Sheep . . . . 11 million
Horses . . . . 5 million
Goats . . . . 2 million
Ducks and geese. . . . 18.4 million slaughtered/year
Cats[3]. . . . . 60 million
Dogs[3] . . . . 57 million
Pocket pets[3] . . . . 12.3 million (rabbits, guinea pigs, hamsters, gerbils, and hedgehogs)

Fish[3] . . . . 12 million fish tanks
Birds. . . . . 8 million[3]–12 million[4]
Reptiles[3] . . . . 7.3 million
Ferrets[3] . . . . 7 million

Ten million Americans belong to animal protection/rights groups. More than 7,000 animal protection/rights groups exist in the United States, with a combined budget of about $200 million. In 1990, PETA boasted a membership of 250,000 and yearly revenues of $9.2 million. Today, it is estimated that PETA has 400,000 members and an annual budget of $13.4 million. PETA is also a multimedia conglomerate that publishes books and magazines, organizes animal rights rock festivals, and produces dozens of videotapes and public service announcements. PETA has had support from celebrities such as Mike Farrell, the late River Phoenix, Paul McCartney, the late Linda McCartney, and k.d. lang. Besides this large number of animal sympathizers, some animal rightists have been conducting terrorist attacks against the research community. Bombings, arson, physical assaults, death threats, vandalism, the release and/or stealing of laboratory animals, and destruction of research equipment and data has markedly increased. According to the United States Department of Justice, there have been at least 313 instances of animal rights violence in the United States.

More than 100 animal rights bills are introduced in state legislatures across the United States every year. Medical schools spend tens of millions of dollars on increased security, litigation, and public education to defend their research. Amendments to the Animal Welfare Act, many of which were passed without public hearings, were pressed for by animal activists and will cost the research community $291 million in capital improvements and $72 million a year to administer, according to the United States Department of Agriculture (USDA). Researchers feel the cost will be much higher. Animal rights organizations claim this expense is only a beginning.

PETA and other groups continue to show photos and documentation of animal cruelty at research institutions, farms, and at family homes with pets (see PETA photos later in the book). In contrast, researchers stress that since 1901, 57 Nobel Prizes in physiology and medicine have been awarded for research involving animals. This research involving animals has led to drugs for treating heart disease, diabetes, and cancer—and has given us aspirin and vaccines against polio, tetanus, diphtheria, smallpox, and whooping cough. In 1997, Dr. Scott J. Hultgren of the Washington University School of Medicine in St. Louis, Missouri, announced research using primates in the development of a new vaccine for the prevention of cystitis in women. Primates are being used because of the potential side effects of the vaccine. PETA and other animal rights groups object to experimenting with these potential side effects on animals, claiming that this action is immoral (for the same reason it would be immoral to test the vaccine on humans).

Because of the controversy listed earlier, a course in applied animal ethics is necessary for all individuals planning to use or currently using animals as part of their livelihood. People of varying viewpoints need to be able to sit down and discuss their concerns without violence or the threat thereof.

DEFINITIONS

While most texts include a brief glossary at the end of the book, a definitions section has been incorporated at the beginning of this text. When speaking any language, we should understand what the other person is trying to communicate so we do not take offense for something not really said or meant. For example, in an attempt to communicate using American Sign Language, a learned professor called a young interpreter a prostitute instead of telling her to smile (as he had intended). This situation occurred because of a simple misreading of a hand gesture and was embarrassing and potentially volatile for an instructor. On the streets of Los Angeles, many innocent individuals have died because they were targets of rival gangs. Misreading of hand gestures (thinking the individual was a member of a rival gang) was one of the excuses given for the justification of these actions.

Language and definitions are extremely important in any discussion. So we can all communicate on the same level of understanding, a common set of terms must be defined or described. Some of these definitions might differ with what you have been taught. Within this text, however, the following definitions will apply:

AAZPA—American Association of Zoological Parks and Aquariums

Abolitionists—Individuals who believe that animals have similar rights as humans. As such, humans must consider whether it is right to use animals for any purpose including food, clothing, entertainment, and research projects. This term is sometimes purposely confused with anti-slavery issues.

Abuse—Physical actions that willfully harm an animal

Analgesia—A state of insensibility to pain, without loss of consciousness

Anesthesia—A state of lack of awareness or sensitivity, with or without loss of consciousness

ALF—Animal Liberation Front

Animal Protectionist—A person who believes animals should be protected from undue suffering, such as hunting, factory farming, nonbiomedical animal research, fur farming, and any animal research that is not essential for saving human and other animal life (this definition, by itself, is controversial).

Animal Rights—A movement that insists animals have moral rights equal to those of humans and is totally opposed to biomedical research using animals, sporting events using animals, using animals for clothing or entertainment, product testing, and the eating of animals.

Animal Welfare—A movement that believes a reduced and minimal number of animals should be used in research—and that those animals used should be treated as humanely as possible. This concept includes proper housing, disease prevention, nutrition, and humane euthanasia or slaughter. This concept implies that humankind has dominion (a power or right) over animals, and as such has responsibility for animal well-being.

Anthropomorphism—The concept of ascribing human traits to animals, gods, etc.

Carcinogenicity—The ability of a substance to cause cancer

Cruelty—Having or showing indifference to, or pleasure in, another's pain or suffering

Culture—The growth of living cells or microorganisms within a controlled, artificial environment.

Cytotoxicity—A substance's capacity to kill or damage cells

Delaney Clause—"No additive shall be deemed to be safe if it is found to induce cancer when ingested by man or animal, or if it is found, after tests which are appropriate for the evaluation of the safety of food additives, to induce cancer in man or animal."

de minimis standard—An amount so small as to be insignificant

Deprivation—Loss of a desired thing, implied cruelty such as limiting an animal's freedom or association with others of its kind, something missing from the animal's environment, and the animals being "stressed," "bored," or "unhappy."

Dissection—To cut apart for purposes of scientific examination (usually refers to either animal or human cadavers)

Domestication—Where taming from the natural environment and breeding, caring, and managing is under the control of humans

Dominionists—Individuals who believe that people can do whatever they want to animals, and humans are the only species with rights. When this interpretation includes neglect and abuse, it is usually deemed socially unacceptable.

Euthanasia—The humane killing of an animal by a method which produces rapid unconsciousness and subsequent death without evidence of pain or distress, or a method which utilizes anesthesia produced by an agent that causes painless loss of consciousness and subsequent death.

Exploitation—Humans having absolute dominion over animals and using the animals as they see fit (animals may be used or abused).

Ethology—The study of behavior of animals.

Humane—Espouses of kind treatment and compassion for humans and other animals

Inherent Value—The idea that since animals are "subjects of life," they have a value or worth that is innate (Tom Regan's inherent value). The subject of a life that is meaningful to that being is also part of this concept.

ILAR—Institute of Laboratory Animal Resources

In situ—In the original place

In vitro—In an artificial environment outside the living organism

In vivo—Within a living organism

Liberation—The concept that animals are not to be put to work in any way, and all use of animals should be eliminated. Animal liberationists have been known to break into research labs and set animals free.

Malthusianism—The belief that population tends to increase faster than food supply, with inevitably disastrous results, unless the increase in population can be checked by war, famine, pestilence and natural catastrophe (named after Thomas Malthus).

Morals and Morality—What an individual or a society believes to be right or wrong

NOEL (No Observed Effect Level)—During the process of toxicity testing on animals, or occasionally on humans, various doses of a pesticide are tested. The dose at which no effect of the type under observation is observed is called the NOEL.

Pain—An unpleasant sensation occurring in varying degrees of severity as a consequence of injury, disease, or emotional disorder

PETA—People for the Ethical Treatment of Animals

Replaceability Argument—Philosophy of activist Peter Singer that states, "Given that an animal belongs to a species incapable of self-consciousness, it follows that it is not wrong to rear and kill it for food, provided that it lives a pleasant life and, after being killed, will be replaced by another animal, which will lead a similarly pleasant life and would have not existed if the first animal had been killed."

Risk Assessment—Compares the harmful effects of an action with the potential benefits

Serendipity—Making a discovery that was unanticipated

Speciesism—The belief that humans are superior to animals, which is a bias similar to racism. Speciesism is a prejudice or bias toward the interests of members of one's own species and against the members of others species.

Standing—One's place or relative position, or one's rank in the community. Moral Standing is the status, rank, or right to do something.

Suffer—To feel or endure pain or distress and to appear at a disadvantage

Suffering—Suffering can result from intolerable emotional pain, as well as from intolerable discomfort. Suffering requires a state of consciousness and a functional cerebral cortex—an emotional state. Pain and/or discomfort causes suffering only when pain and/or distress is at a degree that an individual animal cannot tolerate. The ability to tolerate pain or discomfort will vary widely from one individual to another—and even within a given individual from time-to-time.

Teratogenicity—The capability of a substance, organism, or drug to cause malformations in an animal or human fetus

Transfection—The introduction of foreign DNA into a cell's genome

Transgenic—An animal in which the genetic makeup has been modified by the addition of a DNA sequence from another species

Utilitarian Philosophy—The right action is that what benefits most individuals. Using animals for the betterment of people is acceptable, as long as the animal is treated with compassion.

Vivisection—Cutting into a live animal. Originally, the term referred to cutting the animal without anesthesia (because anesthesia had not yet been developed). Today, the term is generally used to refer to all types of animal experimentation, including research and testing on living animals for medical or scientific research.

Xenotransplantation—The transplantation of animal organs into humans

Chapter 1
EVALUATION

1. What is the difference in the following statements?
   a. Animals have the right to be treated humanely.
   b. Animals have no rights, but humans have an obligation to treat them humanely.
2. A puppy is often treated as an equal family member in many homes. If the house were burning, however, and a baby brother or sister were endangered as well as the puppy, which one would you choose to rescue first? If you had only enough time to rescue one, would you be morally wrong to rescue the puppy?
3. Debate animal rights versus animal welfare philosophies from the following occupational perspectives (should there be a difference?):
   a. Acquired Immune Deficiency Syndrome (AIDS) researcher
   b. Cattle rancher
   c. Caretaker for a research laboratory
   d. Cosmetologist
   e. Horse trainer
   f. Mother with three children
   g. Pet owner
   h. Puppy breeder for a pet store
   i. Severely ill neighbor awaiting open heart surgery
   j. Vegan
   k. Vegetarian
   l. Veterinarian
   m. Zoo keeper
4. The letters PETA stand for ______________________.
5. ______________ is the humane killing of an animal accomplished by a method which produces rapid unconsciousness and subsequent death, without evidence of pain or distress.
6. ______________ is the concept of ascribing human traits to animals.
7. ________________ is the philosophy in which the individual believes using animals for the betterment of people is acceptable, as long as the animal is treated with compassion.
8. ______________ means one's place or relative position, or one's rank in the community.
9. ______________ are people who believe that individuals can do whatever they want to animals, and that humans are the only species with any rights.

10. ____________ is the movement that believes a reduction in and a minimal number of animals should be used in research, and those animals that are used should be treated as humanely as possible. The concept implies that humankind has dominion over animals, and as such has responsibility for animal well-being.
11. __________________ is the movement that insists that animals have moral rights equal to those of humans. This concept is totally opposed to biomedical research using animals, sporting events using animals, using animals for clothing, entertainment, and product testing, and eating animals.
12. _______________ is the belief that humans are superior to animals—a prejudice or bias toward the interests of members of one's own species and against the members of other species.
13. The ________________ sets standards including provisions for housing, feeding, watering, sanitation, ventilation, the use of anesthesia for painful procedures, and the availability of veterinary medical and post-operative care at the federal level.
14. The __________________ of 1985 requires the National Institutes of Health (NIH) to create guidelines for animal use and to develop a research plan for alternatives to the use of animals.
15. __________________ is the study of behavior of animals.
16. _______________ means cutting into a live animal (originally meant "without anesthesia," because it had not been developed yet).
17. ___________ is the percentage of rats and mice that are used in research.
18. _____________ of non-human primates are used in research.

Discussion Question: *(modified from Tannenbaum, Veterinary Ethics)*

You are either a veterinarian or a registered veterinary technician. A client brings into your clinic a one-year-old toy poodle for the second time in a month. The last time the animal was in the clinic, it was diagnosed with a moderate case of flea allergy dermatitis. Medicine was given—as well as flea powder—and instructions were given on vacuuming, dusting the premises, and spraying the outside yard thoroughly for several weeks to prevent recurrence of the problem. After re-examining the dog, you conclude that not only has the dog's condition deteriorated, but you believe the medication was not properly applied as directed. The client tells you, for the first time, that the animal has been impossible to housebreak—and the client wants you to put the dog to sleep.

What do you do? Defend your actions. Discuss the effect of your decision on your business, your livelihood, the client, and the patient. In your discussion, be sure to include the possibility of having one client per day such as this come into your clinic. Would that change your answer to the question, "What do you do?" Is your choice morally correct?

Endnotes

1. Hamilton, Denise. 1995. "Mixing Kids, Animals . . . and Issues." *Los Angeles Times*, 26 November.
2. Shapiro, L.S. 1996. "A Fly Is A Carrot Is A Pig Is A Dog."
3. American Animal Hospital Association North American Pet Owner Survey. 1994.
4. American Veterinary Medical Association. 1991 statistics.

Recommended Readings

Cohen, C. "The case for the use of animals in biomedical research." *New England Journal of Medicine*. 315: 865–870. 1986.

Frey, R.G. (and Singer, P., in Comments). "Moral Standing, The Value of Lives, and Speciesism." *Pacific Division Meeting of Society for the Study of Ethics and Animals*, San Francisco, March 1987.

Orlans, Barbara. *In The Name of Science*, 1993. Oxford University Press, Inc.

Quinn, D. Ishmael. *An Adventure of the Mind and Spirit*. Bantam Books. 1995.

Regan, T. and Singer, P. *Animal Rights and Human Obligations*. Prentice-Hall, Inc. 1989.

Rohr, Janelle. *Animal Rights: Opposing Viewpoints*, 1989. San Diego: Greenhaven Press.

Tannenbaum, Jerrold. *Veterinary Ethics: Animal Welfare, Client Relations, Competition and Collegiality*, 2nd Ed., 1995. Mosby Publishing.

CHAPTER 2

Philosophy Behind the Animal Rights/ Welfare Movements

When we feel strongly about something, we assume that we know the real truth—and therefore do not have to consider the arguments on the other side. This idea is true especially when our feelings are based on cultural or religious conditioning, selfishness, or individual prejudices. An example of this passionate, philosophical "right" can be demonstrated with the Right to Life versus Pro-Choice philosophies and camps.

OBJECTIVES

The student should:

1. Differentiate between Eastern and Western philosophies on animal rights versus human rights.
2. Describe any similarities between traditional American Indian philosophy and the major religions of the modern era.
3. Describe philosophical similarities and differences regarding the care of animals between Judaism, Christianity, Catholicism, Buddhism, Hinduism, Islam, and Jainism.
4. Compare the philosophical similarities of Socrates, Bentham, Descartes, and Mill with the more modern philosophers of Regan, Singer, Cohen, and Frey.

The only way in which a human being can make some approach to knowing the whole of a subject is by hearing what can be said about it by persons of

every variety of opinion, and by studying all modes in which it can be looked at by every character of mind.

(Dr. D.M. Coffey, Western Kentucky University, 1995)

Philosophy's original definition is Greek—a study of the search for the wisdom of life. Others would have us define philosophy as an examination of people's moral responsibilities or social obligations. For individuals studying animal ethics, there is no more important area of study than philosophy. In other words, the animal welfare/rights movement is a movement based on differing philosophies. Although this text is written for the animal science, horse science, veterinary science, and other science students interested in the care of animals, a *brief* understanding of basic philosophical ideas is critical to the understanding of animal ethics. In the past decade, more than 100,000 students per year in the United States have studied philosophy covering the subject of "how we should treat animals." The demand for such a course has not only increased substantially, but the push for a more "applied approach" has been requested by students who actually use animals within their chosen vocations. Thus, this text combines a basic background in theoretical ethics and concentrates more on the applied aspects of this controversial philosophy.

Socrates described moral philosophy as "how we should live" and why. The question of who or what has moral standing, or who or what can be a member of "the moral community," has been discussed by many philosophers on many sides of the argument. Tom Regan espouses inherent value in animals and thus extends moral rights to them. Carl Cohen argues that the capacity for free moral judgment differentiates humans from animals, and thus animals cannot have rights.[5] Therefore, the differences in philosophies continue the discussion and hopefully demonstrate that there is no one correct philosophy.

The modern animal welfare/rights movements became more active in this country and throughout the world in the 1970s. Several books published during that time introduced terminology and set the stage (or rationale) for the philosophy behind animal rights. Ingrid Newkirk and Alex Pacheco, cofounders of PETA, consider Peter Singer (an Australian philosopher) to be the "father" of the modern animal liberation movement. One of his books in particular, *Animal Liberation: A New Ethic for Our Treatment of Animals*, is considered to be the original text of the modern animal rights movement. Singer's book compared the prejudice in using animals for experiments and eating—and not using humans for the same acts—was no less objectionable or biased than slavery or other forms of racism. If species differences are only racial differences, Singer argues, then it is plainly wrong to identify such differences in the treatment of animals and humans. Equal consideration of interests, such as freedom from pain, is what Singer describes in his book when discussing the "rights" of animals. According to Singer, pain is pain, no matter which species is experiencing the pain. Singer uses the term 'speciesism' when referring to the prejudice many humans have in favor of their own species. Singer compares speciesism to racism and sexism. He compares the boundary of one's own group as being superior and requiring special consideration from pain and suffering. Singer insists that if we reject racism and/or sexism, we must then reject speciesism.

Tom Regan's book, *The Case for Animal Rights*, is also widely recognized as influencing the beginnings of the modern animal rights movement. Regan argues that animals have moral rights based on the concept of "inherent value." Inherent value places a burden on others not to interfere with or disturb that inherent value. The non-human animals used in our research laboratories and on our farms are fundamentally like ourselves, according to Regan. They have a biography, not just a biology. State action can only be justified if the action does not violate the rights of its subjects and the "general good" does not excuse homicide, expropriation, or torture. People have a clear right not to be killed, tortured, or even robbed—even if it helps others to achieve the best (utilitarian) outcome. According to Regan, there is no identifiable characteristic all human beings have that is not also possessed by some non-human beings. Therefore, according to Regan, animals must have the same moral rights as humans. Regan argues for empty research cages instead of larger and cleaner ones, and he advocates no farms raising livestock—instead of more traditional, commercial farms. Many people in the American, European, and Canadian communities bought these philosophies (those of both Singer and Regan), and from that point sprang the animal rights movement.

The understanding of less-recent philosophers that influenced Singer, Regan, Cohen, Frey, and others, however, is essential. The French philosopher Réné Descartes (the father of modern philosophy) defended the use of animals in experiments. He insisted that because animals could respond to stimuli in only one way, "according to the arrangement of their organs," they therefore lacked the ability to reason and think. This idea, according to Descartes, made animals similar to a machine in nature and function. Descartes argued that animals could only learn by experience and could not learn by "teaching-learning" the way humans could. Descartes engaged in dissection of different organs, obtaining his specimens from local slaughterhouses and invented the term *embryogeny* for what is now called *embryology*. Followers of Descartes were known to kick their dogs just to hear the machine "creak."[6] Descartes' assertion that animals did not feel pain helped to ease the consciences of many experimenters of the time—and for years to come.

In 1780, British philosopher Jeremy Bentham stated in his *Introduction to the Principles of Morals and Legislation* that the question "is not, can they reason? nor, can they talk? but can they suffer?" Because both animals and humans share the ability to suffer, he considered animals equal to humans. This concept of preventing suffering can be applied to bees, insects, fish, worms, flies, and perhaps to trees. Later in this text, we will describe how Buddhist and Native American philosophies include other life forms (other than human and animal) in their "shared respect for life." Bentham was one of the first to call "man's dominion" tyranny, rather than legitimate government.

Utilitarianism was first explicitly described by Bentham (1748–1832). *Utilitarianism* offers a direct moral view, which concentrates all ethics on questions of pain and pleasure and happiness and misery. This view could thus include animals, as well as humans. The term was further refined by philosopher John Stuart Mill (1806–1873), who advocated the concept of "the greatest good for the greatest number." The basic theory that grew out of their philosophy was as follows: in deciding whether an action is morally right, one sums up the total amount of good the action will bring about and weighs those benefits against the total amount of harm that will be caused. The total good outweighs the individual harm. Utilitarian philosophy embraces what is best for the majority in a situation—that the right action is what benefits the most individuals. Such a theory is in opposition to *egoism*, which is the view that a person should pursue his or her own self-interests, even at the expense of others. According to Mill, acts should be classified as morally right or wrong only if the consequences are of such significance that a person would wish to see the agent compelled, not merely persuaded and exhorted, to act in the preferred manner. In other words, under utilitarianism, the right thing can be done from a bad motive. Both Bentham and Mill balanced pleasure higher than pain in assessing happiness.

CULTURAL AND RELIGIOUS PERSPECTIVES OF ANIMAL USE

In general, Eastern philosophies hold the belief that man is equal to others. The philosophies adopted by many animal activists are thus more related to those of Eastern philosophies.

> ***"Combine the internal and the external into one and regard things and self as equal."***
>
> (Chin-ssu-lu, Chan, *Reflections*)

Most Western philosophies, if not all, teach that man is dominant and is a creation of God:

"Have dominion over the fish of the seas, the birds of the air, and all living things that move on the earth."

(Genesis 1:26–28)

Ingrid Newkirk, founder and president of PETA, has expressed her goal of a world where the lamb will lie down with the lion. The magazine *Animals' Voice* has published a handbook explaining how readers can make their voices heard in church. More recently (spring 1999), PETA has tried to influence "the Christian community," claiming that Jesus was a vegetarian. Billboards in several states (including Texas) have tried influencing the church's traditional preaching of people having dominion over animals.

Most contemporary theologians interested in the animal rights/welfare movement fall into one of three categories:

1. *Traditional orthodoxy*—Judeo-Christian doctrine of kindness toward animals; a prophetic denunciation of animal sacrifice and the eventual ending of the practice by both Jews and Christians
2. *Traditional notions of God*—Theology of creation in which animals have rights
3. *Most radical*—Takes up the questions about God raised by the animal rights movement and answers them with notions taken from contemporary, "process" theologians

AMERICAN INDIAN BELIEFS ON ANIMAL CARE

The destruction of the American bison herds was, in part, a deliberate action taken by the United States Army as a means of bringing the American Indians under control—and to support the expansion of the railroad system throughout the West (see Figure 2-1). Bison tore up the railroad tracks and chased away the cattle brought in by Europeans and American caucasians. Most Native American historians, however, will tell you that the killing of the bison was part of the final solution in ridding the West of Native Americans. The army hired hunters to kill all the bison. They then piled the bison bones in plain sight as an insult to the Native American in this final assault against them.

Animal rights activists may condemn the practices that Native Americans used when they killed the bison themselves. The bison killed by American Indians, however, were killed with great respect and were used only for food, clothing, and shelter. When Native Americans killed a bison, they gave thanks to the bison's spirit. They used every part of the

Figure 2–1 The American bison. *Courtesy of the American Bison Association*

animal they killed. The meat was dried into jerky or was finely pounded and mixed with dried berries and fat to make pemmican. The skins were used for clothing, moccasins, containers, and covers for their teepees. The hair stuffed their pillows and saddlebags. The tendons became their bowstrings and thread. From the hooves, they made glue. They carried water in the bladders and stomachs. The hooves, bones, and horns also became tools and utensils. Even the "buffalo chips" were burned for fuel. To give the bison honor, the Indians painted the skull and placed the skull facing the rising sun. The original population of bison in North America was estimated at 50–80 million. By 1900, the American bison numbered fewer than 2,000.

To all of the Native American people, every creature and every part of the earth was sacred. They believed that to waste or destroy nature and its wonders was to destroy life itself. Chief Seattle (1790–1866) responded to Washington D.C.'s request to buy the lands of Native Americans with this now-famous response:[7]

How can you buy the sky?
How can you own the rain and the wind?
My mother told me,

Every part of this earth is sacred to our people.

Every pine needle. Every sandy shore.

Every mist in the dark woods.

Every meadow and humming insect.

All are holy in the memory of our people.

My father said to me,

I know the sap that courses through the trees as I know the blood that flows in my veins.

We are part of the earth, and it is part of us.

The perfumed flowers are our sisters.

The bear, the deer, the great eagle, these are our brothers.

The rocky crests, the meadows, the ponies—all belong to the same family.

The voice of my ancestors said to me,

The shining water that moves in the streams and rivers is not simply water, but the blood of your grandfather's grandfather.

Each ghostly reflection in the clear waters of the lakes tells of memories in the life of our people.

The water's murmur is the voice of your great-great grandmother.

The rivers are our brothers. They quench our thirst.

They carry our canoes and feed our children.

You must give to the rivers the kindness you would give to any brother.

The voice of my grandfather said to me,

The air is precious. It shares its spirit with all the life it supports.

The wind that gave me my first breath also received my last sigh.

You must keep the land and air apart and sacred,

As a place where one can go to taste the wind that is sweetened by the meadow flowers.

When the last Red Man and Woman have vanished with their wilderness, and their memory is only the shadow of a cloud moving across the prairie, will the shores and forest still be here?

Will there be any of the spirit of my people left?

My ancestors said to me, this we know:

The earth does not belong to us. We belong to the earth.

The voice of my grandmother said to me,

Teach your children what you have been taught.

The earth is our mother,

What befalls the earth befalls all the sons and daughters of the earth.

Hear my voice and the voice of my ancestors.

The destiny of your people is a mystery to us.

What will happen when the buffalo are all slaughtered?

The wild horses tamed?

What will happen when the secret corners of the forest are heavy with the scent of many men?

When the view of the ripe hills is blotted by talking wires?

Where will the thicket be? Gone.

Where will the eagle be? Gone.

And what will happen when we say good-bye to the swift pony and the hunt?

It will be the end of living and the beginning of survival.

This we know: All things are connected like the blood that unites us.

We did not weave the web of life,

We are merely a strand of it.

Whatever we do to the web, we do to ourselves.

We love this earth as a newborn loves its mother's heartbeat.

If we sell you our land, care for it as we have cared for it.

Hold in your mind the memory of the land as it is when you receive it.

Preserve the land and the air and the rivers for your children's children, and love it as we have loved it.

In 1994, a white buffalo named Miracle was born on a ranch in Wisconsin. According to Native American tradition, the birth of a pure white buffalo signals an end to an era. The new era will unite all of the peoples of the earth (different races, red, black, yellow, and white). In Judaism, the red heifer Melody, who was born in 1997, signals the coming of the Messianic Age—when the Temple can be rebuilt.[8] According to Biblical law, the cow's ashes are used for purification, and therefore the ashes are a prerequisite for the renewal of Holy Temple service. In the Book of Numbers XIX, a new era would begin just before the year 2000. Twenty-five leading Biblical experts visited Kfar Hasidim (where the red heifer was born) and concluded the heifer met Biblical requirements. These requirements are described as a red heifer without spots or blemishes—which had "not yet been broken to the yoke" (a virgin heifer). Two hairs of another color are enough to disqualify the animal.

This coincidence is intriguing. Or, is it ancient prophecy that is being fulfilled? Animal rights activists are denouncing the red heifer, because the animal would have to be sacrificed at the age of three (in the year 2000). Moslems object, because the Dome of the Rock (the third-holiest Moslem shrine) was built on top of King Solomon's Temple (where the rededication would take place).

BUDDHIST PHILOSOPHY

Early in the development of the Eastern religions (*Jainism, Hinduism, and Buddhism*), animal sacrifice was no longer practiced. People's feelings against unnecessary destruction of life led to widespread vegetarianism. All Jains and most religious Buddhists are strict vegetarians. They go out of their way to avoid injuring any living thing. One concept preached is that the souls of people are reborn as animals, and vice-versa. Buddhism does not believe in a creator god. Buddhism historically comes from Hinduism, which also shares the belief of rebirth of the soul in human or animal form. Buddha trained himself to be kind to all animate life and taught that it was wrong to kill any living creature. He observed that the key to a new civilization is the spirit of Maitri, the friendliness toward all living things.

There are two main branches of Buddhism. *Theravada* is the form that developed in South Asia and is thought to be closest to the teachings of Gautama Buddha.[9] Mahayana Buddhism is the second and newer sect of Buddhism and is actively practiced in Japan, Korea, and parts of China.

Theravada Buddhism allows the eating of meat, as long as its followers are not responsible for killing the animal—and if the animal was not especially killed to feed them. Meat supplied by Christian or Muslim butchers or fishermen is generally permitted.[10] Many Buddhists criticize the Theravada teaching on meat-eating:

". . . wherever living beings evolve, men should feel toward them as to their own kin, and looking on all beings as their only child, should refrain from eating meat . . . "[11]

In Buddhism, there are five basic *precepts* which serve as moral guidance to practicing Buddhists:

"He who destroys life, who utters lies, who takes what is not given to him, who goes to the wife of another, who gets drunk with strong drinks—he digs up the very roots of his life."

(Dhammapada, 246–247)[12]

The first precept that teaches not to kill and not to hurt is accepted by most Buddhists. The controversy, however, lies in whom Buddhists are forbidden to kill or hurt. In other words, are non-human animals included in this first precept? Some have argued that the protection of "living beings" extends to both plants and animals, because they are alive as well. Hence, it is no worse eating a rabbit than eating a carrot.

The Buddha manifested a complete compassion and respect for all life forms. He taught that those who wished to follow his path should practice loving kindness to all beings, not only to humans. This idea meant an obligation to protect animals and vegetation.[13] The Buddha saw all beings in the universe as equal in nature. So is being a vegetarian the true sign of compassion? Is eating meat against Buddhist philosophy —and evidence of lack of compassion for life?

"Adolf Hitler was a vegetarian; the Dalai Lama, the embodiment of compassion eats meat by his doctor's orders."[14]

HINDU AND JAIN PHILOSOPHY

Hinduism is not as strict (in regards to vegetarianism) as Jainism and Buddhism, and Hinduism allows for limited animal sacrifice as part of religious ceremonies. Cows are considered the most sacred of all animals in Hinduism. Peacocks are also considered sacred. The monkey is protected and is honored over much of India. Proper treatment of animals is important as the Hindu passes toward salvation.

Modern Hindus are still taught that the human soul can be reborn into other forms, such as animals and insects. *Ahisma*, a doctrine of non-violence or non-killing (from Sanskrit; a = without, and himsa = injury) is taken from Hindu, Buddhist, and Jainist philosophies.

Most Americans perceive India as a vegetarian society. In 1993, the government of India conducted its own survey, which showed that 88 percent of the nation's people were non-vegetarian.[15] Even vegetarian Hindus, such as the Bishnoi, are not vegans. For a more descriptive definition of veganism, see chapter 9. The Bishnoi eat dairy products with their vegetable, flour, sugar, and lentil diets. The Bishnoi are also known to be strong ecologists. Several hundred Bishnoi were killed while hugging trees in an effort to prevent deforestation of the Khejare forests of India. Protecting this life form, the tree, was equally important to the Bishnoi.[16]

The kind of food that the Hindu is permitted to eat varies with the caste (class) and geographic locality. The Brahmin will not eat any animal food or drink any alcoholic beverage. Others permit the eating of both sheep and goat meat.[17]

Jainism is one of the world's oldest religions. The Jains have no gods. They instead revere 24 early sages (Jinas). One of its most recent sages, Mahavira (599–527 B.C.E.), preached the shepherding of nature: "A wise man should not act sinfully towards earth, nor cause others to act so, nor allow others to act so . . ." Mahatma Gandhi was schooled by a Jain monk. One of Gandhi's own personal philosophies was the first Jain principle of ahimsa. Gandhi argued for ahimsa throughout his adult life. The Jains teach that every living organism is endowed with a unique, precious soul, or *jiva*, which is a life force which must be respected—even revered. Hence the phrase, "a fly is a carrot is a pig is a dog" developed. According to Jain philosophy, all souls are interconnected (*parasparopagraho jivanam*). Most Jains are vegetarians. Most will not wear leather, keep pets, engage in agriculture, or participate in the exploitation of animals (or, to some extent, even plants). Jains would never condone the keeping of cattle nor any other animal (including pets). They believe that animal husbandry is a form of animal exploitation.

Jains believe that all organisms are endowed with one to five senses. Human beings have five of these senses. A mango fruit has but one sense. Jains are only allowed to eat one-sensed beings in strict moderation, with frequent fasting. This philosophy is a tenet to Jain spiritual ecology. Mahavira catalogued 800,000 organisms and assigned them senses. At approximately the same time in history, biologists in Greece had recognized only 600 different species. Five classifications exist in Jainism:

- Five senses: humans, animals, birds, and heavenly and hellish beings
- Four senses: flies, bees, etc.
- Three senses: ants, lice, etc.
- Two senses: worms, leaches, etc.
- One sense: vegetables, water, air, earth, fire, etc.

The five senses are touch, taste, smell, sight, and hearing. Much more pain results if a life of the higher forms (more than one sense) is killed. All non-vegetarian food is made by killing a living being with two or more senses. Therefore, Jainism preaches strict veganism and prohibits non-vegetarian food. Jainism also preaches strict non-violence. Non-violence is observed in action—but also in speech and thought. One is advised to remain quiet if speaking the truth causes pain, hurt, or anger. Jains are taught to eat food for survival and *not* for taste.

JEWISH (ORTHODOX)

The Jewish attitude toward animals and animal care has always been governed by the consideration that animals are part of God's creation, and that *"His tender mercies are over all His works"* (Psalms 145:9). Judaism espouses the concept that everything created by God was created to serve humankind. In other words, animals may be used for food, for beasts of burden, and for sport, provided that they are managed and are slaughtered in a humane manner. Jewish tradition demands compassion and mercy for animals. Judaism teaches that domesticated animals should be afforded a complete rest on the Sabbath. Many biblical passages describe caring for animals. Rebecca, in proving that she was fitting as a wife for Isaac, brought water not only for Eliezer, the servant of Abraham, but to his camels (Genesis 24:14). In Proverbs (12:10), a righteous man is described as one who pays attention to the needs of his beast. Jonah was admonished by God, who claimed that Jonah should have pity on Nineveh because of the more than six score thousand persons and also much cattle. (Jonah 4:11). In Leviticus (22:28), it is forbidden to slaughter an animal and its young on the same day. This policy was ordered to prevent a young animal from being slaughtered in front of its mother. In Deuteronomy (22:6–7), there is a commandment to move the hen off its nest before gathering her eggs. Also, this commandment forbids eating eggs on which a hen has sat—and eating the hen of which the young need their mother. If the mother is let go on her own accord (leaves the nest) and is not pained by seeing her young taken away, these eggs may then be collected and eaten.

Jewish dietary rituals have nothing to do with health but are simply there to teach the observant Jew that all life is sacred. A confusion has always existed here among Jews and non-Jews alike because of the word "unclean," as it is applied to non-kosher food. The word "unclean" refers to spiritual cleanliness, not biological cleanliness.

People have special responsibilities to build their lives on a higher plane than that of lesser creatures. *Kashrut* is one of the major ways in which an observant Jew trains himself or herself in self-discipline. Until a person can control himself or herself, the observant Jew is taught that one cannot control his or her environment.

All fruits and vegetables can be eaten. In addition, the flesh of animals possessing a cloven hoof and who also chew their cud (Deuteronomy 14:4–15), fowl including chickens, goose, squab, duck, or turkey, and fish with fins and scales are permitted. Items that are forbidden include birds of prey, insects and all other animals of the sea (shellfish [sic] are not permitted).

Meat and fowl must be slaughtered by a ritual slaughterer called a *shochet* (Deuteronomy 12:21). The actual words used in this section of the Old Testament are:

> ***". . . then thou shalt kill of thy herd and of thy flock, which the Lord hath given thee, as I have commanded thee, and thou shalt eat within thy gates, after all the desire of thy soul."***

Moses had previously taught the people a method of slaughtering animals. In Leviticus III:17 and VII:26, it is prohibited to consume flesh containing blood. The purpose was to tame man's instincts toward violence by weaning him from blood and implanting within him a distaste —a horror—for all bloodshed. The kosher method of slaughter (*shechitah*) causes the maximum effusion of blood, with the remaining blood being extracted by means of the washing and salting of the meat.

The actual slaying of animals for food was relegated to a body of pious and specially trained men called *shochetim.* The flesh on an animal that died of itself or was torn by beasts is strictly forbidden. Any animal that was ritually slaughtered but was found to contain injuries or organic diseases, whether the imperfections were obvious prior to or were determined by inspection of the animal after slaughter may not be eaten. Animals not killed strictly in the prescribed manner are forbidden food, as well.

Judeo-Christian ethics teach that God created humans, and that humans have the responsibility for all other living things:

> ***"What is man that you are mindful of him? You made him steward over the works of your hands; you put everything under His feet: all flocks and herds, and the beasts of the field, the birds of the air, and the fish of the sea."***
>
> (Psalms 8:4, 6-8)

Jewish tradition and law grants humans the ability to use and manipulate non-human creatures for their benefit. People were given the responsibility to take care of the non-human species but were still allowed to use them:

> ***"Let us make man in our image, after our likeness. Let them have dominion over the fish of the sea, the fowl of the air, and the cattle, and over all the wild animals, and over all the creatures that crawl on the ground."***
>
> (Genesis 1:26)

"God blessed them, saying: 'Be fertile and multiply; fill the earth, and subdue it. Have dominion over the fish of the sea, the fowl of the air, and all living things that move on the earth."

(Genesis 1:28)

"Unto Adam also and to his wife did the Lord God make coats of skins, and clothed them."

(Genesis 3:21)

Until the laws of kashrut came about, it was permissible to hunt, as seen when Isaac talks to his eldest son, Esau:

"Now therefore take, I pray thee, thy weapons, thy quiver and thy bow, and go out to the field, and take me some venison."

(Genesis 27:3)

Jewish law permitted eating what was slaughtered:

"And make me savory meat, such as I love, and bring it to me, that I may eat."

(Genesis 27:4)

"Notwithstanding thou mayest kill and eat flesh in all thy gates, whatsoever thy soul lusteth after, according to the blessing of the Lord thy God which He hath given thee: the unclean and the clean may eat thereof, as of the roebuck, and as of the hart."

(Deuteronomy 12:15)

"When the Lord thy God shall enlarge thy border, as He hath promised thee, and thou shalt say. I will eat flesh, because thy soul longeth to eat flesh; thou mayest eat flesh, whatsoever thy soul lusteth after."

(Deuteronomy 12:20)

After Noah saved the animals from extinction, God made a concession to man by giving him the right to consume meat, provided that the animals were slaughtered in a humane (kosher) manner (Genesis 1:29 and 9:3).

Although Jewish law permits the eating of meat—and Jewish tradition encourages it—there is also a strong emphasis on animal well-being. Jewish law on animal welfare includes the following tenets:

1. It is forbidden to muzzle an ox when it is treading out grain. It is inhuman to allow an animal to go hungry while he is surrounded by food (Deuteronomy 25:4). Jewish law extends this prohibition of muzzling the ox to workmen employed on production of articles of food (they must not be prevented from eating the food).
2. Animals of different size or of great disparity in strength (an ox and a donkey, for example) are not permitted to be harnessed together. The weaker or smaller animal would endure much suffering.
3. The *Talmud* (*Kitzur Shulchan Aruch* by Shlomo Ganzfried, Chapter 42:1) demands that a person should not sit down to eat before he or she has fed his or her flock (Berachot 40a and Gittin 62a). Today, this idea is further interpreted to mean that a person should feed his or her animals before feeding himself or herself.
4. Hunting as a sport has been outlawed by Jewish tradition for many centuries. The slaughter of animals for food has been carefully regulated to cause the least pain to the animal. This regulation exists not only to protect the animal but to build into the Jewish character a revulsion for cruelty and brutality. The Shocket's purpose is to prevent any unnecessary pain to the animal, so not a single nick or scratch mars the cutting edge of the knife that is used.
5. People have no right to take an animal's life for trivial or unworthy purposes. By minimizing the cruelty to animals, men and women become more sensitive to all suffering. Through this sensitivity, they will learn to empathize with, and so limit, the suffering of all life.
6. The Sabbath is a day of rest for animals, as well as for humans.
7. Castration or emasculation of animals is prohibited (Leviticus 22:34).
8. Prohibition of cruelty to animals (*Exodus* 23:5 and *Mishneh Torah, Hilchot Rotze'ach* 13:1, 13:8, and 13:9) is embraced, even if the animal belongs to one's enemy.
9. Sabbath laws are waived by the duty to alleviate an animal's suffering (milking, feeding, delivering a newborn, helping an animal in distress, etc.) (*Maimonides, Mishneh Torah, Hilchot Shabbat* 25:26).
10. Jewish law prohibits the buying of an animal or a bird unless the buyer can properly provide for it (*Jerusalem Talmud, Ketubot* 4:8).

The following paragraph lists exceptions to cruelty to animals, according to Jewish law:

Anything that is necessary for medical or other useful purposes is excluded from the prohibition of cruelty to animals (Joseph Karo, *Shulchan Aruch, Even Ha'ezer* 5:14). Jewish scientists, doctors, and veterinarians are permitted to test the effects of a new drug on an animal—such as a dog or a cat—prior to applying this medication to a human, to prevent injurious, fatal, unknown side effects in people (Deuteronomy 20:19 and J. Reischer, *Responsa Shevut Yaakov*, p. 3, no. 71). Thus, most rabbinical opinions today agree that animal experimentation for medical research is permissible, as in the question, "What right have you to assume that the pain of animals counts more than the pain of sick people who might be helped [by animal experimentation]?" (Weinberg, *in Responsa Chelkat Yaakov*, p.1, no. 31, idem, *Responsa Seridei Esh*, p. 3, no. 7). One still needs to avoid cruelty to animals where possible, however (*Kiddushin* 82a-b).

The prohibition of cruelty to animals only applies to a live animal. Thus, anatomical dissection on a dead animal is permissible (because the animal feels no pain), whereas vivisection on a live animal would be not allowed unless the action was required to reduce pain and suffering in a human (I. Jakobovits, "The Medical Treatment of Animals in Jewish Law," *J Jewish Studies* 7(1956):207–220).

Although the Old Testament clearly outlines dominion over other life forms, the Hebrew text strongly demands humane treatment of animals:

> ***"A righteous man regards the life of his animal, but the tender mercies of the wicked are cruel."***
>
> (Proverbs 12:10)

Both Jewish and Christian ethics differentiate the value of human versus non-human life:

> ***"If anyone takes the life of a human being, he must be put to death. Anyone who takes the life of someone's animal must make restitution, life for life. Whoever kills an animal must make restitution, but whoever kills a man must be put to death."***
>
> (Leviticus 24:17–18,21)

A description in one family's recent history outlines this troubling differentiation between the value of a human and an animal:[18]

> ***"One of the most vivid memories my grandfather described to me was when he was a young high school boy in South Africa. On the very same day, two trials were being held. One was for a man who shot and killed another man (a black man). The other was for a man who shot and killed a dog. Both individuals were fined. The man who shot the dog was given a stiffer fine. My grandfather told me***

> ***that is why he became a lawyer—to help stop this ridiculous value system that placed greater emphasis on animal life over human (some human) life."***

A generation later in another country, Nazi Germany, a similar story can be told (see Chapter 3, "History Behind the Animal Rights/Welfare Movement").

Judeo-Christian ethics teach that God created humans, and that humans have the responsibility for all other living things:

> ***"What is man that you are mindful of him? You made him steward over the works of your hands; you put everything under his feet: all flocks and herds, and the beasts of the field, the birds of the air, and the fish of the sea."***
>
> (Psalms 8:4, 6–8)

In the *Midrash,* the sages write that Moses' special care for his sheep caused God to choose him as leader of the children of Israel: "If you tend the sheep with such compassion, you will surely be a compassionate leader of my people Israel." King David was also a shepherd who was praised for his compassion toward animals: "One who exerts himself to take care of each sheep's needs will certainly take good care of my people Israel."

According to Leo Baeck, "In an act without parallel in civilization, the Bible placed animals under the protection of laws devised for man." Jewish tradition and law, as previously mentioned, grants humans the ability to use and manipulate non-human creatures for their benefit. People were given the responsibility to take care of the non-human species but were still allowed to use them.

> ***"Let us make man in our image, after our likeness. Let them have dominion over the fish of the sea, the fowl of the air, and the cattle, and over all the wild animals, and over all the creatures that crawl on the ground."***
>
> (Genesis 1:26)

> ***"God blessed them, saying: 'Be fertile and multiply; fill the earth, and subdue it. Have dominion over the fish of the sea, the fowl of the air, and all living things that move on the earth."***
>
> (Genesis 1:28)

CHRISTIAN PHILOSOPHY

Early Christianity professed that animals had no souls. Later, societies admitted that animals experienced pain and had feelings. Humans are unique in that they possess language skills, upright locomotion, use of the thumb, and reasoning skills to solve complex problems.

Christianity has traditionally taught that one can eat what was killed:

> ***"It contained all kinds of four-footed animals, as well as reptiles of the earth and birds of the air. Then a voice told him, 'Get up, Peter. Kill and eat.'***
>
> (Acts 10:12-13)

Jesus helped his disciples catch fish:

> ***"When He had finished speaking, he said to Simon, 'Put out into deep water, and let down the nets for a catch.' When they had done so, they caught such a large number of fish that their nets began to break."***
>
> (Luke 5:4-6)

Jesus fed fish flesh to literally thousands of people:

> ***"'How many loaves do you have?' He asked. 'Go and see.' When they found out they said, 'Five and two fish.' . . . Taking the five loaves and two fish and looking up to heaven, He gave thanks and broke the loaves. Then He gave them to His disciples to set before the people. He also divided the two fish among them all. They all ate and were satisfied, and the disciples picked up twelve basketfuls of broken pieces of bread and fish. The number of people who had eaten was five thousand."***
>
> (Mark 6:38, 41–44)

Jesus hand-fed fish flesh to his apostles after his resurrection:

> ***"When they landed, they saw a fire of burning coals there with fish on it, and some bread. Jesus said to them, 'Bring some of the fish you have just caught.' Jesus said to them, 'Come and have breakfast.' Jesus took the bread and gave it to them, and did the same with the fish."***
>
> (John 21:9–10, 12–13)

To prove that he was physically, bodily resurrected, Jesus ate fish flesh himself:

> ***"And while they still did not believe it because of joy and amazement, He asked them, 'Do you have anything here to eat?' They gave Him a piece of broiled fish, and He took it and ate it in their presence."***
>
> (Luke 24:41–43)

More important than anything else, Jesus demonstrated that one human life can be more important than a thousand non-human lives:

> ***"For Jesus said to him, 'Come out of this man, you evil spirits!' The demons begged Jesus, 'Send us among the pigs: allow us to go into them.' He gave them permission. So they came out and went into the pigs, and the whole herd rushed down the steep bank into the lake and died in the water. The herd being about two-thousand in number."***
>
> (Mark 5:8, 10-13)

Christianity teaches that humans are expected to behave differently:

> ***"But these men blaspheme in matters they do not understand. They are like brute beasts, creatures of instinct, born only to be caught and destroyed, and like beasts they too will perish."***
>
> (2 Peter 2:21)

Christianity teaches that the human species is physically created differently:

> ***"All flesh is not the same: Men have one kind of flesh, animals have another, birds another and fish another."***
>
> (1 Corinthians 15:39)

Christianity also teaches that man has dominion over animals:

> ***"God of my fathers and Lord of mercy . . . by Your wisdom You have formed man, to have dominion over the creatures You have made, and rule the world in holiness and righteousness"***
>
> (Book of Wisdom 9:1, 2–3).

> ***"You have given him dominion over the works of Your hands; You have put all things under his feet, all sheep and oxen, and also the beasts of the field, the birds of the air, and the fish of the sea, whatever passes along the paths of the sea"***
>
> (Psalms 8:6–8).

Some people will have you believe that Christian philosophy compares the animal rights movement with that of foolish actions:

> ***"Some became fools through their rebellious ways and suffered affliction because of their iniquities. They loathed all meat and drew near the gates of death."***
>
> (Psalms 107:17–18)

Others compare animal rights movement to the devil's actions:

> ***"The Spirit clearly says that in later times some will abandon the faith and follow deceiving spirits and things taught by demons. Such teachings come through hypocritical liars, whose consciences have been seared as with a hot iron. They forbid people to marry and order them to abstain from meats, which God created to be received with thanksgiving by those who believe and who know the truth."***
>
> (1 Timothy 4:1-3)

Most Judeo-Christian ethics teach that in a religiously plural nation, it is permissible to be a vegetarian and still hold strongly to one's religion:

> ***"One man's faith allows him to eat everything, but another man, whose faith is weak, eats only vegetables. The man who eats everything must not look down on him who does not. And the man who does not eat everything must not condemn the man who does."***
>
> (Romans 14:2–3)

The earlier quote from *Romans* (in the New Testament) should be a basis for understanding the dialogue of those people who may have differing ideas. Hopefully, in this country we are sophisticated enough to allow differing opinions within a peaceful discussion. To summarize in one word, the world needs *tolerance.*

Still, there are others whose passions are so great that they feel they must lie, degrade, exaggerate, and instill fear for their cause to be understood and to be accepted by others. The one aspect of American life that is cherished more than anything else is the freedom to disagree, which includes the freedom to knowingly have a strong difference of opinion and still feel safe to work, walk, and live in one's neighborhood. More than anything else, America abhors violence as a means of civil disobedience. The earlier section from *Romans* indicates that Christianity embraces this idea as well.

"JUMP OF THE GOAT" FIESTA

In a northern Spanish village (Manganeses de la Polvo Rosa), residents have traditionally hurled a goat from the top of a church tower in an annual ritual to honor their patron saint. Hundreds of people dress up in fancy clothes to witness the tossing of the goat out of the 50-foot-tall belfry. Today, because of animal welfare concerns, the goat is caught in a canvas sheet held open by the villagers below. Although the ritual was officially banned by the governor of the province (Zamora province) in 1992, authorities claim they are unable to enforce the ban. The ritual started in the 19th century, when goats used to sneak into the church tower to eat food left out for the doves. The legend claims that when the priest shooed them away, the goats would jump out of the belfry.

Animal activists say they have seen goats fall to their death when they have missed the canvas below. The town's mayor, Demetrio Prieto, denies that claim. Villagers violently protested the banning and claimed that no one will take that ritual away from them.

THE NEW PAPAL ENCYCLICAL VERSUS TRADITIONAL CATHOLIC DOCTRINE

Saint Thomas Aquinas (1225–1274 C.E.) believed that only humans had rational souls, and animals had sensitive souls which were not rational. Because there are some similarities between human and animal life, Aquinas believed that cruelty to animals would lead to cruelty to humans. "God's purpose in recommending kind treatment of the brute creation is to dispose men to pity and tenderness toward each other," he said.[19]

Cardinal Newman (1801–1890 C.E.) wrote that humans "have no duties toward the brute creation; there is no relation of justice between them and us… they can claim nothing at our hands; into our hands they are absolutely delivered." This statement was widely regarded as the "official" position of the Catholic Church regarding animals.

The newer encyclical *Evangelium Vitae—the Gospel of Life* (1995)—shows a change in the emphasis of animal value:

> ***"To defend and promote life, to show reverence and love for it, is a task which God entrusts to every man, calling him as His living image to share in His own lordship over the world."***

"As one called to till and look after the garden of the world (cf. Genesis 2:15), man has a specific responsibility toward the environment in which he lives, towards the creation which God has put at the service of his personal dignity, of his life, not only for the present but also for future generations. It is the ecological question—ranging from the preservation of life to 'human ecology' properly speaking—which finds in the Bible clear and strong ethical direction, leading to a solution which respects the great good of life, of every life. In fact, the dominion granted to man by the Creator is not an absolute power, nor can one speak of a freedom imposed from the beginning by the Creator himself and expressed (Genesis 2:16-17) shows clearly enough that, when it comes to the natural world, we are subject not only to biological laws but also to moral ones, which cannot be violated with impunity."

Humans are distinct in the "living" world because they are bearers of the image of God. Hence, the new edict proclaims that animals are not our moral equals, but we do have certain duties to them. Humankind has been given the task of caring for God's creation. Humans must respect the good of all life—of every life. Animal rights activists would argue that one cannot respect life and turn around and eat life. Currently, even with the new edict, one can do both according to Catholic doctrine.

INTERNATIONAL NETWORK FOR RELIGION AND ANIMALS

Eighty-three-year-old Virginia Bourquardez, founder of the *International Network for Religion and Animals*, announced in December 1995 that she was leaving the Roman Catholic Church in which she was born and baptized. She said that she could no longer "assent to the teaching in the new catechism of the Catholic Church that animals were created for human use . . . ." She further stated that "at this point in my life, I cannot countenance the fact religion as a whole so permits the terrible suffering of animals." Bourquardez founded her animal rights group in 1985. Where Judeo-Christian tradition often falls short, according to religious animal-rights activists, is in its emphasis on human control over animals—as opposed to the need to respect their divine value.

Bourquardez particularly objected to the statement in the new catechism that *"animals, like plants and inanimate beings, are by nature destined for the common good of past, present and future humanity."* The Rev. Jeffrey Sobosan, a Catholic theologian at the University of Portland in Oregon, interprets the catechism to mean that animals share in the common good with human beings. Animals are God's creatures, and while it is legitimate to use animals for food, clothing, and reasonable medical experimentation, "it is contrary to human dignity to cause animals to suffer or die needlessly," he says.

Sobosan continues describing the new catechism as one that advances the idea of mutual care of human beings and animals. Even when animals are used for food, there "always is the idea the proper attitude is one of gratitude and thanksgiving for the life of the animal."

ISLAM PHILOSOPHY

Moslems are taught that God has given people power over animals, but to treat animals badly is to disobey God's will. Moslems believe that the world belongs to God and that people are accountable to Him for their treatment of the world. It is wrong, therefore, to hunt merely for pleasure, to use an animal as a target, to use its skin, to cause animals to fight each other, to incite them to act unnaturally in entertainment, or to molest them unnecessarily. The prophet Mohammed taught that animals should be killed only out of necessity—and that to do otherwise was a sin.

According to the Moslem belief of respect for life,

> ***"There is not an animal (that lives) on earth,***
> ***Nor a being that flies***
> ***On its wings, but (forms Part of) communities like you.***
> ***Nothing have we omitted***
> ***From the Book, and they (all)***
> ***shall be gathered to their Lord***
> ***In the end."***
>
> (*Qur'an S.Vi.*38)

> ***"In our pride we may exclude animals from our purview, but they all live a life, social and individual, like ourselves, and all life is subject to the Plan and Will of God."***[20]

According to the Moslem belief in man having dominion over animals,

> ***"And cattle He has created***
> ***For you (men): from them***
> ***Ye derive warmth,***
> ***And numerous benefits,***
> ***And of their (meat) ye eat."***
>
> (S.xvi.5)

The other benefits have been described as, "From wool, and hair, and skins, and milk. Camel's hair makes warm robes and blankets; and certain kinds of goats yield hair which makes similar fabrics. Sheep yield wool, and llamas and alpaca for similar uses. The skins and furs of many animals yield warm raiment or make warm rugs or bedding. The females of many of these animals yield good warm milk, a nourishing and wholesome diet. Then the flesh of many of these animals is good to eat."[21]

Halal (permitted) slaughtering of animals is similar to that of the Jewish kosher slaughter. Observant Muslims are permitted to eat anything that is indicated as "kosher."[22] The blood must be drained, and swine and other similar animals are prohibited.

The following passage describes the Muslim belief on the use of work animals:

> ***"And (He has created) horses,***
> ***Mules, and donkeys, for you***
> ***To ride and use for show;***
> ***And He has created (other) things***
> ***Of which ye have no knowledge."***
>
> (S.xvi.8)

> ***"And they feel, for the love***
> ***Of God, the indigent,***
> ***The orphan, and the captive."***
>
> (S.lxxvi.8)

"It has also been held that 'captives' include dumb animals who are under subjection to man; they must be properly fed, housed and looked after; and the righteous man does not forget them."[23]

The Moslem religion has restrictions on the types of foods and on the preparation of these foods, and these laws are similar to those found in kosher laws:

> ***"You are forbidden carrion, blood, and the flesh of swine; also any flesh dedicated to any other than God. You are forbidden the flesh of strangled animals and of those beaten or gored to death; of those killed by a fall or mangled by beasts of prey (unless you make it clean by giving the deathstroke yourselves); also of animals sacrificed to idols."***
>
> (S.v.3)

> ***"The Prophet said, Eat what is slaughtered (with any instrument) that makes blood flow out, except what is slaughtered with a tooth or a nail."***[24]

Animals that were not explicitly killed for food cannot be eaten (dead meat and carrion, for example) (*S.ii.173*).

The Moslem belief on pets is as follows:

There is no mention on the treatment of pets in the *Quar'an*. Some Muslims go to the *hadiths* (written traditions) for matters concerning their pets. Several hadiths are included here:

> ***"If a dog drinks from utensil of anyone of you, it is essential to wash it seven times."***[25]
>
> ***"Five kinds of animals are mischief-doers and can be killed even in the Sanctuary: They are the rat, the scorpion, the kite, the crow and the rabid dog."***[26]
>
> ***"A woman entered the (Hell) Fire because of a cat which she had tied, neither giving it food nor setting it free to eat from the vermin of the earth."***[27]

There are many hadiths that talk about the mistreating of animals. Whether the animal is a snake or a dog, the animal is still forbidden. Muslims do not agree, however, on the matter of pets. Some say that dogs cannot be pets and can only be used for hunting and protecting livestock, because a hadith exists that issues this command. Others speak of the Prophet ordering the slaughter of dogs. More detailed hadiths explain that the dogs that were slaughtered were ill with rabies and were causing the death of livestock, besides posing a threat to human life as well.

While some Muslims consider keeping dogs as pets to be *haram* (prohibited), others disagree. Most, if not all, recognize that dogs are a problem when one wants to pray, because Muslims must be clean to pray. Dogs sometimes like to lick people, and this action would make the Muslim unclean for prayer. Therefore, it is generally accepted that dogs should stay outside. The same is not said about cats, because they do not normally try to lick their masters and are more reserved.[28]

Chapter 2
EVALUATION

1. __________________ is the view that a person should pursue his or her own self-interests, even at the expense of others.
2. __________________ is the philosophy where an action is determined to be morally right and outweighs any harm it might cause.
3. __________________ philosophies embrace the idea that people are equal to other life forms.
4. __________________ philosophies teach that man dominates God's creations.
5. Cows are considered the most sacred of all animals in the __________ religion.
6. ______________ is a doctrine of non-violence or non-killing.
7. All but __________ are forbidden food items according to kosher laws (birds of prey, insects, pork, hunted deer, hunted cattle, shark, lobster, turkey, and horses).
8. The ______________ religion forbids the eating of all meat and animal products.
9. The spirit of ________________ in Buddhism means the friendliness toward all living things.
10. ___________, a doctrine of non-violence or non-killing, means without injury and comes from Hindu and Buddhist philosophies.
11. _____________ meat cannot include birds of prey, shellfish, or hunted animals.
12. __________________ is the country where the "Jump of the Goat Fiesta" takes place.

Discussion Questions:

1. What is the major difference between Eastern and Western philosophies regarding animal use?
2. Why did the United States Army feel that killing all the bison would lead to the final solution of ridding the West of Native Americans?
3. Describe the reverence the Native Americans had for the bison they killed. How can one kill an animal and still have respect, honor, and love for that animal?
4. List at least three examples of traditional Jewish laws regarding animal welfare (how an observant Jew should treat an animal).
5. The prophet Mohammed taught that animals should be killed only out of necessity, and to do otherwise was a sin. How is necessity defined? If there are alternatives to eating meat, does the necessity still exist?
6. The new Catholic catechism, "Animals, like plants and inanimate beings, are by nature destined for the common good of past, present, and future humanity" has been described by some as permitting the terrible suffering of animals. Others

describe the catechism as an advance in the idea of mutual care of both humans and animals. Why do the different interpretations exist? How do you read this new catechism? (Catechism is defined as a book that describes the teachings of the Roman Catholic Church.)

7. Should the army or national police enforce the ban of the "Jump of the Goat Fiesta"? What if the villagers violently revolt again? How many human lives should be used to suppress another's religion to ensure the safety of an animal?

ENDNOTES

5. Cohen, C. 1986. "The Case for the Use of Animals in Biomedical Research." *New England Journal of Medicine*, 315(14):865.
6. Friend, T. H. 1990. "Teaching Animal Welfare in the Land Grant Universities." *Journal of Animal Science*, 68:3462.
7. Ventura County (California) Indian Education Consortium.
8. Hamilton, K. et al. 1997. "The Strange Case of Israel's Red Heifer." *Newsweek*, 19 May.
9. Thittila, B. U. 1956. *The Path of the Buddha: Buddhism Interpreted by Buddhists*. Ed.: Kenneth W. Morgan. New York, NY: The Ronald Press Company.
10. De Bary, William T. 1969. *The Buddhist Tradition in India, China, and Japan*. The Modern Library: New York, NY.
11. *Lankavatara Sutra*. 1923. Ed. Bunyiu: Nanjio, Kyoto: Otani University Press, p. 245.
12. Dhammapada is the moral law portion of the Pali scriptures.
13. Ven. T. T. Q., Buddhist monk and chief editor of a leading Buddhist magazine in Vietnam.
14. Wheeler, Kate. 1997. "Tricycle: The Buddhist Review." Vol. 14, No. 2. *Everyday Mind*, November.
15. Sing, K. "The Life of the People of India." ASI, Government of India: New Delhi
16. Curwen, M. 1993. "Call to Women." Great Britain, April.
17. Johnson, E. B. 1969. "Cultural Patterns in Asian Life." Field Ed. Publications: San Francisco, CA: p. 40.
18. As told by L. S. Shapiro, 1969.
19. Ryder, R. D. 1989. *Animal Revolution: Changing Attitudes Towards Speciesism*. Cambridge, MA: Basil Blackwell, Inc.
20. Note 859 in *The Holy Qur'an*, translation by Abdullah Yusuf Ali, 1934. Dar Al Arabia Publishing.

21. Note 2024 in *The Holy Qur'an,* translation by Abdullah Yusuf Ali, 1934. Dar Al Arabia Publishing.
22. Madani, B. Islamic Ref. Center.
23. Note 5839 in *The Holy Qur'an,* translation by Abdullah Yusuf Ali, 1934. Dar Al Arabia Publishing.
24. *Bukhari Hadith,* Book 7, No. 414, narrated by Rafi bin Khadij.
25. *Bukhari Hadith,* Book No. 173, narrated by Abu Huraira.
26. *Bukhari Hadith,* Book 4, No. 531, narrated by 'Aisha
27. *Bukhari Hadith,* Book 4, No. 535, narrated by Ibn 'Umar.
28. Madani, B. Islamic Ref. Center.

CHAPTER 3

History Behind the Animal Rights/ Welfare Movement

The first recorded use of animals in experiments to study bodily functions (and later to find cures for disease) occurred approximately 2,300 years ago. Veterinary medicine was already quite progressive in India around 1,000 B.C.E. A philosopher and scientist, Erisistratus, used animals in his experiments in the year 300 B.C.E. in Alexandria, Egypt. The Greeks practiced dissection on pigs and many other animals. Aristotle (384–322 B.C.E.) also made observations of animals, which enabled him to establish the sciences of zoology and comparative anatomy.

Objectives

The student should:

1. Outline the beginnings of animal use in the teachings of human anatomy, the study of human disease and search for treatments, and the preventative measures for both human and animal disorders.
2. Outline the beginnings of the protest, organization, and formation of the first groups' protection for animals used in farming, transportation, food agriculture, and research.
3. Describe the controversy among the largest animal rights organizations in their fund raising, fund spending, and use of funds in defense of animals.
4. Outline the formation of and amendments to the Animal Welfare Act (AWA).
5. Describe the differences in animal welfare and rights legislation in Europe from the legislation passed within the United States.

6. Describe the principles of scientific methods established in 1865.
7. Describe how animals are used today compared to how they were used and treated prior to the passage of the AWA.
8. Compare and contrast the various animal rights and welfare organizations as to their militancy, fund-raising and fund-spending practices.
9. Compare the philosophies of prominent Nazi leaders with those leaders in the animal rights community who advocate similar laws concerning vegetarianism, animal research, and the placing of animal rights equal to or greater than certain rights of humans.

One hundred years later, the Greek physician Galen used experimental surgery on animals to demonstrate for the first time that veins carried blood and not air, that kidneys produced urine, that visual images were conveyed by the optic nerves, and that speech skills can be destroyed by neck wounds. Galen is considered to be the founder of experimental physiology. Galen conducted anatomical investigations of monkeys, pigs, and dogs to help people understand human anatomy.

William Harvey (1578–1657 C.E.) was a physician who participated in dissections of most any animal that he could obtain. Harvey's work enabled him to discover some of the basic anatomical and physiological mysteries of the circulatory system.

In 1656, Christopher Wren performed the first intravenous injections in dogs. Wren used quills for needles. Although Wren was known as one of the greatest architects of his time (he designed 53 churches in London), his earlier interests were in science. Wren worked with physicist and chemist Robert Boyle (1627–1691 C.E.) to inject narcotics into a dog's bloodstream.

Boyle and Robert Hooke (1635–1703 C.E.), a philosopher, used mice in experiments to confirm that air contained substances necessary for life. During this same period (1665), Richard Lower performed the first known blood transfusions using dogs.

Although there was some discussion, no laws existed that were designed to protect animals used in these early experiments from cruel treatment. The first law in the modern-day world that protected farm animals from cruel treatment was *The Body of Liberties*, passed by the Massachusetts Bay Colony in 1641. This law did not affect animals used in research. In 1828, the New York state legislature passed its first anti-cruelty law:

> ***"Every person who shall maliciously kill, maim, or wound any horse, ox, or other cattle, or sheep, belonging to another, or shall maliciously and***

> ***cruelly beat or torture such animal, whether belonging to himself or another, shall upon conviction, be adjudged guilty of a misdemeanor."***

In 1820s London, Lewis Gompertz founded the first *Society for Prevention of Cruelty to Animals* (SPCA). Gompertz quoted Psalms (145:9), "The Lord is good to all; and His mercies are over all His works," as justification for the need to support his organization. SPCA's main focus was, and still is, to stop painful animal research. In 1870, the British Association for the Advancement of Science developed its first guidelines for conducting physiological experiments. These guidelines included steps to reduce experiments on animals and to minimize suffering which was not clearly visible.

The *American Society for the Prevention of Cruelty to Animals* (ASPCA) was formed in New York in 1866 by Henry Bergh. The association's main purpose was to look after the welfare of disabled horses and mules and save them from abandonment. The ASPCA was the first humane society in the United States. Henry Bergh saw that no one was prosecuted under the New York State Anti-Cruelty Law, so he drafted a new act: "an act for the more effectual prevention of cruelty of animals." This act had ten sections and today serves as the basis for anti-cruelty laws in forty-one states and the District of Columbia.

Several other anti-vivisectionist and humane organizations were formed following the introduction of the ASPCA to American society:

1877—American Humane Association

1883—American Antivivisection Society

1899—American Humane Education Society

1910—New York Antivivisection Society

1952—*Animal Welfare Institute* (AWI)

1954—*Humane Society of the United States* (HSUS)

One of the main focuses of animal rights organizations is the use of animals in teaching anatomy and other basic sciences, as well as in research areas. Some of the organizations mentioned earlier demand the complete elimination of all animals in research. Other organizations, such as the HSUS and the AWI, originally asked for humane treatment and restraints on unnecessary usage of animals. Today, the HSUS describes its purpose as "promoting animal rights." The HSUS has established a loan program to provide both students and instructors with alternatives to classroom animal dissection and live animal experimentation. The HSUS has recommended the use of video and computer simulation programs, charts, and models.

Anatomy instructors, especially in the veterinary teaching programs, have used models, charts, and computer simulations as adjuncts to their existing dissection programs. During the late-1980s and into the early-1990s, there was an 80 percent reduction in animal use in physiology and microbiology laboratories at Cornell University's College of Veterinary Medicine.[29] Much of this reduction was due to the availability of alternatives and changes in teaching methodology. Still, others have argued that to teach surgery skills and detailed anatomy, an animal cadaver is a "must" prior to cutting into live flesh (in a veterinary hospital). Human anatomy instructors, of course, would prefer using human cadavers for their programs—and have (cost permitting) always preferred this "alternative" to using animals.

Demonstrations show that the number of cadavers used in animal anatomy courses can be reduced to one cadaver for every four students, without reducing the students' mastery of anatomy.[30] Anatomy lessons can be reinforced with computer simulation models, charts, X-rays, skeletons, and other graphics. When using human or animal cadavers, students should be briefed on the respect for life that they will be studying.[31] The use of the cadaver should be used to minimize mistakes made on live humans or live animals. The knowledge and skills obtained should focus on those procedures that will eventually be used to save many more lives —not just to teach the "geography" of the body. No horseplay or disrespect for the "once live animal/human" should be tolerated. These simple guidelines have been adopted by leading colleges and universities that use cadavers as part of their educational process.[32]

Once the student shows competency with cadavers, computers, and non-living beings, the students can then move into live surgery—first watching and then assisting in normal, everyday procedures. The only surgeries performed in many veterinary and veterinary technician programs are the common spaying and neutering of animals. Thus, the cadavers become important for introducing the student to other systems' "geography," relationships and differences found within the same species, and special pathologies which can only be appreciated by necropsy.

In human medicine, the most educated and affluent members of our society demand the most experienced team of physicians (surgeons, surgical technicians, etc.). In animal medicine, one can assume that most animal lovers would *not* want anyone touching their animal in surgery who has not made all of their major "mistakes" on a cadaver. A gradual introduction to live animal surgery, under the direction of a licensed veterinarian, is usually a requirement before sitting for any state boards.

For the non-medical or non-veterinary student, alternative educational tools have shown to be quite useful for teaching anatomy. As previously mentioned, however, for the more advanced student who plans on eventually participating in surgery (whether animal or human), cadavers are usually the preferred alternative to live human or animal specimens. In high school science classes, the need for animal dissection is probably less justified. Films, computer simulations, models, and charts could feasibly make up the majority of most secondary science laboratories. Advanced secondary classes have shown some benefit to visiting college anatomy labs, as well as participating in necropsies at local animal shelters or hospitals.

Most of the alternative resources for dissection are quite expensive. No national organization has been willing to provide these alternatives "free of charge" to the schools needing them. The difference between a classroom set of cat cadavers and cat models is $350 for a set of ten cadavers, versus $3,350 for a set of ten models.[33] The models can be reused more often than the cadavers and thus have a longer "shelf life." The initial outlay of an additional $3,000, however, is prohibitive for most science classrooms.

The HSUS has tried to portray itself as a moderate animal welfare organization. In more recent years, however, the HSUS has taken on a more radical position and has been at odds with several law enforcement agencies over its fundraising and accounting methods. The HSUS has been criticized for not using its enormous annual budget to set up and maintain animal shelters and spaying and neutering clinics—and to provide free alternatives to animal dissection for school districts unable to pay for the more expensive substitutions.

John A. Hoyt, chief executive officer for the HSUS, has asked his members to "fight the well-financed and powerful agribusiness and research industries." Others have chastised this David versus Goliath depiction with a strong charge of hypocrisy. Two nationally respected publications have questioned the HSUS's financial practices.[34] Critics charge that Mr. Hoyt earns more than $250,000 a year in salary and benefits from this non-profit animal humane society. In addition, according to these same sources, the HSUS pays Mr. Paul G. Irwin, its president, an additional $209,051 per year in salary and paid $85,000 for renovations to his Maine cabin. The HSUS also allegedly pays salary supplements of $41,000 to Hoyt and $33,000 to Irwin over a three-year period. In 1982, Hoyt received a $100,000 interest-free loan from one HSUS board member and subsidized overseas trips for his wife from another member of the board. Irwin collected $15,000 in executor's fees from the estate of an HSUS board member without notifying the board of directors in advance (as mandated by the HSUS code of ethics). Irwin owns five homes, including a $768,500 residence in Darnestown, Maryland. He owns three expensive cars: a Mercedes, a Lincoln Town Car, and a Corvette. Thus, the claim by HSUS officials of hardships in fighting "well-financed agriculture and research organizations" does not bring much sympathy among those who know about the financing on the "other side." Many have asked why these high-priced salaries were necessary and why this money could not have been spent directly on reducing the suffering of the animals the organization claims to be helping.

The Better Business Bureau's *Philanthropic Advisory Service* (PAS), European law enforcement authorities, and a Canadian judge all have listed financial improprieties with the HSUS. A Canadian judge ordered the HSUS to return $1 million it allegedly took from its Canadian counterpart (*The Toronto Sun*, January 8, 1997). The *Humane Society of Canada* (HSC) claims that the HSUS illegally withdrew the million dollars in a money-laundering operation. The Canadians are suing for $5 million. Many ask whether these financial irregularities and misrepresentations of HSUS fund spending affect the society's capacity to serve the public in teaching moral responsibilities for the care of animals.

With a $40-million-per-year budget, some think that a dent could be placed in the overburdened animal shelter system. Our national shelters take care of some 15 million animals annually. Of these, approximately 11–13 million animals must be put to sleep each year. Some believe these animal shelters are greatly under funded. Most do not have enough money to adequately feed and care for their animals, thus requiring the animals to be prematurely destroyed before finding adoptive caretakers.

The HSUS, the nation's largest animal advocacy organization, is financially able to fund at least one well-run animal shelter in each state, pay for spaying, neutering, and feeding of these animals, and still have surplus cash. Yet, the HSUS supports no animal shelters. Most individuals who donate money to the HSUS (because of its advertising) believe they are helping reduce the suffering of these 15 million pets. Critics have argued that if indeed animal rights groups claim to really care about the welfare of animals, then why do they not contribute to the reduction of animal suffering in local humane shelters? Many believe millions of dollars should be given to reduce the food shortages, overcrowding, open sewage pits of animal waste, rodent, ant, and cockroach infestation, and lack of medical treatment in the shelters. In 1995, PETA's Ingrid Newkirk described a shelter which was so crowded that "people with animals have been turned away at the door." If these conditions are so bad and are recognized as such by the well-funded animal rights organizations, animal welfare enthusiasts ask the question, "Why don't they help fund the repair, instead of paying these lavish salaries and concentrating on small, isolated cases with much less impact on animal welfare?"

In defense of the HSUS, its educational staff has excelled at addressing alternative educational material for teaching courses in the life sciences. Effective non-animal and/or non-invasive alternatives have been adopted by many institutions as a result of the HSUS's publications, petitioning, and protests.

Most animal welfare organizations recognize the necessity to use animals in some research to find prevention and treatments for various diseases. Most also realize that some type of work on a cadaver—and eventually a living animal—may be necessary for the advanced medical student. The main concern is that excess animals should not be used, that animals should not be cut apart for repetitious work that could be demonstrated with other non-live animal means, and that anesthesia and analgesics should be used when appropriate so that animals do not suffer needlessly.

Several scientific organizations have formed since World War II. Their main purpose in forming was to improve the quality and care for animals used in research and to educate the public about the benefits of their research:

1946—The *National Society for Medical Research* (NSMR) was formed to improve the public's perception on the use of animals in biomedical research and in education.

1950—The *American Association for Laboratory Animal Science* (AALAS) was organized. This organization developed uniform standards of laboratory animal technician training. AALAS is primarily concerned with the production, care, and study of laboratory animals. In 1998, AALAS boasted a membership of 8,500 that consisted of clinical veterinarians, technicians, technologists, educators, researchers, and animal producers, whose stated goal was "dedicated to the humane care and treatment of laboratory animals."

1952—The *Institute of Laboratory Animal Resources* (ILAR) is a coordinating agency and serves doubly as a national and international resource for compiling and disseminating information on laboratory animals. ILAR promotes the humane care of laboratory animals within the United States.

1957—The *American College of Laboratory Animal Medicine* (ACLAM) establishes standards for the training and experience required for veterinarians concerned with the care and health of laboratory animals. ACLAM provides certification by examination to trained veterinary professionals.

1967—The *American Society for Laboratory Animal Practitioners* (ASLAP) was organized to provide training and encourage research in basic and clinical problems relating to laboratory animal practice. Membership is restricted to veterinarians who also exchange experiences and knowledge related to laboratory animal care.

Some additional historical events in the animal welfare movement are outlined in brief:

1906—The *American Animal Transportation Act* was passed to protect animals traveling by rail who were destined for slaughter.

1958—The *Humane Slaughter Act* was passed and was later amended to include the humane handling of animals prior to and during slaughter.

1966—Congress passed Public Law 89–544 (The Laboratory *Animal Welfare Act*, or AWA). President Lyndon B. Johnson signed the original law in 1966, and the law has been amended three times since then. The original law regulated dealers who handled dogs and cats, as well as laboratories that used dogs, cats, hamsters, guinea pigs, rabbits, and primates in research.

1970—The first amendment to the AWA (P.L. 91–579) was passed, which added the regulation of other warm-blooded animals when used in research, exhibition, or wholesale pet trade.

The *Horse Protection Act* (HPA), later amended in 1976, was also passed in 1970 to regulate horse show business, showing, and the sale of horses whose gait is altered by pain in their legs. Congress declared that the process of soring horses, either by chemical or mechanical practices or a combination thereof, is cruel and inhumane. Soring is defined as the application of any chemical or mechanical agent on any limb of a horse, or any practice inflicted upon the horse, that can be expected to cause it physical pain or distress when moving. The soring of horses is aimed at producing an exaggerated gait similar to the gait obtained by conventional training methods (but over a shorter period of time). Consequently, the people who exhibit sored horses sustain an unfair performance advantage over the people who exhibit non-sored horses.

The HPA prohibits the showing, sale, auction, exhibition, or transport of sored horses. The HPA is administered by the USDA through the *Animal and Plant Health Inspection Service* (APHIS). The HPA established the *Designated Qualified Person* program (DQP) to appoint people qualified to detect sored horses. Regulatory policy, procedure, and methods of inspection are reviewed throughout the year, with representatives of the horse industry being used to enforce and strengthen training programs. Farriers and members of the *American Association of Equine Practitioners* (AAEP) are usually licensed as DQPs.

Licensed DQPs receive inspection assignments at various shows and sales though their USDA-certified organization. Affiliation by show or sale management with a certified *Horse Industry Organization* (HIO) permits show management to fulfill its inspection responsibilities.

In 1993, of the 97,597 horses examined by DQPs, 1,198 were turned down because of non-compliance with the HPA (1.227 percent). The APHIS inspected an additional 25,527 horses and turned down 564 horses (2.209 percent). Criminal penalties of up to $3,000 and one year in prison can be assessed against individuals who knowingly violate the act. Each additional violation may result in fines of up to $5,000 and imprisonment for up to two years.

1976—The second amendment to AWA was passed (P.L. 94–279), which prohibited animal fighting ventures and regulated the commercial transportation of animals.

1985—The Improved Standards for Laboratory Animals Act (part of the Food Security Act) was passed as the third amendment to the AWA. The act required the following actions:

1. Issue of additional standards for use of animals in research
2. Exercise of dogs
3. An adequate physical environment to promote psychological well-being of non-human primates
4. Establishment of the *Institutional Animal Care and Use Committees* (IACUC) at the research facilities that set up standards regarding:
    a. The minimization of pain and distress
    b. The use of anesthetics and analgesics
    c. The use of tranquilizers
    d. The consideration of using alternatives to painful procedures

1989—The *Farm Animal and Research Facilities Protection Act* (Rep. Charles Stenholm, D-Texas) finally passed in the Senate in August 1992.

1990—The Food, Agriculture, Conservation, and Trade Act enables the following actions:

1. Authorizes the secretary of agriculture to seek an injunction stopping a licensed entity from continuing to violate the AWA while charges are pending
2. Requires the secretary to issue additional regulations pertaining to random-source dogs and cats (animals that have been obtained from animal and humane shelters for use in laboratories)
3. Covers animals in zoos, circuses, roadside menageries, transportation of animals, research facilities, and dog and cat breeders (see Figure 3–1)

1990—The PL 101–624 *Food, Agriculture, Conservation, and Trade Act,* Section 2503, Protection of Pets, was passed November 28, 1990. This act establishes a holding period for dogs and cats at shelters and other holding facilities before sale to dealers. The act also requires dealers to provide the recipient with written certification regarding each animal's background. Specific items included on the certificate

Figure 3–1 Circus bear act. *Courtesy of PETA*

are mechanisms of enforcement, injunctions, and penalties for violation.

1992—*The Animal Enterprise Protection Act,* Title 18–U.S. Criminal Code (P.L. 102–346) includes farms, zoos, aquariums, circuses, rodeos, fairs, auctions, packing plants, and commercial or academic enterprises that use animals for food or fiber production, agricultural research, or testing. The law imposes up to one year in jail for causing damage of $10,000 or more on an animal enterprise; up to ten years in jail if an intruder injures a person during an attack on an animal enterprise; or up to a life sentence in jail if an intruder kills another person during an attack on an animal enterprise.

1993—The *Federal Register,* Vol. 58, No. 139, July 22, P. 39124, issued a final rule on random source dogs and cats.

The final rules amending the regulations under the AWA require pounds and shelters to hold and care for dogs and cats for at least five days (including one weekend day) before providing them to a dealer. Dealers must provide valid certification to anyone acquiring random-source dogs and cats from that dealer. This law does not prohibit the shelter from having the animal euthanized, however, due to lack of funds or space before that time period is over.

In February 1997, Rep. Charles Canady (R-FL) tried to push through H.R. 596 through the 105th Congress. This "Pet Safety and Protection Act of 1997" would have stopped any biomedical research involving dogs and cats that were not specifically bred for that purpose. Although fewer than 1 percent of the 5–10 million abandoned dogs and cats killed in shelters every year go to Class B dealers and then to research, this bill, if made law, would threaten a necessary supply of random-source dogs and cats needed for medical research, for education, and for product testing. (See the Section 2 reprints on the use of pound animals).

EUROPEAN ANIMAL RIGHTS/WELFARE HISTORY

"From the late-17th century through the 18th century, a strong tradition emerged in England and France of animal experimentation based on the notion that animals are incapable of feeling pain."[35]

Two French physiologists, François Magendie (1783–1855 C.E.) and his student Claude Bernard (1813–1878), established live animal experimentation as common practice. Because of widespread public awareness of their experimentation, an antivivisection movement began in England (although their experiments were conducted in France). In 1876, the *Cruelty to Animals Act* was passed in England. This legislation was the first in the world to regulate the use of animals in research. The morality of animal experiments was at issue. This act drastically reduced animal experimentation. In response to the act, the *Association for the Advancement of Medicine by Research* was formed in 1882. Its purpose was to promote research and to seek a "just implementation" of the Cruelty to Animals Act.

During this time, anesthetics (which were not discovered until 1846) were *not* widely used in animal experimentation. (To be fair, however, anesthetics were not used widely in human medicine during this time, as well.) Hence, the amount of pain inflicted on animals was immense. Members of the scientific community expressed equal hostility to the work being conducted by Magendie and Bernard. Even 150 years ago, scientists believed and taught that animal experimentation was justified only if such research led to remedies for disease and pain—and if these experiments were performed with the least possible pain.

Bernard felt that "the science of life can be established only through experiment, and we can save living beings from death only after sacrificing others" (Bernard, 1865). He further stated, "It is essentially moral to make experiments on an animal, even though painful and dangerous to him, if they may be useful to man." Bernard did have limits to his experimentation, however. He would *not* use monkeys, because they resembled human beings.

In assessing Magendie's lack of compassion, we must also realize that had he waited until anesthesia was discovered, "the delay in the critical discovery of the function of spinal roots and the use of this finding may have been more costly to humans in the end."[36]

When judging Bernard, Magendie, and others of the 19th century, we must take into consideration the limitations of the time. In the 1860 publication *The Lancet*, Alfred Perry, an English physician, wrote about what he personally witnessed at the French veterinary school near Paris (Alfort Veterinary School):

> ***"Every week, old and worn-out horses and mules are provided, and the students of the two senior classes commenced, soon after nine in the morning, with slighter operations of bleeding from the neck and feet, nicking the tail . . . etc. At midday . . . (they) proceeded to perform the more serious operations of firing, lithotomy, neurotomy (respectively, burning with a hot iron, removal of stones, and dissection or cutting of a nerve) . . . and other operations equally painful. This lasted till near five in the afternoon when the classes were dismissed, and the animals,***

if not already dead from pain and loss of blood were dragged into the yard and destroyed." [37]

The veterinary professor defended this exercise, because "it accustomed the students to the shrinking of the animal when touched by the instruments, and it made them cool at operating." In addition, after much protest by British animal protection organizations, the French Academy of Medicine defended the practice of the veterinary school. While the British passed the subsequent 1876 Cruelty to Animals Act, the French did not pass any legislation controlling animal experiments until 1963.

RELATIONSHIP WITH TODAY'S USE OF ANIMALS

How does all of this history of animal experimentation relate to what is going on today?

Today, most animal experiments fall into one of three general categories:

1. Biomedical and behavioral research
2. Education
3. Drug and product testing

Biomedical research experiments follow the principles of scientific method established by Claude Bernard in 1865. These methods include the following requirements:

- Controlling all variables so that only one factor—or set of factors—is changed at a time
- Replication of results by other laboratories

Both of these requirements must be met for the experiment to be considered scientifically valid. While biomedical research advances our knowledge of how biological systems function and is used extensively by the medical establishment, behavioral research is directed toward learning factors that affect human and animal behavior.

Educational research, in many instances, is limited to dead animals and provides a training means to students in human and veterinary medicine, physiology, and general sciences. As pointed out earlier, some work with dead animals (cadavers) for veterinary students is required for the same justification in using human cadavers for human medical students. Numbers used, methods of obtaining animals, and how the animals were euthanized are all matters of concern for both animal rights and animal welfare organizations.

In drug and product testing, animals are used to determine the efficacy and safety of new drugs, the toxicity of chemicals to which humans and animals may be exposed, and the continued safety and efficacy of new batches of products already on the market. Most of these experiments are conducted by commercial firms to fulfill government regulations.

TODAY'S ANIMAL RIGHTS ORGANIZATIONS

More than 400 animal welfare and animal rights groups exist in the United States, with budgets believed to exceed $200 million. These organizations are classified as moderate, militant, or terrorist.

Moderates—Moderates play a small role in the animal rights movement. While many in this group appear to take an anti-science stance, it is more likely that they simply lack the basis for scientific understanding and have a more dogmatic approach to their positions. The moderate animal rights movement is not the same segment of society trying to improve the environment of animals used in research, food production, and sports. This separate group would fall under animal welfare, as defined earlier in the text.

Militants—Militants in the animal rights movement are more aggressive in method and are definitely more vocal. Many have goals of making the nation a vegetarian nation. Others see their movement as a part of a larger social revolution intended to remodel western society by changing its values, institutions, and laws. Their ultimate goal (according to Peter Singer) is to alter all of man's relationships with animals and to put an end to all forms of exploitation, whether for research, business, pleasure, fur, sport, transportation, or food. Their present concentration of effort on animal experimentation and factory farms was a strategic decision based on the belief that these would be the easiest causes for which to generate public support.

One of the larger militant animal rights groups is PETA. In 1995, it was estimated to have more than 400,000 dues paying members in the United States with chapters in many states. Founder Ingrid Newkirk states their philosophical belief that "a rat is a pig is a dog is a boy," and that human treatment of animals represents "super-racism" and "fascism." In October 1995, PETA announced the planned relocation of its 100-person headquarters to Seattle from the Washington D.C. area. As of December 1998, they have stayed in Washington D.C.

Terrorists—The leading radical organization in the United States is the ALF. Scotland Yard has also identified ALF as an international underground terrorist organization active in the United Kingdom, France, and Canada. As already mentioned, ALF has claimed responsibility for burning laboratories under construction in California, destroying equipment, records, and facilities across the country, and stealing ("liberating") research animals. Death threats and harassment of scientists and their families are included in their activities. At the University of California—Davis school of veterinary medicine, ALF caused more than $4 million in damage to a laboratory under construction. In 1991, ALF claimed responsibility for setting a fire to a Seattle-area animal food warehouse, for staging an arson attack on an experimental mink farm at Oregon State University, and for breaking into veterinary research labs at Washington State University.

Although PETA correspondent Kathy Guillermo claims that "PETA neither supports nor commits 'terrorist' acts," evidence released by the FBI contradicts this statement. In 1992, FBI director Louis Freeh cited one example where PETA sent $45,200 to defend Rodney Coronado, a member of the ALF who pleaded guilty to arson at the mink research facility at Michigan State University. Freeh claimed that the amount spent to defend Coronado was 15 times the amount PETA spent on animal shelters nationwide during that same year. The attack on that lab interfered with toxicology research that was designed to help both humans and marine life in the Lake Huron area.

In PETA's publication *Animals' Agenda* (March 1990), political scientist Kevin Beeday writes, "Terrorism carries no moral or ethical connotations. It is simply the definition of a particular type of coercion . . . It is up to the animal rights spokespersons either to dismiss the terrorist label as propaganda or make it a badge to be proud of wearing." The encouragement in PETA's publications and the financial support of those who commit these terrorist acts seem to indicate that PETA may not be the non-violent organization that most of its members believe it to be.

The following list describes several objectives of animal rights organizations in general:

1. Banning the use of animals
2. Impeding, curtailing, and controlling the conduct of research using animals by doing the following acts:
    a. Attacking laboratories
    b. Raising the costs of conducting experiments using animals by increasing regulations
    c. Establishing the rights of humans to act as guardians of animals or agents in the enforcement of animal welfare
    d. Diverting funds from research projects

e. Gaining the right to examine *all* animal research protocols
f. Prohibiting *all* animal research considered to be cruel or unnecessary by animal rights organizations

The following list describes some of the organizations' recent accomplishments:

1. Bills introduced in several state legislatures that ban the use of animals in biomedical research
2. Bills introduced in several state legislatures that establish the fact that animals have rights
3. In 1987, more than 85 separate bills were introduced in state legislatures. Today, there are close to 100 separate bills introduced every year across the country.
4. Massachusetts passed a bill which prohibited the use of pound animals for research and testing within its borders. More than 15 million unwanted dogs and cats are euthanized each year in North America alone. More than a dozen states ban the selling of these animals to laboratories. This bill forces labs to then buy animals bred specifically for research.
5. New York, in 1987, passed a bill that prohibits the release of any dog or cat from any shelter—except for adoption or return to its owner.
6. The social price of being a researcher is too high, causing a 27 percent drop in the number of college graduates with degrees in biomedical science between 1977–1987.
7. The Downed Animal Protection Act of 1992 was introduced because of a home video taken by a Minnesota animal activist that showed the handling of downer animals in the south Saint Paul stockyards.
8. PETA's recruiting of stars such as Paul McCartney and k.d. lang and the use of these stars' names in PETA's endorsements and fund-raising activities.

The following list describes where animal rights groups have *not* been successful (as of yet):

1. Attempts to divert 30–50 percent of *National Institutes of Health* (NIH) research money into the development of alternatives to such experiments using animals. In the 96th Congress (1980), H.R. 4805, the Research Modernization Act, was introduced. The act would have required a diversion of 30–50 percent of all federal funds supporting research and involving the use of animals to be spent for the develop-

ment of alternative methods for such research. This bill would have included funds coming from the *Food and Drug Administration* (FDA), the Department of Defense, the National Science Foundation, the NIH, the *Environmental Protection Agency* (EPA), the Department of the Interior, the *Veterans Administration* (VA) and military medical centers. This bill failed.

2. An attempt to seek legal sanctions to act on behalf of animals. The Federal District Court, the United States Court of Appeals (May 1994), and the United States Supreme Court denied the Animal Legal Defense Fund, the HSUS, PETA, and other animal rights groups the right to assume guardianship over animals. The courts said that the plaintiffs lacked the constitutional "standing to sue." They further stated that if they were to grant guardianship, this "might unleash a spate of private lawsuits that would impede advances by medical science in the alleviation of suffering." This decision overturned a lower court's ruling to include rats, mice, and birds under the AWA.
3. PETA has not been successful at convincing the general public that the Oscar Mayer Wienermobile is a "bad thing." The Wienermobile is a 27-foot-long, six-cylinder vehicle that travels from town to town promoting Oscar Mayer hot dogs. The vehicle is considered truly American. Attacking the vehicle and those around it, which are usually children, proved to be bad public relations for PETA, at least as reported in the summer of 1997. Most reported that it looked like PETA was attacking the children visiting the vehicle, rather than protesting for veganism. Headlines in the *Los Angeles Times* read, "PETA Gets Roasted While Hot-Dogging." The *Los Angeles Times*, usually a paper that is supportive of the animal rights movement, claimed that PETA's publicity ploy ignited "flames of indignation" among the children and parents they were targeting for support.

Animal rights organizations emphasize three arguments:

1. Animal research is immoral, regardless of the benefits.
2. Animal research has no value in treating human disorders.
3. Research is unacceptably cruel.

The moral argument of animal activists is that rights of animals are based on their capacity to feel pain—that is, suffering. Thus, anyone who causes a sentient being to suffer in any way is violating its rights—thus acting unethically. Using this analogy, murder would not be prohibited as long as it were done without warning and without pain. Peter Singer asserts that "if the experimenters would not be prepared to use a human infant, then their readiness to use non-human animals reveals an unjustifiable form of discrimination" (the idea of speciesism).

The argument against the animal rights interpretation of "rights" is that rights are based on the capacity to think or to reason. Philosopher Carl Cohen contends that rights are a moral concept, where certain actions are right and others are not. Thus, rights presuppose beings who are capable of grasping moral principles (beings who are rational). The concept of rights prohibits people from initiating force against others. To quote Edwin Locke (professor in the College of Business and Management, University of Maryland, College Park), "Rights protect people from other people in the name of human life. The concept is totally inapplicable to animals, which do not possess reason, cannot grasp moral principles, and cannot volitionally direct their actions." The concept of rights, Locke continues, "is not formulated by focusing on infants, the senile, the comatose, or the mentally ill. These fall into the special category of developing or sick humans." Locke uses another example of a special application of rights: that children have the right to life, but they do not have the right to vote or sign contracts.

No decent human being would want to gratuitously cause pain or suffering to an animal simply for the purpose of causing pain. Human action toward animals should be governed by sympathy, not by rights, argues Locke. Locke further emphasizes that the issue is not that man is morally superior to animals, but that animals are simply outside the realm of morality. Locke concludes that animal rights advocates are anti-life, their hatred of people is openly and loudly proclaimed. They must be fought, he counters, and stopped in the name of morality.

> ***"The idea of animal rights may set out to ensure that animals are treated as human beings, but by blurring the essential distinction between the two, it lends itself just as readily to the suggestion that human beings may be treated as animals . . . Thus, the doctrine that purports to elevate the status of all living things is, in the end, a doctrine that debases the status of mankind, and endangers our essential freedoms."***
>
> (Constance Horner,
> United States Under Secretary of Health and Human Services)

See the section in this book on Nazi Germany for examples of animals being given higher moral and legal status over some human populations. In more recent times, we have learned how our own government used African-American males in the infamous "Tuskegee Experiments" involving syphilis research. These men were valued far less than research animals.

While it is true that finding or developing an animal model for a disease can be extremely difficult, animal models are excellent sources of fundamental knowledge. "Most fundamental cellular processes are similar across species. Regulation of gene expression, mechanisms of nerve action potential, and macro-molecules that govern these processes have been highly conserved in evolution," according to Dr. W.T. Grenough, Department of Psychology, The Beckman Institute, University of Illinois, Champaign-Urbana.

Ever since Darwin unfolded his theory of evolution, scientists have justified using animals' similarities to humans as a means of finding cures and preventative measures—and as a means of understanding how our body works or does not work. Darwin and others who followed him showed that there were greater similarities than dissimilarities between humans and animals.

The goal of human disease research is to select the species that most closely resembles the biological function that one is studying in humans. The basic biological principles characterize a wide range of species. Computer models and cell and tissue cultures help to formalize a theory, but they do not tell you how a toxin or treatment works where several systems work synergistically (e.g., when a substance's effect upon a portion of the body is only seen after the substance is metabolized by, or acts upon, or is stored in, another organ). Even determining the concentrations of a substance to which a given body cell will be subjected in real life is dependent upon "a complex array of factors, such as how much enters the body through the skin or the lungs or the intestine, how much is broken down by the body's detoxifying systems in the liver, and how much simply passes through the body to be excreted." Computer simulations are suited only to assembling information already obtained (largely by laboratory work) and are largely useless in relation to entirely new medications or potential toxins. While computational chemistry has made great strides in its capacity to analyze known (and some unknown) toxins, live animal studies, which include interactions of organs and their systems, is a requirement for understanding mechanisms, treatments, and preventative measures.

Many of the non-animal tests still come from animal tissue. An example is the non-animal Ames Test. This test uses mouse liver cells to test for carcinogenicity and is praised by animal rights advocate Richard Ryder, author of *In Defense of Animals*. Ryder, however, fails to concede that the non-animal tests he promotes fail to screen out two carcinogens: diethylstilbestrol and vinyl chloride. The non-animal tests are generally less expensive, but they have not yet be proven to be as reliable as the scientifically accepted animal tests.

Diabetes researcher Mark Atkinson, M.D., developed a blood test for early diabetes detection, and that test indicated that researchers seeking cures for diabetes had failed in their initial attempts at human-based testing. Therefore, the researchers had to return to animal-based research to answer questions about immunological failings and detailed surgical techniques.

COMPARISON WITH NAZI GERMANY

Few people today suggest that Nazi philosophies are either acceptable or normal. It is well-documented that the Nazis treated human beings with extreme cruelty. The Nazis also promoted animal protection and environmental conservation. The documentation of the extensive measures taken by Nazis to ensure the humane care and protection of animals can be explained by the psychological assessments of many prominent Nazi political and military leaders, who felt affection and regard for animals but enmity and distance toward humans. The Nazis abolished moral distinctions between animals and people by viewing people as animals. The result was that animals could be considered "higher" than some people. As pointed out earlier, this philosophy was practiced in South Africa early in this century and in the United States during the 1930s Tuskegee Experiments.

At the end of the 19th century, kosher butchering and vivisection were the foremost concerns of the animal protection movement in Germany. In 1933, the Nazis passed a set of laws regulating the slaughter of animals and banning vivisection. Hermann Goering ordered an end to the "unbearable torture and suffering in animal experiments" and threatened to "commit to concentration camps those who still think they can continue to treat animals as inanimate property." Mercy killing of animals (painless death) was required. Hitler himself promoted vegetarianism and denounced Christianity as a "symptom of decay," because Christianity did not advocate vegetarianism. The German composer Richard Wagner respected race and animal protection (which was evident in his opera, "Parsifal"). The Nazis were more concerned with the suffering of lobsters and crabs being killed in restaurants and the shoeing of horses without the use of anesthesia than the millions of people they tortured and murdered every day. The rough mistreatment of an animal could result in a punishment of two years in prison—plus a fine. The Nazis preached that the mistreatment of animals was evidence of a fundamentally antisocial mentality or evidence of Jewish blood (Staff, I. 1964; *Justiz im Dritten Reich: Eine Dokumentation.* Frankfort: Fischer Bucherei). The Nazis thus clearly demonstrated the results of giving equal or greater "rights" to animals instead of humans—the willingness to kill and torture a human in defense of an animal.

The preamble to these new laws of 1933 clearly defines the superior rights given to animals over certain groups of people:

"The animals protection movement, strongly promoted by the National Socialist government, has long demanded that animals be given anesthesia before being killed. The overwhelming majority of the German people have long condemned killing without anesthesia, a practice universal among Jews though not confined to them, . . . as against the cultivated sensitivities of our society."

The purpose of the Law for the Protection of Animals was "to awaken and strengthen compassion as one of the highest moral values of the German people" (Giese and Kahler 1944; *Das deutsche Tierschutzrecht: Bestimmungen zum Schutze der Tiere*, Berlin: Duncker and Humbolt).

One of several controversial statements made by the cofounder of PETA can be used in an analogy of thoughts that prevail in modern times, when equal rights are given to animals: "Six million Jews died in concentration camps, but six billion broiler chickens will die this year in slaughterhouses" (Ingrid Newkirk, *The Washington Post*, November 13, 1983).

The German philosopher Friedrich Nietzsche rejected intellectual culture and reason and praised the animal instinct in man. Nietzsche played up the beast in man as a type of secret idol (the blond beast). He thought and described man as a predator. Man was domestic only on the surface. Nietzsche claimed that the humanitarian behavior toward fellow humans was insincere. Hitler, who identified with Nietzsche's philosophy, emphasized that the new German should emulate certain animal behaviors such as obedience and faithfulness and the strength, fearlessness, aggressiveness, and even cruelty found in beasts of prey.

Hitler's cruel training of his SS personnel is well-documented. After 12 weeks of his soldiers working closely with a German shepherd, Hitler had each of his SS soldiers break their dogs' necks in front of an officer so the soldiers could earn their stripes. Doing so, he thought, would instill teamwork, discipline, and obedience to the Fuhrer (Arluke and Sax, *Anthrozoos*, Vol V, Number 1).

Hitler demanded, "I want violent, imperious, fearless, cruel young people . . . The free, magnificent beast of prey must once again flash from their eyes . . . I want youth strong and beautiful . . . , and athletic youth . . . In this way I shall blot out thousands of years of human domestication. I shall have the pure, noble stuff of nature." These new Germans were to be part animal, renouncing a certain side of their humanity. The compassion normally reserved for humans was to be redirected toward animals, and the cold aggressiveness of animal instinct became the model German.

Today, we can find greater protection of companion animals—and greater punishment for those who abuse these animals—than for those doing similar crimes against people (see the section at end of this text). It is not uncommon in the State of California for an abused child to be returned to the abusing parent, after the parent goes for a counseling session and receives a court warning. When a citizen abuses a companion animal, however, the animal is rarely returned to its "owner," and a stiff fine and sometimes a jail sentence is invoked. Why is there so much more legal protection for the animal than for the child? Where does this pervasive philosophy come from that gives the animal greater protection?

German zoologist Ernst Haeckel attacked Christianity for putting man above animals and nature. He believed that man and animal had the same nature, as well as moral status. He supported "racial hygiene" through euthanasia, citing biological fitness of animals as essential to animal societies. Equating animals to humans, the Nazis also practiced this racial hygiene with humans—exterminating those they felt were biologically unfit (mentally retarded children and adults, those with physical deformities, homosexuals, etc.).

David DeGrazia in his *The Moral Status of Animals and Their Use In Research: A Philosophical Review* (see its entirety in Section 2 of this text) reviews and discusses the ethical concerns of granting animals moral status. In the German empire, animals were considered sentient objects and were accorded love and respect as a sacred and essential element in man's relationship with nature. According to Goering, "Animals are not, as before objects of personal property or unprotected creatures, with which a man may do as he pleases, but pieces of living nature which demand respect and compassion . . . For the protection of animals, the education of humanity is more important than laws."

Philosopher Richard Wagner urged the smashing of laboratories and the removal of scientists and "vivisectors." Wagner portrayed the vivisector as both evil and Jewish. Wagner also expressed views on the biological purity of Aryans and that the human race had become contaminated and impure through the mixing of races and the eating of animal flesh. "Regeneration of the human race" was linked to animal protection and vegetarianism," he said.[38] Wagner believed people had become corrupted by the blood of slaughtered animals. What was misleading about Wagner's statements and the statements of others was that Jews were not permitted to eat blood, nor to hunt. Animals, fish and fowl were divided into categories of which animals were to be eaten and which were not (Leviticus XI). These laws, which Jews had been following for generations, were laws created and followed not because of health reasons (although there are many healthful reasons that can be cited here)—but out of the respect for life—a compassion for other life forms.

Moses had previously taught the Jewish people a method of slaughtering animals. In the Old Testament (Leviticus III:17 and VII:26), it is prohibited to consume flesh containing blood. The purpose was to tame man's instincts toward violence by weaning him from blood and implanting within him a distaste—a horror—of all bloodshed. The Kosher method of slaughter (shechitah) causes the maximum effusion of blood with the remaining blood being extracted by means of the washing and salting of the meat. If any people were to be admired for eliminating blood from their diets, it should have been the Jewish people—and not the Germans that Wagner and others admired.

The degeneration that Wagner described in people who ate the blood of slaughtered animals was then spread, according to Wagner, through the mixing of races. Emulating Wagner, Hitler, and other elite Nazis (such as Rudolf Hess) became vegetarians. Hitler felt that civilization could be regenerated through vegetarianism.

The concept of creating the super race was central to the Nazi ideology and was characterized by the quest for racial purity and the elimination of "inferior races." Germans were to be treated as farm animals, bred for the most desirable Aryan traits, while ridding themselves of weaker and less-desirable animal specimens. Distinctions between humans, animals, and the larger, "natural" world were not to make up the basic structure of life. That which was regarded as "racially" pure—and that which was polluting and dangerous—were central to the Third Reich's philosophy. According to Hitler, non-Aryans were subhuman and should be considered beneath domestic animals. The Nazis' phylogenetic hierarchy had certain "races" beneath animals.

Chapter 3
EVALUATION

1. Animals have been used in experiments for more than ________ years.
2. _____________ used animals in experiments, helping him to create the sciences of zoology and comparative anatomy.
3. _____________ used animal experimentation to demonstrate that veins carried blood.
4. The first known blood transfusions by Dr. Lower in 1665 were performed on _____ (animal specie).
5. ____________________ was the first law in the modern-day world to protect farm animals from cruel treatment.
6. __________________ is the largest American animal rights organization—boasting a $40 million a year budget.
7. The number of animal shelters supported by this organization in the United States: ___________
8. Number of companion animals taken in by American animal shelters annually: ____________
9. Number of companion animals euthanized annually in the United States: ___________
10. The _____________ is the legislation that included the humane handling of animals prior to and during slaughter.
11. What year was the American Animal transportation Act passed?
12. Define soring.
13. The ____________ prohibits the showing, sale, auction, exhibition, or transport of sored horses.
14. ________________ was the year the Animal Welfare Act first became law in the United States.
15. Which animals were originally covered by the Animal Welfare Act?
16. What is a DQP?
17. The ________________ Act makes it a felony and imposes a fine of up to $10,000 and a jail sentence of up to a year for individuals who injure a person during an attack on an animal enterprise.
18. The _____________ requires animal shelters to hold and care for dogs and cats for at least five days before providing them to a dealer.
19. If indeed animals have equal moral and legal rights to humans, as some animal rights activists claim, then should we not be spaying and neutering the human species as we do to companion animals to correct the overpopulation we currently have? If your answer to this question is yes, then which humans do we not castrate and "spay"? Who decides which humans are able to breed? Who chooses the

mating of these humans left to breed? Answer these questions, keeping in mind that humans and animals have the same moral and legal rights. How is this different from what Nazi Germany did in requiring the "mating" of individuals, who chose, who kept the offspring, etc.? What you are willing to do to one specie you should be willing to do to the other, unless you are a speciesist. If you disagree with this analogy, justify your disagreement.

20. Why is it considered cruel to *not* euthanize an animal who has a terminal illness and is suffering, yet it is considered barbaric (and illegal) to suggest such treatment to a human?
21. Why did Hitler denounce Christianity as a "symptom of decay?"
22. What types of laws did the Nazis pass to protect animals in 1930s Germany?
23. The goal of human disease research is to select the species that most closely resembles the biological function to the species being studied. If that is the case, then why do we not use the animal closest to the one we are studying: humans? Justify and explain why or why not? Cite examples where human research included humans—and today is looked at as "inhumane." Compare and contrast these thoughts to current charges from animal activists, when animals instead of humans are used in similar experiments.
24. Most animal experiments used today use rodents. Is it still speciesism to use mice in research to find cures or prevention for diseases in cattle, dogs, cats, or horses? Justify this use. Justify not using the mice. Remember the definition of speciesism.
25. What are the objectives of most animal rights organizations?
26. What are the objectives of most animal welfare organizations? What are the differences?
27. Cite recent accomplishments of animal rights groups. In which areas have these groups not been successful?
28. What is the leading, radical, terrorist animal rights group in both the United States and Great Britain?
29. What are the two methods established by Claude Bernard in 1865 that are still used today in experiments using the principles of scientific method? Will the cloning of animals help reduce the numbers of animals (replications) required in the scientific method? (See the later section on cloning.)
30. Dr. Orlans' description of the mistreatment of animals at the French veterinary school (Alfort) in 1860 was and is disturbing. How could we allow such treatment of animals? Why do we generally not have the same disdain for similar treatment of humans? At the exact same time throughout the world, human patients were strapped down, and surgery was performed without the use of anesthesia. During the American Civil War, soldiers had limbs amputated, bullets removed, and flesh sutured—all without anesthesia. Where is the same condemnation for this treatment? After all, anesthetics had been used to reduce pain during surgical procedures since 1846 (diethyl ether). Animal anesthesia (application of pressure on the nerves, freezing of the part to be operated on, and inhalation of fumes from the

extracts of narcotic plants) had been given to animals since about 1800. So why did the French elect *not* to use these pain reducing methods at Alfort?

31. Can one ever justify such painful experiments on animals in modern times? How about in people?
32. The ________________ was the first legislation in the world to regulate the use of animals in research.
33. What was Bernard's justification of using animals in research?
34. Why did Bernard feel that monkeys should *not* be used in animal research?
35. Compare and contrast the use of animals in biomedical research as outlined in the "White Paper" by the AMA, literature from the Americans for Medical Progress, and information sheets provided from PETA (all can be found in Section 2). Why is there such a difference in perception?
36. Veterinarian Adrian Morrison outlines "misunderstandings" of the animal rights movement in the United States (see her article in Section 2). Compare and contrast her view and other scientists' views of the need to use animals to find cures and preventions for disease—and the views of PCRM and PETA, which claim the opposite.
37. Compare and contrast the German empire's description of the moral status of animals with that of modern-day animal rights activists. Describe the Nazi phylogenetic hierarchy. Where were blacks, Jews, Catholics, homosexuals, and gypsies in relation to animals?
38. Describe the "regeneration of the human race" as it was linked to animal protection and vegetarianism.

ENDNOTES

29. Dr. Franklin M. Loew, Dean, College of Veterinary Medicine, Cornell University.
30. Pierce College Pre-Veterinary Program.
31. Policy Statement, "Dissection of Animals in the Classroom," Educators for Responsible Science.
32. Pierce College, Cal Poly SLO, Oregon State University, Georgetown University, and Tufts University.
33. Carolina Biological Supply Company, record number 1776.
34. The Washington Post, September 25, 1996; U.S. News and World Report, January 22, 1996.
35. Orlans, B. 1993. In the Name of Science. Oxford University Press, p. 4.
36. Sechzer, Jeri. 1983. "The Ethical Dilemma of Some Classical Animal Experiments." The Role of Animals in Biomedical Research. Annals of NY Acad. Sciences, Vol. 406, pp. 5–12.

37. Orlans, B. 1993. In the Name of Science. Oxford University Press, p. 17.
38. Viereck, P. 1965. Metapolitics: The Roots of the Nazi Mind. New York, NY: Capricorn.

RECOMMENDED READINGS: (From Section 2)

1. "Animals in Research." *JAMA*, Vol. 261(24): 3602–3606. 1989.
2. HSUS-PETA Connections
3. Loeb, J.M. et al, *Human vs. Animal Rights. In Defense of Animal Research. JAMA*, 262(19): 2716–2720. 1989.
4. Policy on Humane Care and Use of Laboratory Animals. AALAS.
5. Policy Statement: Dissection of Animals in the Classroom (Educators for Responsible Science).
6. "The Humane Care & Treatment of Laboratory Animals." NABR. 1995.
7. "Use of Animals in Biomedical Research. The Challenge and Response." An American Medical Association White Paper, 1992 (revised).
8. "Use of Animals in Medical Education." *JAMA*, Vol. 266(6): 836–837. 1991.
9. World Medical Association Statement on Animal Use in Biomedical Research.

CHAPTER 4

The Three Rs as Applied Practically to Animal Use

In 1959, *The Principles of Humane Experimental Technique*, written by Russell and Burch, described a method of minimizing unnecessary suffering and use of animals in laboratories via the implementation of the three Rs. The authors pointed out that the three Rs were not simply a humane goal, but they were also a scientifically sound goal. The three Rs include *replacement*, *reduction*, and *refinement*.

OBJECTIVES

The student should:

1. Outline the three Rs and give examples of how each could be implemented.
2. Describe the procedures of experimental design for new medical treatments.
3. Outline a database search for a topic on animal ethics.
4. Describe the functions of the IACUC.
5. Describe the individuals serving on the IACUC.

Typically, in *in vivo* (whole animal) research, a scientist will select a group of animals, use half of them as a control group, and administer the substance to be tested to the experimental group while administering a placebo to the control group. The scientist will then observe differences between the two groups to come up with some kind of conclusion as to the effect (or lack thereof). In many cases, but not all or even most, the experimental animals need to be sacrificed to examine their tissues. This final assessment of a product is needed to determine risk or safety of its use. While no one wants to eliminate risk assessment (no parent would give their child a *new* cough syrup or antibiotic that had not undergone extensive safety testing), most would agree on the need to develop alternatives to the traditional whole-animal models used. Lower species do not precisely match humans, but a great deal can be learned from models of particular systems or organs that mimic those of humans. For example, thousands of dogs died as medical scientists learned to perform open-heart surgery.

Scientists now have available a variety of animal and non-animal methods to conduct biomedical research, but many questions can still be answered only through the study of a whole living system. Cell structures and tissue studies, rather than animals, are "adjuncts, but not alternatives," to whole-animal testing. What this idea means is that these non-animal tests complement, but do not always or completely replace, animal studies. Animals, including humans, are so complex that they cannot be duplicated by using adjuncts. Computer models, to be effective, must be based on actual animal laboratory research. Cell cultures and computer models cannot show what happens inside the brain or how a new drug will affect the heart, the nervous system, or a developing fetus. The continued use of laboratory animals is still essential; however, a reduced number of animals are required due to refinement in statistics and actual procedures used in many research projects. With the inclusion of cloned animals, fewer and fewer research subjects will be required (see the section in this book on cloning). In addition, continued efforts to improve alternative non-animal methods are ethical necessities.

Today, there is an international consensus within the research community to speed up the search for scientifically valid alternatives to whole-animal test models. Research companies around the world are spending millions of dollars searching for new, scientifically accepted alternatives. Currently, our knowledge does not permit us to use a single-cell culture to mimic the complicated physiological actions that take place in a whole-system animal (one that includes the interactions of the endocrine, immune, circulatory, neurological, and digestive systems). In addition, international law prohibits the use of human beings as subjects for such experiments.

Researchers do become attached to the animals they use in their research. Thus, the dichotomy exists between the love of science (in finding a cure or a treatment to a disease) and the love of the life with which the researcher is working. During the early '90s, one researcher describes his feelings while assisting in the development of a product that would speed up the healing of burned epithelial cells (skin).[39] Initially, tissue cultures and computer models were used. Eventually, the researchers had to use whole-animal models. Toward the end of the experiment, the animals used (pigs) had to be sacrificed to examine the various body organs for "possible side effects." One intern had grown attached to three of the pigs used in this experiment and questioned the need for their sacrifice—when the computer model, the tissue cultures, and the live animals all showed benefit without any "harm." Upon completion of the necropsy (examination of an animal's body after death), the need for their sacrifice became apparent. The pig's skin did heal much more quickly, which would be of great value to both human and animal burn patients. The intern remembered the dog that was severely burned in the 1994 Northridge (California) earthquake. One local doctor, an expert in human burn healing, initially worked on this canine, and then the dog was transferred to the University of California-Davis Veterinary College. With our current knowledge and skill, there is only so much one can do. The canine did not survive.

The intern further commented that he was hoping, as were the researchers involved with this study, to find a new product that would have helped this patient and others. The veterinary pathologist conducting the necropsy quickly discovered that the product was destroying the liver. The side effect was an unacceptable one—a side effect that no computer and no tissue culture could have predicted.

No researcher wants to kill or cause pain to the animals in his or her laboratory. Of course, there are exceptions (although few in number) to this observation. These exceptions (animal abusers) should *not* be employed by the research community. Animals used in research are not strays off the street. They are primarily specially bred and raised animals that are expensive. All of the researchers that this intern worked with seem to have grown attached to these "assistants" in their labs. Non-animal methods are cheaper and easier to use, and they do not have the uncomfortable feeling attached to them of having to cause one animal harm to reduce harm in another.

Russel and Burch stated that "good science and animal welfare are not incompatible." When an animal is stressed, experimental results may be misleading or erroneous. By following the three Rs, a more precise science—as well as a more humane one—should result. The three Rs *do not* mean the elimination of animals in research. They mean (when appropriate) the replacement, reduction, and refinement of the experiment. Replacement can mean eliminating the need for whole-animal testing by replacing with an *in vitro* (tissue) culture. Replacement can also include using mice instead of dogs (animals lower on the phylogenetic hierarchy). In the modern western world, humans are at the top of the phylogenetic hierarchy. Humans are not equal to animals. Dogs are not equal to mice. Mice are not equal to cockroaches. *A fly is not equal to a carrot, which is not equal to a dog, which is not equal to a pig.*

Reduction of animals might mean using three animals in a typical Draize test instead of six. Six have traditionally been used over the years (for statistical significance). Once again, with the incorporation of cloned animals, we should be able to greatly reduce the number of animals used in most experiments. Refinement might include a better method of providing post-operative analgesics to the animal to minimize the pain or discomfort. In general, refinement alternatives improve the design or efficiency of an experiment, lessening the distress to the animal.

Continued research is needed to find non-whole-animal methodologies, while still protecting both the animal and the human population that animal testing provides. Johns Hopkins' *Center for Alternatives to Animal Testing* (CAAT) is one of the leading centers promoting such research. There are other research centers smaller in size and scope doing similar work (including one at the University of California-Davis).

For an example of how the three Rs are practically implemented, see "Review of Concepts Case Study" in the next section.

PUBLIC UNDERSTANDING OF SCIENCE AND THE IACUCS

ANALYZING PUBLIC POLICIES

In 1985, the United States Congress passed two laws that required facilities conducting biomedical research on animals to establish *institutional animal care and use committees* (IACUCs). These laws were the *Health Research Extension Act* (HREA) and the *Improved Standards for Laboratory Animals Act* (ISLA). The ISLA was an amendment to the AWA. Both amendments require the IACUCs review proposals for research and periodic inspections of research facilities, among other activities.

Both HREA and AWA amendments mandated that IACUCs include at least one member who is not otherwise affiliated with the facility. The AWA, in addition, requires that the *non-affiliated members* (NAMs) be representative of the general community interests in the proper care and treatment of animals. Much controversy exists as to who these NAMs should be. Research scientists typically appoint or approve of RVTs (Registered Veterinary Technicians), veterinarians, or scientists from other institutions who do *not* have a conflict of interest with the research at hand. Animal protectionists would prefer to have members of their organizations on the IACUCs. Research scientists fear sabotage of their work, among other factors, and have tried to prevent such inclusion. Several successful IACUCs, however, have included such noted animal protectionists as Barbara Orlans from Georgetown University.

The main purpose of the IACUC is to ensure ethical and sensitive care and use of animals in research and that scientific reliance on live animals is minimized. Each institution which falls under authority of the AWA and/or the *Public Health Service* (PHS) receives PHS support for research and teaching involving laboratory animals and must operate a program with clear lines of authority and responsibility, a properly functioning IACUC, procedures for self-monitoring, adequate veterinary care, a program of occupational health, sound animal husbandry practices, and appropriate maintenance of facilities for housing animals.

IACUCs must have at least five members, including a veterinarian with program responsibilities, a scientist experienced in laboratory animal research, a non-scientist, and an individual who has no other affiliation with the institution besides membership in the IACUC.

The IACUC must do the following tasks:

1. Have the full support of the institutional official responsible for the program
2. Evaluate the entire program every six months
3. Prepare a report on the evaluation and the inspection of the facilities, which is filed with the institutional official
4. Make recommendations to this official concerning deficiencies, with a proposed timetable for corrections
5. Have the authority to suspend PHS-supported research activities

The IACUC has an obligation to review all research projects proposed for PHS support prior to receiving funding. A written report of this review confirms that the project will be conducted in accordance with PHS policy, the *Institutional Animal Care and Use Committee Guidebook*, and the AWA. At least one member of the committee must review each proposal, but all members must have prior opportunity to request full committee review. The IACUC has the authority to approve, require modifications before approval, or withhold approval of proposals submitted. *No activity* involving animals can begin unless the activity is *first* approved by the IACUC.

FEDERALLY MANDATED IACUC FUNCTIONS

1. Review, at least once every six months, the research facility's program using USDA regulations and the Guide as a basis.
2. Inspect, at least once every six months, all of the animal facilities, including animal study areas/satellite facilities, using USDA regulations and the Guide as a basis.
3. Prepare reports of IACUC evaluations and submit the reports to the institutional official.
4. Review and investigate legitimate concerns involving the care and use of animals at the research facility (resulting from public complaints and from reports of non-compliance received from facility personnel or employees).

5. Make recommendations to the institutional official regarding any aspect of the research facility's animal program, facilities, or personnel training.
6. Review and approve, require modifications in (to secure approval), or withhold approval of those components of proposed activities related to the care and use of animals.
7. Review and approve, require modifications in (to secure approval), or withhold approval of proposed significant changes regarding the care and use of animals in ongoing activities.
8. Suspend an activity involving animals when necessary, take corrective action, and report the incident to the funding agency and to the USDA.

FEDERAL CRITERIA FOR GRANTING IACUC APPROVAL

- Activities must be in accord with USDA regulations and PHS policy.
- Pain/distress must be avoided or minimized. If pain or distress is caused, appropriate sedation, analgesia, or anesthesia will be used. The attending veterinarian must be involved in planning. The use of paralytics without anesthesia is prohibited. Animals with chronic or severe, unrelievable pain will be painlessly euthanized.
- Surgery must meet requirements for sterility and pre/post-operative care. One animal cannot be used for several major operative procedures from which it will recover—without meeting specified conditions.
- Euthanasia methods must be consistent with USDA regulations and AVMA recommendations.
- Animal living conditions must be consistent with standards of housing, feeding, and care directed by a veterinarian or a scientist with appropriate expertise.
- Alternatives must be considered to painful procedures, and consideration of alternatives must be documented if animals experience pain or suffering.
- A written narrative of methods or sources must be provided.
- A limited amount of duplication is an essential part of the scientific method as scientists confirm and expand on findings of other scientists. Scientists must provide assurance, however, that activities do not unnecessarily duplicate previous efforts.
- Personnel must be appropriately qualified.
- Deviations from requirements must be justified for scientific reasons, in writing.

ROLE OF THE VETERINARIAN ON THE IACUC

Institutions using animals for teaching and research are required by law to have an attending veterinarian associated with their animal care and use program (unless only rats and mice are used, *and* they receive no federal funds).

The veterinarian's chief responsibility is to provide for the health and welfare of the animals. The veterinarian is the spokesperson for the animals being used. He or she must coordinate with the technical staff to ensure adequate, daily animal husbandry. Immunization against disease-causing agents, surveillance of animal colonies for specific infectious microbial agents, prophylaxis utilizing pharmaceutical agents, isolation and quarantine of incoming animals, and separate housing of animals according to species and source falls under the consultation and/or administration of the veterinarian. The veterinarian is responsible for monitoring animal health and providing adequate diagnosis and treatment of animals when illness or injury dictates veterinary medical care. The veterinarian also has the responsibility to assure proper usage of anesthetic and analgesic agents appropriate for the species and surgical procedures being used.

In most institutions, the veterinarian and his or her staff is also responsible for the training and instruction to institutional personnel on humane methods of animal maintenance and experimentation. USDA regulations require such training in all institutions using animals in research and teaching.

REQUIRED CONTENTS FOR AN INSTITUTIONAL TRAINING PROGRAM

(Source: USDA Regulations, 9 CFR Part 2, Subpart C, Section 2.31. *Federal Register,* August 31, 1989)

Training and instruction of personnel must include guidance in several areas:

1. Humane methods of animal maintenance and experimentation, including the following:
   i. The basic needs of each species of animal
   ii. Proper handling and care for the various species of animals used by the facility
   iii. Proper pre- and post-procedural care of animals
   iv. Aseptic surgical methods and procedures
2. The concept, availability, and use of research or testing methods which limit the use of animals or minimize animal distress

3. Proper use of anesthetics, analgesics, and tranquilizers for any species of animal used by the facility
4. Methods whereby deficiencies in animal care and treatment are reported, including deficiencies reported by any employee of the facility. No facility employee, committee member, or laboratory personnel shall be discriminated against or be subject to any reprisal for reporting violations of regulations or standards under the AWA.
5. Utilization of services (e.g., the National Agricultural Library, the National Library of Medicine) that are available to provide information about the following topics:
   i. Appropriate methods of animal care and use
   ii. Alternatives to the use of live animals in research
   iii. Unintended and unnecessary duplication of research involving animals
   iv. The intent and requirements of the AWA

MONITORING RESEARCH AND TEACHING ACTIVITIES USING FARM ANIMALS

This monitoring requires the IACUC to deal with routine problems and ethical dilemmas of daily agricultural traditions. Monitoring is complicated by administrative procedures found in most colleges of agriculture. The PHS policy and USDA regulations do not cover food and fiber research (only biomedical research). The welfare of the animals should not differ.

A major goal of farm animal research is to reduce the labor and feed required to produce food and fiber, thus increasing efficiency of production. Many of these innovative methods decrease the activity and alter the behavior of the animal. Ethical questions are compounded with economic, bioethical, and environmental questions.

The *Guide for the Care and Use of Agricultural Animals in Agricultural Research and Teaching (Agri Guide)* was prepared specifically for farm animal research activities with special consideration given to current practices and issues in commercial agriculture.

The IACUC for farm animal institutions should include the following members:

1. A scientist from the institution with experience in agricultural research or teaching involving farm animals
2. An animal scientist with appropriate training and experience in the management of agricultural animals—and with recognized, highly

professional credentials as verified by the scientific and professional societies in animal, dairy, or poultry science

3. A veterinarian who has appropriate training and experience with farm animal medicine and is appropriately licensed
4. A non-scientist affiliated with the institution
5. A person not otherwise affiliated with the institution
6. Other members as required by the institutional needs and applicable laws, regulations, and policies

Although not currently required by law, the monitoring of food and fiber animal research and teaching activities can significantly benefit an institution by improving the overall quality of the animal care program and improving the image of the college with the community at-large.

REVIEW OF CONCEPTS CASE STUDY

(reprinted with permission from AWIC, the National Agricultural Library, Beltsville, MD)

SITUATION

Dr. Stanley Braeger has always used pigs and dogs in his advanced trauma life support training course. All procedures are conducted on anesthetized animals. When the training session is complete, all animals are euthanized. He plans to continue using the same animal models in future semesters. He presents his protocol to the IACUC, which asks that he search for "alternatives."

LEGISLATION

The committee chairperson explains that Dr. Braeger's animal models are covered by the AWA and refers him to Title 9, Code of Federal Regulations, Chapter 1, Subchapter A—Animal Welfare. Because the protocol involves the use of anesthesia to relieve pain, she cites the following passage from Section 2.31:

> ***"The principal investigator has considered alternatives to procedures that may cause more than momentary or slight pain or distress to the animals and has provided a written narrative description of the methods and sources, e.g., the***

Animal Welfare Information Center, used to determine that alternatives were not available . . . ."

Although not required by law, the *Animal Welfare Information Center* (AWIC) is mentioned as one source of guidance (others include the Johns Hopkins Center for Alternatives to Animal Testing, the University of California Center for Alternatives, etc.). Dr. Braeger contacts a technical information specialist at AWIC.

ALTERNATIVES

AWIC and the scientific community currently use the three Rs—reduction, replacement, and refinement—to define alternatives. The concept was conceived and developed by W. M. S. Russell and R. L. Burch (1959) in their book, *The Principles of Humane Experimental Technique* (currently being published by the Universities Federation for Animal Welfare, Hers, England). Reduction means reducing the numbers of animals used in the experiment without compromising results and maintaining statistic integrity (e.g., transgenics and new techniques). Replacement refers to substitution, either wholly or in part, of the animal model with a non-animal model (e.g., computer simulation or in vitro methods), or using an animal that is lower taxonomically. Refinement includes the use of techniques that can reduce or eliminate pain and distress in the animal model, such as the use of positive reinforcement training or more efficacious anesthetics or analgesics. The information specialist and Dr. Braeger discuss this concept in more detail so that the broadened definition of alternatives is clearly understood.

LITERATURE SEARCHES

Now that Dr. Braeger understands the "alternatives" concept, he asks how to determine whether they exist—and what "written narrative description" means. The AWIC staff member tells him that *Regulatory Enforcement and Animal Care of the Animal and Plant Health Inspection Service* (APHIS/REAC) requires that the availability of alternative methods be determined by a literature search. The documentation is provided by indicating which databases were searched, which keywords were used, and a printout of the literary citations during the search.

The fastest and most comprehensive way to do a literature search is by using an online service such as DIALOG (3460 Hillview Ave., P.O. Box 10010, Palo Alto, CA 94303-0993; phone: 800-334-2564). The online service contains several hundred databases, including Medline and Embase (medical databases), which can be selected and searched simultaneously. AWIC can help Dr. Braeger select databases and develop a search strategy, so he can run the search at his university library or department. If he did not have access to the online services, AWIC could run the search for him, but AWIC would have to charge him for all online costs that exceed $25.

DATABASE SELECTION

Each database contains bibliographic citations from professional journals which often include abstracts—and, occasionally, the full text of the article. There is some overlap of journals covered among databases, but each contains materials unique to that database. Dr. Braeger describes his protocol to the information specialist. Based on the description, the specialist recommends not only biomedical databases but also a few databases that cover computer simulations and bioengineering.

The following list shows the databases selected by the specialist:

MEDLINE. . . . . TOXLINE
EMBASE . . . . HEALTH PERIODICALS DATABASE
PASCAL . . . . INSPEC 2
BIOSIS PREVIEWS . . Ei Compendex Plus™
SCISEARCH . . . . MICROCOMPUTER INDEX™
NTIS. . . . . Federal Research in Progress
TRIS . . . . Nursing and Allied Health

SEARCH STRATEGY

The search strategy is listed in the following table. Next to each search "set" is the number of citations retrieved from the selected databases ("Items"), followed by the keywords searched ("Descriptors"). "?" indicates a truncated word. For example, "TEACH?" retrieves "TEACH," "TEACHES," "TEACHING," AND "TEACHER." "RD" removes the duplicates from overlapping databases, retrieving only unique citations.

Set	Items	Descriptors
S1	3947	(trauma or life(w)support)/Tl,DE and (train? or teach?)Tl,DE
S2	68	S1 and (alternative? or computer(N)simulate? or model? or cadav? or videodisc)
S3	54	RD

S1 is a broad search using keywords that relate to life-support training. If the keyword or keyword combination is found in the title, abstract, or descriptor of an article indexed in the database, the citation will be retrieved. The addition of "TI,DE" reduces the number of citations by retrieving only those with the keywords appearing in the title and descriptors.

Because S1 produced 3,947 citations, the search is narrowed in S2 by adding keywords which offer potential alternatives. When duplicates are removed in S3, 54 citations remain from the original search set. Because there are so few citations, it is not necessary to limit the search more. All animal and non-animal model alternatives are cited, including those that use pigs and dogs.

DOCUMENTATION

Dr. Braeger then prints out his search. He submits a copy to the IACUC if requested and keeps a copy in his records to document that he has considered alternatives. This process meets the documentation requirement requested by the AWA. He must review the literature he has found and discuss with the IACUC whether any of the alternatives in the citation list are relevant to his research. As can be seen from the sample citation list attached, potential alternatives are available—such as the use of cadavers or videodiscs.

Remember that every search is unique. Selection of databases and the search strategy will differ depending upon the protocol and the questions being asked.

SAMPLE CITATIONS

```
(item from file: 73)
7770940  embase nO: 90199937
Animal cadaveric models for advanced trauma life
support training.
Eaton B.D.; Messent D.O.; Haywood I.R.
Royal College of Surgeons of England, London  United
Kingdom
Ann. R. Coll. Surg. Engl. (United Kingdom), 1990, 72/2
(135-139)
(item from file: 155)
06085558  87059558
Interactive videodisc to teach combat trauma life
support
Henderson JV; Pruett RK: Galper AR; Copes WS
J Med Syst   Jun 1986,  10(3) p271-6, ISSN 0148-5598
```

Chapter 4

EVALUATION

1. In 1959, _______, written by Russell and Burch, described a method of minimizing unnecessary suffering and use of animals in laboratories.
2. ______________ means (whole-animal) research.
3. CAAT is located at which institution? ____________________________
4. ____________ was the year the AWA was initially passed by Congress.
5. CAAT stands for ____.
6. USDA stands for ____.
7. In 1985, the United States Congress passed two laws that required facilities conducting biomedical research on animals to establish IACUCs. These laws were:
   a. ____.
   b. ____.
8. IACUC stands for _______________________________________.
9. NAM stands for __________________________.
10. The __________________ has an obligation to review all research projects proposed for PHS support prior to their receiving funding.
11. _____________________ must prepare reports of animal facility inspections and submit them to the institutional official.
12. Institutions using only rats and mice in research and/or teaching, while receiving federal funds, must have a(n) ____________ associated with their animal care and use program.
13. A(n) ______________________ is the spokesperson for the animals being used in any research facility using animals.
14. In most institutions, the ______________________ is responsible for the training and instruction to institutional personnel on humane methods of animal maintenance and experimentation.
15. Humane methods of animal maintenance and experimentation would include the following:
   a. ____.
   b. ____.
   c. ____.
   d. ____.

16. Two commonly used sources for researchers to obtain information on appropriate methods of animal care and use, alternatives to the use of live animals in research, and unnecessary duplication of research would include the following:

    a. ____.

    b. ____.

17. The ________________ was prepared specifically toward farm animal research activities with special consideration given to current practices and issues in commercial agriculture.

18. P.I. stands for ________________ on any research project.

19. ______________________ refers to substitution, either wholly or in part, of the animal model with a non-animal model.

20. __________________ is a non-animal test used to test for carcinogenicity—but is made from mouse liver cells.

21. ________________ and ________________ are two methods used to reduce the numbers of animals used in experiments without compromising results and statistical integrity.

22. APHIS stands for ______________________________________.

23. List the three Rs:

    a. ____________________, b. __________________, and c. ________________

24. ______________________ is an experiment done within a whole living organism.

25. ____________________ is an experiment done in an artificial environment outside the living organism.

26. HREA stands for _____________________.

27. ISLA stands for ______________________________, an amendment to AWA.

28. The minimum number of members on an IACUC is __________________.

29. The IACUC must have at least one member who is a licensed ______________.

30. The IACUC must have at least _____ member(s) who has/had no other affiliation with the research institution besides membership in the IACUC.

31. It is federally mandated that the IACUC must review, at least every _______, the research facility's program using USDA regulations and the Guide as its basis.

32. The IACUC must inspect, at least once every ____________, all of the animal facilities, including animal study areas.

33. The _______________ is the animal's representative (spokesperson) on the IACUC.

Discussion Questions

1. Discuss reasons why research scientists should fear placing animal protectionists on their IACUCs. Discuss reasons why research scientists should place animal protectionists on their IACUCs. Is there a happy compromise? Why or why not?

2. How might the ethical considerations of farm animals differ from that of traditional research animals? Should there be an added consideration? How about economics?
3. Do you agree with the finding of the IACUC which reviewed Dr. Braeger's protocol in the previous section? How would you have decided? Why?
4. Many animal activists have demanded that a member of their organization should be on every IACUC to represent their point of view. Should we then also include a member of iiFAR (see the iiFARsighted Report in Section 2) on every IACUC to represent those humans who would suffer the most, should a particular research project not proceed? What complications would result if both were on the IACUC? What benefits would result?

ENDNOTES

39. Shapiro, L. S., biotechnology intern, 1994.

RECOMMENDED READINGS (From Section 2)

1. iiFAR sighted Report
    a. Animals in Diabetes Research
    b. Meet our Members
    c. Cats and Medical Research
2. "Making Animal Tests the Scapegoat for Rare Side Effects." *FBR Facts* Vol. 11, No. 4.
3. "Historical Revisionism and Intellectual Dishonesty." *FBR Facts* Vol. 11, No. 3.

CHAPTER 5

The Basic Legal System Affecting Animal Welfare

Contributed by Robyn Abrams, MS, JD

OBJECTIVES

The material covered in Chapter Five was organized, summarized and placed in non-legalistic terminology so that the student of animal ethics should:

1. Be able to easily access the location of various articles of the AWA
2. Cite references as to how the act is enforced
3. Demonstrate understanding of the IACUC and its workings

SUMMARY OF REGULATORY STRUCTURES

THE ANIMAL WELFARE ACT

For more than 30 years, the USDA has enforced the AWA. When the law was passed in 1966, it required that minimum standards of care and treatment be provided for certain animals bred for commercial sale and for research and for animals transported commercially or exhibited to the public. At that time, the law mainly concerned itself with the acquisition and treatment of cats and dogs; birds, rats, mice, and domestic farm animals were excluded. The act was extended in 1970 to include other warm-blooded animals. Also, rules were developed to ensure that animals would be given an adequate amount of anesthesia and analgesics. In 1976, the law was extended to cover live animal transport, and in 1979, standards for marine mammals were adopted. In 1985, the reach of the act was extended by requiring registered institutions to establish IACUCs, which would not only oversee animal care—but for the first time would examine the use of the animals and whether alternatives are available—even if analgesics are used. In addition, the amendments required institutions to address the psychological well-being of primates and the exercise and socialization of laboratory dogs. Although these requirements establish acceptable standards, they are not always the ideal. Regulated businesses are encouraged to exceed the minimum standards.

The AWA requires that all individuals or businesses dealing with animals covered under the law must be licensed or registered with APHIS. *Regulatory Enforcement of Animal Care* (REAC) is the unit of APHIS that directly monitors animal care facilities through unannounced visits for compliance with the regulations. REAC inspectors, called *veterinary medical officers* (VMOs), complete an "Animal Inspection Report" within a certain time. If deficiencies remain uncorrected at the next unannounced visit, legal action can be taken, fines can be levied, or a suspension of the facility's registration rights can occur. The specific rules and regulations pertaining to the implementation of the AWA are published in the Code of Federal Regulations–CFR—Title 9 (included within this book).

Section 1: Statement of Policy

Congress created this act to:

1. Ensure that animals intended for use in research facilities or exhibition or for use as pets are provided humane care and treatment
2. Assure the humane treatment of animals during transportation in commerce
3. Protect the owners of animals from their theft by preventing the sale and use of animals which have been stolen

Congress further finds that:

1. The use of animals is instrumental in certain research and education for advancing knowledge of cures and treatments for diseases and injuries which afflict both humans and animals.
2. Methods of testing that do not involve animals continue to be developed, which are faster, less-expensive, and more accurate than traditional animal experiments for some purposes.
3. Measures that minimize or eliminate the unnecessary duplication of experiments on animals can result in more productive use of federal funds.
4. Measures that help meet the public concern for laboratory animal care and treatment are important in assuring that research will continue to progress.

Section 2: Definitions

When used in this act:

a. The term "person" includes any individual, partnership, firm, joint entity, or other legal entity.

b–d. further description of "person."

e. The term "research facility" means any school (except elementary and secondary), institution, organization, or person(s) that uses or intends to use live animals in research, tests, or experiments and that (1) purchases or transports live animals in commerce, or (2) receives funds under a grant, award, loan or contract from a department of the United States government, for the purpose of carrying out research, tests or experiments.

f. failure to comply = grounds for denial of license.

g. The term "animal" means any live or dead dog, cat, monkey (nonhuman primate mammal), guinea pig, hamster, rabbit or other such warm-blooded animal that is being used or intended for use for research or exhibition. This term here excludes horses not used for research and other farm animals such as poultry and livestock used for food or fiber or other production purposes such as breeding, animal nutrition improvement, etc.

(F) The term "dealer" means any person who, in commerce, for compensation or profit, delivers for transportation or transports except as a carrier, buys or sells, or negotiates the purchase or sale of:

1. Any dog or other animal whether alive or dead for research, teaching, exhibition, or use as a pet, or
2. Includes all premises, facilities, or sites where dealer operates or keeps animals.

Exceptions:

1. Any retail pet store, unless that store sells any animals to a research facility, an exhibitor, or a dealer
2. Any person who does not sell or renegotiate the purchase of any wild animal, cat, or dog, and who derives no more than $500 gross income from the sale of other animals in one year

Sections 3–12 deal with licensing requirements (complying with standards in section 13) of dealers, research facilities, etc., record-keeping requirements as to purchase, sale and previous owners of animals, and humane marking requirements for live dogs and cats.

ANIMAL WELFARE ACT REGULATIONS–9 CFR

PART 1—DEFINITION OF TERMS

1.1 Supplemental definitions to Animal Welfare Act

1. "Animals"—this term excludes: birds, rats of the genus Rattus and mice of the genus Mus bred for use in research.
2. "APHIS"—Animal and Plant Health Inspection Service, USDA
3. "Attending Veterinarian"—a person who has graduated Veterinary School or who has received equivalent formal education; has received training and/or experience in the care and management of the species being studied and who has direct or delegated authority for activities involving animals at a facility.
4. "Department"—United States Department of Agriculture
5. "Farm Animal"—any domestic species of cattle, sheep, swine, goats, llamas or horses which are normally and have historically been kept and raised on United States farms, or used as food or fiber or for improving animal nutrition, breeding, management, or production efficiency, or for improving the quality of food or fiber. This term also includes animals such as rabbit, mink and chinchilla, when used solely for meat or fur, and animals such as horses and llamas when used as work and pack animals.
6. "Handling"—petting, feeding, watering, cleaning, loading, transferring, restraining, treating, training, moving, or any other similar activity with respect to an animal.
7. "Housing facility"—any land, premises, shed, barn, building, trailer, or other structure intended to house animals
8. "Primary enclosure (P.E.)"—any structure or device used to restrict an animal to a limited amount of space, such as a room, pen, run, cage, pool, hutch, tether. In case of tether (dogs on chains), it includes the shelter and area within reach of tether.
9. "Indoor Housing"—any structure with environmental controls intended to house animals and a) having temperature and humidity control, b) an enclosure created by a continuous connection of floor (foundation) walls and roof, and c) at least one door for entry and exit.
10. "Outdoor Housing"—structure or land intended to house animals which is not #7 or #9 and temperature can't be controlled within set limits.
11. "Sheltered Housing"—facility which provides the animals with protection from the elements and protection from the temperature

extremes at all times, such as runs or pens totally enclosed in a barn or building or connecting inside/outside runs.

12. "Exotic Animal"—any animal not identified in definition of "animal" that is native to a foreign country, such as lions, antelope, water buffalo and species of foreign domestic cattle such as a yak.
13. "Weaned"—the animal has become accustomed to take solid food without nursing, for at least five days.
14. "Painful Procedure"—any procedure that would reasonably be expected to cause more than slight or momentary pain or distress in a human being, such as pain in excess of that caused by injection or other minor procedures.
15. "Paralytic Drug"—a drug which causes partial or complete loss of muscle contraction, with no analgesic or anesthetic properties.
16. "Principal investigator"—research facility employee or associate responsible for a proposal to conduct research and for the design and implementation of the animal research.

PART 3—STANDARDS

Subpart A—Specifications for Humane Handling, Treatment, and Transportation of Dogs and Cats

Facilities and Operating Standards

3.1 Housing (General)—Consideration is given to:

- Structure—kept in good repair, protect animals from injury, restrict other animals from entering.
- Surfaces—made of materials easily sanitized, be free of jagged edges etc., regularly maintained.
- Food or bedding storage within—area must be free of any trash, weeds, waste material, etc.
- Cleaning—hard surfaces that animals contact must be spot-cleaned daily and sanitized, floors made of absorbent bedding, dirt, sand, gravel, grass, etc., must be raked or spot cleaned to ensure all animal's freedom to avoid contact with excreta; replace surfaces if necessary to eliminate odors, insects, etc.
- Water and electric power—housing facility must have reliable electric power for adequate heating, cooling, ventilation and lighting, and provide running potable water for dogs' and cats' drinking needs, cleaning and other husbandry requirements.

- Storage—store food and bedding in a way to protect it from spoilage, contamination and vermin infestation. Store supplies off floor and away from walls to allow better cleaning around supplies. Store food with regard to maintaining nutritive value, open food must be kept in leak-proof containers with tight lids to avoid contamination, etc.
- Drainage and waste disposal—facility operators must provide for frequent collection, removal, and disposal of animal and food wastes, bedding, debris fluid wastes, dead animals, etc., in a manner that minimizes contamination and disease risks; housing must have disposal and drainage systems so that waste is rapidly removed and animals stay dry, constructed with traps to prevent backflow of gases and sewage and much more.

3.2–3.5 Indoor, Sheltered, Mobile, Travel Housing

The regulations then go through these same considerations for cats and dogs, but they describe indoor housing, sheltered housing, mobile or traveling housing, and outdoor housing. Specifics are provided, such as exact range of acceptable temperatures for each type of housing, specific space requirements for different animal sizes, and specifics as to which kinds of outdoor conditions and which kinds of animal type (short hair, etc.) restrictions are placed on such housing. Additional requirement specific to cats, then dogs, are enumerated—such as space requirements if the dog or cat has a litter, compatibility considerations (aggressive cats must be housed separately), and dog tether length, construction, access to shelter, etc.

Animal Health and Husbandry Standards

3.7 Compatible grouping:

1. any dog or cat that is overly aggressive must be housed separately.
2. puppies and kittens less than four months old may not be housed in the same P.E. with any adult dog or cat other than their mothers.
3. any animal suspected of having a contagious disease must be isolated from healthy animals in the colony.

3.8 Exercise for dogs:

Dealers, exhibitors, and research facilities must develop, document, and follow an appropriate plan to provide an opportunity to exercise. The plan must be approved by the attending veterinarian. The frequency, method, and duration of exercise must be determined by the attending veterinarian, and at research facilities, with approval by the Committee (IACUC). Exercise plans should consider providing positive physical contact with humans that encourages exercise through play. If a dog is housed without sensory contact with another dog, it must be provided with positive human contact at least daily. Forced exercise, such as swimming, treadmills, or carousel-type devices, are unacceptable to meet the exercise requirements. A research facility may be excepted from these requirements if the principal investigator determines that for scientific reasons set forth in the committee-approved research protocol, it is appropriate for certain dogs to exercise.

3.9 Feeding

Dogs and cats must be fed wholesome, uncontaminated, palatable, and sufficient food at least once a day. The normal condition and weight of the animal must be maintained. The food must be protected from rain, etc., and located to minimize contamination by pests and excreta.

3.10 Watering

If potable water is not continually available to cats and dogs, it must be offered to them as much as necessary to ensure their health and well-being, but not less than two times daily for at least one hour each time.

3.11 Cleaning, sanitation and pest control

Excreta and food waste must be removed from P.E. daily. If water or steam is used, the animals must be removed first unless the enclosure is large enough to ensure that animals are not distressed or wet by the process. Used P.E. and food and water receptacles must be sanitized at least once every two weeks. An effective program of pest control (insects, external parasites, birds) must be established so as to promote the health and well-being of the dogs and cats.

3.12 Employees

The employees providing for the animals' husbandry and care must be supervised by an individual who has the background and knowledge in proper husbandry of cats and dogs to supervise others.

Transportation Standards

3.13 Various rules regarding transport by carriers and intermediate handlers are listed, such as:

1. cats and dogs must not be handed over to a carrier more than four hours before departure.

2. affixed to each carrier must be the name and address, where the animal came from, its tag or tattoo number, and when it was last fed and watered (within last four hours).
3. When a P.E. containing a cat or dog arrives at the animal holding area at a terminal facility, the carrier must attempt to notify the consignee upon arrival. If the consignee cannot be notified within twenty-four hours, the animals must be returned to the consignor. The carrier must continue to provide proper care, feeding, and housing until the animal is returned.

3.14 Primary enclosure used to transport live cats and dogs: The P.E. must be constructed so as:

1. The interior has no sharp edges that could injure animals in it
2. The animal can be quickly removed from it in case of emergency
3. Made of non-toxic materials
4. Proper ventilation insured
5. Method(s) by which animal does not have to come into contact with its own wastes
6. Cleaned and sanitized with each use
7. Puppies and kittens can only be transported in the same P.E. with their mothers
8. Must be large enough so animal can turn around normally while standing, to stand and sit erect and lie in a natural position
9. Only one animal older than six months per P.E. when traveling by air. No more than two puppies or kittens between eight weeks and six months can be put together in the P.E. if shipping by air. When shipped by surface vehicle, no more than four dogs or four cats over eight weeks old (of comparable size) can be put in the same P.E.
10. Food and water instructions must be attached to each P.E.

3.15 Primary conveyances (P.C.) (rail, air, marine, motor vehicle)

The animal cargo space used to transport dogs and cats must be made and maintained in a manner that at all times protects the health and well-being of the animals:

1. Situate that engine exhaust from P.C. does not reach animal
2. Animal cargo space must have sufficient air for normal breathing at all times
3. Situated in cargo space that animals are totally protected from the elements

4. Air cargo space must be properly pressurized when above 8,000 feet
5. Auxiliary ventilation (fans, blowers) must be used when temperatures reaches 85°F. The ambient temperature cannot exceed 85°F for more than four hours (nor fall below 45°F).

3.16 Food and Water

Each dog and cat more than sixteen weeks old must be offered food at least every twenty-four hours. Puppies and kittens less than sixteen weeks old must be offered food at least every twelve hours. Water must be offered at least every twelve hours. Verification of these feedings must be made and signed for at each feeding period.

3.17 Care in transit

The operator of ground transportation, or an accompanying person, must check on the animals at least every four hours to ensure proper air, temperature, etc. If any animal seems in distress, veterinary care at the nearest veterinary facility must be sought. If air transport is used and cargo area is not accessible, the carrier must observe the animals whenever they are loaded and unloaded and to check air pressure, etc. when in the air.

3.18 Terminal facilities

All concerns and rules regarding an animal's welfare and health, such as ambient temperature, etc., apply to this Section.

3.19 Handling

Any handler who moves the animals within, to, or from an animal holding area must make sure to provide animals with shelter from the sunlight and extreme heat, rain and snow, and cold temperatures. The P.E. must not be tossed, tilted or dropped.

Subpart F —Specifications for the Humane Handling, Care, Treatment and Transportation of Warm-Blooded Animals Other Than Dogs, Cats, Rabbits, Hamsters, Guinea Pigs, Non-Human Primates, and Marine Mammals

Under the AWA, biomedical research performed with the use of farm animals is covered by 9 CFR, the standards and regulations part of the Act. Subpart F sets forth somewhat general standards, similar in many respects to the standards set out in the prior sections—but even more general. For example, rather than setting out specific temperature ranges as in the dog and cat section, the regulations just state: "The ambient temperature shall not be allowed to fall below nor rise about temperatures compatible with the health and comfort of the animal." 9 CFR, Subpart F 3.126(a). In discussing space requirements, the regulations—rather than providing specific measurements as it does in the primate sections—state here that "... provide sufficient space to allow each animal to make normal postural and social adjustments with adequate freedom of movement. Inadequate space may be indicated by evidence of malnutrition, poor condition, debility, stress, or abnormal behavior patterns" Subpart F 3.128.

The PHS also sets forth standards of care for common species of livestock in the *NIH Guide for the Care and Use of Laboratory Animals*. This guide provides standards of care for livestock under the experimental settings of biomedicine, but not for agricultural production research. These standards of care will be set forth when the NIH guidelines are reviewed.

Around 1985, various groups of animal scientists and members of the government and veterinary communities attempted to fill the gap in standards for agricultural production by developing the "Guide for the Care and Use of Agricultural Animals in Agricultural Research and Teaching." The "Ag Guide" is meant to provide responsibility and institutional accountability and compliance guidelines for farm animals used in agricultural research and teaching. This guide has been adopted by a majority of agricultural institutions, and the American Association for Accreditation of Laboratory Animal Care requires compliance with the Guide for accreditation of farm animal research facilities.

The Canadians have published similar documents and guides for its agriculture enterprises. Agriculture Canada, as an example, has published the following items:

1. Recommended code of practice for care and handling of pigs
2. Recommended code of practice for the care and handling of special fed veal calves
3. Recommended code of practice for the care and handling of dairy cattle

The Canadian Federation of Humane Societies, the Canadian Society of Animal Science, the Canadian Veterinary Medical Association, the Dairy Farmers of Canada, the Canadian Society of Animal Science, the Canadian Council on Animal Care, the Agricultural Institute of Canada, the Ontario Trucking Association, the World Society for the Protection of Animals, the Western Veal Producers Association of British Columbia, the Quebec Department of Agriculture, the University of Guelph, the Agriculture Canada Animal Health Division Research Branch, and other organizations contributed to the drafting of these codes. This fact clearly demonstrates that even when a large group of individuals with varying viewpoints come together, a meaningful and productive result can occur.

In 1993, the USDA and Purdue University published *Food Animal Well-Being*. The conference proceedings and deliberations detailed methods of "assuring an appropriate climate for food animal agriculture to prosper, retaining economic viability without sacrificing social acceptability." The review group met again in 1997 to revise its guidelines. The participants again included animal protectionists, animal scientists, professors, veterinarians, philosophers, and food producers to make sure that more than one side was heard and understood. The well-being of farm animals used in the production of food and fiber for human consumption and the potential for animal distress and pain during experimentation related to food and fiber production was discussed.

The result of these conferences was that all USDA/Cooperative State Research, Education, and Extension Service (CSREES)-sponsored research projects are protocol-reviewed individually for proper well-being standards by IACUCs when submitted, and the resulting animal well-being assurances are also monitored by the CSREES scientists. The Ag Guide provides the standards that are used to assess the well-being of the animals in question in the protocol reviews.

Because these three documents (the AWA, the NIH Guide, and the Ag Guide) deal with farm animals for different purposes in contrasting environments, it is important that farm animal facilities familiarize themselves with all three.

THE NIH GUIDELINES

The NIH is the major United States government agency funding lab animal research. Institutions must meet NIH requirements, in addition to AWA regulations, to get that funding. NIH policies originally dealt only with the care and maintenance of lab animals and not the experiments themselves. In 1985, NIH updated its policy statement to include regulations over the experiments themselves in the wake of several news stories concerning the use of lab animals in certain questionable experiments. NIH also conducts spot checks to ensure compliance. When the institution is found to be noncompliant, the funding is suspended. An office of NIH, Office for Protection from Research Risks (OPRR), develops, monitors, and exercises compliance oversight relative to NIH policy on humane treatment of research animals. NIH policy is implemented through the "GUIDE" (Guide For the Care and Use of Laboratory Animals). The Guide, intended as a source of information on common lab animals housed under a variety of circumstances, covers any live vertebrate animal used or intended for use in research, research training, experimentation, biological testing, or related purposes. IACUC committees, through an institutional official, must report the results of the committees bi-annual inspections at least annually to OPRR.

NATIONAL SCIENCE FOUNDATION (NSF)

This agency is a federal agency outside of NIH that requires its grantees to comply with NIH standards. The main difference between NIH and NSF is that one-fourth of the animals under NSF grants involve field studies of free, living, wild animals.

FOOD AND DRUG ADMINISTRATION (FDA)

Since 1979, all product studies submitted for FDA approval are required to follow "Good Laboratory Practice" regulations. These regulations, are not directly enforceable legally, such as the USDA regulations, nor are they directly related to financial support such as NIH. They are secondary regulations under the Federal Food and Drug and Cosmetics Act and must be followed to get a new product released. The regulations are intended to ensure the quality and integrity of the safety data filed.

I. Transportation

The transportation, care and use of animals should be in accordance with the AWA (7 U.S.C. 2131 et seq.) and other applicable Federal laws, guidelines, and policies.

II. Housing

It is essential to properly manage animal facilities because:

- It affects the welfare of the animals to be used in research.
- The validity of research data can be affected. The animal scientist must minimize variations that can modify the animals' responses to experimentation.
- It can affect the health and safety of the animal care staff.

Housing is an important element of the physical and social environment of the animal.

A. System itself should:

1. Provide a comfortable environment
2. Provide adequate space which permits freedom of movement and normal postural adjustments and a resting place appropriate to the species
3. Provide an escape-proof enclosure that safely confines the animal

4. Provide easy access to food and water
5. Provide adequate ventilation
6. Meet animals' biological needs (body temperature, urination and defecation, and reproduction, if appropriate)
7. Keep animals dry and clean
8. Avoid unnecessary physical restraint
9. Protect animals from known hazards

B. Social Environments

Although there is little objective evidence for defining what is adequate care in terms of social environment, some guidelines can be followed. Attention should be given to whether the animal is communal or territorial, so as to know whether to house the animal in a group or singly. Age, sex, and rank must be considered in group housing. Proper population density must be considered, because it can affect reproduction, metabolism, behavior, and immune responses. It is important to keep the group composition as stable as possible, especially for highly social mammals such as non-human primates and canines, because the introduction of new members or mixing up members can alter behavioral and physiological functions.

C. Space Recommendations

Again, although there is little definitive, objective data on space requirements, successful experience and professional judgment should be used. Minimum space requirements are provided in Figure 5–1. Because of all the complex factors affecting caged animals, however, no system is defined as perfect.

D. Activity

Although the form of activity an animal has in its "natural" state is restricted in the laboratory environment, the amount of exercise does not necessarily have to be restricted. For example, supplementary activity can be provided through the use of treadmills, leash walking, access to a run, or releasing the animal from its cage into an animal room. For large farm animals, loafing areas, exercise lots, and pastures are suitable. For animals with specialized locomotor patterns, such as brachiating non-human primates, ropes, bars, and perches, are important, especially if the animal is held for long-term (greater than three months). In determining the type and degree of exercise, breed, temperament, age, history, physical condition, the nature of research, and the expected research duration must be

Figure 5–1 *Minimum space recommendations for laboratory animals, courtesy of the USDA*

Animals	Weight	Type of Housing	Floor Area/Animal		Height[a]	
	g		in²	cm²	in	cm
Mice	<10	Cage	6.0	38.71	5	12.70
	10–15	Cage	8.0	51.62	5	12.70
	15–25	Cage	12.0	77.42	5	12.70
	>25	Cage	15.0	96.78	5	12.70
Rats	<100	Cage	17.0	109.68	7	17.78
	100–200	Cage	23.0	148.40	7	17.78
	200–300	Cage	29.0	187.11	7	17.78
	300–400	Cage	40.0	258.08	7	17.78
	400–500	Cage	60.0	387.12	7	17.78
	>500	Cage	70.0	451.64	7	17.78
Hamsters[b]	<60	Cage	10.0	64.52	6	15.24
	60–80	Cage	13.0	83.88	6	15.24
	80–100	Cage	16.0	103.23	6	15.24
	>100	Cage	19.0	122.59	6	15.24
Guinea pigs[b]	≤350	Cage	60.0	387.12	7	17.78
	>350	Cage	101.0	651.65	7	17.78
	kg		ft²	m²	in	cm
Rabbits[b]	<2	Cage	1.5	0.14	14	35.56
	2–4	Cage	3.0	0.28	14	35.56
	4–5.4	Cage	4.0	0.37	14	35.56
	>5.4	Cage	5.0	0.46	14	35.56
Cats	≤4	Cage	3.0	0.28	24	60.96
	>4	Cage	4.0	0.37	24	60.96
Dogs[c]	<15	Pen/run	8.0	0.74	–	–
	15–30	Pen/run	12.1	1.12	–	–
	>30	Pen/run	24.0	2.23	–	–

Figure 5–1 *Continued*

Animals	Weight	Type of Housing	Floor Area/Animal		Height[a]	
	kg		ft²	m²	in	cm
	<15	Cage	8.0	0.74	32	81.28
	15–30	Cage	12.1	1.12	36	91.44
	>30	Cage	c	–	c	–
Nonhuman primates[d]						
Group 1	<1	Cage	1.6	0.15	20	50.80
Group 2	1–3	Cage	3.0	0.28	30	76.20
Group 3	3–10	Cage	4.3	0.40	30	76.20
Group 4	10–15	Cage	6.0	0.56	32	81.28
Group 5	15–25	Cage	8.0	0.74	36	91.44
Group 6	>25	Cage	25.1	2.33	84	213.36
Pigeons	–	Cage	0.8	0.074	e	–
Quail	–	Cage	0.25	0.023	e	–
Chickens	<0.25	Cage	0.25	0.023	e	–
	0.25–0.5	Cage	0.50	0.046	e	–
	0.5–1.5	Cage	1.00	0.093	e	–
	1.5–3	Cage	2.00	0.186	e	–
	>3	Cage	3.06	0.285	e	–
Sheep and goats						
1–4/Pen	<25	Pen	10.0	0.93	–	–
	25–50	Pen	15.0	1.39	–	–
	>50	Pen	20.0	1.86	–	–
5/Pen	<25	Pen	8.5	0.79	–	–
	25–50	Pen	12.5	1.16	–	–
	>50	Pen	17.0	1.58	–	–
>5/Pen	<25	Pen	7.5	0.70	–	–
	25–50	Pen	11.3	1.05	–	–
	>50	Pen	15.0	1.39	–	–

Figure 5–1 *Continued*

Animals	Weight	Type of Housing	Floor Area/Animal		Height[a]	
	kg		ft^2	m^2	in	cm
Swine						
1–4/Pen	<25	Pen	6.0	0.56	–	–
	25–50	Pen	12.0	1.11	–	–
	50–100	Pen	24.0	2.23	–	–
	100–200	Pen	48.0	4.46	–	–
	>200	Pen	60.0	5.57	–	–
5/Pen	<25	Pen	6.0	0.56	–	–
	25–50	Pen	10.0	0.93	–	–
	50–100	Pen	20.0	1.86	–	–
	100–200	Pen	40.0	3.72	–	–
	>200	Pen	52.0	4.83	–	–
>5/Pen	<25	Pen	6.0	0.46	–	–
	25–50	Pen	9.0	0.84	–	–
	50–100	Pen	18.0	1.67	–	–
	100–200	Pen	36.0	3.34	–	–
	>200	Pen	48.0	4.46	–	–
Cattle	<350	Stanchion	16.0	1.49	–	–
	350–450	Stanchion	18.0	1.67	–	–
	450–550	Stanchion	22.0	2.04	–	–
	550–650	Stanchion	24.0	2.23	–	–
	>650	Stanchion	27.0	2.51	–	–
1–4/Pen	<75	Pen	24.0	2.23	–	–
	75–200	Pen	48.0	4.46	–	–
	200–350	Pen	72.0	6.69	–	–
	350–500	Pen	96.0	8.92	–	–
	500–650	Pen	124.0	11.52	–	–
	>650	Pen	144.0	13.38	–	–
5/Pen	<75	Pen	20.0	1.86	–	–
	75–200	Pen	40.0	3.72	–	–

Figure 5–1 *Continued*

Animals	Weight	Type of Housing	Floor Area/Animal		Height[a]	
	kg		ft²	m²	in	cm
	200–350	Pen	60.0	5.57	–	–
	350–500	Pen	80.0	7.43	–	–
	500–650	Pen	105.0	9.75	–	–
	>650	Pen	120.0	11.15	–	–
>5/Pen	<75	Pen	18.0	1.67	–	–
	75–200	Pen	36.0	3.34	–	–
	200–350	Pen	54.0	5.02	–	–
	350–500	Pen	72.0	6.69	–	–
	500–650	Pen	93.0	8.64	–	–
	>650	Pen	108.0	10.03	–	–
Horses	–	Tie Stall	44.0	4.09	–	–
	–	Pen	144.0	13.38	–	–
Ponies	–	–				
1–4/Pen	–	Pen	72.0	6.69	–	–
>4/Pen	≤200	Pen	60.0	5.57	–	–
	>200	Pen	72.0	6.69	–	–

[a]From the resting floor to the cage top.

[b]Space recommendations are comparable to the current regulations of the Animal Welfare Act. Mothers with litters require more space (CFR. 1984a).

[c]These recommendations may require modification according to the body conformation of individual animals and breeds. Some dogs, especially those toward the upper limit of each weight range, may require additional floor space or cage height to ensure compliance with the regulations of the Animal Welfare Act. These regulations (CFR. 1984a) mandate that the height of each cage be sufficient to allow the occupant to stand in a "comfortable position" and that the minimum square footage of floor space be equal to the "mathematical square of the sum of the length of the dog in inches, as measured from the tip of its nose to the base of its tail, plus 6 inches, expressed in square feet." If dogs are housed in group pens or runs, only compatible animals should be housed together.

[d]The designated groups are based on approximate sizes of various nonhuman primate species used in biomedical research. Examples of species included in each group are:

Group 1—marmosets, tamarins, and infants of various species
Group 2—capuchins, squirrel monkeys, and similar species
Group 3—macaques and African species
Group 4—male macaques and large African species
Group 5—baboons and nonbrachiating species larger than 15kg
Group 6—great apes and brachiating species

considered. Although cages might be necessary for post-surgical care, isolation of sick animals and metabolic studies of less than three months, pens, runs, etc., are encouraged when a longer "stay" is expected.

ANIMAL ENVIRONMENT

The environment within which the animal is kept should be appropriate for the species and its life history. For example, commonly used lab animals such as rats and mice are readily adaptable to lab caging systems, whereas less-commonly used animals might have more specific needs (i.e., voles' reproductive behavior can be disrupted if their cages are changed too often).

A. Micro- and Macro-environments

The micro-environment of an animal is the physical environment immediately surrounding the animal, such as the temperature in the cage of primary enclosure. The macro-environment is the physical conditions in the room or secondary enclosure. There are differences between these two environments, and they can be influenced by cage design. Experimental exposure (in a micro-environment) to rodents of elevated levels of temperature, humidity, or concentration of ammonia can increase their susceptibility to infections and toxic agents, thus affecting research results.

B. Temperature and Humidity

Because temperature and humidity can greatly affect an animal's metabolism and behavior, it is important to regulate them closely. The optimum temperature for an animal is that in which an animal responds most favorably and is usually determined by maximum production in farm animals (gains in weight, milk, wool, work, and eggs) and feed efficiency.

C. Ventilation

Ten to fifteen room air changes per hour provide adequate room ventilation. The air should not be re-circulated, however, unless it has been treated to remove particulate and toxic gaseous contaminants. Consideration must also be given to the control of relative air pressure in animal housing. For example, non-human primate housing and areas of quarantine and soiled equipment should be kept under relative negative air pressure, whereas pathogen-free animal housing and clean equipment should be kept in a relatively positive air pressure environment.

D. Illumination

Precise lighting requirements for maintenance of physiological stability and good health are not known (for all animals). In the past, lighting levels were used that unknowingly caused retinal damage to the eyes of albino mice if used over a certain time period. Therefore, it is important especially with albino animals to provide variable intensity controls to ensure light intensities consistent with the needs of the animals and the working personnel. A time-controlled lighting system should be used to provide a regular, diurnal lighting change.

E. Noise

Noise controls should be considered in facility design, because it can be annoying to both personnel and animals alike. Noisy animals, such as dogs and non-human primates, should be housed away from cats, rodents, and rabbits, because the noises produced by the former can be disturbing to the latter. Moreover, continued exposure to acoustical levels higher than 85 dB can have both auditory and non-auditory effects, including eosinopenia and reduced fertility in rodents and increased blood pressure in non-human primates. Appropriate hearing protection programs must also be established for personnel.

FOOD

Feed animals daily according to their particular requirements, unless the experimental protocol requires otherwise. The food should be palatable, uncontaminated, and nutritionally complete. Lab animal diets should not be manufactured or stored in facilities used for farm feeds or any products containing additives such as rodenticides, insecticides, hormones, antibiotics, fumigants or other potential toxicants.

Contaminants in food can have dramatic effects on biochemical and physiological processes, even when the amounts are present in concentrations too low to cause clinical signs of toxicity. For example, some contaminants induce the biosynthesis of hepatic enzymes, which can alter an animal's response to drugs. Consider shelf life, temperature, humidity, vitamin durability, light, oxygen, etc. in the storage (length, place, etc.) of food.

BEDDING

Bedding should be of a type not readily eaten by animals, free of toxins, and absorbent. The bedding material should be plentiful enough to keep the animal dry between cage changes. Be aware that bedding materials containing aromatic hydrocarbons from cedar and pine can induce the biosynthesis of hepatic microsomal enzymes.

WATER

Unless the experimental protocol requires otherwise, animals should have continuous access to fresh water. The water should be periodically monitored for pH, hardness, and microbial or chemical contamination. Treated or purified water can be used to minimize such contamination.

SANITATION

A. Cleanliness

Sanitation is essential in an animal facility. Bedding should be changed as often as necessary to keep animals dry and clean. Litter should be emptied from cages in areas other than the animal room itself to minimize exposure of animals and personnel to aerosolized waste. In some instances, however, frequent cage cleaning is not appropriate. For example, it would be counterproductive to clean a cage and thus remove pheromones if reproduction was a research goal. This section reviews in detail the frequency and types of cleaning sanitation of cages, runs, equipment, etc., that are recommended.

B. Waste Disposal

Waste should be removed regularly and frequently. A safe and sanitary method of waste disposal is incineration. The federal government and most states and municipalities have laws controlling waste disposal. Compliance with these statutes is an institutional responsibility. Hazardous wastes must be rendered safe by sterilization or containment before being removed from the facility.

C. Vermin

Programs should be created to control, eliminate, or at best prevent infestation by pests such as flies, wild or escaped rodents, and cockroaches. The screening of openings, sealing of cracks, and elimination of breeding sites will go far to prevent their entry. The use of pesticides should be avoided if at all possible, and relatively non-toxic compounds such as boric acid should be used to avoid harm to animals and personnel (and any negative influence on experimental results).

IDENTIFICATION AND RECORDS

Methods of animal identification include collars, bands, plates, colored stains, ear notches, tags, tattoos, and freeze brands. Records on experimental animals are essential and can be limited or detailed, depending on the type of animal and type of research. The records should include the source of the animal, the strain or stock, the name and location of the responsible investigator, and any pertinent dates.

NIH GUIDELINES FOR FARM ANIMALS

Farm animals (cattle, sheep, goats, swine, and fowl) can be housed in a lab setting, as discussed in the general NIH guidelines, or on a farm. Animals housed on farms, however, can be housed under less-strict conditions. The rest of this section deals with farm animals involved in biomedical research.

A basic requirement of housing is protection against environmental extremes. In determining the proper degree of care, the animal's well-being should be considered—and not merely the animal's ability to survive and produce under adverse conditions, such as climatic extremes and increased population densities.

Many different types of holding systems exist on farms, such as minimum protection (climatic housing) to total environmental control. The type used depends on factors such as species, breed, age, climate, pelage or feathers, and research or husbandry goals. Climatic housing, which is most suitable in moderate climates, should provide animals with the opportunity to protect themselves from wind, rain, and sun. In temperate climates, animals can be placed inside pastures and paddocks without shelter other than trees, wind fences, and sun shades. Controlled housing can be used for all animals but is especially useful with sick or young animals. The micro- and macro-environments must be regulated (as described in the previous section). This section discusses in detail suggestions for housing material and construction.

Although there is no standard temperature at which livestock must be maintained, the thermal environment has a strong influence on farm animals. Factors such as species, breed, age, weight, feeding level, activity of the animal, opportunity for evaporative cooling, experience, and husbandry should be considered in determining the proper temperature. For maximum performance and physiological stability, animals should be maintained as much as possible in their *thermo-comfort zone*—the temperature range at which animals show no particular preference for any particular location to keep cool or warm, and the temperature range in which physical temperature regulation is employed. Below the "lower critical temperature" of an animal, the chemical regulation mechanism of the body is no longer able to cope with cold—and the body temperature drops, resulting in death. Above the "upper critical temperature" of an animal, the animal becomes heat stressed—and physical regulation comes into play to cool them. In this case, fertility drops, as does feed efficiency and economic stability. Hence, from an economical and animal welfare concern standpoint, livestock should be raised as close to their thermo-neutral comfort zone as possible.

To avoid undue competition for feed, animals fed in groups should be provided adequate trough space. Feeding space is determined by the size and number of animals eating at one time. When the animals are more restricted in their movement, food should be provided in a rack, which results in a decrease of disease agents and parasites. In cold climates, water sources should be equipped with heating devices to prevent freezing.

III. General Husbandry

IV. National, local, and international law

A. Nearly all other industrialized countries outside of the United States have established laws concerning the use of animals in research, including the 31 members of the Organization for Economic Cooperation and Development (OECD).

B. Canada relies on a "voluntary" system administered by the *Canadian Council on Animal Care* (CCAC). The CCAC establishes national standards and guidelines and conducts inspections. Representatives of major interest groups, including animal protection organizations, sit on the committee. Before the United States came up with the IACUCs, the CCAC suggested that all institutions set up animal care committees, so therefore, Canada's oversight structure has the same feel as that of the United States.

C. Rather than rely on local institutional committees, Britain utilizes a system of personal and project licenses that is overseen by a *home office inspector.* Britain is part of the *European Community* (EC) and must therefore meet the minimum standards established by the EC 1986 Directive on animal research. The Directive requires that member countries collect comprehensive statistics on animal research activities, promote alternatives, establish standards to minimize pain and suffering, and establish standards for adequate care and housing.

D. In Australia, there are different state laws dealing with animal research—as well as a national Code of Practice and an Australian Council on the Care of Animals in Research and Teaching. The Code requires that all experiments be approved by an institutional "Animal Experimentation Ethics Committee," which consists of several members. One of the members must be an independent person not involved in animal experimentation in any way.

E. Many OECD countries have established formal government or government-supported forums where opposing views can be heard. Sweden and the Netherlands also have some formal consultation mechanisms. Nearly all OECD countries encourage the development, promotion, and use of alternatives.

California does not have a specific statute to protect research facilities. State law does, however, permit the release of unclaimed pound animals for research. State law requires shelters selling animals for such a purpose to post a sign stating that, "Animals turned in to this shelter may be used for research purposes (California Civil Code, Section 1834.7 (West 1985))." The regulation of research facilities falls under the department of health, which is permitted to make reasonable rules and regulations. These regulations include the control of the humane use of animals for the diagnosis and treatment of human and animal diseases, for research in the advancement of veterinary, dental, medical, and biologic sciences, for research in animal and human nutrition, and for the testing and diagnosis, improvement, and standardization of laboratory specimens, biologic products, pharmaceuticals and food, sanitation, and record keeping, and for the humane treatment of animals by persons authorized by the board to raise, keep, or use animals under the provision (California Health and Safety Code, Section 1660 (West 1990)). Although the law permits the sale of shelter animals that are to be euthanized, none of the shelter animals in Los Angeles County are sold. They are euthanized and then are disposed of. The research facilities in the Southern California area purchase specifically bred animals for their research. Part of their reasoning is that they need a history of the animals to be used, and a secondary reason is because of the pressure from the animal rights community to not allow the use of shelter animals.

PATENTING OF ANIMALS

(See the section on ethics of animal use, "Patenting and Copyrighting of Animal Forms.")

Chapter 5

EVALUATION

1. Who enforces the AWA? ____________________
2. When the AWA first became law, which animals were its main concern?
   ____
3. APHIS stands for ______________________________.
4. ________________________ is the unit of APHIS that directly monitors animal care facilities through unannounced visits for compliance with the regulations.
5. Which animals are excluded in the AWA?
   a. __________ b. ____________ c. ____________
6. According to the USDA, which one of the following is *not* a domestic farm animal: cow, sheep, goat, pig, horse, llama, or water buffalo?
7. Who determines the frequency, method, and duration of exercise required for dogs in research facilities, for dealers, and for exhibitors? ______________
8. When shipping dogs and cats, food and water instructions must be attached to each __________________.
9. When shipping dogs and cats, water must be given at least every _____ hour.
10. When transporting animals, the accompanying person must check on the animals at least every _____ hour to ensure proper air, temperature, etc.
11. Which three documents should farm animal facilities be familiar with in order to stay in compliance with standards for agricultural production, and thus partially qualify for accreditation?
    a. ________________ b. __________________ c. ______________
12. _______________________ is the major United States government agency funding lab animal research.
13. _________________ is the branch of the NIH responsible for the monitoring of compliance to standards of humane treatment of research animals.
14. One-fourth of the animals under _________ grants involve field studies of free living wild animals.
15. In California, the regulation of research facilities falls under __________________.

Discussion Questions

1. Justify why farm animals' housing conditions are less strict than other research animals. Why or why not?
2. Why is there no standard temperature set for livestock, unlike other animals? Should there be? Why or why not?

Recommended Readings

1. *Regulation of Biomedical Research Using Animals.* National Association for Biomedical Reasearch
2. *The Use of Dogs and Cats in Research and Education.* National Association for Biomedical Research

CHAPTER 6

Specific Cases of Animal Welfare

Chapter 6 reviews specific cases of animal welfare that have made national and international news. Statistics, data, and photos from both the research and animal rights communities are included in this chapter. These, and other cases, have had a tremendous impact on legislation attempted and passed, both locally and nationally, that restrict the uses of animals by society.

Objectives

The student should:

1. Outline a brief history of each case cited and explain the case's relevance to societal changes in accepting the use of animals in research and in food production.
2. Describe events using animals that even the research community condemned.
3. Describe events where members of animal rights organizations broke the law, endangered human life, and destroyed millions of dollars of property in the "defense of animals."

LOUISIANA STATE UNIVERSITY CAT EXPERIMENTS (DOD) 1983–1989

From 1983–1989, Dr. Michael Carey, a United States Army neurosurgery researcher, conducted experiments at Louisiana State University (LSU) to study the kinds of brain wounds that he had personally observed in combat soldiers. Dr. Carey became interested in head injuries while he was a Mobile Army Surgical Hospital (MASH) neurosurgeon in the Vietnam War. He argued that his experiments were crucial for improving survival and recovery from gunshot, shrapnel, and other missile-induced brain wounds. Half of all single-wound deaths in combat are caused by head wounds, Carey said, and there had been no improvement in the post-operative recovery rate since World War II. In addition, in the United States, 16,000 people die each year from gunshot wounds to the head. Up until this time, fewer than 30 papers in the world had been written on the use of anesthetized animals to study brain wounds and a quarter of those publications came from Dr. Carey's team at LSU. His work was the only scientific work in the world studying the kind of brain wounds soldiers receive in combat.

During this six-year period, 125 anesthetized cats each year were shot through the head to aid Dr. Carey in his research (see Figure 6–1). In 1987, after animal rights organizations discovered Dr. Carey's cat experiments, letters and phone calls to Sen. Bob Livingston of Louisiana began, demanding that the Army-sponsored research be shut down. Sen. Livingston, who sat on the House of Representatives subcommittee which allocated money to the Army research, called for *Congress' General Accounting Office* (GAO) to look into Dr. Carey's animal research. The GAO asked for a panel of experts to convene and review the research. While the review was being conducted, Dr. Carey's experiments were suspended.

The head of the GAO expert medical panel of eight neurosurgeons and other experts (including a veterinarian and an anesthesiologist) was Dr. John Jane, chairman of the neurosurgery department at the University of Virginia. The panel said that Dr. Carey's research was "unique, worthwhile, and should be continued." Dr. Roy Schwarz, senior vice president of medical education and science for the American Medical Association (AMA), described Dr. Carey's work as "important research. It's critically important,

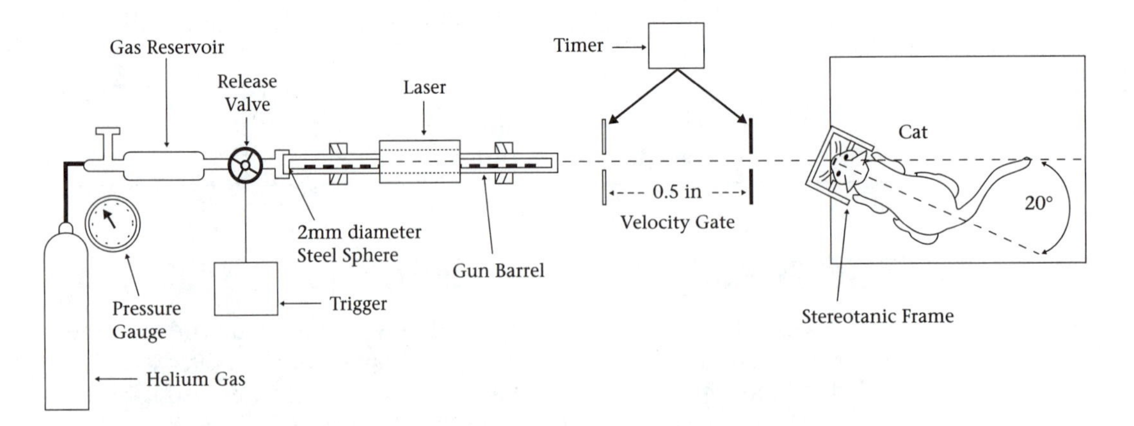

Figure 6-1 Schematic illustration of experimental layout. *Courtesy of PCRM*

and it needs to go on." The panel concluded that the experiments involved no pain, and post-operative care appeared adequate.

A team of eighteen neurosurgeons, neurologists, and trauma experts brought in by the Physicians Committee for Responsible Medicine (PCRM), an animal rights organization, found that, in addition to obvious humane concerns, many differences between the structure of a cat's brain and a human brain for this particular type of experiment would lead to problems in extrapolating the results to humans. Dr. Carey countered that the cat has been the most-studied laboratory animal for the past 100 years, and that more knowledge about the cat's brain is available than knowledge about the human brain. Many of the mechanisms and chemicals that make a cat's brain work also make a human's brain work.

PCRM insisted that the experimental wounds did not sufficiently resemble battle field injuries (the ratio between bullet size and the size of the brain was different in the cat and human models, and the brain injuries were not as "clean" as in experimental wounds—shrapnel, etc.), and much of the study had already been thoroughly researched. PCRM cited work made by Victor Horsely, one of the fathers of neurosurgery, in the May 31, 1894 issue of the British publication *Nature*. They claim that Dr. Carey's work simply replicated this earlier study.

The PCRM was not accepting of the GAO panel review and issued its own rebuttal, "The Army's War on Cats." PETA and PCRM, using the Freedom of Information Act, requested and then released Dr. Carey's proposal for a second contract. PCRM alleged that the cats used were in pain, and that their crying disrupted work at LSU. "The crying of these animals was so loud that the people down the hall could not even work. They couldn't even see patients" (*PCRM Update*, September–October 1989). All of these allegations were proven *not* to be the case; rather, they were simply rumors repeated enough that they wound up being printed in *CAT* magazine (and thus were believed). The person making the original charge admitted to making the claim simply because she personally considered the experiments to be "horrendous and not valuable," as reported on *60 Minutes* on January 24, 1992.

Due to the public outcry, the GAO decided to seek a second opinion—this time from a group of veterinary anesthesiologists. None of the members of this group of veterinarians on the second panel had ever done any research on brain injury. None of the veterinarians was asked to comment on the issue of pain that was thought to be severe in the experimental cats. The veterinarians simply criticized Dr. Carey for the method by which the cats were anesthetized. They claimed that Dr. Carey's method skewed his research. After reviewing the issues raised by the second GAO panel, the Pentagon ended the experiments in 1989. The GAO issued a report noting poor anesthesia techniques, poor documentation of post-operative care, and the omission of data from approximately one-third of the cats tested. The GAO claimed that these omissions interfered with the reliability of any results. The Army's poor monitoring of the experiments was also noted. Dr. Carey was also criticized for a lack of reportable productivity, despite having been given six years and $2.1 million by the Army. Dr. Carey claimed that the most important data that his research was showing was "that we were studying a drug which made the injured brain get better faster, and get better to a greater extent, get better further. And we were just this far from showing that this drug would really work when our research funding was shut off." Dr. Carey claims the funds were shut off because of animal rights activists and the pressures they put on Congress.

Dr. Jane accused the GAO of bias, saying, "I think what they (the GAO) did was take our report, which they didn't like, and try to get other people to express concerns." In addition to Dr. Jane's panel, other groups of scientists and consultants reviewed Dr. Carey's work (which had been already peer-reviewed by the Department of Defense) and gave their approval. An LSU committee, an Army investigatory team, and the American Association of Neurologic Surgeons, as well as three anesthesia experts to whom Carey sent his papers for opinions, all said they favored his research.

In 1989, Dr. Carey was targeted by animal rights activists. Threatening calls were made to his house, and similar-sounding letters were sent to his home. While Dr. Carey, still a colonel in the Army Reserves, was recalled to active duty in Operation Desert Storm, his wife, Dr. Betty Oseid Carey (a pediatrician), continued to receive threatening phone calls at her home.

In May 1994, the AMA gave the Medal of Honor to Drs. Michael Carey and Betty Jean Oseid Carey for their continued, four-year crusade (since his work was terminated) on behalf of brain-wound research and a "defiant and unflinching stand against animal rights extremists" (*60 Minutes*, January 24, 1992).

PENNSYLVANIA HEAD INJURY LABORATORY (PRIMATES)—1984

In May 1984, five people from the ALF broke into a University of Pennsylvania medical school laboratory, vandalized equipment, and stole 60 hours of videotape depicting the activities of the lab filmed by research personnel. Head injury research had been conducted in the lab—specifically, experiments on baboons designed to produce non-impact damage to the brain and spinal cord. Dr. Thomas Gennarelli was trying to mimic whiplash in auto accidents, which is the jerking damage resulting from the brain being slammed against the side of the skull. Baboons were studied to determine the type and extent of damage suffered—and the effects of this damage on the animals' subsequent behavior. The videotapes stolen were given to PETA. PETA condensed and edited the videotapes from 70 hours down to 20 minutes and widely distributed them to the media. Many people became concerned about the animal care standards in the laboratory and the utility of the research.

In this case, there was alleged misconduct on the part of laboratory workers. Animal research scientists have also deplored the workers' actions, claiming their acts were indefensible. This documented abuse brought punishment from the regulating authorities. Dr. Thomas Gennarelli's funding was cut off, and the research institution was fined.

Although NIH investigations of the incident took over twelve months to complete, the NIH ultimately did suspend the research, concluding that the laboratory had failed to comply with animal care standards. The NIH had originally funded the research. During this time, media interest in the story was so great,—and the criticism of the research itself was so loud—that even *Science* magazine, the official journal of the American Association for the Advancement of Science, commented ruefully that "from a public relations standpoint, some scenes on the tapes—which were made for documenting the research, not for public viewing—range from embarrassing to disastrous."

The edited tapes made it clear that the laboratory workers were not just "documenting" the research. Posing before the camera, the laboratory workers held dazed baboons in silly "say cheese" poses, dangled them by crippled limbs, laughed when they struggled, etc.

THE HUMANE FARMING INITIATIVE (HFI)—1988

The animal agriculture industry was relatively unaffected by the public's perception after the 1904 release of Upton Sinclair's *The Jungle*. The book shocked the nation with its detailed and graphic descriptions of the conditions in the slaughterhouses of Chicago. In 1964, Ruth Harrison published *Animal Machines—The New Factory Farming Industry*, an exposé which listed the abuses of animals in Great Britain's intensive farming systems. This book also deeply shocked the public. Harrison called attention to the new animal housing systems that she claimed restricted the natural behavior of animals. The book led to the establishment of the Brambell Commission in Great Britain. Brambell released five essential freedoms for animals and in 1965 established a code of practice for animal agriculture in Britain. Others continued to intensify the debate of "factory farming," leading to attempted legislation against such practice.

The HFI was a state referendum in Massachusetts—a proposal written by an activist group known as *Coalition to End Animal Suffering and Exploitation* (CEASE), a radical reform group believing in animal rights. The initiative included provisions such as: prohibiting the suffocation or grinding of chicks; requiring veal producers to provide larger spaces, more bedding, and more food for calves; general requirements minimizing the pain from castration and dehorning of livestock; and a directive to the Massachusetts commissioner of agriculture to study six areas, including humane transport, healthful diets, and scientific advisory reviews.

The main groups involved were: CEASE, the *Massachusetts Farm Bureau* (MFB), a conservative group consisting of the producers and holding a traditional utilitarian view, and the MSPCA, a moderately liberal animal welfare group. The actual voters were the consumers and taxpayers in this power-play. Their interests were decisive. CEASE believed that the current animal agricultural situation was lacking in morality and needed sweeping reform. The MSPCA wanted to move animal welfare legislation through more normal channels, rather than through a referendum vote, and respected the legitimacy of the existing interests. The Massachusetts Farm Bureau felt that the existing legislation adequately protected farm animals and that new regulations would impose needless costs on farmers. The voters, it turned out, were more concerned with how the regulations would impact the small farmers as opposed to increased consumer costs (early polls were 55 percent in favor of the initiative, but the final result was 71 percent against the initiative). The family farm still occupies an almost mythical place in American culture. The MFB was successful in convincing consumers that the effect of the initiative on farmers would be devastating. They referred to the initiative as the "Anti-Family Farm Bill," which was particularly effective in Massachusetts, because there are many small farmers and non-farm rural residents living in Massachusetts—as opposed to a state where the population is less in contact with agriculture or views agriculture as a large-scale enterprise. The initiative did exempt small-scale farmers from some of its more costly provisions, and the big farming enterprises would have absorbed much of the costs or passed them on to the consumer. Ultimately, the animal products industry and some consumer protection groups (that were worried about increases in food prices) were much more powerful politically than CEASE—and so had the greater impact on public policy. Although the referendum provided for much discussion, the loudest voice won.

Several other states have more recently tried to pass similar legislation. One advantage the animal rights groups have is that few people understand animal agriculture today. As a society, we are too far removed from the farm to really understand the management, the husbandry, and the science of food-producing animals. Major farm organizations have done a poor job of reaching the voting and consuming public with education on their needs. The HSUS, PETA, and other groups have been much more effective in this arena.

SILVER SPRINGS, MARYLAND

On September 11, 1981, in an unprecedented raid on a scientific laboratory (the Institute for Behavioral Research, founded by B. F. Skinner, et al), Montgomery County police seized seventeen monkeys being used for neurological research. For the first time in United States history, police raided a scientific research laboratory because of purported cruelty to animals. This case has become a landmark case that has set legal and political precedents that still affect animal research throughout the United States. The infiltration of the laboratory by Alex Pacheco, the subsequent police raid, and various news stories helped to make PETA the fastest-growing and one of the largest animal rights organizations in the United States.

Alex Pacheco was a 23-year-old college student at the time and a veteran, militant animal rights protester. He applied and was hired by researcher Edward Taub. Pacheco was an outspoken vegetarian and was known to denounce farmers who castrated their livestock without using anesthesia. He went into Dr. Taub's laboratory not to get a job, but rather for purposes of "infiltration." Together with Ingrid Newkirk, Pacheco founded PETA as a group designed to be more militant than the average chapter of the Humane Society in promoting animal rights.

The research that Dr. Taub was working on in 1981 involved monkeys to gather information on the rehabilitation of human stroke victims. Dr. Taub received a $180,000 grant from the NIH. The process Dr. Taub used involved *deafferentation.* This operation on the spinal cord affected sensory nerves leading to one arm—severing them, and thus severing all sensory communication between the arm and the brain (see Figure 6–2). The left limb was left completely numb but still was able to move. Dr. Taub was trying to apply information from deafferented monkeys with the rehabilitation of humans who had suffered strokes or spinal cord injuries. Dr. Taub felt that his research could ultimately affect the treatment of as many as 50,000 Americans each year. In both deafferented monkeys and humans who have suffered stroke or spinal cord damage, neurological damage resulted in a lack of sensation in the limbs. Doctors assumed, at this point, that those limbs could never be used again. Dr. Taub thought that both the monkeys and stroke victims could be trained to use the limbs, and he was granted research money by the NIH for this work.

Pacheco was opposed to all animal research that caused pain to the animal. Pacheco admitted to staging photographs (as exposed in *Washingtonian* magazine) and exaggerating his charges to build PETA's reputation throughout the country as a strong defender of animal rights. According to court testimony, Pacheco took the photographs that allegedly showed unsanitary conditions at the laboratory. None of the visitors or the staff members who were in the laboratory at the same time these pictures were taken, however, recalled seeing such conditions. USDA Veterinarian A.G. Perry had made an unannounced inspection of the colony during this time, and under oath he reported that he saw no such conditions. In Dr. Taub's first trial, Pacheco admitted to staging two of his photographs. Although Pacheco exaggerated his charges, some feel he was just in decrying the filth and sloppiness that occasionally occurred in Dr. Taub's laboratory. Others felt he was just simply

Figure 6-2 Restricted primate subject. *Courtesy of PETA*

because animals were used in the experiments. The laboratory was dirty, according to Pacheco, and he also claimed that the colony room was cleaned haphazardly by a couple of college students who were not concerned with the welfare of the animals in the room. Pacheco brought other primate researchers and sympathetic veterinarians into the laboratory to gather support for his claim of a "miserable and unhealthful environment." Dr. Taub was on vacation when all of these visits were taking place, and Pacheco felt that the laboratory was not properly cared for in his absence.

On the night before the raid, Pacheco and Newkirk brought in one final visitor: Dr. Richard Weitzman, a veterinarian. Weitzman felt that although the laboratory was dirty, the animals looked well-fed and healthy and were in no danger of dying. Weitzman asked Pacheco why he simply did not point out these conditions to Dr. Taub and have him fix them? The next morning, a warrant was obtained by Pacheco and Newkirk to search the lab and take the monkeys. The monkeys mysteriously disappeared and then reappeared again after authorities refused to look at any improprieties. About a week after returning the monkeys, one of them was killed as a result of mixing them together during a cage cleaning.

Seventeen misdemeanor counts of cruelty to animals were brought upon Dr. Taub, and he was initially found guilty of six counts of cruelty to animals for failing to provide adequate veterinary care to six of the monkeys. He was fined $3,000. Dr. Taub was acquitted of the other eleven charges against him. Dr. Taub did not feel he did anything wrong, and thus he appealed the verdict. In a second trial, he was acquitted of five of the convictions. The second court let stand the sixth conviction of inadequate veterinary care of a monkey named Nero, whose wounds caused the monkey to have his arm amputated. The judge fined Dr. Taub $500 and told him, "I hope and trust this will not deter you from your efforts to assist mankind with your research" (Taub et al., *Archives of Physical Medicine and Rehabilitation*, in press, and Peter Carlson's "The Strange Ordeal of the Silver Spring Monkeys," from *The Washington Post*, December 26, 1996). Dr. Taub appealed the final verdict and then was acquitted of all charges.

Moorpark College (a two-year community college in Southern California) tried to obtain some of the monkeys to use in its animal caretakers program. The NIH refused to release the monkeys to Moorpark's Gary Wilson, charging that the college would have inadequate veterinary care for the animals.

While this custody battle over the monkeys continued, other neuroscientists from the *National Institute for Mental Health* (NIMH) discovered that the brain could reorganize itself throughout adulthood. Scientists had deafferented one finger on a cat's paw and then examined the cat's brain, which had reorganized itself so that the tiny portion of the cortex that had received input from that finger was now receiving the input from elsewhere. The neuroscientists now wanted to know whether the monkeys' brains had also reorganized—and if so, how much? The scientists felt this information was vital to understanding the human brain's response to stroke and spinal cord injuries. PETA won an injunction prohibiting both the euthanizing of the three monkeys in question and their use in this final experiment. While arguments were going back and forth, two of the monkeys died in their cages in great pain. The third monkey was anesthetized, experimented on, and euthanized after the scientists received a vacated order from another judge.

Today, several scientists are using the knowledge that Dr. Taub gained by experimenting with deafferented monkeys to help rehabilitate human stroke and spinal cord injury victims. Today, scientists are much more careful of the management, healthful conditions, and sanitation of the laboratory than they were in 1981.

THE "BABY FAE" CASE—1984

In October 1984, a baboon heart was transplanted into a 12-day-old infant, "Baby Fae," who had a hypoplastic left-heart syndrome. The transplant was conducted at Loma Linda University Medical Center. Three weeks later, "Baby Fae" died of kidney failure.

The transplant created much debate. Critics of using an animal's organ in a human questioned the fact that the hospital made little attempt to get a human infant heart and that "Baby Fae" had been used in a clinical experiment. Loma Linda spokespersons said that the procedure was experimental therapy, but the experiment offered the infant her only chance at long-term survival. They also said that attempts had been made to find a human heart. They insisted that human hearts are rare. The available data, however, indicated the chances that "Baby Fae" would survive with a baboon heart for more than six months were not good, leading many newspapers to portray the infant as just another experimental animal. At this point, some animal activists used the media criticism to underscore their point that the baboon was needlessly killed. Although public opinion about animal research showed that the public was strongly influenced by the amount of pain and distress they thought the animals experienced—and the perceived importance of the research—the activists' arguments were not received too well with the public or the media. In the *Boston Herald,* a cartoon appeared, capturing the public's rejection of the animal rights argument. On the left-half of the cartoon was a picture of "Baby Fae" with the caption, "Born with half a heart"; on the right-half was a picture of the activists with the caption, "Born with half a brain.

In comparing the "Baby Fae" case with the "head trauma" case, one can see that the public and media viewed these cases quite differently in terms of the animal activists' reactions to those cases. The public and media regarded the animal activists' criticisms of the "Baby Fae" case to be unsupported and misdirected, whereas the activists' condemnation of the "head trauma" cases were consistent with those of the public and the media. The contrasting response could be explained in terms of a balancing formulation; that is, how these cases were viewed in terms of their benefit to society, versus suffering on behalf of the animal in question. Specifically, the suffering of the heart-donor baboon appeared minimal, whereas the baboons in the videotape seemed to be in great distress.

These cases provide a good analysis of how the public's perception of the "costs" and "benefits" of the research plays a role in the support of the research. This point is also evidenced by the fact that a majority of the public is opposed to testing new cosmetics on animals, while only about 15 percent oppose the use of animals in new drug research, especially for diseases such as cancer and AIDS.

> ***"Even if animal research resulted in a cure for AIDS, we'd be against it."***
>
> Ingrid Newkirk, PETA Founder, *Vogue,* September 1989 and *The Washington Times,* December 26, 1996

AN AIDS PATIENT RECEIVES A BABOON BONE MARROW TRANSPLANT

In a related case, on December 14, 1995, a 38-year-old man named Jeff Getty received a bone marrow transplant from a baboon in an extremely risky experiment to fight AIDS. Doctors hoped that the baboon's bone marrow cells, which are resistant to HIV (the virus that causes AIDS), would migrate to Getty's own bone marrow and produce immune cells to fight the virus. Getty's own immune system was weakened temporarily by chemotherapy and radiation treatment. The treatment was necessary to prevent his body from rejecting the baboon bone marrow. Months later, it was determined that the transplanted cells did not work in fighting off the virus. The experiment was the first test of the facilitated bone marrow in a human recipient of a cross-species transplant. *Facilitator cells* are a special type of white cell that is believed to be crucial in promoting the transplant and removing the cells that are the major cause of transplant rejection. Animal activists condemned this "experimental procedure," while scientists rallied around a potential breakthrough in the fight against auto-immune disorders.

DRAIZE TEST

In 1921, the *Journal of the American Medical Association* quantified the results of the lack of testing for eye irritation, reporting numerous cases of human blindness and disfigurement—as well as one death—resulting from the use of a synthetic aniline dye called *Lash-Lure.* Lash-Lure was applied by operators in beauty salons to darken eyebrows and eyelashes.

Dr. John Draize, a scientist for the United States government, standardized a scoring system of a pre-existing test for ocular irritation in 1944. With this test, a liquid or solid substance is placed into one of the rabbits' eyes, and changes in the cornea, conjunctiva, and iris are observed and scored. The rabbit's eyes are inspected at 24, 48 and 72 hours—and at four and seven days. The Draize test has proven quite accurate in predicting human eye irritants, particularly for slightly to moderately irritating substances which are difficult to identify using other methods. Rabbits are used in both skin and eye tests, because they are the best models available for predicting potential human toxicity. Almost anything that is put into or around the eye is first tested by rabbits before humans are allowed to use the product.

The FDA is the federal agency responsible for ensuring that products are safe and effective. Whenever possible, scientists use validated, non-whole animal techniques, including tissue cultures, egg embryos, and biomedical assays. Many in vitro tests are being used as reduction and refinement alternatives while undergoing *validation,* the process by which the suitability of a particular test is assessed for a specific purpose (with its reliability and reproducibility verified). The number of rabbits used in the cosmetics industry had been reduced by 87 percent in the 1980s by using alternative methods, outlined as follows. The Draize test and skin irritant test, however, continue to be considered among the most reliable methods currently available for evaluating the safety of a substance introduced into or around the eye or placed on the skin. Non-animal tests are widely used as screening tools. Their results do *not* necessarily demonstrate the safety of a substance. The effects of a substance on a specific cell or tissue might differ significantly from its effect on a specific organ, such as the eye.

Most toxicologists and ophthalmologists are reluctant to repudiate the Draize test, because no one test has been developed as of yet that will measure all the necessary variables described by Draize.

On September 9, 1991, Governor Pete Wilson of California vetoed Assembly Bill 110, which would have banned the Draize eye and skin irritancy tests for personal care and household cleaning products. At the urging of former United States Surgeon General C. Everett Koop, Gov. Wilson stated the "overwhelming judgment of the scientific community... that these alternatives cannot now or in the foreseeable future completely replace the use of animals for testing..." and that "to guarantee the safety of new products for consumers, we must continue to rely on tests based on the reactions of a living organism. The information gathered from these tests helps people to use these products properly and helps health professionals to treat cases involving the accidental or deliberate misuse of these products" (Shapiro, L.S., *A Fly is a Carrot is a Pig is a Dog,* 1996). In addition to Dr. Koop, the FDA, the EPA, the United States Consumer Products Safety Commission, the Department of Health Services, the American Association of Poison Control Centers, the

California Association of Ophthalmology, the American Academy of Dermatology, the American Diabetes Association, the Center for Alternatives to Animal Testing at Johns Hopkins University, and the Stanford University Medical Center all urged the governor to veto the bill.

The humane objections to the Draize Test are obvious. What should also be obvious are the humane objections for not testing these products to guarantee human and animal safety. Substances tested using the Draize method are not as corrosive and as highly irritating as many animal rights activists claim (see the following description). Still, the mildly irritating substances that are put into rabbits' eyes create ethical concerns that should be addressed using the three Rs as a guide.

WHEN TESTING PRODUCTS USING THE DRAIZE TEST

1. pH test—Anything below 2.0 or above 12 is not tested further.
2. Primary Dermal Irritation Test (anything causing severe irritation is not tested any further)
3. Non-irritating to moderately irritating solutions are then tested further using the Sequentially Performed Eye Test.

Using the three Rs, the Interagency Regulatory Alternatives Group (composed of representatives from the FDA, Consumer Products Safety Commission, and the EPA) has recommended reducing the number of animals used per eye test to three (was at six). Further development of legally accepted alternatives are being researched—and will, hopefully, fully replace the Draize Test.

EMBRYO COLLECTION, TRANSFER, SPLITTING, AND CLONING

A great deal of progress has been made during the past 30 years through better management of an animal's natural reproductive processes. Farmers routinely use artificial insemination, synchronization of estrus, superovulation, and embryo transfer to improve the reproductive potential of their animals. Non-surgical techniques of recovery of embryos have been practiced for both the cow and the mare. Two-way flow catheters enable the flushing of the embryos from the uterus into a collecting receptacle. A small balloon near the end of the catheter prevents the flushing fluid from escaping through the cervix. Embryos are recovered from donor animals in the *morula,* or early blastocyst stage. These embryos (usually in the eight-cell stage) can then be transferred to the oviducts of the recipients, frozen for future use, or split into eight identical embryos (see Figure 6–3).

Since 1981, animal scientists (such as Jon Gordon) have been introducing foreign DNA into the mammalian genome. Transgenic mice have been routinely produced to aid in the study of gene function. A *transgenic animal* is an animal whose genetic makeup has been modified by the addition or removal of a specific DNA sequence. This modified, chromosomal DNA is then transmitted to all future generations.

The animal rights movement has protested the use of animals for this purpose. The scientific community has insisted that the primary value of the transgenic animal model to society will be as a scientific tool for elucidating the mysteries of gene regulation. Thus, this new scientific tool could be used to find a prevention and/or treatment for many genetic disorders—and other disorders—that plague both human and animal kind. Currently, proposed transgenic livestock projects include the following:

1. *Enhancement of livestock production traits*
   a. Increasing wool production in sheep
   b. Reducing the time it takes to get livestock to market weight
   c. Decreasing the amount of feed required to get livestock to market weight
   d. Producing leaner carcasses

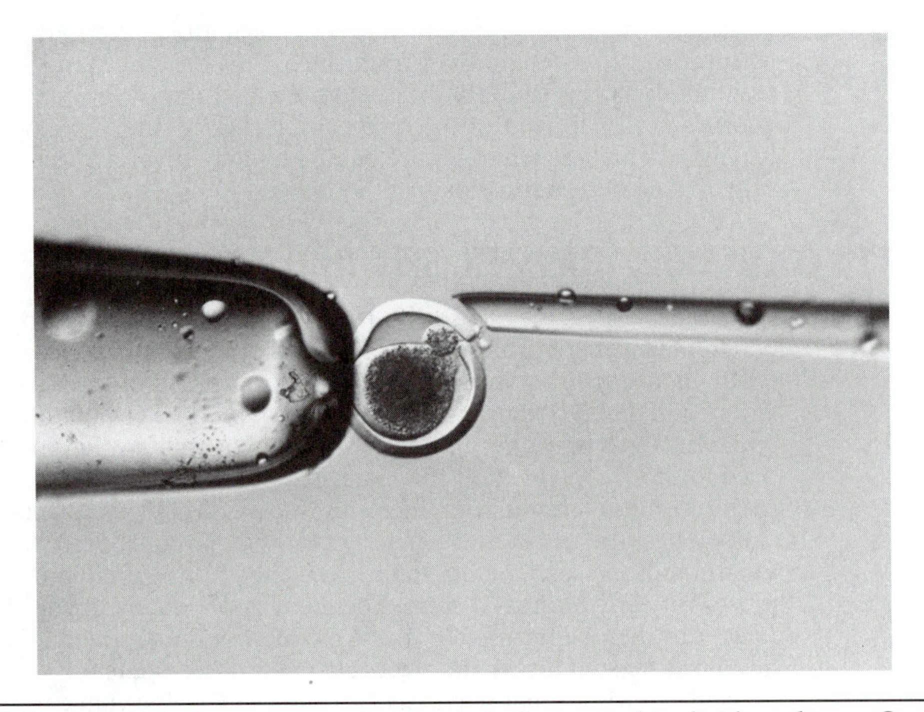

Figure 6-3 In the cloning process, microscopic tools are used to divide embryos. *Courtesy of Dr. Carol L. Keefer, ABS Specialty Genetics*

2. *Increase disease resistance*
   a. Introduction of genes that encode the necessary information to produce specific antibodies to fight off disease in various livestock species
   b. The introduction of genes that encode viral envelope proteins

   Both of these experimental transgenics might have the capacity to produce strains of livestock that are genetically immune to a variety of viruses. Livestock professionals and university personnel claim this experimental work would greatly reduce animal suffering and produce a much cheaper food. Animal activists argue that this idea is "messing with Mother Nature" and would cause untold, catastrophic effects.
3. *Creation of new animal products*
   a. A variety of blood-born coagulation and anti-coagulation factors (tPA, blood clotting Factors VIII and IX, and Protein C) have already been successfully produced in the mammary glands of goats, pigs, and sheep.
   b. Human hemoglobin has been produced in the blood of pigs.
   c. Human lactoferrin is currently being researched for production in bovine milk. When isolated, this product can be used as an additive for infant formula. Lactoferrin has strong bacteriostatic properties. Scientists are also working on increasing bovine lactoferrin in the cow to help prevent some forms of mastitis. This could greatly reduce the need to use antibiotics for the dairy animal.

   Again, many animal activists argue that we are "using," and thus "abusing," animals when we turn them into "pharmaceutical factories." Scientists counter that the high cost of alternate production processes and the reduced availability of human blood (from which many of these products are currently extracted) warrant continued research in this area. Use of large animals as "bioreactors" for the production of these and other drugs cannot be replaced with cell or bacterial culture systems because of their complexity. In addition, extracting from tissues draws concerns about contamination of known and unknown agents.
4. *Creation of human genetic disease models*

Currently, there are a number of experiments taking cells from a patient and transfecting these cells with the desired gene—then reintroducing these cells into the patient. Eventually, scientists hope to simply inject DNA directly into the tissue to be engineered. What about animal models and human models? Does one model exclusively help the other? Are both related to disease prevention, treatment, and cures for each other?

"OH DOLLY, OH DOLLY"

"Mary Had a Little Lamb"(s) revisited . . .
Mary had a little lamb, its fleece was slightly grey,
It didn't have a father, just some borrowed DNA.
It sort of had a mother, though the ovum was on loan,
It was not so much a lambkin, as a little lamby clone.
And soon it had a fellow clone, and soon it had some more,
They followed her to school one day, all cramming through the door.
It made the children laugh and sing, the teachers found it droll,
There were too many lamby clones, for Mary to control.
No other could control the sheep, since their programs didn't vary,
So the scientists resolved it all, by simply cloning Mary.
But now they feel quite sheepish, those scientists unwary,
One problem solved, but what to do, with Mary, Mary, Mary . . .

Anonymous

In February 1997, the highly respected journal *Nature* confirmed that researchers at the Roslin Institute near Edinburgh, Scotland, had successfully taken a cell in an adult ewe's mammary gland and cloned the cell, thus creating a new lamb (named Dolly). Dolly is a genetic carbon copy of her biological mother. She is her mother's identical twin. The cattle industry has been cloning for years. What is different here is that Dolly was cloned using cells from an adult sheep. Previous cloning was limited to embryonic cells, where an embryo was simply split—creating identical twins. In Dolly's case, a cell from an adult ewe's mammary gland was taken and "put to sleep" in order to later cause simultaneous cell division. The cell was later put into an emptied, unfertilized ovum (its genetic material removed). A pulse of electricity was used to fuse the two cells together and to start the cell division. A week later, the embryo was placed into another ewe's uterus. Rather than going through the specific biology of how this technique became perfected, a philosophical discussion is demanded for this scientific "advancement."

Both the benefits and the real and hysterical concerns of society need to be addressed. Soon after Dr. Ian Wilmut, research scientist from the

Roslin Institute in Edinburgh, Scotland, announced his findings, President Bill Clinton charged a federal commission with the task of investigating the legal and ethical implications of cloning. All types of hysteria involving cloning and its effect on society, the security of our nation, and the environment began to appear in the popular press. The press reported that Saddam Hussein ordered Iraqi scientists to clone himself. Hussein has been an assassination target for many years. His son, Uday, had a double (a surgically altered person) made so that attempted assassins would be less likely to attack him. Could the world allow a cloned Hitler or a cloned Saddam Hussein? These are questions that were being asked on talk radio after Dolly's appearance in the press.

In this country, noted science critic Jeremy Rifkin argues that "it's a horrendous crime to make a Xerox™ of someone." Rifkin and others ask, "Where will this experimentation stop?" A baby created to provide transplant material for the original is not being advocated yet. How about creating a team of Michael Jordans? What most of the articles did not explain is that clones cannot replicate anyone emotionally, culturally, or morally —only physically. All of the person's physical illnesses will be duplicated. You cannot create another Hitler, another Saddam Hussein, another Einstein, or another Jordan simply by cloning. Saddam Hussein is 60 years old. Another 60 years of raising his clone in the exact same manner that he was raised would be necessary to create an identical individual.

What about endangered animals? Should we clone the elephant and other endangered animals? Is it wrong to deny these animals a chance to continue to exist on this planet? Some animal activists feel that it is wrong to "mess with Mother Nature." Others feel that anything we can do to save a species should be tried.

So why bother with cloning animals? From an ecological as well as economical basis, cloning could be beneficial to agriculture. Currently, approximately 9.5 million dairy cows produce enough milk to feed our nation. Their average milk production is slightly greater than 16,000 pounds per cow each year. A handful of cows have produced 60,000 pounds of milk each year. Cloning these high producers could create a much smaller national herd, eating less feed, growing on less land, creating much less manure waste. Thus, from an ecological and economical basis, cloning is a benefit to the dairy farmer, consumer, and ecologically minded public. Cloning can also speed up the clock for other genetic advances (leaner beef, more wool growth, and more disease resistance). What about the negative aspects of this idea? The price of milk, most likely (due to supply and demand), would be much lower—forcing the small family farmer out of business. Stress to the cow would be much less, because only high-producing cows showing minimal stress would be cloned. If farmers were to clone only the top producers and cease breeding everything else, however, we would risk loosing the gene pool necessary for all the genetic variations found in the world's animals. These variations might prove necessary in the future to fight new diseases or to improve production.

Would the industry as a whole clone two million, 60,000 pounds of milk per year cows? Obviously not, and hopefully not. The acceptance of artificial insemination took half a century to take hold in the United States dairy industry. *Bovine Somatotropin* or *Bovine Growth Hormone* (BST) is not used in most dairy herds, and even in those herds where BST is used, it is not used in all cows. Still, the use of this new technology should be reviewed, critiqued, and analyzed each step of the way to safeguard our food supply, our ecology, and our cost of producing food and fiber.

In a February 1997 telephone poll for TIME/CNN of 1,005 adult Americans, 65 percent of those polled felt the federal government should regulate the cloning of animals. Seventy-four percent thought that it was against God's will to clone human beings, and only seven percent said they would clone themselves if they had the chance. Who would serve on this federally regulated committee? Sen. Tom Harkin, a Democrat from Iowa, argued before the entire Senate hearing on cloning that it was wrong for President Clinton to issue an order to stop all federally funded human embryo research —and that it was wrong for Sen. Christopher Bond, a Republican from Missouri, to propose that the ban be permanent. Sen. Harkin compared the federal government's attempts to those efforts in the 17th century to punish Galileo, who advanced Copernicus' theory that the Earth orbits around the sun and not the reverse. Sen. Bond countered that there "are aspects to life that should be off-limits to science. We must draw a clear line. Humans are not God, and they should not be allowed to play God. It is morally repugnant" (*Time*, March 10, 1997).

An eighteen-member National Bioethics Advisory Commission to the President has called for laws to forbid human replication through cloning, but the commission did not address the issue of experimentation with cloned human cells that go no further than a laboratory dish. Most have agreed, including Dr. Ian Wilmut, that human cloning leading to the birth of a child should be strictly forbidden by United States laws. The commission, however, has stated that human embryo research (including cloning) that stops short of producing a child should not be addressed by federal laws. The commission has also recommended that laboratory research using cells of humans and animals should continue. Scientists feel that cloning research could lead to the production of new skin for burn victims or new organs for people who need transplants. The commission said, "Professional and scientific societies should make clear that any attempt to create a child by somatic cell nuclear transfer and implantation into a woman's body would at this time be an irresponsible, unethical, and unprofessional act."

In other words, the somatic cell nuclear transfer and implantation technique employed with creating Dolly would be considered unethical for creating a human child—but would be "okay" for creating an animal "child." Most scientists have agreed with this conclusion, but many animal activists have not.

REPLACEMENT ORGANS

On July 23, 1997, scientists at the British Association of Pediatric Surgeons in Istanbul, Turkey, announced they had grown replacement organs for sheep, rats, and rabbits using the animals' own cells and using lab molds to help the tissue take the shape of various organs. Dr. Anthony Atala and Dr. Dario Fauza, two Harvard researchers, claim to be the first to have grown animal tissue from a variety of organs (heart, kidneys, and the urinary bladder). The organs were built with tissue that was taken from both grown and fetal animals and was later transplanted into those animals. Scientists hope that eventually this technique will be used to make "spare parts" for people and will be especially useful in correcting common birth defects.

According to Dr. Thomas McDonald, a Cornell University researcher, this new technique will be helpful in organ transplants, because the biggest obstacle to successful transplantation is the body's rejection of foreign parts. In human surgery, typically a mismatched tissue is used to repair defects (taking a piece of intestine to repair or patch a hole in the urinary bladder, for example). The researchers described how quickly tissue can be grown in an incubator. Scientists claim that a square centimeter of cells could produce enough tissue within two months to cover two football fields.

Although this new research using animals was developed for human benefit, many animal scientists are excited about the possibilities of using this new technique to benefit animal patients as well. Clear benefits exist of using one's own DNA, compared to the current usage of donated tissues and organs from cadavers or the experimented use of *heterograft* (xenotransplant). Although pig heart valves have been used in humans for years, many individuals have raised the ethical concerns of using animal parts in humans.

On September 20, 1996, proposed guidelines for xenotransplants were presented to the public by the NIH, the FDA, and the Centers for Disease Control and Prevention, to create discussion and generate comments. Recommendations included screening animals carefully to minimize the risk of cross-species transmission of animal disease.

GENETIC MAPPING

Throughout the country and in several foreign nations, researchers are trying to determine where the genes for desirable traits lie along the strands of DNA on each chromosome. Scientists are developing tests that farmers could use to determine whether their cattle, sheep, pigs, horses, and chickens carry the desired genes. Genetic engineering would enable farmers to create new breeds or strains by adding or knocking out one or more genes in their prized animals. The possibility of bringing back a gene from a foreign country that does not permit exportation of its herds—and the creation of strains of animals resistant to parasites and diseases which possess a better feed-to-gain ratio—is frightening to many. But these concepts are exciting to others who see endless opportunities to use this technology to produce more (and cheaper) sources of food and fiber.

Chapter 6
EVALUATION

1. ______________________ is an animal rights organization purported to represent physicians in favor of animal rights.
2. ________________ is the United States Army neurosurgeon who researched head injuries in cats.
3. GAO stands for ___________________________________.
4. The Pennsylvania Head Injury researchers were trying to mimic which type of common human injury? ____________________
5. Which type of animals were used in this research trial? __________
6. The HFI was a state referendum in ________________.
7. ____________________ is a loss of the sensory nerve fibers from a portion of the body.

Discussion Questions

1. Discuss the need or lack of a law such as the HFI. What effect would it have on the price of food? Should that be a consideration? If so, why? If not, why not?
2. Discuss the importance of Dr. Taub's work on spinal cord injuries and strokes. Should this type of work ever be researched using animals? If so, why? If not, then what alternatives are there to replacing the animals and still conducting the research? Or should the research simply be abandoned? Actor and director Christopher Reeve was a loyal PETA animal rights supporter. Since his recent spinal cord accident, he has had a change-of-heart when he realized the need for animal research in order to find a cure for his predicament. He has since proposed increased support of spinal cord injury research. Discuss this change-of-heart. Are people who are in favor of this type of research more likely to have a personal gain? Are people who are opposed to this type of research more likely not to have experienced a tragedy similar to that of Reeve?
3. Was PETA justified in refusing to allow the euthanasia of injured animals in the Silver Springs case? Defend your position on an ethical basis.
4. Compare and contrast the Baby Fae case with that of Jeff Getty receiving a bone marrow transplant from a baboon. Should either of these transplants have taken place? Defend your position. Why did the public and media generally support Baby Fae but not the research in the Silver Springs case? Read PETA's *Xenografts: Frankenstein Science* in Section 2 when analyzing your conclusion.
5. The Draize test has long been a symbol of protest by the animal rights movement. Describe some alternative tests that legally and/or scientifically (biologically) equate the accuracy of the test. If you find none, can the test be conducted with fewer animals? Be sure you can describe this test, how the animal is used, the pain or suffering

(if any) the animal feels, and ways to alleviate this pain. What are the risks if this test is completely eliminated?

6. Defend or justify your position in allowing the cloning of Dolly and not the cloning of a human child—as outlined by the National Bioethics Advisory Commission in 1997.

Recommended Readings

1. PCRM Update, Sept.–Oct. 1989. Physicians Committee For Responsible Medicine.
2. *Xenografts: Frankenstein Science.* PETA
3. *Protecting Laboratory Animals.* USPHS
4. *Public Health and the Role of Animal Testing.* USPHS

CHAPTER 7

Why Animals Are Needed in Research

In 1832, the Warburton Anatomy Act legalized the sale of bodies for dissection in England in an attempt to end the practice of stealing bodies from cemeteries. Similarly to today, people who died and had no family or friends to claim their body were (and are) regularly handed over to be used as human cadavers. Many universities have traditionally used animals to teach human anatomy simply because they were unable to obtain human cadavers, and many have ended their use of cats to teach human anatomy due to the ease in obtaining human cadavers. Even local area high schools are obtaining human cadavers to teach human anatomy.[40] Scientific arguments for using animals in experiments have also been based on the ethical inadmissibility of using living human beings for most research, necessitating the use of animals as surrogates. Some have also argued that many of America's poor are often victimized by unscrupulous and/or inadequately trained physicians—and that their bodies, in life as in death, are often treated as "teaching or research material."

Objectives

The student should:

1. Outline the current common uses of animals in education and research, and cite claimed needs for such.
2. Outline the procedures required in developing a treatment for a new disease.
3. List the major benefits for both humans and animals in using animals in biomedical research.
4. List major advancements in both human and veterinary medicine that have been achieved by using animals in research.

5. Describe the American Veterinary Medical Association's (AMVA) policy in the use of animals.

Many animal activists argue that non-human species are useless in research because of their inherent differences from people. These same individuals are often advocates of computers, cell cultures, and human surrogates. But which method of observation—non-human species or cell cultures—will give you a better assessment as to the effect, methods, and hazards of a medicament to a human? Is it morally or ethically sound to produce a new medicine that has not been thoroughly tested in another living being? Would a caring parent permit his or her child to take new cough syrup that has not been thoroughly tested for safety on another living creature? If animals are equal to people, then do we seek poor, volunteer children (perhaps from third-world countries) to take money in exchange for being "human guinea pigs?" Many animal activists do not want to answer these questions. Others simply say, "What you are not willing to do to a human child, you should not do to an animal child."

What is the normal procedure in testing a new product? What is the normal procedure in finding a cure for a new or currently researched disease? (Please see Figure 7–1.) Unlike what is presented in many emotionally filled mailings, animals are *not* the first step in assessing a treatment or preventative measure in disease control. Animals are, however, eventually used in most (if not all) new treatments for humans. Prior to the AWA, there was little regulation as to the care of these animals. Many laboratories provided excellent care, and others were less than humane. Because of the latter, the need for the AWA and its amendments was and is still necessary. Also, due to the lack of humane practices, the use of mandatory IACUCs (unannounced government inspections and detailed records or all animal research facilities) was required.

The record-keeping and required IACUCs have created a better science, as well as a more humane treatment of animals used in research. Because scientists and physicians continue to use animals in their search to find preventative and treatment measures for both animal and human diseases in biological research, these regulations need to be maintained.

THE BENEFIT TO HUMANS OF USING ANIMALS IN BIOMEDICAL RESEARCH

Virtually all medical knowledge and treatments—certainly, every medical breakthrough in the last century—has involved research with animals. Seventy-six Nobel prizes have been awarded in medicine or physiology in this century. Fifty-four Nobel prizes were based on animal research. Medical

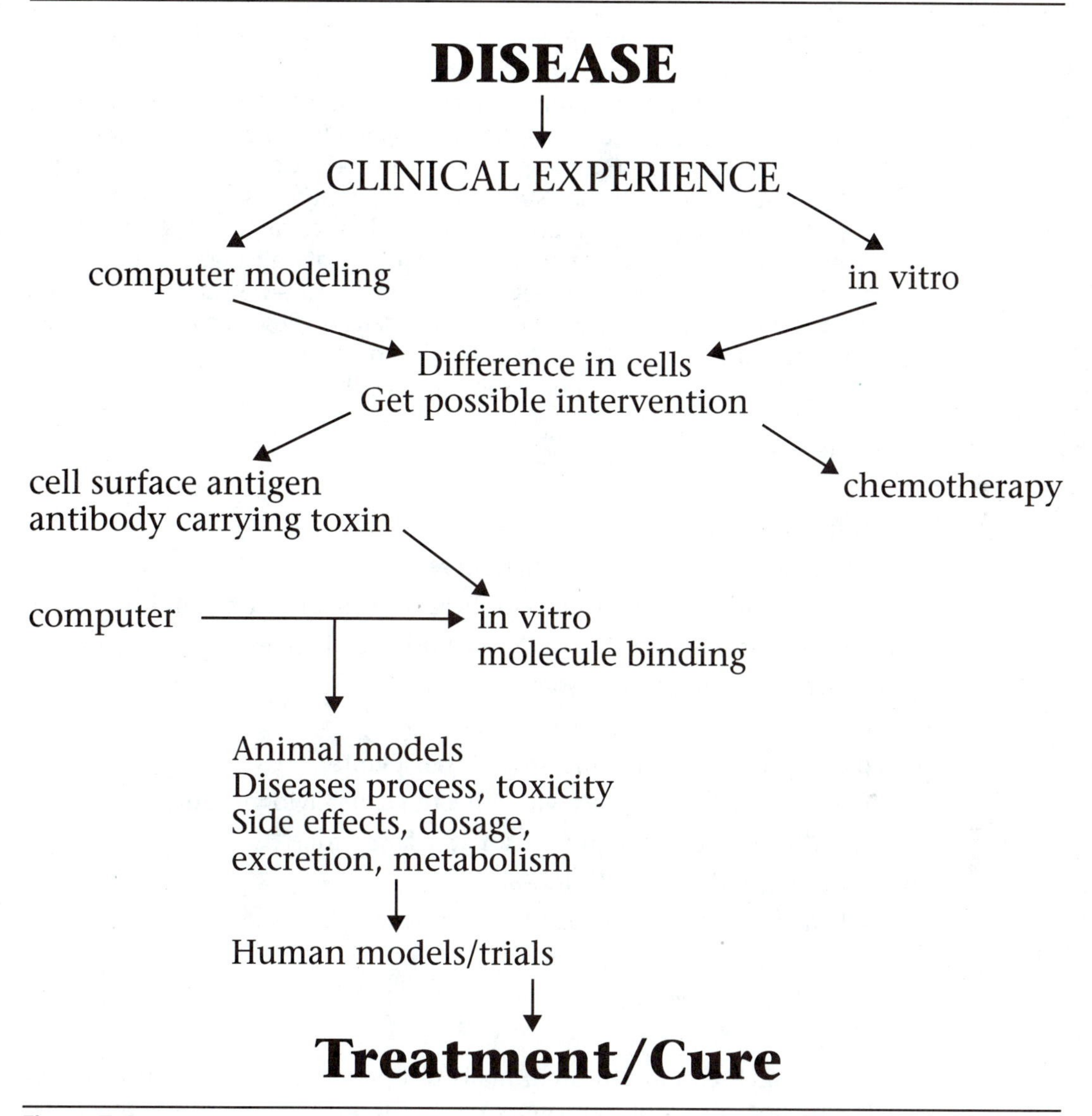

Figure 7–1 Process of developing a pharmaceutical product

practitioners use animals to gain knowledge needed to understand life processes. Biomedical researchers conduct experiments to formulate hypotheses and make predictions. As tests are made and predictions are agreed upon, the sharing of these research findings is vital for human as well as animal well-being. Researchers use animals to save both human and animal lives, extend our life expectancy, and improve upon the quality of life for both humans and other animals.

Researchers use many animals to gain insight into how different parts or systems in our bodies adapt to different research techniques. The use of specific animals for certain purposes is necessary, because animals have unique parts in their systems that are similar to those of humans. As mentioned earlier, between 17–22 million animals are used in research and testing in the United States. Ninety percent of these animals are rats, mice, and other rodents specifically bred for such use. Dogs and cats make up 1–1.5 percent of research animals. Primates represent fewer than 0.5 percent. Research institutions use fewer than one percent of all the animals sent to pounds. The following list describes some examples of how some animals have specific uses in research:

armadillo—leprosy
bear—osteoporosis
chinchilla—middle ear infections
cockroach—central nervous system repair
dog—hip and joint replacement, heart transplants, hemophilia
ferret—gastro-intestinal disorders, endocrine diseases
mouse—auto-immune system
opossum—embryonic development
pig—heart, lung, skin, and kidney transplants
primate—hormone levels, viral infections, behavioral studies
rabbit—tissue damage from heart attack or stroke
rat—drug addiction withdrawal
salamander—replacing dead heart muscle
shark—cancer resistance
woodchuck—hepatitis

Research conducted on animals has provided humans with the following benefits: joint replacement, surgical procedures, arthroscopy, osteoarthritis treatment, organ transplants (kidney, liver, cornea, bone marrow, skin), vaccines for cholera, diphtheria, Hepatitis B, influenza, meningitis, mumps, the plague, pneumonococcal pneumonia, polio, rabies, rubella (measles), tetanus, and typhoid and yellow fever.

Specific animals which have made contributions to an improved human lifestyle include the following.

PARTS FROM ANIMALS

HOGS

Heart valves—In 12 years, 250,000 lives have been saved.

Skin—Burns and learned skin stretching; gelatin for capsules and pills

Thyroid—Metabolism in humans to treat low calcium and phosphate levels and regulate heartbeat

Pancreas—Insulin; five percent of diabetics are allergic to synthetic insulin

Other—Cortisone, blood fibrin, and estrogen

CATTLE

Adrenal glands—Norepinephrine used to relieve symptoms of hay fever, asthma, and some allergies; heart stimulants used by dentists to prolong the effects of local anesthetics

Blood—Helps blood clotting, skin grafting

Liver—Treating anemia

Pancreas—Insulin for diabetics; glucagon counteracts an insulin shock.

Other—Hormones (TSH, ACTH, estrogen, thyroid extract, and cholesterol as building blocks for sex hormones)

PROCESSES

RATS

Aging process, toxic levels of chemicals, diabetes, obesity, hypoglycemia, measuring alcohol consumption, regulation of cholesterol, effect of penicillin, cancer, infections, tissue rejection, blindness prevention, Alzheimer's disease aging process, and reproductive problems

PRIMATES

Hepatitis, AIDS, genital herpes, Parkinson's disease, Hepatitis-B vaccine, malaria, children's vision, pregnancy, and fetal development

RABBITS

Processes of vision, blood pressure, tuberculosis, and cystic fibrosis

DOGS

Cardiac bypasses; electrocardiograph; heart disease; open heart surgical techniques; kidney, liver, heart, and lung transplants; diabetes; muscular dystrophy; anemia; aging; sinus; using pins in setting fractures; cardiac pacemakers; arteriosclerosis; hemophilia; hepatitis; nutrition; glaucoma; cataracts; emphysema; spinal surgery (disks); trauma and shock.

CATS

Feline leukemia, brain damage, strokes, breast cancer, leukemia vaccine, cataracts, spinal cord injury, nerve impulses, and arteriosclerosis

Innovations are made possible by biomedical research using animals:

Artificial hearts
Artificial hips and joints
Artificial limbs
Blood pressure measure
Cardiac catheters
Pacemakers
CT scans
Electrocardiograph (ECG)
Electroencephalograph (EEG)
Electron microscope
Kidney dialysis machine
Magnetic Resonance Imaging (MRI)
Radiation therapy
Ultrasound
X-rays

Ongoing animal-based research promises to show similar benefits. Research with Rhesus monkeys and chimpanzees underlies the recent release of a vaccine for Hepatitis B. Rhesus monkeys are also being employed in testing of experimental treatments for Parkinson's disease and AIDS researchers are attempting to determine why chimpanzees can be infected with the HIV virus, yet they almost uniformly survive the infection. Former Surgeon General C. Everett Koop has suggested that "we'd be in absolute, utter darkness about AIDS if we hadn't done decades of basic research in animal retroviruses."

BENEFITS TO ANIMALS OF USING ANIMALS IN RESEARCH

For pets as well as people, preventative medical care is the key to good health. Nutrition and breeding research ensures that puppies, kittens, calves, foals, and kids grow into strong animals. Medications that kill harmful parasites in animals, such as heartworms and hookworms, were created by animal research. Although a heartworm or hookworm is a life form, it is not, by Western thought, given equal consideration to that of our pets (a form of speciesism?). Vaccines that keep our animals safe from deadly illnesses, such as feline leukemia, rabies, distemper, and infectious hepatitis, were created by animal research.

In 1978–1979, hundreds of thousands of puppies and young dogs died by hemorrhaging and vomiting from an outbreak of parvovirus, a new disease. Using the latest techniques of virology, researchers were able to quickly adapt an existing feline vaccine to protect puppies from this sickness. The development of this life-saving vaccine would have been impossible without animal research.

A major cause of heart failure in cats was wiped out when researchers traced the deterioration of heart muscle (called *feline dilated cardiomyopathy*) to an amino acid deficiency. Cat food manufacturers quickly fortified their foods with taurine to guarantee that thousands of housecats no longer died of this disease.

Devastating genetic problems such as hip dysplasia, deafness, and hemophilia can now be avoided through careful breeding. Current chromosome analysis and genetic research technology helps prevent the spread of crippling traits in pedigreed animals.

New veterinary techniques adapt human medical breakthroughs for pets and livestock. Artificial hips and joints give pets with injuries, arthritis, or hip dysplasia the freedom to romp again. Aging animals can have cataracts removed or even receive hearing aids for deafness. Animal blood banks are saving the lives of pets hurt in accidents or pets undergoing surgery.

One of every ten foals is born prematurely. These foals now stand a chance of surviving because of equine neonatal centers, modeled on facilities for premature infants. Admitted with their mothers, the foals receive drugs, respirators, and 24-hour monitoring until all risks are gone.

Pacemakers extended the lives of hundreds of dogs and cats last year. The first dog implanted with a pacemaker in 1967 thrived so well that when the battery ran low after five years, a second pacemaker was inserted. Because pacemakers are expensive, they are usually donated to veterinarians by the families of deceased heart patients. Two families then benefit from the contribution made by animals to research.

Kidney failure is another leading cause of death in cats and dogs. A cat's kidney is a bit larger than a human thumb, so veterinarians are beginning to use modern microsurgical techniques and the latest immunosuppressive drugs for successful organ transplants. Perhaps the new cloning techniques being developed for producing human replacement organs will soon be used to also help our companion animals needing such transplants.

Some of you may laugh at this proposition. According to the veterinary college at Davis, California in 1976, however, we were "only 20 years behind human medicine" in offering medical care to our pets. In 1997, we were "only two years behind" human medical offerings. Surgeries such as kidney transplants in dogs (not for research purposes, but to save the dog's life), total hip replacements, and corneal transplants are being performed on our companion animals—and dental suites that offer routine preventative care and are equipped to perform minor orthodontics can be found at many, if not most, of the schools of veterinary medicine in this country. Animal research has benefited the animal patient, as well.

Cancer is responsible for almost half of the deaths in pets older than age ten. Animals suffer from bone, skin, and breast cancer and leukemia. As animal medicine grows more sophisticated, veterinarians, neurologists, radiologists and surgeons are able to develop treatments for animals, including radiation therapy, chemotherapy, cryosurgery, and treatment for hypothermia. These treatments can give years of life to pets with cancer.

Before effective vaccines and veterinary treatments were available, a farmer would often have to destroy both healthy and sick animals to halt a contagious disease. Today, a number of drugs and vaccines safeguard the animals we grow and the food we consume.

VACCINES

Dogs—Rabies, distemper, parvovirus, infectious hepatitis, parainfluenza, leptospirosis, canine adenovirus, bordetellosis, and coronavirus

Cats—Leukemia, rabies, rhinotracheitis, panleukopenia, calicivirus, and FIP

Horses—Strangles, tetanus, encephalomyelitis, and rabies

Poultry—Marek's disease, Newcastle disease, fowl cholera, duck hepatitis, hemorrhagic enteritis, fowl typhoid, fowl pox, encephalomyelitis, erysipelas, infectious bursal disease, mycoplasmosis, pasteurellosis, and reovirus

Cattle—Leptospirosis, brucellosis, salmonellosis colibacillosis, clostridial diseases, pasteurellosis, haemophilus somnus infection, pinkeye, anthrax, parainfluenza 3, bovine viral diarrhea, and rabies

Sheep—Anthrax, bluetongue, contagious ecthyma, clostridial disease, rabies, and pasteurellosis

Swine—Erysipelas, actinobacillosis, clostridial disease, leptospirosis, pseudorabies, transmissible gastroenteritis, atrophic rhinitis, pseudorabies, salmonellosis, and tetanus

Research on diseases, breeding, and nutrition protects the health of cattle, hogs, sheep, chickens, and humans. Tuberculosis is no longer a threat to most of the world's population.

The AVMA's official policy on the use of animals is described as follows: The carefully controlled use of animals contributes greatly to improving the health and welfare of both animals and human beings. Therefore, the AVMA believes that there is ample justification for the prudent and humane use of animals in veterinary medical education and biomedical research, provided that:

1. The institution which conducts such teaching and/or research has met all legal requirements and guidelines pertaining to the acquisition, care, and use of animals for these purposes;
2. The individual investigators have thoughtfully examined the need for such animals and have appropriately selected the species and carefully determined the number required to meet the needs of the protocol;
3. Adequate safeguards are used to ensure that only unidentified, unowned, and specifically authorized animals are obtained from animal shelters and humane societies for these purposes; and
4. Preventative measures are taken to guarantee the health of animals obtained from such facilities before they are used in teaching or research situations.

MEDICAL ADVANCES THROUGH THE USE OF ANIMAL RESEARCH

More than eighty medicines developed for humans are today used to heal pets, farm animals, and wildlife. Anesthetics, tranquilizers, and pain killers enable veterinarians and wildlife biologists to treat wild or frightened animals without pain or stress.

Modern diagnostic technologies such as ultrasound, MRI, and CT scans are transforming the nature of veterinary medicine. Veterinarians today can diagnose and remove delicate brain tumors, help with difficult pregnancies, and correct birth defects.

Advances in reproduction technology have aided both humans and animals. The same techniques that help childless couples conceive are being applied to breeding programs for endangered species such as the Florida panther and black-footed ferret.

During the Persian Gulf War, new techniques researched on dogs to replace expected limb loss were developed and are now being used to treat severe osteoarthritis, hip dysplasia, and limb loss in canines.

The largest-selling animal health product in the world is being used to save people in Africa from river blindness. Research found that Ivermectin, the drug that kills parasitic worms in livestock and heartworms in dogs, also attacks the tiny worms that harm millions of people.

GENERAL SUPPORT FOR ANIMAL RESEARCH

SUPPORT FROM THE MEDICAL ESTABLISHMENT FOR ANIMAL RESEARCH

Surveys of physicians confirm that 97 percent support the use of animals in research. One animal rights group that purports to speak for organized medicine is the PCRM. PCRM claims 60,000 members, but only 3,000 are physicians—fewer than one-half of one percent of all the physicians in the United States. (See the PCRM solicitation form.) PCRM shares its office space with PETA. PCRM President Neal D. Barnard, a psychiatrist, is the medical advisor to PETA. Dr. Barnard claims that anti-AIDS drugs have been developed using cell cultures, and that 80 percent of cancers are potentially preventable by diet modification and the cessation of tobacco use. Both of these claims are false (Sources: "The importance of animals in biomedical and behavioral research," a statement from the Public Health Service, March 1994; "Use of animals in biomedical research, The challenge and response." AMA White Paper, 1989; Balch, C.M., et al., "The vital role of animal research in advancing cancer diagnosis and treatment," *The Cancer Bulletin*. Vol. 42(4):266–269 (1990); Horton, L. "Physicians and the politics of animals research in the 1990's." *The Cancer Bulletin*. Vol. 42(4):211–219 (1990)).

At the 1990 annual meeting of the AMA, the California Medical Association urged the AMA to condemn PCRM "for misrepresenting the critical role animals play in research, teaching, and testing, and for obscuring the overwhelming support for such research which exists among practicing physicians . . ." (according to the AMA). They also stated that PCRM "erroneously reports that a growing number of physicians question the continued need to use animals in research," and that the group "has supported a campaign of misinformation against important animal research on HIV/AIDS and responsible animal users." This resolution unanimously passed the AMA House of Delegates.

Dr. Roy Schwarz, AMA's senior vice president for medical education and science, states, "The medical community overwhelmingly agrees that without biomedical research, the battles against HIV/AIDS, Alzheimer's disease, cancer, heart disease, stroke, multiple sclerosis, and hundreds of other major health threats will be markedly compromised, if not completely lost."

SUPPORT FROM THE GENERAL PUBLIC FOR ANIMAL RESEARCH

All major public surveys have shown overwhelming support for continued use of animals in medical research. A 1989 Gallup survey reported that 77 percent of Americans believe the use of animals in research is necessary for progress in medicine. Polls reported in the October 1989 issue of *Parents* magazine showed that 80 percent of its readers believed that animals had rights. Eighty-five percent also believed, however, that it was morally acceptable to use animals for the benefit of people. Only five percent of those polled thought that it was wrong to kill for food, and two-fifths of those who said it was wrong admitted that they still ate meat.

Professors Hoban and Kendall of North Carolina State University showed similar results with studies of consumer attitudes on food production and biotechnology in agriculture.

At the same time, the public believes that animals should be treated humanely. Researchers should be required to obey laws that establish standards for animal research. Efforts should be made to reduce the number of animals used, or to find alternative methods whenever practical or possible. The general public believes that humans have responsibilities or obligations that require us to treat animals humanely. In other words, the general public, including most scientists, support the three R's as outlined by Russell and Burch (*The Principles of Humane Experimental Technique*, 1959).

Chapter 7
EVALUATION

1. The _________________ legalized the sale of bodies for dissection in England, in an attempt to end the practice of stealing bodies from cemeteries.
2. ___________ animals are used in research and testing each year in the United States. (quantity)
3. _____ percent of these animals are rats, mice, and other rodents.
4. _____ is responsible for almost half of the deaths in pets older than ten.
5. _____ percent of physicians in the United States support the use of animals in research.
6. One physician-based animal rights group that does not support the use of animals in research is ___________________________.
7. In a 1989 Gallup survey, _____ percent of Americans believed the use of animals in research was necessary and should continue.

Discussion Questions

1. Without the use of animal surrogates for medical doctors to practice on, are our nation's poor more liable to become "teaching or research material?" What are our alternatives? Defend your position.
2. Outline the process of development of a pharmaceutical product. Relate this outline to the human AIDS disease. Show in your discussion the relationship, if any, of HTLV-III (the Macaque Monkey), HTLV-II, FLV (feline leukemia), and FTLV (T-Lymphotropic Lentivirus) with that of the HIV virus. Show how the studying of these related animal viruses and the search for treatments for those disorders will or will not speed up the discovery of prevention, earlier diagnosis, and treatment of human AIDS.
3. Why is it sometimes necessary to use animals other than mice and rats (i.e., a bear, a dog, or a pig) as human surrogates for different research techniques?
4. In 1978–79, hundreds of thousands of puppies and young dogs died of parvovirus, a new disease. Quickly, researchers developed a life-saving vaccine by using research animals. Many of these animals were purposely given the virus, and they died by hemorrhaging and vomiting. Discuss the ethical implications of this research. Should the research have been conducted? Why? Why not? How does this situation relate to using animals as surrogates for discovering preventative measures for humans? Is there a difference? Why? Why not?
5. Read Compton and Taylor's "Current Ethical Issues Surrounding Animal Research," Loeb et al.'s *Human versus Animal Rights*, *iiFARsighted Reports on Animal Research*, and PETA's "Companies that Don't Test on Animals," "Animal Experimentation:

Sadistic Scandal," and "AIDS: Contagion and Confusion." Compare and contrast the drastically different views on the need to use animals in research (benefits and risks).

Endnotes

40. Simi Valley High School, Simi Valley, CA

Recommended Readings (From Section 2)

1. Compton, L., and C. Taylor. *Current Ethical Issues Surrounding Animal Research. Drug Information Journal,* Vol. 26:85–94. 1992.
2. *Animal Rights: What's at Stake.* AMPEF.
3. *Testimony to the Value of Animal Research.* AMPEF.
4. *Lives Saved Thanks to Animal Research.* AMPEF.
5. *The Vital Role of Animals in AIDS Research.* AMPEF.
6. *AIDS: Contagion and Confusion.* PETA.
7. *Animal Experimentation: Sadistic Scandal.* PETA.

CHAPTER 8

Ethics of Animal Use

Rainbow Bridge

Just this side of heaven is a beautiful place called Rainbow Bridge. When an animal dies who has been especially close to someone here on Earth, that animal goes to Rainbow Bridge. There are meadows and hills for all our special animal friends so they can run and play together. There is plenty of food and water and sunshine, and they are warm and comfortable. Those who were ill and old are restored to health and vigor; those who were hurt or maimed are made whole and strong again, just as we remember them in our dreams of days and times gone by. They feel no pain or suffering, only comfort.

The animals are happy and content but for one small thing: they miss someone very special to them who had to be left behind.

They all run and play together, but the day comes when one suddenly stops and looks anxiously into the distance. The bright eyes are intent; the eager body quivers. Suddenly, he begins to break away from the group, flying over the green grass, his legs carrying him faster and faster. YOU have been spotted, and when you and your special friend finally join, you cling together in joyous

reunion, never to be parted again. The happy kisses rain upon your face; your hands again caress the beloved head, and you look once more into the trusting eyes of your animal friend, so long and painfully gone from your life, but never ever absent from your heart.

Then you cross the bridge to heaven together . . . to live together forever in peace, joy, and love.

This anonymously written poem captures the feelings we bestow our pets—our companion animals. Many individuals have been fortunate enough to also share these feelings for livestock. Many consider their livestock not only a business, but a way of life—a part of the family. The strong attachment or feelings the farmer has towards his or her stock can best be demonstrated by the dedication he or she has in delivering young calves, foals, lambs, and piglets by breathing into their mouths and giving them the first breath of life. They also show this dedication by doctoring these animals when they are injured or ill and by crying for them when they accidentally die or are shipped to market at the end of their productive lives. Farm animals are more than just a means of making a living. They are a farmer's friends. The farmer believes that his or her stock has been entrusted to his or her care.

How does this idea relate to animal ethics? Our personal experiences, or the lack thereof, guide our feelings and our passions in life. When 98 percent of the voting public has no direct connection to animal agriculture, it can only be expected then that the support for the farmer is not understood nor received. The animal rights community understands how removed most Americans are from animal agriculture, animal research, the fur industry and other animal enterprises. Because of this "ignorance" perceived in these areas, the main focus currently employed by animal rights activists is factory farming, animal research, and wearing of fur. The other areas, companion animal "ownership" and "veganism," would not be so readily accepted by most Western philosophies (see the following information).

Objectives

The student should:

1. Define speciesism. Discuss its ramifications in modern America (pros and cons).
2. Describe how personal experiences, or the lack thereof, guide our beliefs in animal ethics.
3. List the biological rights of domestic animals as proposed by animal welfarists.

4. Describe why non-domestic animals can't be guaranteed these same biological rights.
5. Describe why many veterinarians and animal nutritionists might consider it cruel to try and make a cat eat a vegetarian diet.
6. Describe how Hollywood has shaped our views on animal welfare/rights.
7. Compare and contrast the classifications of various philosophies of individuals who are active in animal treatment and behavior.

PHILOSOPHICAL ISSUES AND CONCEPTS INVOLVED IN ANIMAL USE

ASSIGNING RIGHTS TO ANIMALS

Humans perceive things differently in different situations. For example, a mouse could be a person's pet, a nuisance, a laboratory animal, or food for a hungry snake. Depending on one's background, different individuals have developed different philosophies—usually falling somewhere within a utilitarian versus animal rights perspective.

"Speciesism" is a word that animal activists use to describe the philosophy that humans are superior to animals, and thus they compare speciesism to racism (where one race is said to be superior to another). These animal activists argue that animals should be treated similarly, if not equally, to humans. Animal rights activists insist that domestic animals—and even wild animals—have rights independent of the values that humans place upon them. In most societies, people have given legal rights to domestic and non-domestic animal species in the form of humane laws, animal welfare laws, and wildlife protection or management laws. In addition to these laws, animal rights activists argue that all species of animals have certain moral and ethical rights that are intrinsic to all living beings.

Animal welfare proponents stress the humane treatment of animals. They generally feel that humans are different from animals and, therefore, humans can have dominion over other animal species. However, the ultimate use of animals is for the purpose of whatever benefits the majority of human-kind. In other words, because only humans are capable of moral and ethical behavior, only humans would have moral and ethical rights. Most animal welfare proponents believe that animals have biological rights independent of people. Biology is governed by nature, however, not humans. Nature does not guarantee survival of a species nor the humane treatment of an individual. Extinction is generally the rule of nature (there are ten times as many fossil species than known species to exist today). Have you ever watched a domestic cat catch and eat a mouse? The cat bites off one of the mouse's legs, letting the mouse think it is getting away, and recatches the mouse, only to torture it as it is being eaten.

Nature can be beautiful, and it can also be cruel. Because humans are capable of distinguishing the difference between right and wrong, we have moral rights. One cannot assign moral rights to an individual who does not have moral responsibilities. When a human being does not abide by his or her moral responsibilities, we take away his or her moral rights (throw him or her in jail, etc.). When an animal poses a danger to a human population, we try and relocate the animal (if possible), knowing that the animal (i.e., a mountain lion) cannot distinguish the moral difference between eating a young human child and eating a young deer. Some animal rights activists argue that there is no difference. If you permit the lion to eat a young deer, why would you not then permit the lion to eat a young child (that is, if we are *all* equal?). Some insist on making the lion lie down with the lamb. Perhaps one day this could occur—but the event will not be "natural."

Animal welfarists support the following biological rights of an animal:

1. The ability to find food to sustain energy requirements
   herbivores—plant material
   omnivores—both plant and meat material
   carnivores—only animal (meat); need amino acids from a meat source
2. Conserving an adequate amount of water
3. Finding or creating shelter from the elements and predators
   a. nests
   b. forming groups (herds, droves, or flocks)
   c. dens
4. The opportunity to reproduce

All four biological requirements listed are needed to preserve a species. These requirements are the biological rights of an animal. They are not moral or ethical rights; rather, they are simply biological rights.

When many species share a common environment, the basic requirements of survival are often met through the interaction of two or more species. This interaction many times involves obtaining food from another species (a symbiotic relationship). In the case of humans, animals have been used for food, and humans have preserved and/or expanded the animal species eaten. In some cases, the competition for limited resources greatly decreased non-domestic animals in favor of humans and domestic animal species. Some non-domestic animals are only being preserved in zoos, wild animal parks, and nature preserves created by humans to manage such animals. These animals have no moral or ethical rights in these parks. They simply have legal protection. Humans have a moral and ethical responsibility to care for and manage the endangered life forms.

Some misguided animal lovers will try to take away the natural instinct of a carnivore and make a vegetarian out of him (feeding a cat a vegetarian diet, for example). This is inhumane, cruel, and unnatural. Humans are territorial animals. Humans—men in particular—are extremely territorial. This statement is true whether we are talking about men living in the inner city of Los Angeles, in the deepest jungles of Africa, in the rural areas of Idaho, or on the plains of Kansas. Biologically, we are no different than other animals. The tiger, the mountain lion, and others have vast territorial claims. Likewise, humans include the entire planet as their territory. One difference is that humans have ethical and moral responsibilities that other territorial animals do not have (because they are not moral or ethical beings). Some humans have sought to extend their moral, ethical, and legal principles, which are necessary to maintain order among people, to domestic and wild animals. When people grant such rights to animals, these rights carry no weight or respect among the animals when people are not present. The animals have no desire to be morally or ethically correct.

Of course, in Hollywood and for those people who actually believe in the fantasy movies they watch have problems separating reality from fiction. *Babe* did not really talk, nor did the dogs and cat in *Homeward Bound*. Most people realize that, but the feeling and the attachment to the animal in the film leaves an almost "human" quality about the animal that most find difficult to shake.

CLASSIFYING ANIMAL TREATMENT AND BEHAVIOR

Philosophers and animal science professionals have grouped individuals who are active in animal treatment and behavior into one of five groups or classifications:

1. Animal Cruelty—These individuals have no problems "punishing" animals. Luckily, there is little intentional animal cruelty in the United States. There is still considerable animal cruelty due to ignorance, however. Starvation is often associated with cruelty.
2. Animal Exploitation—Some people abuse animals and use them only for their benefit. Exploiters have absolute dominion over animals. These people have no problem with dog fighting, bull fighting, cock fighting, poaching, trapping, and using animals for target shooting.
3. Animal Welfare—Expresses the responsibility to protect animals from harm and ascribes to the idea of humane ethics. They will use animals for food and research but will minimize suffering and advocate a fast, painless death.
4. Animal Rights—People who support this concept believe that the rights of animals should be guaranteed and that animals should not

be eaten or used for research or sport. They are opposed to vivisection, hunting, and trapping. They are also opposed to killing, except to reduce suffering.

5. Animal Liberation—These people are the most extreme activists. They try to eliminate all types of animal use. They have been known to commit terrorist acts including arson, bombings, and murder. They believe their cause is so grand that it is higher than the law (see the previous section on religious philosophies).

MORAL OBLIGATIONS OF MAN TOWARD ANIMALS AND THE CHANGING STATUS

Animals play a major and essential role in testing and researching new animal and human products—and in education for both human and animal medical students. The AVMA recognizes that humane care of animals in research, testing, and education is an integral part of those activities. The use of these animals is therefore a privilege that carries with it a unique professional, scientific, and moral obligation. The AVMA encourages the proper stewardship of animals, but AVMA also defends and promotes the use of animals in meaningful research, testing, and educational programs.

DISSECTION AND ALTERNATIVES

Although computer models, coloring books, and plasticized models enhance a student's ability to comprehend anatomy, "hands-on" dissection experience enables the use of touch, smell, and sound senses that are sometimes missing from simulations. Many students enrolled in biology and anatomy classes are not interested in becoming surgeons (human or veterinary) or surgical assistants. In those cases, the instructor should employ a growing array of alternative learning devices to accommodate students who oppose dissection on religious, ethical, and/or environmental grounds. For the same reason that no right-thinking person would ever fly with a pilot who has only flown a computer model, no real animal lover would allow a veterinarian to operate on their pet after having experienced only computer and plasticized models.

Several high school, college, and university anatomy classes have traditionally used the cat for dissection exercises. Many have now switched to human cadavers for human anatomy classes and use cat cadavers for animal anatomy courses. In both cases, the cadavers' lives were not ended for the purpose of dissection; both the humans and cats had died previously and instead of burying or burning the bodies these cadavers are utilized to educate future doctors and veterinarians so that other lives may be saved. For basic junior high and high school anatomy classes, plasticized and computer models may be all that is necessary, while a field trip to a pathology lab may be helpful for advanced students.

What happens when a student does not want to dissect an animal? Can he or she still learn the material? The student would have to work a little harder in memorizing anatomical parts, their order, and so on, but it can be done. Usually, professors will group a student who cannot dissect, for one reason or another, with students who can. The student can then simply observe the exercise. A student who refuses to participate in a dissection needs to understand, however, the likelihood of becoming a surgeon or surgeon's assistant is greatly reduced without the experience that he or she is missing. Some argue whether or not it is ethical to pass a student without "hands-on" experience in dissection. Are these students capable of becoming surgeons or surgical assistants? Evidence, in small numbers, from several veterinary and human medical teaching hospitals, is inconsistent.[41]

EMERGING FIELDS OF ANIMAL WELFARE CONCERN AND THEIR POTENTIAL FOR ANALYSIS (TRANSGENIC ANIMALS AND ARTIFICIAL BREEDING METHODS)

Conventional methods of animal breeding have partially been replaced in recent years and are likely to increase rapidly in the near future—with the new possibilities for *genome manipulation*. A genome is the total gene complement of a set of chromosomes found in animals, including humans. Genetic engineering technology has now enabled us to manipulate an animal's genome by the addition of new genes and the removal of existing genes. The addition of new genes can enhance the productivity and the welfare of animals. For example, we can now select double-muscling in beef cattle. Scientists are also working on transferring genes in cattle, with the result of an increased growth rate and modification of the final body form, thus producing an animal that is larger and reaches its mature size more quickly and with less feed. BST is currently being used in many dairy herds throughout the world. This naturally occurring hormone enables a dairy farmer to produce up to 20 percent more milk, with only five percent more feed. Thus, the farmer can produce more milk with fewer cows, taking up less room and consuming less feed (according to the American Dairy Science Association and American Society of Animal Science).

Resistance to specific parasites and diseases can be increased by the addition of a gene which encodes a natural, non-toxin protein from plants. This protein is deleterious to insect survival. These new techniques in agricultural science could help alleviate blowfly strike, lice infestation, tick predation, and perhaps internal parasites in sheep.

The secretions of the mammary gland can be changed so that cow or sheep milk contains proteins of value for human or animal therapy (i.e., alpha-S1 anti-trypsin and the anti-hemophilic factor IX).[42]

Any modification of an animal's normal genetic makeup should be checked carefully to assess the effect on the welfare or well-being of the animal. Sometimes these modifications of genes could actually improve the animal's well-being. Genes exist that could be replaced to better the welfare and productivity of an animal. An example of this concept is also found in sheep. The limitation of wool growth by the supply of the amino acid *cysteine* can be changed by isolating the functional genes for the enzymes *serine transacetylase* (SAT) and *O-acetylserine sulfhydrylase* (OAS) from some other organism and transfer them to sheep in a suitably modified form. This process would enable the sheep to synthesize its own cysteine. From a welfare viewpoint, sheep containing this unique genetic makeup would appear to be at a significant advantage over those sheep who did not. Food utilization is improved in these transgenic animals (according to the American Society of Animal Science).

With the use of BST, nutrients are directed away from other body tissues toward the mammary gland. Basal metabolism and maintenance requirements are unaffected. A slight increase in body temperature results with the higher milk yields. Just like human trials of new products, dairy cows were given sixty times the commercial dosage of BST during a two-week period and up to six times the dosage for two consecutive lactations. Both cows and their calves were proven to be healthy. After considerable research covering more than two decades, regulatory agencies gave approval for commercial use.

When using artificial insemination or embryo transfer, care must be taken that the resulting offspring is not too large for parturition with that recipient dam, which is the female who receives the transferred embryo; in other words, she is the surrogate mother, not the biological mother. Artificial insemination has been used successfully for decades in the dairy industry and more recently in beef, swine, goat, and sheep industries. Embryo transfer—and now embryo splitting—is a more recent advancement in the animal sciences.

I. Patenting and Copyrighting of Animal Forms

1. Patent Competitiveness and Technological Innovation Act of 1990 (September)

   It is not an infringement (does not violate the law) for a person to reproduce a patented transgenic farm animal by conventional means, or to sell the animal, or the reproductive material (i.e., sperm, ova, or embryos) of the animal in the farming operation.

2. April 1992

   Congress finds that the patenting of animals raises serious economic, environmental, and ethical issues not yet addressed by Congress, and that the granting of numerous animal patents could expose patent holders to revocation or alteration of their patents and expose the federal government to potential financial liability for restitution. Therefore, a five-year moratorium will be imposed in which no vertebrate animal, including a genetically engineered animal, shall be considered patentable.

3. The Patenting of the Obese Mouse

   Since 1953, subject matter patentability permits patenting an animal in the United States as long as there has been human intervention in the preparation of that animal (not the same as the patenting of a gene).

Chapter 8

EVALUATION

1. _____ percent of the voting public in the United States has no direct connection to animal agriculture.
2. _____________ is a word that animal activists use to describe the philosophy that humans are superior to animals (and thus we compare the term to racism).
3. A(n) __________ animal is one whose genetic makeup has been modified by the addition or removal of a specific DNA sequence.
4. List the four biological rights of an animal:
   a. ________________
   b. ________________
   c. ________________
   d. ________________

Discussion Questions

1. What are the moral implications of increasing the genetically-encoded disease resistance of an animal? What are the benefits to the animal, to the farmer, and to the consumer? For more information on this topic, see the D. Lo, et al. paper, *Expression of Mouse IgA by Transgenic Mice, Pigs and Sheep*. Eur. J. Immunol. 21: 1001–1006 (1991).
2. We can now cause modifications that are neutral to the animal but are of benefit to people. For example, we can alter the keratin proteins that make up wool fiber and thus change the properties of the wool so produced (producing a wool of superior quality). What, if any, are the ethical implications of such research?
3. Accepting the fact that research has demonstrated human safety, product efficacy, animal safety, manufacturing quality, and environmental assessment, which ethical considerations should be decided before a dairy farmer uses BST in his or her milking herd?
4. What is the difference between animal cruelty and animal exploitation?
5. What is the difference between animal welfare and animal rights philosophy in the use of biotechnology?
6. The AVMA describes the use of animals in research as a privilege (not a right). What does this idea mean morally, ethically, and legally?
7. Some animal activists insist on making all animals vegetarians. This act will cause many problems in cats, including feline dilated cardiomyopathy. Discuss the ethical implications of this unnatural diet to the feline. Is it inhumane? Why? Why not?

Endnotes

41. Georgetown University, UC Davis, Oregon State University, Michigan State University
42. A.J. Clark, et al.—Expression of human anti-hemophilic factor IX in the milk of transgenic sheep. *Biotechnology* 7:487–492. 1989.

Recommended Readings (From Section 2)

1. DeGrazia, D. *The Moral Status of Animals and Their Use in Research: A Philosophical Review.* Kennedy Institute of Ethics Journal, Vol. 1(1): 48–70. 1991.
2. *Animals in the Classroom: Lessons in Disrespect.* PETA.
3. *Companion Animals: Pets or Prisoners?* PETA.
4. *Policy Statement Dissection of Animals in the Classroom.* Educators for Responsible Science, 1992.

Suggested Additional Reading

1. *Man and Beast He Preserves: Humans and Animals in the Bible and in Jewish and Christian Tradition,* James V. Parker, 25 pages. Available for $5—prepaid from James V. Parker, 4327 NE Glisan, Portland, OR 97213.

CHAPTER 9

Animal Environments and Agri-Ethics

Although many animal activists would love to completely end all animal agriculture within the United States, that situation is an improbability. Other nations have been successful at curtailing agriculture production and slowing down or stopping certain advancements in animal agricultural science. In 1987, Sweden passed legislation providing for the fining and imprisonment of violators of their new *Farm Animal Welfare* laws. The new laws phase out layer cages (as soon as a viable alternative can be found) and orders the swine industry to stop using sow stalls and farrowing crates and to provide more space and straw bedding for slaughter hogs. The legislation forbids the use of genetic engineering, growth hormones, and other drugs for farm animals, except those used for veterinary therapy.

Objectives:

The student should:

1. Describe the five freedoms guaranteed to farm animals in Great Britain.
2. Describe the importance of agriculture to the gross national product of the United States.
3. Describe items other than meat that are normally obtained from animal agriculture.
4. Describe items other than meat that are normally obtained from animal agriculture.
5. Compare food safety, both animal and plant, and include methods, under law, that helps protect the consumer from food contamination.
6. Describe the various types of vegetarianism practiced today.

7. Describe methods of assessing and improving animal environments.
8. Describe the AVMA's official position on transportation of animals, castration, docking, and ear cropping as well as their position on caging laying hens.

In Great Britain, the United Kingdom Farm Animal Welfare Council has demanded *five freedoms* as essential guarantees that farm animals are being treated humanely. These freedoms appear in the following list:

1. freedom from hunger and thirst
2. freedom from discomfort
3. freedom from pain, injury, and disease
4. freedom from fear and distress
5. freedom to display normal behavior

Of course, non-domesticated animals (animals in more natural surroundings) would not fall under this legislation and would not be free from these natural stressors. Farmers who want to stay in business should realize that these basic *five freedoms* are good husbandry practices that, if followed, greatly reduce animal stress, increase animal comfort, and thus productivity in most cases.

The United States makes up only .3 percent of the world's agricultural labor, and yet the country produces 11 percent of the world's pork (see Figure 9–1), 25 percent of the world's beef, 15 percent of the world's feed grains, and 11 percent of the world's grains. Sixteen percent of the nation's

Figure 9–1 Pigs in farrowing crates

gross national product (GNP) is connected to United States food agriculture. Figure 9–2 displays the gross dollar value of the major animal agriculture products in the United States.

Animal products are valuable to the United States economy. As fewer and fewer urbanites understand livestock production, however, the animal welfare gap between city folk and farmers widens. Only two percent of the total population in the United States is involved with farming, and less than 25 percent live in rural areas where they are exposed to farm animal husbandry. Most Americans today have difficulty in associating meat with the animals from which that meat comes. Shopping at a butcher shop in the 1960s was totally different from picking up cellophane-wrapped meat packets in 1990s supermarkets. Thus, the attitudes of the American public and its perception of what is acceptable farm husbandry have greatly changed.

Animal rights activists feel that animals should be accorded the same moral protection that humans have. Therefore, besides the five freedoms previously mentioned, they believe that animals should have the right to live. Thus, trying to convince an animal rights supporter that there have been improvements in the methods for raising animals, who will be eventually killed and eaten, is fruitless. Most animal welfarists simply want animals to be treated in a more humane way. Their understanding of animal husbandry must be improved if farmers are to have a say in how urbanites pass legislation that affects their livelihood. Hence, farmers need

Figure 9–2 *Gross dollar value of the major animal agricultural products in the United States*

	(Dollars in Millions)
Beef	30,550.7
Chicken	8,385.3
Lamb	666.7
Pork	11,094.8
Turkey	2,344.0
Mutton	84.0
Dairy Prod.	18,325.6
Eggs	3,886.8
Mohair	19.4
Wool	46.8
Mink Pelts	70.9
TOTAL	**75,475.0**

(Source: USDA, 1993 Agricultural Statistics)

to improve some of their husbandry methods *and* educate the animal welfarists on how and why these methods are being used. In other words, we need to better communicate with each other.

NATURAL AND SYNTHETIC FIBERS

Besides using animals for biomedical research and food production, animals also provide products such as hair, hides, and pelts. Wool is a necessary part of survival for many people throughout the colder areas of the world. Fur is a natural product of many animal species. Leather is used for shoes, jackets, and seat covers. Synthetic fibers have replaced some of these animal by-products in modern Western countries, but not to any degree in third-world countries. The synthetic fibers are usually made from non-renewable petroleum used in plastics and synthetic polyester. These materials can be more expensive in third-world countries and are not part of the normal, recyclable chain.

ANIMALS' ROLE IN FOOD PRODUCTION

While some individuals believe that it is healthier to eat a diet devoid of meat, milk, and eggs, (a vegan diet), others argue the contrary. The best possible diet for a person is one that is well balanced in the four (or five) basic food groups. The proteins of animal origin—meat, milk and eggs—are highly digestible and rich in essential amino acids that are lacking in most plant proteins. Meat also contains minerals and vitamins needed daily in the diet. The problem in America is that we tend to go from one extreme to another. Many (if not most) Americans consume a diet too rich in fat and too low in fruit and vegetables. A diet absent of fat is equally as dangerous to one's health. Farmers, geneticists, and nutritionists have been working in recent years to create more dairy products that are low in fat, eggs that are lower in cholesterol, and meat that is leaner, thus lowering fat and cholesterol content.

The pork industry, for example, has responded to consumer demands by producing a pig that is 72 percent muscle, 13 percent bone, and only 15 percent fat. These percentages compare to a market hog 40 years ago, which was 50 percent fat, 40 percent muscle, and 10 percent bone.

Many philosophy courses will argue or discuss whether people should eat meat. Others (see articles by Dr. Temple Grandin in Section 2) prefer to discuss the most humane methods of animal husbandry that should be practiced in raising animals for food. This argument should be addressed not from the human standpoint but from that of the animals. In other words, because we are castrating, dehorning, vaccinating, restraining, transporting, and eventually slaughtering these animals, what is the most morally or ethically acceptable way to perform these tasks? Not everyone will agree, obviously. We, as a society, however, need to come to some type of agreement as to a line we should *not* cross. Rep. Charles Stenholm, a strongly pro-agriculture member of the United States House of Representatives, has told his agricultural audiences that by the year 2000 there will most likely be United States federal legislation governing farm animal welfare.

The earlier description of the French medical students learning anatomy on a screaming horse is and should always be totally unacceptable. The AWA provides better protection from pain to research animals than is provided to common household pets (companion animals). No veterinarian prescribes analgesics to dogs and cats going home after a spay or neuter, yet the AWA mandates such relief in most similar type operations for research animals.

Many livestock operators compare the pain of administering an anesthetic with the various "painful" procedures of castrating and dehorning because of the time it takes to administer the anesthetic medication. Whichever is the least painful and least prolonged methods is usually adopted.

Historically, anesthesia for farm animals was rarely considered. Today, America's farmers are much more highly educated and more capable of mitigating many of the necessary and painful procedures involved with animal raising. Many times the risk of providing anesthetic to the animal is too great to consider, and the pain must be minimized in another manner. In addition, local anesthetics have inherent risks themselves. Besides the pain of injection and the stinging sensation of the anesthetic, infection can easily spread with the anesthetic as it permeates the tissue. Ruminant animals have an additional risk when given general anesthesia, due to gasses produced in the rumen that can be absorbed across the ruminal wall and into the blood.

Veterinarians and many animal caretakers argue about the need to use anesthesia during dehorning procedures. Calves are dehorned to reduce the risk of injury to other animals and to the people working with them. Dehorning is more easily accomplished and with the least discomfort to the calf if the procedure is performed at an early age (less than two months of age). When dehorning occurs after this age, many veterinarians suggest the administration of a local anesthetic.

Local anesthesia by cornual nerve blockade minimizes pain to the calf without causing excessive stress. The cornual nerve is a branch of the zygomatic temporal nerve. A small amount (usually between 3–10 ml) of two percent lidocaine can be administered with an eighteen-gauge, 3.75 cm needle.

Admittedly, this procedure is impractical on the open range in the beef industry. With lock-up stanchions in closely confined dairy operations, however, cornual nerve blocks should become standard procedure in calves older than two months of age. For calves older than six months of age, a sedative analgesic may be additionally provided to the animal (see case study on calf dehorning).

FOOD SAFETY

Many animal activists will argue against the consumption of animal products due to contamination of meat, eggs, and milk with pesticides, herbicides, hormones, antibiotics, etc. Again, this subject should be addressed in an animal science course, and thus we do not have time to cover that aspect of food production in this book. We cannot forget the fact, however, that imported fruit and vegetables are rarely inspected for residues of pesticides, herbicides, or other carcinogens. A detailed inspection is conducted on meat, milk, and eggs produced in the state of California and in most other states as well. Misuse of drugs given to animals could cause *traces* (residues) in the edible parts of slaughtered animals. In high doses, it is possible that these traces could be hazardous to consumers. On a federal level, the Food Safety and Inspection Service monitors the processes to assure a safe food supply. The service sets specific regulations on the amount of residues of hormones and other additives that might be left in meat.

Philosophical discussions involving the moralistic views of animal activists who believe it is unethical to eat meat have academic merit. Those who argue from a purely health viewpoint, however, can be considered hypocritical for the reason mentioned earlier, especially if the individual smokes, consumes alcohol, eats fatty foods such as peanut butter, chocolate, avocados, coconut, and palm kernel oil, or eats imported fruit and vegetables loaded with herbicides and pesticides.

How much hormone residue is too much? Scientists have studied rats to find the exact amount of a particular substance that will cause no observable physiological effect in the animal, which is called the *No Observed Effect Level* (the NOEL). Once the NOEL is found, then the *Acceptable Daily Intake* (ADI) is calculated. An ADI is one-hundredth of a NOEL. This value is the "hundred-fold" safety level. The ADI indicates that the acceptable level of intake is one-hundredth of the amount of material that causes no observable effect on change in growth in laboratory animals. Compared to a human, that amount is extremely small. The Food and Drug Administration (FDA) is the federal agency that is primarily responsible for monitoring the use of animal drugs and maintaining safe food for consumers. In addition to the NOEL and ADI regulations, the FDA sets a withdrawal time for all medications prior to the slaughter or milking of the animal.

NATURAL VERSUS RESIDUES

Many plants contain hormone levels that far exceed the maximum acceptable "residues" in meat. As an example, a four-ounce portion of coleslaw has about 500 times the estrogen level found in a steak from an implanted steer. This estrogen in the coleslaw is naturally occurring plant estrogen. Naturally occurring plant estrogen can be more harmful in these large amounts. Sheep and cattle, as an example, have been proven to abort when consuming large amounts of naturally occurring plant estrogens found in plants such as alfalfa and soybeans. Figure 9–3 summarizes the comparison of estrogen levels in various foods.

According to former Surgeon General C. Everett Koop, "You and I consume every day about 45 micrograms of potentially (carcinogenic) man-made pesticide residues in our (total daily intake of) food. But there are 500 micrograms of naturally occurring carcinogens in one cup of coffee, 185 micrograms in a slice of bread and 2,000 micrograms in a glass of cola." Government tests show that beef contains no violative residues of any potentially harmful compounds, including pesticides.

THE DELANEY CLAUSE

The Delaney Clause, passed by Congress in 1958, stated that no carcinogens shall be added to the supply of food (Shapiro, L.S., *A Fly is a Carrot is a Pig is a Dog*, 1996). Since 1958, any food additive found at any level to "induce cancer" in experimental animals or people was prohibited. The United States was the only country in the world with a regulation so restrictive. Originally, this clause measured carcinogenic amounts in parts per million. In 1996, the clause was being used at any measurable level. Scientists in 1996 were capable of routinely detecting substances at levels as minute as one *part per trillion* (ppt)—the equivalent of one second in 32,000 years. Experts on both sides of the animal welfare/rights issue believe that some tests in which laboratory animals were being fed extremely high levels of a substance over their lifetimes and developed cancer may have little, if any, relevance to safety levels in food consumed by humans. Scientists argued that negligible risk ("de minimis" risk) of a substance was so low that it was considered insignificant.

Figure 9–3 *Comparison of estrogen levels in various foods*

Food	Estrogen levels (nanograms)
Beef	1.2–1.9
Milk	11
Potatoes	225
Ice Cream	520
Cabbage	2,000
Soybean Oil	16,800,000

One of the arguments that scientists used in defense of this change was in the comparison of an implanted versus non-implanted steer. A three-ounce serving of beef from an implanted steer contains 1.9 nanograms of estrogen, while the same size serving from a non-implanted steer contains 1.2 nanograms. The difference is seven-tenths of a billionth of a gram (or .0007 of a millionth of a gram), the equivalent of one blade of grass in one and one-third football fields.

In 1997, in a bipartisan effort, Congress passed H.R. 1627, which eliminated the Delaney Clause banning even minute traces of carcinogens in processed foods. Congress replaced this clause with legislation where only substances that pose a "reasonable certainty of no harm" would be allowed. This phrase is generally understood to mean, "creating a lifetime risk of developing cancer of no more than one in one million." The controversy of using animals to determine this risk still continues.

ARGUMENT FOR VEGETARIANISM

Many vegetarians argue that a vegetarian diet is more ethical and healthier. From an ethical view, each individual has a line that they will *not* cross. If the eating of animal flesh is unethical for any individual, then a vegetarian diet is the correct choice for that person.

We as a society do *not* have a right to try and force all of our ethical beliefs on society as a whole, however. Certain areas of behavior, such as what you read or do not read, cohabitation, how much charity you give, what social groups you belong to, whether you smoke (as an adult), whether you swear, drink alcohol, or help an elderly lady cross the street are personal ethical choices that our system of government relegates to the discretion of the individual. Many believe we should not mandate these behaviors in a free society, while others disagree.

Many Americans came to this country from a totalitarian society, where they were told which books they could or could not read, to whom they were allowed or not allowed to give charity, whether they were allowed to go to school, what jobs they were allowed to hold, and to what degree they could criticize their non-elective government officials. Let us hope that our society never becomes that restrictive. Our personal ethics should be used as examples, so that others can understand who we are, perhaps even admire our choices (whether you eat meat and why), and decide, based on that admiration, to experiment with those beliefs.

Some ethical beliefs need to be adapted by society as a whole, however. Hearing the views of others that might differ from our own is still important and will help us try to "get along" as a civilized society.

The second argument that vegetarian diets are more healthful is not a very good argument. Both healthy and unhealthy vegetarian diets exists. Healthy diets exist that include animal products, and there are unhealthy diets also containing animal products. Many degrees of vegetarianism exist as well. Some vegetarians eat no red meat but eat fish and poultry. Other vegetarians do not eat meat at all but still consume dairy products and eggs. Vegetarians fall into the following general categories:

Semi—eat poultry, fish, and dairy products but little or no red meat or pork

Ovo-lacto—eat animal products from milk and eggs, but no meat

Pesco—eat fish but no red meat

Fruitarians—eat fruits, nuts, olive oil, and honey

Vegans—do not eat or use any animal products

Vegetarians can have healthy diets. It is essential that vegans know enough about amino acids, which foods are high-quality protein foods, which blends of proteins are required, and which blends of vitamins and minerals are required to balance the diet. Most corn products are low in lysine. Most legumes are low in methionine and cystine. The mixing of corn and beans tends to balance out the amino acid needs. Other examples of blending diets can create healthy vegetarian diets. Once again, however, this is *not* a nutrition text; hence, further discussion in this regard should include a registered dietitian.

NON-ANIMAL USES OF ANIMAL BY-PRODUCTS

While many animal rights activists and strictly observant Jews would never wear certain lipsticks because they contain *carmines* (which are derived from pulverized insects) as well as animal fat, most American women never give their lipstick a second thought. Inedible tallow and grease by-products are used in soaps and animal feeds. Fats also provide sources for floor waxes, paints, inks, plasticizers, cosmetics, synthetic rubber, food emulsifiers, candles, lubricants, and pharmaceuticals. Farmers save millions of dollars each year by using manure instead of chemical fertilizer for their vegetable and fruit crops. Jains, because of their sincere belief that all animals and most plant life is deserving of protection, would not use the benefits of any of these items.

> ***"To believe in something and not live it is dishonest."***
>
> Mahatma Gandhi

What this statement means is that if you condemn others for eating meat because "eating meat is murder," then you would be dishonest if you drove in a car with leather upholstery, wore leather shoes, used any lubricants or pharmaceuticals made with animal by-products, etc. Sure, it would be less comfortable and more difficult to do without some of these items, but it would be much more honest. The Jains and many Hindus have proven it possible, that is, if you admire and wish to live that lifestyle. Many in our society condemn an action without thinking about the consequences or the alternatives.

FARM ANIMAL WELFARE IN PRACTICAL TERMS

People often demand cheaper food and fiber. In the United States, we are fortunate in that we pay less than 10 percent of our disposable income for food. Many countries in Europe (such as Poland and France) pay up to 60 percent. Other than animal feed, labor is the single most expensive item in producing meat, milk, and fiber. The high cost and low availability of experienced farm labor has forced the agriculture industry to develop systems that save on labor. These systems usually require extensive capital outlay but pay for themselves in the long run.

PORK PRODUCTION

In the swine industry, an example of this high-intensity system is a series of farrowing crates situated over a manure handling system that removes the waste products by a flush system. This environmentally friendly design keeps room humidity, temperature, and, most importantly, ammonia levels to a minimum. The purpose of the farrowing crate is to prevent the mother pig (sow) from accidentally rolling over and killing her small litter of piglets. In nature, this incident quite commonly happens. In addition, the individual housing eliminates the competition for food between sows and the bullying and fighting that usually results. To produce cheap pork, the cost of labor must not only be reduced, but the number of piglets weaned must be maintained. The more piglets that become injured and/or die, the greater the cost of production.

From an animal well-being standpoint, one must compare the benefits of producing cheaper pork, having a lowered mortality of piglets, and fewer injuries to the sows with the confinement housing that is required. Many small pork producers have been able to compromise in this regard. Switching from open-pasture housing of the sows, these farmers noticed a tremendous improvement in weight gain, fewer parasitic infections, fewer injuries to the sows (fighting), and much lower piglet mortality. Sows are generally kept in the crates for 28 days. During this time, sows and piglets can be more closely examined. Many farmers exercise their sows daily in open lots as they clean out the crates. Better sanitation practices can be maintained this way. Labor costs make this practice prohibitive in most larger facilities.

In England, Holland, and on one farm in Colorado (Bell Farms), sow units are still set up outside. Typically, much larger areas of land are required, and the operations must be farther removed from urban sprawl. Environmental concerns of water contamination, parasites and other microbial transfers between the environment and animal, and odor complaints by neighbors are what primarily changed this once widely practiced rearing in America to totally confined operations.

The National Pork Producers Council, the association that represents United States pork producers, has an animal welfare committee composed of veterinarians and animal scientists from across the country. This committee studies new and improved methods of handling, feeding, breeding, and caring for swine. They have published and distributed a Swine Care Handbook for the purpose of educating swine producers on the most humane methods to care for hogs in modern swine operations. More than $400,000 was spent researching housing design, building ventilation, swine nutrition, and general methods of maintaining a profit-generating yet humane business.

The committee works with the Pork Quality Assurance program in developing guidelines to ensure that consumers are guaranteed a wholesome and safe product. Limits on medicines and guidelines on their administration, health needs of the pig, and maximization of the comfort of the animal are outlined in these guidelines.

RAISING VEAL

Certain methods for raising dairy calves for veal have undergone considerable condemnation. Three general classifications of veal exist. *Bob veal* (or baby or drop calf veal) refers to calves that are one-month-old or less and thus have only received a milk-based diet. *Grain-fed veal* calves are fed milk, grain, some hay, and other processed feeds. Some refer to this type of veal as *baby beef. Special-fed veal* calves are fed milk-based feeds and are usually slaughtered at three to four months of age. This veal is also known as *white veal*, or *milk-fed veal*, and is the most controversial.

Most people in the veal industry use individual pens to raise their calves. Those raising replacement calves for the dairy industry use similar facilities. The purpose of using the individual pens is similar to using individual incubators in a human hospital. To prevent the spread of disease from one calf to the next, and to prevent the natural "sucking impulse" of one calf on another, the calves are isolated from direct contact with each other.

The pens used should be large enough to provide for natural resting positions throughout their growing cycle (or they should be moved to larger pens as they grow). Most pens are constructed of wood or metal with slatted floors, an open front, a back, and a partially covered top and slotted sides. This design provides for optimum ventilation and visual contact between the calves. The flooring should not cause discomfort, deformities, or injury to the calf. The crate should be disinfected on a regular basis. Each and every calf should be monitored three to four times a day.

Group pens are generally discouraged while the animals are consuming milk. As long as milk is being given, the calves will want to suck on each other, thus increasing disease transmission and injury to the penile and umbilical regions. Urine sucking also increases with group housing. With dairy replacement calves, once they are weaned, they are usually placed together in group open housing units of ten or more calves.

Light in totally enclosed barns should be adequate enough to observe all calves for a minimum of eight hours per day. Natural light within the barns (windows) is encouraged. Many operations have proven that they can be successful at raising grain-fed veal in outdoor, individual housing, using straw bedding and receiving top-dollar for their finished product. Buyers of veal calves prefer that the housing and care provided is the most humane. Economics, however, is a great factor for the buyers. Promotion of "ethically raised calves"—clean and healthier calves—could be a bonus to the buyer and consumer alike. Death rates can reach levels of two percent, which is sufficiently lower than the average (15 percent).[43] Low death rates can be credited to the reduction of stress, maintenance of optimum sanitation, and other enriched environmental criteria demanded by many veal operators. Veal calves raised in Canada undergo similar restrictions for humane care. In other words, veal can be and is, in many parts of this and other countries, raised in a humane manner.

CALF WELFARE

Many urbanites do not understand the husbandry needs of cattle. One such example in the dairy industry is the condemnation of separating calf from cow shortly after birth. This separation is required for the health of both the calf and the cow and is much more humane. Consider a human mother who is giving up her newborn for adoption. Which would be more cruel: allowing the mother and baby to bond with each other and letting the mother nurse her new baby, and then ripping the two apart from each other three days later, or separating the two from one another immediately after birth?

A tremendous difference in stress levels exists between cows allowed to nurse their calves and those who were bottle-fed from the first feeding. When a calf is allowed to stay with its mother for even one day and is then forced to separate, excessive bawling and refusal of feed occurs on the calf side. Refusal of feed and milk let-down on the cow side occurs. Instead, dairy farmers convince the newborn that they are its parent and convince the cow that the milker is its calf. All are happy and very little stress is involved.

Colostrum, by legal definition, is milk from a cow fifteen days before and five days after birth. This milk cannot be legally sold for human consumption, but because it is rich in antibodies, it is thus always given to the calf. Dairy cows, however, have been bred for generations to produce much more milk than any one calf needs. Calves are not smart enough to stop nursing when they have had enough. Like any young child, they will eat and eat and eat, so long as it tastes good. Calves will often overeat and develop *scours* (diarrhea) from overconsumption. Uncorrected scours is the most common cause of death in young dairy calves. Most dairy farmers, therefore, regulate the quantity, quality (monitor the colostrum IgG levels), temperature, and sanitation of the colostrum and the milk that follows to each and every replacement heifer (the calf being raised). They raise these calves in isolation pens until they are consuming enough grain and hay to be weaned. At that time, they are placed in group housing. Group housing is much cheaper from a labor standpoint, but it is much crueler in that, at a very young age, it is associated with an increase in mortality, a competition for feed, injuries resulting from "horseplay" and bullying, and a greater increase in the spread of infectious diseases. For this reason, group housing should be avoided until after weaning.

UNDERSTANDING THE ENVIRONMENTAL FACTORS AFFECTING WELFARE

Consumers have benefited greatly from the modernization of animal agriculture. A little less than 10 percent of an American's disposable income is used for food purchases. The public concern about the treatment of farm animals, however, has also increased. Modern farming methods disrupt social structures in a number of ways. Groups in which animals are maintained are generally far larger or smaller than those of a more "natural" population. Selection for breeding is decided by the farmer and does not normally permit the mixing of sexes in herds or flocks, except when the farmer decides to breed.

Early weaning of piglets, artificial incubation, and the rearing of chicks deprives the animals of normal parent-offspring bonding. The greatest public concern, however, is that of long-term confinement, namely the caging of laying hens and the crating of veal calves and pregnant sows. The primary reason behind sow confinement is to decrease piglet mortality, and the primary reason for hen confinement is for ease of egg collection and disease control. Before laying hens were placed in cages, it was quite common to see death rates as high as twenty percent or higher due to disease. Today's modern poultry farms are well sanitized, and disease outbreaks of this magnitude are a thing of the past.

Another point that is secondary to the reasons listed earlier is that animals are confined to permit increasing numbers to be housed together in a single building. The housing of large numbers of hens in well-insulated, well-ventilated buildings is advantageous for the egg farmer, because it largely eliminates the need for costly supplemental heating during cold winters. In addition, on a modern day farm, a single worker may be responsible for 250,000 broiler chickens each year, and thus does not have the ability to care for birds that are spread out over large distances.

While some veal producers raise their calves in elevated crates for seventeen weeks on an all-milk substitute, iron-deficient diet, this is not necessary. A large percentage of veal producers raise veal in isolation pens (to protect the animal from disease), but they supply the calf with whole milk and grain (a well-balanced diet). The calf has room to stand up and walk around. Many leave half of the pen covered (to provide protection from the sun and rain) and half of it exposed to allow increased ventilation. Dairy replacement heifers are treated in the same manner. Marketing veal raised in this "more humane" manner is not a problem.[44]

METHODS OF ASSESSING AND IMPROVING ENVIRONMENTS

A necessary component to understanding animal well-being is understanding animal behavior. Can the animal express its "natural" behavior patterns in an "unnatural" environment? Are current confinement housing systems appropriate environments from the animal's viewpoint? How much pain and suffering do animals experience when they receive treatment such as beak-trimming or dehorning? Is boredom stressful to an animal? How about the backyard horse that is all alone all day long while its owner is at work? Is the horse bored? Is that boredom stressful? Is that stress cruel? These questions are much more than simply questions of whether the animals are being abused, treated cruelly, or neglected. These questions need to be addressed.

More research is needed to assess farm animal cognition, behavior, pain, and stress. By maximizing farm animal well-being, the animal should also become more productive to the farmer. In most cases, domestic, commercially raised farm animals "outproduce" wild animals by far. Wild animals are subjected to considerable stress. No caretakers are available to modify the weather, to provide adequate nutrition during times of drought or floods, or to protect them from predators, diseases, and parasites. Many animal abusers (out of ignorance) release their domestic animals into the wild when they can no longer take care of them, move from the area to a location that does not permit animals, or have a litter "they did not want." Domestic animals suffer at an even greater rate than non-domestics when placed in the wild.

In farm animals, all of these "normal" stressors can be reduced considerably with proper management. Still, as farmers try to spend less money on labor and feed (the two largest expenses of producing livestock), these shortcuts may cause a decrease in welfare to the animal. Continued research on improving the economics of production without decreasing the welfare of the animal is needed.

PROBLEMS ASSOCIATED WITH SOCIAL GROUPINGS, ISOLATION, CLEANING, AND OTHER HUSBANDRY PROCEDURES

DISCUSSION QUESTIONS

1. List common husbandry procedures that increase stress in farm animals. How might these procedures still be carried out but with minimal stress? How might some of these procedures be replaced with others that are less stressful? Should they be mandated by law, suggested, or advocated by peer pressure? What is the best approach? Why?
2. When introducing a bull to a herd of cows, does this situation cause stress? If so, should bulls not be permitted to commingle with the cows? If so, does this cause stress? Which is the greater stress? Is the answer the same in all cases? Before deciding how much stress is too much, what other considerations should be made?
3. Can isolating a young calf for six to eight weeks to prevent illness and death justify the isolation? If your answer is no, should we then not isolate human infants in incubators? If this situation is not the same, justify objectively your conclusion.

COMPANION ANIMALS AND THEIR "HARMFUL" ENVIRONMENTS

Several routine procedures with companion animals have been deemed cruel or inhumane by animal rights activists. These procedures include declawing, ear cropping, euthanasia, and debarking.

Declawing keeps cats more manageable around the house and causes less destruction. The AVMA's official position is that declawing of domestic cats is justifiable when the cat cannot be trained to refrain from using its claws destructively. Declawing outdoor cats can be considered inhumane if the cat can no longer fend for itself or run up a tree when being pursued.

Ear cropping and tail docking makes an animal look better, and the procedure keeps standards of the breed high. The AVMA has recommended to the American Kennel Club (AKC) and appropriate breed associations that action should be taken to delete any mention of cropped or trimmed ears from breed standards for dogs and to prohibit the showing of dogs with cropped or trimmed ears if such animals were born after a reasonable date. Tail docking has some justification for dogs used in breeding, security work, or for protection.

Animal rights activists charge that animals deserve to have a life of their own without human intervention, and thus they are opposed to euthanasia of unwanted animals. Euthanasia is done, however, to control overpopulation and to prevent suffering. The AVMA supports euthanasia of unwanted animals by properly trained personnel using acceptable humane methods, because euthanasia is more humane than letting the animals suffer slow deaths due to disease, starvation, and exposure.

In July 1997, the Los Angeles Department of Animal Regulations was called to a home in Van Nuys, California. There, 589 cats and twenty-nine dogs were found in filthy, disease-ridden conditions. Several dead animals left to decay were also found in the house. Animal control officers had to euthanize thirty-nine animals immediately for humane reasons. Additional animals were showing signs of starvation. Friends of the woman who operated this "rescue" operation claimed that she loved animals but that she was simply overwhelmed with the numbers she was trying to save. She did not want animals taken to the pound to undergo euthanasia. Instead, she caused dozens of these animals to *suffer*. Each animal had a living space of only 1.3 square feet. Which is more cruel: taking excess animals (twelve to thirteen million each year in the United States) and having them humanely put to sleep, or having them wind up either in the conditions just described, set loose in the wild to starve to death, or left to be eaten by a predator or hit by a car?

PUPPY MILLS

Commercial pet production and dealers are covered by the Animal Welfare Act (AWA). Animal activists charge that mass production does not consider the physical and mental needs of the animals involved and that there is little regulation taking place. The AVMA supports increased USDA, Animal and Plant Health Inspection Service (APHIS), and Regulatory Enforcement and Animal Care (REAC) funding for adequate enforcement, along with appropriate statutes and regulations to better ensure humane care of small companion animals in breeding facilities and pet shops.

Much of the breeding of purebred animals risks the animal's health simply to obtain a desired appearance. The inbreeding of dogs, such as the German shepherd, is thought to have increased their incidence of hip dysplasia. While hybridization and selective breeding for traits are a proven method of improvement, not all companion animal breeders understand genetic technology or are willing to choose animals inferior to their standards from their breeding pool. Breed registries must recognize that ensuring the health of their respective animals ensures their continuation and the continuation of the breed they represent.

TRANSPORT AND SLAUGHTER

The federal government has specific laws regarding the time an animal can be transported on the road without adequate feed, water, and exercise. The *Packers and Stockyard Administration* also regulates stockyards regarding the treatment of animals. The USDA regularly inspects processing plants to ensure that the quality of the slaughter method is humane and that the quality of the product meets government standards.

OFFICIAL AVMA POSITION OF TRANSPORTATION OF ANIMALS

Adequate protection should be provided in adverse environmental conditions. Time limits on the railroads, as specified by state and federal regulations, must be observed. Similar consideration should be given for feed, water, rest, and protection with all other types of transportation. Excessive crowding and deprivation of food and water must be avoided. The AVMA supports government regulations pertaining to the humane slaughter of food animals and supports continued research on improved techniques for humane slaughter.

Care must be observed in the loading and the unloading of animals to avoid any potential injury or stress to the animals. Dr. Temple Grandin (a leading expert in animal behavior and animal welfare) has created a series of designs to minimize stress to the animal during loading and unloading and during general restraint (see several of Dr. Grandin's reprint articles in Section 2). Established protocols for the handling and separating of sick and injured animals is a must. These policies and procedures should be enforced to minimize the transmission of infectious diseases. The sorting, grouping, and penning in the sales yards must be performed without animal abuse and with minimal stress.

TRANSPORTING PIGS

In transporting pigs, it has been shown that simply placing a boar among females being transported markedly decreases the amount of aggression among the females. If two boars get together, however, then fighting takes place between the boars, and pushing and bruising of other animals may occur. Some breeds of hogs are more resistant to stress during transport.

TRANSPORTING HORSES

Horses cause a major transportation challenge because of their long necks and size. Horses should *not* be transported on cattle trucks. Single-deck trucks with air ride suspension are recommended for the transportation of horses.

TRANSPORTING CATTLE

Federal law imposes a 48-hour transportation limit on cattle. No more than 32 hours should elapse, however, before allowing a "stress stop" for loading, unloading, feeding, and watering. Stressed cattle are easily susceptible to disease and exhaustion, resulting in problems for the producer.

Horned cattle cause twice as much bruising than non-horned animals. Equipment and flooring designs need to eliminate the sounds of hooves clanging on metal floors or plates. Light in the wrong place in a chute, a shadow grate, etc., can also cause stress on cattle more easily. Dr. Grandin has designed many facilities with this idea in mind.

SLAUGHTER

Recently, PETA stirred up some additional controversy with a print advertisement that played upon the concluded O.J. Simpson trial. PETA's graphic advertisement showed a silhouette of a gloved hand holding a knife in a stabbing pose. At the top of the advertisement were the words, "If the (bloody) glove fits..." The rest of the advertisement described how people who eat meat are paying professionals to kill ("Imagine if you were a professional killer, paid to slit throats"). The *Los Angeles Times* and the *Los Angeles Weekly* refused to run the advertisement.

The process of slaughtering most animals in the United States starts with stunning the animal, which renders the animal "brain dead." Most stunning is done with captive bolt guns. About ninety-seven percent of animals are stunned correctly and without any pain. Sometimes the guns are not well maintained, however. Stunner maintenance of at least thirty minutes per day and replacement of broken seals is recommended to service the gun.

CASUALTY ANIMALS AND LAMENESS

Even the best producers, the most careful transporters of livestock and the most well-run stockyards will occasionally have disabled livestock. Just like humans, animals sometimes slip and fall, breaking their bones. Paralysis occurs in calving (a disorder known as *obturator paralysis*). When an animal is severely injured, most of the time it is more humane to immediately slaughter the animal. If the animal is in extreme distress or if the condition is irreversible, the animal should be moved humanely and directed to a federally or state-inspected slaughter plant, slaughtered on the farm (if possible), or immediately and humanely euthanized. If the animal is slaughtered on the farm, it *cannot* be sold for human consumption. Many state laws also prohibit injured animals from being slaughtered for human consumption (depending on the injury). Non-ambulatory animals should never be sent through intermediate marketing channels. They should either be euthanized or shipped directly to a federally or state-inspected slaughter facility. When immediate euthanasia is *not* possible, pain relief should be provided in the interim.

Humane livestock care is recognized as good business by both ranchers and farmers. The greatest concern for the care of farm animals is from the individual who raises them. Rarely will you find a producer who does not have a deep sensitivity for his or her animals' well-being, because economic survival depends on them. Productivity is *not* the *only* indicator of animal welfare. Animals under stress *do not* gain weight or produce eggs or milk efficiently. Today's food animals live in a highly improved environment that eliminates many natural stresses.

Humane livestock care maximizes the animal's performance and minimizes veterinary expenses. Humane livestock care reduces injuries and assures a higher-quality food and fiber for the consuming public. Although farmers and ranchers are the ultimate animal welfarists, they have a need for much improvement in the care of their animals. The philosophical differences as to the degree of improvement should be discussed further in both animal science curricula and in courses involving animal ethics. Remember to discuss these matters with facts, rather than emotions. Who saw what, when, and where? Many of the charges simply do not hold water, are greatly exaggerated, or at one time happened many years ago and perhaps in another country. When poor husbandry is taking place, it is the job of industry, as well as the general public, to initiate change. Farm advisors, county extension agents, and university personnel are recommended resources to start diplomatic changes among farmers.

Whenever animal agriculture is guilty of animal abuse or neglect, the industry itself must become proactive in alleviating the problem. By its very nature, animal agriculture sometimes creates situations where moral and ethical considerations need to be addressed. Two of our nation's leading animal behaviorists, Dr. Grandin of Fort Collins, Colorado, and Dr. Janice Swanson of Kansas State University, have listed four areas where animal agriculture needs to improve its practices:

1. Handling of non-ambulatory animals
2. Processing of live animals
3. Transporting of livestock
4. Addressing the psychological well-being of confinement

We will briefly address these concerns and note that the degree of concern and improvement will vary due to the differences in people's philosophical values.

NON-AMBULATORY OR "DOWNER" ANIMALS

Downer animals are animals that are not able to stand due to calving problems, broken bones, or paralysis. They might become non-ambulatory at the processing plant or before they even leave the farm. Laws vary from state to state as to what to do with a downer animal. Many states require a veterinarian to euthanize the animal or allow the farmer, stockyard personnel, or rancher to perform that task himself or herself. After the animal is put down, do you bury it, burn it, or call the rendering truck? What if no rendering truck is available?

Most cases (about seventy-five percent) of downer animals can be prevented. Most are dairy animals. Broken legs represent a small portion of all downer animals. Drs. Grandin and Swanson recommend the following preventative measures:

1. Changing attitudes in management
2. Improving calf handling
3. Eliminating rough handling and prodding
4. Preventing abuse on non-ambulatory animals (downers)
5. Not overloading trucks
6. Proper management and adjustment of equipment
7. Maintaining an optimal number of animals in a loading area
8. Keeping the rest area open

Grandin and Swanson further recommend that in preparing to handle "downer" animals in their location, one should complete the following tasks:

1. Establish policies on the proper treatment of non-ambulatory animals.
2. Have only trained personnel move non-ambulatory animals.
3. Take prompt action when an animal becomes non-ambulatory.
4. Gently roll the animal to move it. *Do not drag* or lift the animal by its limbs unless there is no other alternative. If it is necessary to attach a belt from a hoist to the limbs, use uninjured limbs.
5. Use equipment and handling devices for moving non-ambulatory animals that are appropriate for the size of the animal (i.e., a modi-

fied handcart with a plywood base is adequate for moving crippled sheep and hogs, whereas conveyer belting is recommended for cattle).

6. If forklifts are used, construct an angled metal pallet to fit over the forks to assist with rolling the animal onto the pallet.
7. If euthanasia is recommended for animals that go down in a truck, at the auction or market the captive bolt stunner or a firearm is recommended. The captive bolt stunner fires a blank cartridge which propels a steel bolt into the animal's brain and kills the animal instantly. Firearms are less recommended because they could injure another victim. Cattle and horses are shot in the center of the forehead, sheep are shot on the top of the head, and hogs are shot in the forehead. Pigs are electrically stunned, rendering them brain dead.

EMERGENCY MEDICAL CARE AND EUTHANASIA FOR CATASTROPHIC INJURIES

In the past fifteen years, the development of new equipment and implants has enabled veterinary surgeons to repair many fractures that were previously considered hopeless. There are still many situations, however, when a horse with a serious fracture cannot be saved. The fracture site might be too contaminated, or the blood supply may be sufficiently compromised to allow for successful repairs. The horse's large size and weight often inhibits treatment as well. Sometimes an animal's temperament will not allow for recuperation. Many times complications occur in the other limbs, due to excessive weight-bearing during the rehabilitation period. In these cases, the humane solution is euthanasia.

The *American Association of Equine Practitioners* (AAEP) takes a definite position on euthanasia. The following criteria should be considered by veterinarians in evaluating the immediate necessity of euthanasia to avoid and terminate incurable and excessive suffering to horses:

1. Is the condition chronic and incurable?
2. Has the immediate condition determined a hopeless prognosis for life?
3. Is the horse a hazard to himself or his handlers?
4. Will the horse require continuous medication for the relief of pain for the remainder of its life?

Once again, the ultimate cut-off of life for a companion animal is most difficult for both the practitioner and for the animal owner. One of the most challenging decisions of the veterinarian is whether or not to put down a suffering animal. Some ask, "Do we have a right to make this decision?"

Others respond by saying, "Do we have a right not to make this decision of life with suffering over a painless death or permanent sleep?" Is it cruel to make the animal suffer? Now compare this idea with a human being in the same predicament.

SHOW ANIMALS AND THEIR BREEDING

Why do people show animals? Is it necessary? Is it cruel to the animal?

Showing animals enables a number of values to be developed, including teamwork, sportsmanship, competition, discipline, and responsibility. Developing pride in ownership, a business sense, and occupational experience are outcomes of producing and exhibiting show animals. Breaking rules, lying, deceiving, and cheating are wrong. The industry must strive for 100 percent of all people involved in these activities to engage in ethical behavior. The majority of those involved in exhibiting livestock and horses do not tolerate unethical behavior. The dairy industry is at least one area that has most consistently shown ethical behavior and practice. The Holstein Association has published in its national newsletter a list of names of people who practiced unethically. These people were blacklisted. No one was allowed to buy or sell animals to or from them. If all livestock agencies would follow suit, we could maintain a showmanship tradition that would be much more respected. Dr. Jeff Goodwin of the University of Idaho, has prepared videotapes and lectures regarding show animal ethics. Dr. Goodwin is one of many who continue to preach, teach, and guide the young about their moral or ethical obligations in caring for livestock.

Exhibitions provide an arena to improve the quality of breeds and classes of animals. Animals are selected for their superior qualities, whether for conformation, meat cutability, or performance. As a result, winning animals and their bloodlines are sought for breeding purposes. Strict guidelines banning drug abuse in animals exist and are enforced. The American Quarter Horse Association, in its 1994 rule book, states, *"Any surgical procedure or injection of any foreign substance or drug which could affect a horse's performance or alter its natural conformation or appearance is prohibited, except for those surgical procedures performed by a duly licensed veterinarian for the sole purpose of protecting the health of the horse. However, no foreign substance or drug which could affect a horse's performance is acceptable, whether or not administered to protect the health of the horse and, on the contrary, is prohibited."* Many question why American human sports events are not as strict at enforcing drug abuse among its competitors.

ASSESSING FARM ANIMAL WELL-BEING

Because of the increasing outcry and concern about the quality of life of farm animals, there has been a tremendous effort by animal scientists to develop clear, interpretable scientific criteria for assessing the well-being of an animal (animal welfare). The indicators which have been commonly used in farm animals are productivity, animal health, physiology, and behavior. One problem with these indicators is establishing a norm or standard to which measures can be compared in an effort to evaluate the animal's welfare. Two of the nation's leading authorities on measuring farm animal welfare are Dr. Joy Mench and Dr. Gary Moberg at the University of California at Davis.

Dr. Mench describes animals as complex adaptive systems. She points out that at any point in time an animal is exposed to a wide variety of different environmental stimuli (food, predators, etc.). At the same time, the animal may be experiencing different internal states such as hunger or sexual drive. These external and internal stimuli could send conflicting messages—eat, flee from a predator, perform a mating display, etc. The integration of these messages in the brain and their comparison with memories of previous experiences results in the behavioral and physiological responses we see, and that process enables the animal to adapt to changes in its environment.

So how do we compare well-being to a specific stressor? Animals do experience both emotional states (fear) and sensations (pain), but these experiences differ in important respects from that of human experience. Emotions and sensations are accompanied by changes in hormone levels, which can be measured, and brain functions, which will mitigate the animal's behavior.

In humans, *cognition* (the individual's ability to form mental representations of the environment and to be aware of its own sensations, emotions, and motivations) is an important element of subjective experience. We can intensify pain if it is associated in our minds with fears about a possible lingering illness. Hence, an animal's level of awareness can be an important consideration in evaluating its well-being.

PUBLIC ACCOUNTABILITY

CLASS DISCUSSION

1. Be prepared to discuss the public accountability of no more animal research and its effect on a paraplegic, a child with AIDS, or a youngster with multiple sclerosis or muscular dystrophy, etc.
2. Who should regulate farm animal welfare? What would be the most effective means of policing, convincing, or changing attitudes in this area of animal use?
3. Who should police the laboratory animal research facility? Are the IACUCs effective? Why? Why not?

4. Compare and contrast the following: existing laws protecting animals with those laws protecting children, the common citizen, etc.

RECOGNITION OF PAIN IN FARM ANIMALS

"Humane treatment of animals does not mean treating animals as humans."

Patrick D. Wall, University College, London

The misunderstanding of "pain in animals" is a severe detriment to the animals because it attempts to assign them specific human values. Each individual seeks to regulate his or her internal environment, which is a crucial condition for free life. One method of assessing pain is to study the animal's efforts to stabilize its internal environment, and then helping it or at least not inhibiting those efforts without a good reason.

If we were to use this method on non-domestic animals, would we not consider them in a constant state of stress?

LARGE ANIMAL ANESTHESIA AND POST-OPERATIVE CARE

How far do we go in assessing and treating animal pain? We need to define and make distinctions between pain and suffering. Pain is defined as one form of perceived threat to the individual. Suffering accompanies the experience of persisting pain and it is an enduring negative emotional state. A human in athletic competition feels pain but does not necessarily suffer. Perceived helplessness seems to be an essential element of suffering.

Veterinary anesthesia has been a respected discipline for over 40 years. Today we have drugs for various types of animal control, ranging from mild sedation to tranquilization and temporary taming, to short-term immobilization and general anesthesia. The administration of these drugs requires a sound knowledge of the animal's behavior and physiology. Certain types of animals react strenuously in fear or rage when someone attemps to control them. Violent responses are stressful to the animal, and chemical restraint of stressed animals cannot be approved on welfare grounds unless a true emergency exists that requires control of the animal. Even then, consideration of the risk to the controlling agent (minimal lethal potential) must be taken. During exertion, there is increased production of carbon dioxide from glycolysis in the vigorously contracting musculature. This carbon dioxide is taken up by the red blood cells and transported to the lungs. If the increased carbon dioxide is not removed by the lungs, the hydrogen ion concentration increases in the blood and the pH decreases, resulting in acidosis. Profound hypoxia can also occur in all tissues.

Previously we spoke of comparing the risk and pain of administering an anesthetic with the pain, time, and risk of the procedure. Weighing the two against each other should determine if and what type of relief should be offered.

FUTURE DEVELOPMENT OF USDA STANDARDS FOR FARM ANIMALS UNDER THE AUTHORITIES OF THE ANIMAL WELFARE ACT

What is being proposed? What should be proposed? How will the proposal affect food prices? Should that be a consideration? Why? Why not? Is the average American willing to pay double or triple for food? Are you? Should cost matter?

INTENSIVE ANIMAL PRODUCTION

Animal rights organizations have charged that confinement of animals has contributed to over-reliance on antibiotics to manage disease and growth, hormones to enhance production and reproduction, and chemicals to control flies, rodents, and other pests. If improperly used, these drugs and chemicals can show up in human food supplies, posing health hazards including allergies, cancer, and anaphylactic shock.

While it is true that high-density production can increase the potential for diseases, it can also provide better opportunities to monitor animals' health and nutrition. Large "factory-style" farming operations have environmental systems that can greatly curtail the spread of disease. These structures are well-ventilated, warm, and sanitary. They provide shelter from predators, bad weather, and extreme climates. Breeding and birth are also less stressful to the animal compared to free-range natural environments. Health management is much easier and more attention can be given sooner to animals when needed. It is also more efficient and economical to clean the large intensive facilities.

Guidelines direct producers to use scientifically developed animal health products only after they have been proven to pose no threat to human health. Growth-promoting antibiotics have undergone rigorous development and testing. All products and pesticides are strictly regulated and are monitored by the government to minimize risks to both animals and humans. While abuses within the system exist, the possible contamination of the human population is minimal compared to the risk of imported farm produce from Central America and overseas where such regulation is lacking. The industry producers and veterinarians, as well as government agencies, need to continue to monitor residue levels and illegal usage to guarantee a safe food supply.

Uptake of pesticides or herbicides by animals can take place if certain safeguards are not enforced and monitored. These situations can happen simply due to human error, such as the heptachlor poisoning of the dairy herds in Hawaii in 1983. Today, much of our fruits and vegetables are coming into the country from nations that do not have as strict environmental laws as we demand. There is considerably more danger of toxic pollution from environmental hazards such as these than from our domestic livestock. Dr. C. Everett Koop, former United States Surgeon General, when commenting on the threat of domestic use of pesticides, said, "Cancer rates in the United States have dropped remarkably over the past forty years. During this period of time, stomach cancer has dropped more than seventy-five percent, and rectal cancer has dropped more than sixty-five percent. The only cancer going up today is cigarette-induced lung cancer."

Animal activists have criticized farrowing crates as being unnatural. Sows cannot walk or turn around in them. They ask, "Would you like to be tied down all your life and just have babies?" Once again, we must be careful in comparing (anthropomorphizing) what is wanted by animals and people. Think about a young infant in an incubator in a hospital. You never see two babies in the same incubator. By isolating them, does that cause their "psyche" to be harmed? The sow is permitted to stand and lie down. The farrowing crate (or sometimes a tether) reduces the death and injury to baby pigs due to crushing. Individual stalls provide better control of feed, reduce aggression, and improve sanitation. Visit a commercial swine unit during the semester and have the operators demonstrate that these crates are actually more humane because they reduce the pain, suffering, or stress among the majority of the animals. Visit other swine operations where "open range" conditions exist and have them show you that their animals are not more prone to various parasitic diseases, that they exhibit less "stress", etc. Also talk to a large animal veterinarian. In other words, try and see first-hand, when you can, before your opinions are etched in stone. Try and see all sides of the issue. This is not to say that the enrichment of animals' lives that will be eventually slaughtered should not take place; it should. The animal's enrichment should be based on the animal, however, and not on human norms. Domestic and wild animals also should not be compared. Their norm is not the same. Researchers, scientists, farmers, and university personnel need to continue to find methods of producing cheap and abundant foods without severely compromising the welfare of the animals being raised for that food production.

The official policy of the AVMA on individual tethers and stalls for sows indicates that it is acceptable when monitored, maintained, and adjusted by responsible personnel. The standards for housing, flooring, environmental control, and stocking density are set forth by the North Central Extension Agricultural Engineers, Midwest Plan Service, in Ames Iowa, and are considered the guide for the humane care of swine in confinement. Environmental deficiencies can be calculated by tail biting, gastric ulcers, poor maternal care, and the loss of young. These environmental stress factors cost the farmer money. Hence, the farmer should want to reduce confounding factors causing such economic loss. Not all farmers are up-to-date on newer methods of animal welfare, and this is where the farm advisors and other such personnel come in handy. Education on better methods from both an economic and welfare standpoint should be made on a regular basis.

For poultry, cages also protect birds from predators, disease, and weather extremes. Producers are better able to control hygiene, environment, livability, feed conversion, and production in "battery cages," compared to free-range chickens. Cleaner eggs with less labor results from this system, as well. The AVMA's current position on the use of cages to house layer chickens is that their use should continue, and that present knowledge is not sufficient to support a radical change or ban of this system.

Animal activists charge that unwanted chicks (mainly those from a pipped egg, when the chick or poult has not been successful in escaping the egg shell during the hatching process) are left to suffocate in egg shells. Some producers have been known to throw unwanted or incompletely hatched chicks into the garbage bins to die (see Figure 9–4). No accurate estimate exists as to how widespread this practice may be. The most humane disposal method for unwanted chicks is to dispose of them while they are young. Smothering unwanted chicks or poults in bags or containers is *not* acceptable and should be condemned. Pips, unwanted chicks, or poults should be killed prior to disposal by an acceptable, humane method such as carbon dioxide euthanasia.

INDUCED MOLTING OF LAYER BIRDS

Molting, or feather loss, is a natural process. Birds in the wild molt. In large commercial poultry operations, managers will bring all animals into a nonlaying period at the same time. Managers can plan ahead for all birds so that cash flow will be steady (see Figure 9–5). The process causes a rejuvenation of the ovarian cycle of all the birds at the same time. The AVMA considers this practice acceptable management, when done under careful supervision and control. The once-practiced, long-term total feed and water withdrawal that resulted in high levels of mortality is *not acceptable.* Under no circumstances should water be withheld for any length of time. A carefully monitored and controlled program that usually includes reducing photoperiod, controlled caloric intake through dietary restriction, and/or reduction in some nutrients essential for egg production (i.e., sodium), is acceptable. Careful monitoring of the birds' weights and weight loss, egg production, mortality, and behavior is needed to assure that proper results are achieved humanely. Molting is not a reliable index of the progress and success of a resting and rejuvenation program. The ultimate goal is to improve

Figure 9–4 Live chicks found in dumpster . . . *Courtesy of PETA.*

Figure 9–5 Today, most eggs are produced by caged layers in large confinement flocks. *Courtesy of the University of Illinois, Champaign-Urbana.*

the bird's ability to produce a high-quality egg with a strong shell. The main goal *is not* to replace the feathers.

VEAL PRODUCTION

Many small, medium, and large veal operations exist across North America that do *not* raise their animals in the following conditions described by PETA: "Veal animals are kept in small crates and continued in separate stalls in the dark with improper nutrition to maintain light meat color. Animals are kept alive by dosing with antibiotics."

Veal calves are allowed to stand, lie down, touch, see, and react to other calves in clean, well-lighted barns or pens (see Figure 9–6). Calves are segregated to allow closer medical supervision and to control deadly bacterial diseases and parasites. Group housing increases the spread of infectious diseases. The main diet of the veal calf is milk and a calf starter (grain). The calf at this age is a non-ruminant and is incapable of digesting any roughage (thus, no hay is given). Heifer calves, not raised for veal but instead raised as replacement animals, are reared in a similar fashion. Human babies are never grouped together in the same incubators and are also not permitted to run amuck (free roaming). Their movement is restricted

Figure 9–6 Individual, portable calf hutches

to maintain sanitation and safety, similar to young calves. Replacement heifers, when weaned, are placed in group housing. Veal calves are slaughtered at weaning and thus are not normally introduced to a roughage diet. The AVMA has issued its "Guide for Veal Calf Care and Production." The guide outlines the proper procedures of veal production and housing that are considered to be the most humane. Those who violate these procedures and raise their calves in a cruel method should be condemned.

A whole industry should not be condemned because of the poor standards of a few. Many poor human parents exist who take drugs while they are pregnant, abandon their children to the streets, etc. Should we condemn all human parents due to the poor performance of the "bad" parents who are few by percentage?

Chapter 9
EVALUATION

1. The ________________________ Administration regulates stockyards regarding the treatment of animals.
2. Federal law imposes a(n) _____ hour transportation limit on cattle.
3. Most stunning of animals to be slaughtered in the United States is done with a(n) ____________.
4. List four areas where animal agriculture needs to improve its practices (according to leading animal behaviorists):
   a. ____________________
   b. ____________________
   c. ____________________
   d. ____________________
5. ____________________ animals are animals not able to stand due to calving problems, broken bones, or paralysis.
6. AAEP stands for the ____________________________________.
7. Today's modern day pork carcass contains ____ percent fat.
8. On a federal level, the ____________________ monitors the processes of food production to assure a safe food supply.
9. NOEL stands for ____________________________________.
10. ADI stands for ____________________________________.
11. The ____________________________ is the federal agency that sets specific regulations on the amount of residues of hormones and other additives that may be left in meat.
12. An ADI equals _____ of a NOEL.
13. Which of the following items has the largest amount of carcinogens?
    a. a cup of coffee
    b. a three-ounce serving of beef
    c. a glass of cola
    d. a slice of bread
14. The difference in the level of hormones in a 3-ounce serving of beef from an implanted steer versus a non-implanted steer is approximately equivalent to:
    a. one blade of grass in 1.3 football fields
    b. .0007 of a millionth of a gram
    c. a needle in a haystack
    d. .7 of a billionth of a gram
    e. all of the above

15. Comparing 3-ounce servings, which of the following foods has a higher level of the estrogen hormone than the average 3-ounce serving of beef? (More than one answer might be correct.)
    a. milk
    b. potatoes
    c. ice cream
    d. cabbage
    e. soybean oil
16. The ___________________, passed by Congress in 1958, stated that no carcinogens shall be added to the supply of food.
17. _________________ are vegetarians who do not eat or use *any* animal products.
18. ___________________ are vegetarians who eat animal products from milk and eggs but no meat.

Discussion Questions

1. Rodents have a value on the planet, but is that value sufficient to give them permission to eat 20 percent of the grain harvest in India? Justify your answer on an animal-rights based philosophy and on an animal welfare-based philosophy. Is there a difference?
2. Are we morally compelled to find ways to control the locust from its natural self-destructive rampages? Why? Why not?
3. Can an animal express its "natural" behavior patterns in an "unnatural" environment? If not, then how do you assess pain, stress, and suffering by observation of behavioral patterns?
4. Discuss the boredom of a backyard horse left in a 20-by-20 foot corral all day while his owner is at work or school. Is this boredom stressful to the horse? Do the behavioral patterns change because of the boredom? Does this mean the animal is under stress and is therefore suffering? If so, is it morally or ethically right to force the horse to endure these conditions? What about your dog or cat?
5. Read the positions enclosed from PETA on companion animals ("Companion Animals: Pets or Prisoners?"). Assess the environment you have in your home for your dogs, cats, birds, fish, or horses, and describe any differences. Do you feel you need to change the environment for your companion animals based on the descriptions from PETA? Explain. Why? Why not?
6. What is the official position of the AVMA on ear cropping? On tail docking? On declawing? What do you think about these procedures? Why?
7. What is the AVMA's official position on the euthanasia of unwanted pets? What is the animal rights movement's stance on this policy (see the PETA factsheet, "Euthanasia: The Compassionate Option")? What would happen if we did not

perform euthanasia on unwanted pets to the population of pets within our borders? Could we afford to feed all of these animals? What about disease control? Here is the devil's advocate question: If people were breeding in such a manner and we could not afford to feed all of the children resulting from such breeding, would you advocate the euthanasia of these children? Why? Why not? (Remember the case in Brazil.)

8. Discuss how breed registries could help in the regulation of "puppy mills."
9. What are some of the reasons large animals who are injured are put down instead of receiving medical assistance? Use the horse with a broken leg as an example.
10. Discuss methods of assessing a farm animal's well-being. Discuss problems with the accuracy of these methods.
11. What is the official AVMA policy on battery cages for house layer hens? Why?
12. Why is it considered better by most hog farmers and veterinarians to limit the movement of sows in farrowing crates? Is there a better method available when producing large numbers of piglets? In answering this question, include the consideration of price (labor, facilities, and weather conditions).
13. What is the AVMA's acceptable method of inducing molting in layer birds? What is an unacceptable method that is sometimes but rarely used?
14. Using hormone implants allows cattlemen to produce lean beef more efficiently and to provide beef at lower prices year-round. It is estimated that hormone growth implants reduce the total cost of beef production by $50–$80 per steer and reduce retail prices of all cuts of beef by 20 cents (30 cents per pound). Anti-beef activists claim this situation causes beef to be full of harmful hormones. Is this true?

Case Study 9–1:

Dehorning Calves by Steven Bursian, William Frey, Faith Gandiya, Leland Shapiro, and Sally Walshaw. (*Permission to reprint has been granted by the editor of The Ag Bioethics Forum.*)

A recent newspaper article from a case study done by the Ag Bioethics Forum reported the visit to a local dairy farm by a self-professed animal rights activist. She said, "I was horrified to see little calves having their tiny horns removed without any anesthetic, while the calves screamed and bawled."

According to a spokesperson for the state dairy cattle association, Holstein cattle (male and female) have horns. As these animals become mature, the horns are hazardous to people and to other cattle. Horns also increase necessary feeder space for each animal. Non-chemical methods of dehorning include electrocautery irons and various cutting and scooping instruments.

A veterinarian who agreed to be interviewed stated that the head of a calf is well-supplied with nerves that carry pain sensation. Older calves may struggle more and may seem more depressed after dehorning than younger calves. Some dehorning procedures appear to be less painful to calves than other procedures.

Case Study 9–2:

Polling, or dehorning, is an issue for both range and feed lot beef operations and on dairy farms. Cattle with horns are more difficult to handle safely, because they require more room in transportation and confinement systems. The presence of horns can also exacerbate problems associated with dominance hierarchies.

Dehorning is done in several ways: by treatment at an early age with a caustic chemical that causes some irritation; by burning the horn bud with a hot iron, also when the calf is quite young, which causes momentary pain because the interior of the horn is innervated; and using a "dehorning spoon," which levers the horn out of the skull (a procedure which becomes increasingly painful and bloody as the calf ages). Cattle with the *poll* (horn-free) gene are born hornless, so it would be possible to breed cattle to have no horns. A dairy and reproductive specialist estimates, however, that introducing the poll gene while preserving other superior traits in Holsteins (the most common dairy breed) would raise the price of milk 4–5 percent.

QUESTIONS

1. Is solving the animal welfare problems associated with dehorning worth a 4–5 percent rise in the cost of milk?
2. What if a number of other animal welfare problems could each be solved at a similar cost, with the aggregate rise in cost associated with virtually eliminating problems involved in management, housing, transportation, and slaughter all being solved for a 50 percent rise in the cost of meat and animal by-products? Would an ideally humane animal agriculture be worth it?
3. If so, should the changes be mandated, and how should the transition be phased in?
4. Would *you* be willing to pay for a 50 percent *real* increase in the cost of your food?
5. If these laws were mandated in the United States, how would our farmers be able to compete with imported animal products from countries such as Mexico where these laws do not exist due to the NAFTA agreement? Would this situation not then put all of our farmers out of business requiring us to buy only foreign-produced products? Remember that with NAFTA, we cannot put our environmental, labor, or ethical concerns as restrictions for commerce.

TOPICS FOR DISCUSSION

1. Invite other students to engage in a role-playing exercise using rights-based, duty-based, and goal-based ethical theories to evaluate the case.

Possible roles:

Animal activist—Justify your opposition to dehorning.

Farmer—Justify the continued dehorning of calves.

2. Ask other students to suggest alternative procedures to avoid the problem of apparent pain in this common agricultural procedure (i.e., a cornual nerve blockade using 3 to 10 ml of 2 percent lidocaine).
3. Should the alternatives you mention be mandated for farmers? Why or why not?

General Problem:

a. To what extent are we morally required to limit pain in animals used for agriculture?

Specific Problems:

b. Should we develop more humane methods for calf dehorning?
c. Should humane dehorning methods be required?

Case Study 9-3

By Gary Varner, *Permission to reprint granted by editor of The Ag Bioethics Forum.*

THE VEGETARIAN MEAL

A professor at a major state university has received federal funding for a program on ethics, which life sciences professors from around the country will attend. The professor is a vegetarian on moral grounds. In making arrangements for the five-day program, he specifies that all of the (optional) lunches will be lacto-ovo vegetarian. Several of the participants are outraged. In fact, one participant sends a long, angry email message to professors across the country and to highly placed personnel in federal funding organizations, including the National Science Foundation, which funded the program.

QUESTIONS

1. Why do you think some of the participants were outraged?
2. Was the professor wrong to design the lunches this way? Should funding for future programs be rescinded because of the meal plan?
3. Suppose the professor in question had believed, on moral grounds, that every meal should include some broccoli, and he had insisted that the lunches all do so. Would this action have angered participants? What if the professor believed, on moral grounds, that people should consume some animal flesh at each meal, and he had designed the menus accordingly?

ENDNOTES

43. Pierce College's average in the 1980s.
44. M. Santoro, Goldenmeat Farms, Canoga Park, CA

RECOMMENDED READINGS: (From Section 2)

1. *Factory Farming: Mechanized Madness*. PETA.
2. Grandin, T. "Principles of Abattoir Design to Improve Animal Welfare." *Progress in Agricultural Physics and Engineering*. Pg. 279–303.
3. ——. *Recommended Animal Handling Guidelines for Meat Packers*. American Meat Institute. 1991.
4. *Living in Harmony with Nature*. PETA.
5. *Veal: A Cruel Meal*. PETA.

CHAPTER 10

Veterinary Aspects of Welfare

Animal activists allege that veterinarians are too closely tied to the animal industry, making them unable to be objective when pointing out animal welfare problems. They claim that the veterinarian's main goal is to increase profits for the producers who pay their fees. Veterinarians are obligated morally, ethically, and philosophically to promote the welfare of all animals, as defined in the AVMA policy. The AVMA affirms that animals raised for food, fur, and/or fiber should be treated and handled humanely with due consideration to their welfare and well-being. Veterinarians often, and should, assume leadership roles to help eliminate cruelty, abuse, and neglect of animals in modern livestock production. Livestock, dairy, and poultry producers also have the duty to recognize and deal with people who are cruel, abusive, and neglectful in their production practices as well as to ensure that those practices that are contrary to animal welfare are abandoned or replaced with procedures that are more humane.

OBJECTIVES:

The student should:

1. Outline the AVMA official policy on methods, which should be followed in the caring of animals that are raised for food production, sports, fur, and/or fiber.
2. List the main categories regarding degrees of pain associated with animals in research.
3. List and then discuss the AVMA official policy on euthanasia.
4. Discuss the assessment of adverse effects on animals that routine procedures such as castration, tail docking, and dehorning might have.

5. Play various roles in the case studies provided at the end of this chapter. The student should try and play roles not normally agreeing with his or her own perspective.

FINANCIAL PRESSURES IN INTENSIVE AND EXTENSIVE SYSTEMS

Do financial pressures dictate the care that an animal receives from the veterinarian? Why? Should they? If not, then who pays? Because of these financial pressures, the veterinarian's job has become more of a consultant, rather than a practitioner who treats illnesses. Is this a correct assessment? What are the alternatives?

Many activists and others condemn the veterinary practitioner for not providing free treatment. Idealistic students have this same mentality. A typical freshman student wanting to become a wildlife veterinarian does not consider, how am I going to make a living? who is going to pay the bills? where do you need to live in order to treat and take care of large numbers of wildlife? and can you raise a family there? Do we, as a society, order the free treatment of animals, as we have with people in similar situations? Do we mess with "Mother Nature?" Where do we draw the line?

Similarly, with livestock, there are many possible treatments for sick or injured livestock. The veterinarian, however, needs to determine who is going to pay for that treatment, or better yet, how payment will be made. If a beef steer nets its owner $300 per year, a veterinarian cannot expect him to pay $500 for a displaced abomsal surgery. In California, a typical urban veterinarian performing a left flank abomasopexy will charge only $600 for this fairly long surgery. For a similar surgery on a horse, the veterinarian would charge $6,000. Why is there a difference? The owners of companion animals can afford to pay more, because they are not running a business where they need to break even. Similarly, the owner of a dairy or beef cow cannot afford to pay expenses to "fix" a cow that are worth more than the cow itself. The veterinarian needs to make determinations based on economics when determining which procedures, if any, to conduct. The cow farmer is a businessperson who depends on the veterinarian to not only heal the cow, but to keep him or her afloat financially.

PAIN RECOGNITION, ASSESSMENT, AND THE PHARMACOKINETICS OF PAIN RELIEF

Research using animals is divided into three categories regarding the degrees of pain associated with that research. Level 1, which is no pain, is associated with routine procedures such as blood sampling and vaccinations; level 2 includes pain relieved with anesthesia; and level 3 is when no anesthesia is given to lessen the pain. According to the USDA, 93 percent of research animals do not experience pain, either because the experiments do not involve pain (58 percent) or because anesthesia or pain killers are used (35 percent). In a small number of studies (7 percent), such as those on chronic pain, which is a major problem for cancer patients and burn victims, anesthesia and/or pain killers cannot be administered. These studies are among some of the most carefully evaluated studies in research, with the quantity and duration of pain strictly regulated and monitored to ensure that those levels are kept to a minimum level. IACUCs are now largely fulfilling the task of alerting investigators to the need for using anesthetics, as well as other refinements, that reduce the pain to the animal.

In 1992, the HSUS submitted a "Petition for Changes in Reporting Procedures Under the Animal Welfare Act" to the USDA. During this same year, the NIH/Arena Guidebook advised "the higher the level of the anticipated distress the stronger must be the justification of the value of the research." Thus, the level of distress must be identified and quantified. The Canadians, British, and Dutch have adopted "categories of invasiveness in animal experiments." Five categories, A through E, are used instead of the three categories described by the USDA. These categories include:

A—experiments involving no living material, tissue cultures, or invertebrate species.

B—experiments that cause little or no discomfort or stress.

C—experiments that cause minor discomfort, stress, or short duration pain.

D—experiments that involve significant but avoidable pain or stress but having the pain or stress alleviated by the use of anesthesia.

E—experiments that involve severe pain without the use of anesthetics. (Orlans, B. *In the Name of Science*. Oxford University Press. 1993.)

Thus, the three previously mentioned categories would be replaced as the following:

Level 1 = Category C, Level 2 = Category D, Level 3 = Category E.

METHODS OF EUTHANASIA AND THEIR ASSESSMENT OF HUMANENESS

Tom Regan, an animal rights philosopher, notes that euthanasia requires more than just killing painlessly or killing with a minimum of suffering.

Regan asserts that the term can be applied to people *or* animals, only if killing them is in their best interest. Veterinarians use the word euthanasia to refer to the killing of an animal painlessly or with as little pain as possible, regardless of whether the animal is healthy or whether killing the animal can be said to be in its best interest.

In veterinary medicine, the distinction between killing an animal and letting one die is of much less importance than in human medicine. Typically, veterinarians do not use the term euthanasia when they allow a patient to die. Few veterinary clients have the will or financial capability to prolong an animal's life when the animal has reached a vegetative state, nor do most veterinarians find it reasonable to utilize limited resources trying to keep an animal alive in such a state. No decent parent would hesitate selling their car or their home if needed to pay for a medical bill that "may" provide a "chance" to keep their child alive. But who would condemn a dog or cat owner for *not* doing the same for their beloved pet who suffers from cancer, even if the treatment of tens of thousands of dollars would cure the illness? So yes, the veterinarian must consider price and affordability with any treatment that he or she prescribes or recommends to a client. Increasingly, human health maintenance organizations (HMOs) are also considering price in their decisions about which medical treatments they will allow. The ethics of human suffering are not always considered.

Additionally, most animal owners and veterinarians would consider it inhumane to keep an animal alive that has an incurable ailment and is in pain or distress. Although there has been much discussion of this issue in human medicine, in most states and countries it is still illegal to perform euthanasia on people. For a suffering animal, though, it would be considered cruel or inhumane *not* to do so.

The choice of which agent to use in euthanizing an animal can be an ethical decision as well. If the wrong agent is given or is improperly administrated, the animal can suffer needlessly. The AVMA, in its "Report of the AVMA Panel on Euthanasia," periodically documents the most common methods with regard to such factors as safety for personnel, ease of performance, rapidity of death, efficacy, and species suitability. The report details the dosage, routes of administration and techniques recommended to assure greatest animal comfort. The report also lists several general criteria for choosing an acceptable product and method:

- Ability to induce loss of consciousness and death without causing pain, distress, anxiety, or apprehension
- Time required to induce unconsciousness
- Reliability
- Safety of personnel
- Irreversibility

- Compatibility with subsequent evaluation, examination, or use of tissue
- Emotional effect on observers or operators
- Compatibility with requirement and purpose
- Drug availability and human abuse potential
- Age and species limitations
- Ability to maintain equipment in proper working order

ASSESSMENT OF ADVERSE EFFECTS ON ANIMALS ASSOCIATED WITH ROUTINE PROCEDURES (CASTRATION , TAIL DOCKING, AND DEHORNING)

The purpose of castration and dehorning is to reduce animal aggressiveness, to prevent physical dangers to other animals and producers, to enhance reproductive control, and to satisfy preferences regarding the taste and odor of the meat. If performed on young animals, these procedures result in minimal discomfort compared to the long-term benefit for all the animals in the group (see Figures 10–1 and 10–2). With older animals, pain

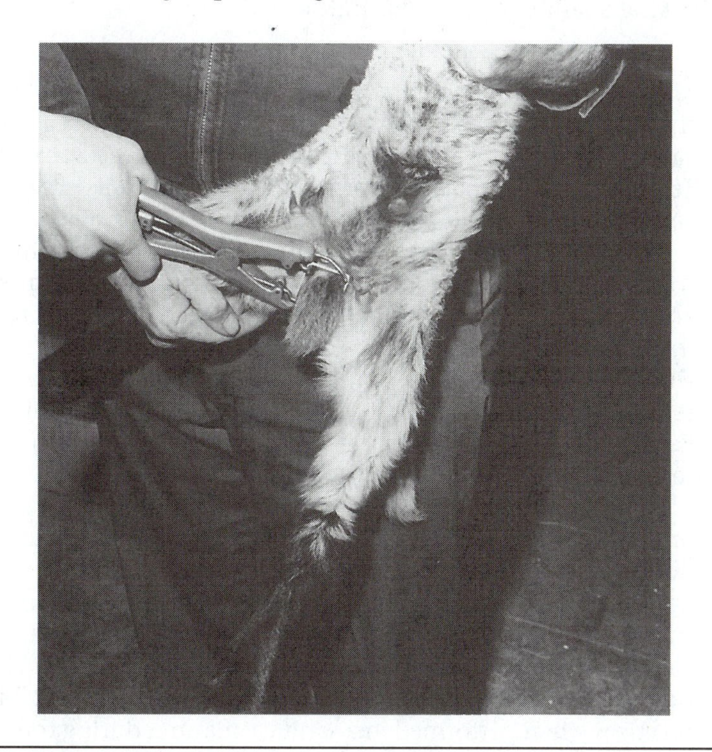

Figure 10–1 Castration of a lamb using a elastrator

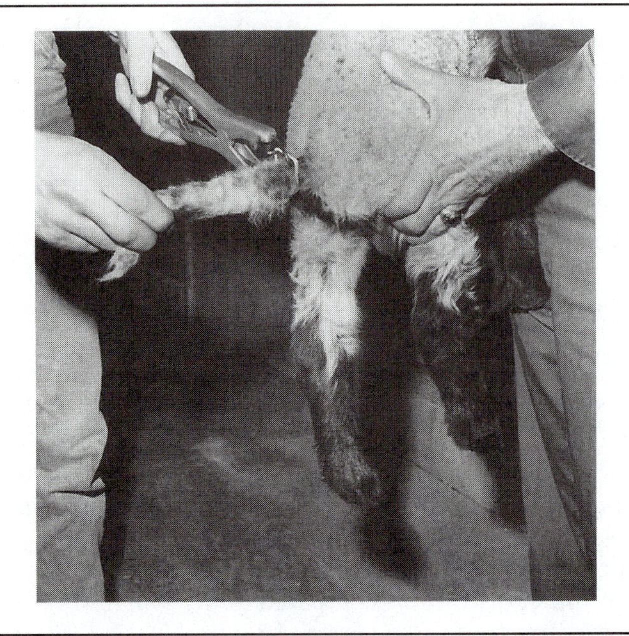

Figure 10–2 Docking a lamb using an elastrator

relief (in the form of a local anesthetic) should be provided. General anesthetic is too risky compared to the procedure itself. If male swine are not castrated at a young age, fighting in herds might become common. In about 10 percent of the males, a hormonal secretion called *boar taint*, causes the carcass to be inedible.

Docking of tails (in sheep) prevents fly and parasite infestation. In some countries, dairy cattle have their tails docked to improve sanitation and to make it easier to attach the milking machines from the rear. Docking piglets' tails keeps them from biting off each other's tails, a situation that creates the potential for illness and spreading disease among the litter. Ear notching provides an inexpensive, easy means of identifying pigs. Ear tags are easily chewed off by litter mates, and tattoos are not easily read without restraining the animal.

The AVMA's official position supports the use of procedures that reduce or eliminate the pain of dehorning and castrating of cattle. These procedures should be completed at the earliest age possible. Research in developing new and improved techniques for painless, humane castration and dehorning is encouraged. It is also recommended that viable alternatives to the castration and dehorning of cattle be developed and applied. Castration, ear notching, and tail docking of piglets are acceptable management practices when performed in a sanitary manner during the first week of life.

BEAK TRIMMING IN POULTRY

Trimming beaks is a common practice in the poultry industry and is done to prevent fighting and stress among poultry while limiting unnecessary feed intake (see Figure 10–3). If done properly, beak trimming removes the sharp tip on the end of the beak. This procedure should only be performed on young chickens and turkeys to prevent or reduce the natural tendencies toward cannibalism, fighting, and feather-picking. The majority of the beak is *not* removed (as purported in many animal rights papers). The AVMA's official position on beak trimming is that it should only be performed by well-trained personnel, and that alternative methods for the control of cannibalism should be sought.

COMMUNICATION SKILLS WITH CLIENTS IN DIFFICULT SITUATIONS (EUTHANASIA, NEGLIGENCE, CRUELTY, ETHICAL PRACTICE, ANIMAL RESEARCH)

Physicians, both veterinary and human, must become involved with controversial issues, many of which will ultimately result in the more humane treatment of their patients.

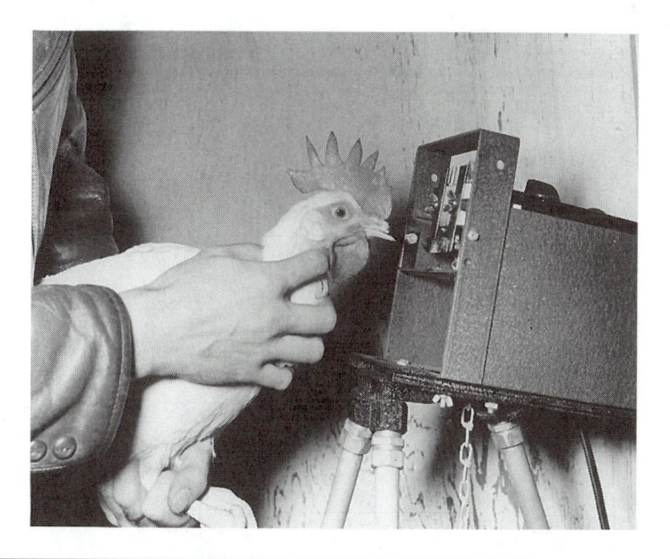

Figure 10–3 Debeaking a chicken

The AMA urges its members to tell their patients the following statements:

1. Research using animals saves lives, extends life expectancy, and improves the quality of human and animal life.
2. Virtually every major medical advance in this century has depended on animal research.
3. At the present time, there are no real alternatives to using animals in research.

Because veterinarians serve two customers, the client and the patient, the interests of patients sometimes conflict with those of the client. When this situation happens, veterinarians must make ethical choices to resolve the conflict. These choices are rarely black and white. Because of the difficult decisions a veterinarian must ethically make, most veterinary colleges now require a course in veterinary ethics as a part of its curriculum. Several scenarios are included here to help you understand this dichotomy. Many of the scenarios were borrowed from the Animal Ethics Institute (organized by Gary Comstock of Iowa State University), and others were modified from *Veterinary Ethics* by Tannenbaum.

Chapter 10
EVALUATION

Several of the veterinary ethics questions were modified from Tannenbaum's Veterinary Ethics, *2nd Ed. 1995, a strongly recommended text for the veterinary student.*

1. Does the veterinarian serve the patient or the client?
2. Should the client be responsible for treatment of the patient (his or her animal) if a significant financial expense is required? If so, how great an expense can the client be expected to bear?
3. Does a veterinarian have a moral obligation to take his or her time and money and attempt to adopt *all* curable animals brought in by clients who can no longer take care of them? If not, then who should care for these animals? If yes, then how are the veterinarian's expenses met?
4. Now that your patient is dead, what do you tell the client? While performing surgery on a dog (the patient was maintained on inhalation anesthesia and IV fluids), your most experienced registered veterinary technicican (RVT), who is monitoring the anesthesia by observing respiration, color, pulse, etc. and by listening through an esophageal stethoscope, is called to assist in another emergency. Your patient is stable and is doing well, so you see no problem and send her into the other room. Ten minutes later, you complete your closure, remove the surgical drape, and find that the patient is dead.
    a. Do you bill the clients for the surgery?
    b. What do you tell them?
    c. Do you share your feelings of guilt with the owners, or keep them to yourself? If the owners are lawyers from a large, well-known law firm, would that change your answer? If so, is that ethical? What if they were "old" clients that you have known for years?
    d. Would your answers to any of these questions be different if you determined that your technician had inadvertently turned off the oxygen instead of the nitrous oxide in her haste to assist your associate with his emergency case?
5. List the three animal categories regarding the degrees of pain associated with research.
6. What is the difference between animal rights philosopher Tom Regan and the AVMA on the term euthanasia as it relates to animals? Some animal rights activists insist on using the word *killing* or *slaughtering* instead of euthanizing an animal. Are they correct? What is the difference?
7. Are there times when it is more humane not to use anesthetic when castrating or dehorning an animal? Explain or defend your position.
8. What is the AVMA's official position on beak trimming?

Case Study: Euthanasia in Veterinary Medicine

By Walt Weirich, *Permission to reprint has been granted by the editor of The Ag Bioethics Forum*

SCENARIO ONE

Jack B. Boomer brings in his dog for a consultation. You watch him drive up in his "Beamer" and roughly extract the dog from the back seat. After he has the dog out on the parking lot, he stops to brush the dog hair off the seat. The dog tangles the leash around Jack's leg and you notice that Jack is wearing well-shined loafers and expensive slacks, but no socks.

When Jack is in your examination room, he tells you that he wants his dog Clarence put down. He explains that the dog has caused him grief one too many times, and he wants Clarence put to sleep. You ask him why and he says that his former girlfriend liked the dog, but his current girlfriend does not. Just when he and his girlfriend get in a certain "mood," the dog interrupts and spoils everything, according to Jack. He will not listen to any of your suggestions and insists that you euthanize Clarence. Clarence is a neutered, four-year-old Jack Russell terrier in excellent health.

What do you do? What are the consequences for Jack if you euthanize Clarence? What are the consequences for you if you euthanize Clarence? What are the consequences for Clarence if you do not euthanize Clarence (remember that Jack is determined).

SCENARIO TWO

Mrs. Jane B. Practical brings in her beloved "Poochie." Poochie is an eighteen-month-old Yorkshire terrier that has recently been acting strangely. Poochie has been depressed after eating to the point of almost appearing to be in a light coma. Poochie has had some "spells" that might just have been seizures. Poochie is small and has never been a vigorous dog. Mrs. Practical asks that you evaluate Poochie to find out what is wrong with her.

You examine Poochie and discover what is almost certainly a portosystemic shunt. You recommend that she take Poochie to the nearby university for surgery and give the dog a special diet that will almost certainly relieve Poochie's symptoms and allow a nearly normal life. Poochie will require special care, but with a little diligence on Mrs. Practical's part, Poochie will be all right.

Mrs. Practical's face goes dark when you tell her this information. She does not want to put Poochie through all of that and asks you to put Poochie to sleep. What do you do? What will happen to Poochie if you refuse to euthanize? How will Mrs. Practical feel if you give her a hard time about her decision? Is euthanasia a valid treatment in veterinary medicine?

SCENARIO THREE

Ima V. DeVoted brings in her aging, mixed-breed Beagle female "Mitzie." Mitzie has several moderate-to large-size mammary tumors in her right chain and some smaller ones on the left. Aspiration of one of the masses reveals adenocarcinoma, with a high score for malignancy. You recommend radical mastectomy as a two-stage procedure. Mrs. DeVoted quickly accepts, and the surgery is done on the right side. Everything seems to go well, but before the second stage of the surgery can be done, regrowth on the right side has already started. The growth is aggressive, and in a short time the new tumor is ulcerated and is draining profusely. Thoracic radiographs reveal metastatic masses within the lung that were not there at the initial examination six weeks prior. Mitzie is in a lot of pain with the rapidly growing tumors. It is clear that nothing more can be done that will correct the problem and make Mitzie more comfortable. You recommend that Mitzie be euthanized. Mrs. DeVoted refuses and stomps out of your office. She subsequently quits her job and stays home with Mitzie full-time.

What is causing Mrs. DeVoted to take such drastic steps for Mitzie, when Mitzie is clearly suffering? What do you tell Mrs. DeVoted to encourage her to change her mind? What is the moral basis for the argument that you might be able to present to convince Mrs. DeVoted of what is wrong with her actions?

SCENARIO FOUR

Johnny Dollar of the Dollar Cattle Farms calls you to see a cow on his farm. The last two weeks before calving, the cow has fallen on the way to the barn and fractured her left femur, and probably her pelvis. Johnny is concerned that she will deliver the calf she is carrying, because the calf could be the basis of a new line of genetically improved dairy cattle. He wants you to do what you can to keep her alive until she calves naturally. After that, he does not care what happens to the cow. He would like to get some money out of her, but the calf is of paramount importance to him.

How will you deal with this situation? Will you acquiesce to his request to keep her alive for two weeks until she calves? Are there alternatives that might be morally correct for the cow?

RECOMMENDED READINGS (From Section 2)

1. *Euthanasia: The Compassionate Option*. PETA.
2. Grandin. T. *Public Veterinary Medicine: Food Safety and Handling*. Euthanasia and slaughter of livestock. JAVMA 204(9): 1354–1360. 1994.
3. *Use of Animals in Biomedical Research. The Challenge and Response*. An American Medical Association White Paper. 1992 (revised).

CHAPTER 11

Zoo, Wildlife, and Utility Animals

The three main purposes for having zoos are entertainment, education, and sensitization to animals, their habitats, and their plights. Animal and other biological scientists study exotic and endangered species using the zoo as their laboratory. Before World War II, most of the great zoos were in Germany and were primarily a collection of animals. Since the Second World War, zoos have emerged throughout the world. Today, they primarily focus on improved breeding programs and technology for saving many endangered species.

Objectives

The student should:

1. List the justifications for and arguments against having zoos.
2. List the justifications for and arguments against permitting the hunting of animals for food.
3. Discuss the history of rodeo and its use in today's modern agriculture industry.
4. List rules enforced to ensure humane treatment of rodeo stock as outlined by the PRCA.
5. Discuss the concerns of animal activists regarding the use of PMU horses in the production of Premarin.
6. Discuss ethical concerns in the care of racehorses.
7. Justify the use of animals that help disabled people (guide, hearing dogs, and primates for paraplegics, etc.)
8. Discuss the pros and cons of using animals in motion pictures and television.

Animal activists and other zoo critics argue that zoos rarely simulate the animal's natural habitat, that many zoo animals become ill, and that nature should take care of itself. Small zoos with individual animals are particularly criticized, because there is usually no long-range plan for restoration into the animal's natural habitat.

If zoos are to be successful propagators of some of the world's rarest life forms, *culling* must be practiced. The killing of these culled, non-domestic animals might be required, which brings about the ethical dilemma of selecting individuals for physical soundness and genetic make-up with the ultimate goal of preservation of the particular species. What is equally disturbing to some people on an ethical level is that someone must weed out surplus exotics or exotics that do not fit the intended goal. Sometimes it is not possible to transfer excess offspring or animals themselves to other zoos. What do you do with all of these "extras." Animals that are not part of a breeding program can be sterilized, or sometimes they are simply euthanized as "surplus animals."

What many people feel is a worse case scenario is the direct or indirect sale of these excess animals to hunting ranches, where hunters pay for the privilege of stalking exotic game in "canned hunts." The *American Zoo and Aquarium Association* (AZA) has incorporated into its code of ethics a prohibition against the sale or transfer of animals for the purpose of sport, trophy, or other kinds of hunting. Animals that can no longer serve their purpose at the zoo and cannot be reasonably placed elsewhere must be euthanized.

Today, in Yellowstone National Park, the American bison is inflicted with the abortive and zoonotic disease, *brucellosis*. Because cattle graze the same range as the bison, cattle farmers want the bison shot before they contaminate their cattle herds. Environmentalists want the bison left alone and to let nature take its course. A more prudent and perhaps ethical approach might be to vaccinate the young bison calves and to test and slaughter only the positive-testing cows and bulls. This process would require the cooperation of the park service, environmentalists, and the ranchers. If we all learned to talk, work together, and stop pointing fingers, much more could be accomplished. As of March 31, 1999, there were only five cattle herds and one bison herd remaining affected in the United States. Three herds in Texas, one herd in Oklahoma, one herd in Florida, and one Bison herd in South Dakota were still infected with brucellosis. Forty-five states, Puerto Rico, and the United States Virgin Islands hold a class-free status. Only the states of Texas, Louisiana, Kansas, Missouri, and Florida remain on the list "non-brucellosis free states." It is expected that shortly after the year 2000, brucellosis will be eradicated from the United States—a long and expensive cooperative fight. It is not known if the current (new) vaccine for brucellosis will be effective at preventing the disorder in bison. Until we learn to work together and try to fight this disease as a team, however, we will never know. The success story in nearly eradicating this costly and deadly disease came about only through the cooperation of national, state, and local agencies, dairy and beef ranchers, and the veterinary community.

RACING, HUNTING, AND WORKING HORSES

Hunting is the pursuit of prey or game, with the intent to possess it. Some view hunting as an environmental management technique. Others view hunting as a barbaric act to harm others. Those who do hunt cite various reasons for their actions:

- Companionship or solitude
- Communal or participation with nature
- Challenge
- Tradition or heritage
- Food
- Population management
- Environmental management

Steel-jawed leghold traps are sometimes used and are thought to be the best mechanism for getting rid of predators, because fences and biological controls often do not work. The AVMA considers the steel-jawed leghold traps inhumane. Another type of trap should be used if predators are a problem.

In several areas within the United States, Amish farmers continue to use horses for cultivation and transportation. Throughout the world, especially in third-world nations, the equine is still used for work. Water buffalo still pull plows through rice paddies. Horses, mules, camels, and llamas all make life more convenient and easier for many of the world's poor by providing labor, transportation of goods, and human transportation. No alternative source of transportation or labor exists for these poor, developing areas of the world.

Many athletic equine events are deeply rooted in practical beginnings. Rodeo events test a cowboy's working skills (Figure 11–1). Horse racing resulted from man's naturally competitive nature to determine which horse is the fastest and has the greatest endurance. Since 1947, the Professional *Rodeo Cowboys Association* (PRCA) has published established animal welfare policies. In 1993, there were more than 60 rules pertaining specifically to animal welfare. To qualify for sanctioning, a rodeo must have a veterinarian on-site or on-call for that event. In 1993, 86 percent of the rodeos sanctioned by the PRCA had veterinarians on-site. The *PRCA*

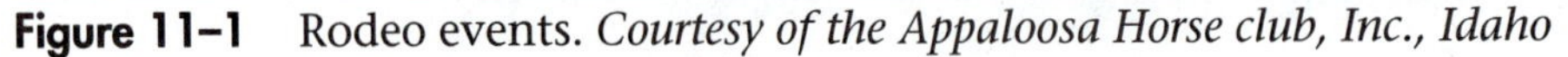

Figure 11–1 Rodeo events. *Courtesy of the Appaloosa Horse club, Inc., Idaho*

Veterinarian Handbook outlines treatments, emergency plans, and evacuation procedures with all species of rodeo livestock. In addition, the American Association of Equine Practitioners (AAEP) has published a book, *Guide to Veterinary Service and Judging of Equestrian Events*, which assists veterinarians in their medical roles at various events.

The following list describes some of the rules that are enforced to ensure the humane treatment of rodeo livestock (modified from the PRCA handbook):

- No locked rowels, rowels that will lock on spurs, or sharpened spurs may be used on bareback horses or saddle broncs.
- Wooden paddles are not to be used by any PRCA member to beat or intimidate animals.
- In calf roping, calves may not intentionally be flipped backward. The contestant must adjust the rope and reins in such a manner that will prevent the horse from dragging the calf. The rope is to be removed from the calf's body as soon as possible after a "tie" is completed. Roping calves shall weigh at least 250 pounds each, and be strong and healthy.
- Placing fingers in the eyes, lips, or nose of steers while wrestling them is forbidden.
- Animals for all events will be inspected before being selected for competition, and no sore, lame, sick, or injured animal, or one with defective eyesight, shall be permitted in the "draw" at any time. Should an animal become sick or injured between the time it is drawn and the

time it is scheduled to be used in competition, that animal shall not be used in competition. Another animal will be drawn for the contestant, as provided in the PRCA rule book. A veterinarian shall be present or on call for every performance and/or section of slack. Failure to do so shall result in a fine to the rodeo committee of $100 per performance.

- No animal shall be beaten or cruelly prodded. Standard electric prods shall be used as little as possible. Animals shall be touched with the prod only on the hip or shoulder area.
- A conveyance must be available and used, if possible, to remove animals from the arena in case of injury. Injured calves shall be removed from the arena in a pickup truck or calf stretcher.
- No sharp or cutting objects shall be permitted in cinch, saddle girth, or flank straps. Only sheepskin-lined flank straps shall be used on bucking stock, and flank straps shall be of the quick-release type. Sheepskin-lined flank straps shall be placed on the animal so the sheepskin-covered portion covers both flanks and the belly of the animal.
- PRCA-approved saddles or riggings must be used in the saddle bronc riding and bareback riding events.
- No stimulants or hypnotics are to be given to any animal used for contest purposes.
- Chutes must be constructed as to prevent injury to stock. Maintenance personnel and equipment shall be stationed at chutes to assist in the removal of any animal, should it become caught. The arena shall be free of rocks, holes, and obstacles.
- Clowns are not to abuse stock in any fashion.
- No small animals or pets are allowed in the arena where restraint is necessary, or where they are subject to injury or attack by another animal.

In 1988, on-site veterinarians at seven of the largest rodeos compiled statistics on all animal injuries. Of the 1,050 bulls used, there were no injuries that year. There were also no injuries in the 861 saddle bronc runs, in the 814 bareback runs, in the 1,633 steer wrestling runs, in the 925 team roping runs, or in the 104 steer roping runs. A total of 12 injuries occurred out of the 1,546 calf roping runs. Five calves had broken legs (all were casted and healed fine); one had a mildly torn shoulder joint ligament; one had a mildly bruised lower spinal cord; and three had mildly torn ligaments in the stifle joints. All recovered from their injuries. Although there were only 0.78 percent injuries in the calf roping event, many feel that this number is too high. Professor Ron Wechsler of Los Angeles Pierce College, and Pierce's Rodeo Team Advisor, believes that they should either eliminate calf roping from rodeo *or* disqualify anyone who jerks a calf over in the event. Professor Wechsler is a strong proponent of animal welfare at all of the rodeos that he manages and supervises. He feels that this policy would send a strong statement to the cowboy, indicating the proper treatment of the calf.

The AVMA condemns spectator events involving animals in which injury or death are the intended results. The AVMA recommends that spectator events involving animals, where injuries could occur incidentally to the practice of the sport, should be practiced in a manner that minimizes injury. These sports would include dog racing, dog sled racing, cutting and reining exhibitions, and field trials. The AVMA recommends that all rodeos abide by the rules to ensure the humane treatment of rodeo livestock, such as the rules established by the Professional Rodeo Cowboys Association and the International Rodeo Association. The AVMA further recommends that all organizations involved with animals used in spectator events should develop appropriate guidelines or standards to ensure the humane treatment of those animals. AVMA also recommends that the organization should support continued research into sport animal medicine to reduce injury of contestants and animals.

PREGNANT MARE URINE (PMU) HORSES

For more than 40 years, farms in North Dakota have produced pregnant mare urine (PMU) as a byproduct of the equine industry. PMU has been prescribed extensively by gynecologists to women enduring complications resulting from menopause (among other medicinal problems). The urine from pregnant mares contains estrogen during the later part (fourth month and beyond) of pregnancy. Wyeth-Ayerst, a pharmaceutical company headquartered in Philadelphia, buys the urine and extracts the estrogen, making the commercially available *Premarin* ®. Premarin, in 1995 alone, brought in close to $800 million in sales (source: NAERIC).

PETA claims that plant-based drugs could achieve the same result for hormone replacement therapy in women. Most researchers do not agree, however. No synthetic or plant substitute exists that is identical to Premarin, according to the nation's most-respected research scientists. A PETA undercover "investigator" collected video footage at several North Dakota ranches in 1994. The film, according to PETA, shows all manner of horse mistreatment. PETA also claims that the horses are abused during the collection process.

A mare's gestation lasts approximately eleven months. Urine is collected from month four through month ten. Each mare spends six months in a collection barn (from October 1 through the end of March). During this time, North Dakota is blanketed with snow. If the horses were not used for collection purposes, they would still be kept in these barns during this time. All the barns are well ventilated.

During the rest of the year, the horses are pastured. Mares receive drinking water eighteen times per day. Amounts are regulated for maximum urine production and for maximum health of the mare. Water is not restricted; rather, it is instead evenly distributed throughout the day. All horses are fed according to nutritional guidelines twice a day for grain and five times a day for hay. Health records are kept on every horse. Each mare is exercised a few hours every couple of weeks.

Animal rights activists claim that the horses being used are treated inhumanely. 49,000 horses exist on 488 PMU farms that are in contract with Wyeth-Ayerst. The farmers have a code of ethics they are required to follow, with frequent inspections. Many of the complaints of animal rights activists about the way the horses are housed are answered by the farmers themselves, the AAEP, and the North American Equine Ranching Information Council. You can review their findings and publications in Section 2 of this text.

MEDICATION OF RACEHORSES

Many horse trainers, similar to human athletic trainers, give their animals pain-killers such as phenylbutazone (bute) and corticosteroids, instead of giving them time off to rest. Horses that continue to run on pain-killers cannot feel how sore they are and can suffer serious injuries as a result. Bute a *non-steroidal, anti-inflammatory medication* (NSAIM) and corticosteroids are prescribed by veterinarians for the treatment of minor injuries sustained by athletic horses. These prescriptions are similar to the medicines given to human athletes for the same reasons. The primary function is to reduce inflammation. As a result of the reduced inflammation, the pain is also lessened. An overuse of steroids can lead to a weakening of the support tissues, such as cartilage and ligaments of a joint. This weakening can cause injury to a horse during strenuous exercise and should be avoided.

Veterinarians should prescribe corticosteroids only for "the therapeutic treatment of specific medical conditions." The administration of NSAIMs and corticosteroids might be preferable to a prolonged period of inactivity for a horse. At the racetracks, these drugs are prescribed by licensed equine practitioners who must comply with state regulations regarding administration, timing, and dosage.

RACING TWO-YEAR-OLDS

A study published in 1990 by *Thoroughbred Times* cites a survey which listed all the horses suffering catastrophic injuries in North America between January 1, 1989, and November 18, 1990. Two-year olds were among the 1,354 horses ages two and older that suffered injury (4.5 percent). In the ten age categories studied, two-year-olds ranked sixth in incidence of catastrophic injuries. Of all two-year-olds that raced in 1989 and 1990 (a total of 29,592), the 61 injuries represent .002 percent of injuries.

The two-year-old population represents approximately 16 percent of the racehorse population in a given year.

According to 1992 figures compiled by The Jockey Club, two-year olds as a group fell into the lowest category for total earnings ($89,672,798), for percentage of all earnings by racehorses (11.6 percent), and for average earnings ($6,613), despite ranking third in number (13,561 and 16.3 percent of all runners in the five age categories studied).

The cost of training a two-year old horse is approximately $17,500 per year. There is little economic incentive to race two-year-olds, given the cost of their training and their relative low earning power; however, there is one extremely important benefit: the experience. Two-year olds are still growing at an extremely rapid rate. Inexperienced trainers pushing two-year olds do not allow the animals' bones to strengthen sufficiently during this time period. Perhaps waiting an additional year would be beneficial to the well-being of the animal. It has been suggested that the equine racing industry consider adopting this additional rule in their animal welfare guidelines (waiting until age three to race).

POST-RACING USE OF HORSES

A large percentage of race horses, particularly fillies and mares, retire from racing and begin living as broodmares. Others retire to second careers that include hunters and jumpers, three-day event horses, polo ponies, and pleasure horses. Some are simply placed in retirement homes. Small numbers are ultimately slaughtered for human consumption outside the United States (greater than one percent). The horse industry supports the introduction of legislation and the enforcement of current laws guaranteeing the humane transportation of horses to federally inspected processing plants.

GUIDE, HEARING, AND WORKING DOGS

Guide dogs assist the physically challenged. Beagle Brigades protect our borders from contaminated food and disease, and they also sniff out illegal drugs. Animals are used in therapy for handicapped and elderly individuals. Therapists use horses, dogs, lambs, and other animals in therapeutic ways for their patients. Police departments use dogs as "quasi" officers. These working "officers" save many lives by hunting violent suspects, lost children, and explosives.

PRIMATES FOR PARAPLEGICS

One of the best ways students might gain an understanding of the benefits that primates provide to human paraplegics is to have a paraplegic come to class with his or her primate assistant. It is difficult for most of us who have two hands and two feet that work to understand the frustration of not being able to take care of our basic needs. Animal activists have condemned the use of primates for such activities. Once again, this area requires much discussion. Should we allow this use? Are we morally correct in preventing it? If we permit it, to what degree would we do so? What requirements, if any, should be placed on the human benefactor? What do we do with the primate assistant when he or she becomes too old, or when his or her services are no longer needed (when the human benefactor dies)?

ANIMALS IN MOTION PICTURES AND TELEVISION

Animals that are used in a film or on TV are usually safe from harm since most animals are handled by trained professionals who have a vested interest in ensuring the health and welfare of the animals in their care. Thanks also to the efforts of the AHA, animals are protected from abuse during filming. Throughout the filming, the treatment of animals is monitored by an AHA representative to assure compliance. The AVMA encourages the humane use of animals by the entertainment media, and strongly opposes the mistreatment or abuse of animals in any manner. The AVMA commends and supports the AHA for its efforts to prevent cruelty to animals in the production of motion pictures in the United States.

Many animal activists claim that this is not enough. They claim that animals are demoralized when asked to perform "embarrassing" tasks. Animals are thought to be "used" and "abused" because they earn millions of dollars for their "masters" at the expense of this humiliated pet.

Most Hollywood movies today have a disclaimer at the beginning and/or end of the film informing the viewing public that "no animals were hurt as a result of the filming of this movie." Through computer-generated graphics, trick photography, and other techniques, animals do not have to be blown up, sent over a cliff, or eaten by a predator to demonstrate the same action on film.

A few years ago during the Super Bowl, PETA protested an Anheuser-Busch commercial shown during halftime. The commercial showed horses playing football. It was all computer-generated, so why the protest? PETA claimed the commercial was denigrating to the horse.

Chapter 11
EVALUATION

Case Studies: *Permission to reprint has been granted by the editor of The Ag Bioethics Forum.*

Case #11–1

Peter Kirk is a midwestern farmer who hunts deer every fall, just like his father and grandfather before him. Each year, Kirk spends several weekends afield, tracking deer and eventually killing one. He cures and freezes the venison, which is treated as a delicacy in his family.

Case #11–2

Rick Pearson is a National Park Service ranger in Yellowstone National Park. One of his jobs is to serve as a marksman during yearly culling of the northern elk herd, which has repeatedly exceeded the carrying capacity of its range. After extensive attempts at trapping and relocation, the park service killed about 4,000 animals (upwards of one-third of the population at the time) during the winter of 1961–62. The hunt was staged in the winter, when the animals are concentrated at lower elevations, and used park service marksmen, and the carcasses were processed on the scene, with the meat being given to area Native American tribes. The initial herd reduction caused a public outcry because the methods seemed so unsporting or cruel, but an influential government report endorsed the technique. Now each winter, several hundred elk are shot under similar circumstances, which has stabilized the population within the carrying capacity of its range.

Case #11–3

Howard Stancer is a successful Hollywood actor who has traveled the world for years trophy hunting. He has heads of nineteen big game animals on his wall and needs only the head of a rare cat to have completed the prestigious "big 20" of trophy hunting. So this summer he is going to a game ranch in central Texas and paying $3,500 to be guaranteed a cat of the species in question. These cats, along with a dozen other exotic species, are bred on the ranch in one-acre pens for the purpose of such "canned" hunts.

QUESTIONS

1. What conclusion would you personally reach about the moral permissibility of hunting in each of these cases? Why?

2. Would you have the same conclusion for Native Americans performing the hunting in case #11–1? Why or why not? If your answer was different for Native Americans, would you have a different conclusion for someone who is half-Native American? How about one-quarter? one-eighth? At what point does this become a racial conclusion?
3. Read and discuss "Living in Harmony with Nature" (PETA factsheet #7). Can most people adopt these suggestions? Would most people want to adopt them? Why or why not? What would you do (if anything) to change this outcome?
4. Read the articles on "Checks and Balances on PMU Ranching" (Section 2) and comment on the sentence, "is that enough to guarantee humane treatment of the horses being used?" Is using the horse in this manner worth the medical benefit to the millions of women who use (need) this product (Premarin)?

Recommended Readings: (From Section 2)

1. *AAEP Officials Inspect PMU Farms*. AAEP Report.
2. *Checks and Balances in PMU Ranching Ensure High-Quality Care for Horses*. NAERIC.
3. Greer, T. *Canada and U.S. Join forces in the War on Animal Rights*. Canadian Rodeo News, Jan. 1995.
4. ——. "Rights stunt backfires." Canadian Rodeo News, Feb. 1995.
5. "Independent Equine Veterinary Practitioners to Conduct 1,450 Herd Health Reviews in 96-97 Seasons."
6. *Living in Harmony with Nature*. PETA.
7. "NAERIC Updates Equine Welfare Committee at Largest AAEP Convention Ever."
8. "North American Equine Ranching Information Council (NAERIC) Seeks to Inform, Educate Public About Mare, Foal Husbandry."
9. North American Equine Ranching Information Council State of Purpose.
10. "PMU Horses Command Record Prices at South Dakota Livestock Exchange."
11. "PMU Ranchers to Benefit from Breeding Enhancement Program."
12. "PMU Ranching's Veterinary Care Standards Surpass Those of Private Ownership."
13. *Recommended Code of Practice for the Care and Handling of Horses in PMU Operations*.

CHAPTER 12

America's Newest Extremists: The Animal Rights Movement

SPECIEISM, RACISM, AND ANIMAL ABUSE

"The ultimate objective of the (animal) rights view is the total dissolution of the animal industry as we know it."

Tom Regan (in Pringle's *The Animal Rights Controversy*)

Tom Regan, Peter Singer, and other philosophers have argued that "speciesism," or the use of species to determine membership in the moral community, is no more morally justifiable than using race, sex, or age to determine who has rights and who does not. William M. Kunstler, Esq., co-founder of the Center for Constitutional Rights, has stressed that speciesists might exploit non-humans simply because humans are more powerful and they believe we will benefit from that exploitation. He further states that discrimination against other disadvantaged groups becomes much easier if one is a speciesist.

OBJECTIVES

The student should:

1. Describe the goals of each of the major animal rights organizations.
2. Discuss the impact of these goals with modern every day life (what we eat, wear, drive).
3. Discuss man's inhumanity to man and compare it to animal abuse charges levied by many leading animal rights organizations.
4. Discuss animal rights terrorism and relate it to "the ends justifying the means."

5. Discuss what would happen if all animal research suddenly stopped. Discuss who would be affected most and how they would be affected.

Our current legal system tries to resolve human and animal conflicts by balancing human and animal interests. In doing so, we seek to compare the incomparable. Human interests are protected by claims or right; however, animals are regarded as property under the law and are thus not regarded as capable of having similar rights. What the law does allow is to judge whether the treatment of animals is humane. Kunstler and others compare laws that supposedly protected slaves from abuse by their masters to laws that supposedly protect animals from their owners. Kunstler claims that in both cases the legal system was and is more interested in serving the interests of the powerful than it is in providing justice to the disempowered. Animal activists further claim that the legal system does not enforce the laws currently in place to protect animals. Although many criminals do go free because of the protections afforded criminal defendants, the same is true for *all* crimes, not just those relating to animal welfare.

PETA'S STATEMENT ON COMPANION ANIMALS

> ***"Although most of us love the animals who share our lives, we treat them with a kind of benign neglect. For instance, we leave them at home while we work, go shopping, or go to school, and when we return we feed them and take them out briefly or clean out their litter boxes, but often neglect them for most of the time we're at home, while we prepare a meal, talk on the telephone, watch television, write letters, sleep, etc. Companion animals eat when and what we want them to eat, and go out or stay in when we want them to. As John Bryant has written in his book*** **Fettered Kingdoms*****, they are like slaves, even if well-kept slaves. Many guardians fail to treat their companion animals with kindness and respect, leaving them outdoors in all kinds of weather, neglecting to provide fresh water at all times, etc."***
>
> PETA's Position on Companion Animals

How extreme do some of the animal extremists go? PETA is opposed to the use of mosquitoes in medical research. Roberta Wright, a PETA official, while interviewed by the *Arizona Daily Star*, was quoted as saying, "Why wouldn't I be (concerned about cruelty to mosquitoes)? They are a part of this planet, too. We don't have any right to take a being we have deemed as inferior and exploit it for our purposes."

In fact, there is more enforcement of cruelty to animals than cruelty to children within our current legal system. In July 1995, Raphael Diaz Rodriguez, a North Hollywood, California resident, was arrested for killing a pet rabbit and for choking his ex-girlfriend. He faced a one-year jail sentence and a $20,000 fine on the cruelty charge, and a one-year and $1,000 fine for the battery charge. Plea bargaining, Rodriguez pled guilty to the battery and was given ninety days in jail plus three years probation. Domestic violence advocates have used this and other similar cases to demonstrate the great disparity in punishment for those who hurt animals and those who hurt humans.

In the July 14, 1997 issue of the *Los Angeles Times*, Arianna Huffington writes a guest column comparing the mourning for a dog and the mourning for two dead boys. The dog, a K-9 cop, was killed in the line of duty. At the dog's memorial, 800 people showed up, including police officers from all over the state of Florida. The dog received a twenty-one-gun salute, and a flag was presented to the dog's partner. Four days after the K-9 cop was killed, two young Native American boys (ages 3 and 5) were strapped into the back seat of a car and were driven into the Everglades to drown. The father of the boys killed them after having a fight with the boy's mother. Huffington points out that there was no outpouring of support for the boy's mother, no large crowds of mourners, and no outcry by the public over the death of these two young children from the same community or neighboring communities that gave such support for the dog.

Earlier in this text, a scene in early 20th century South Africa described the value of a human being (a black man) being placed below the value of an animal (a dog). Have we not changed our value system in 100 years? Can we still justify giving greater value to a pet than certain classes or races of people? If so, what does this statement say about our society?

The term speciesism implies that we do not believe that "a dog is a mouse is a chicken is a pig is a person." This phrase was coined by Ingrid Newkirk to demonstrate that we are all equal, and thus we all deserve the same protection under the law.

> ***"We even oppose the use of honey, because bees die in the process of gathering it."***
>
> Carol Burnett, PETA (*Marin Independent Journal*, June 25, 1989)

Pet ownership is an "absolutely abysmal situation brought about by human manipulation," says Ingrid Newkirk of *Washington Magazine* (August 1986). Newkirk continues this discussion in the August issue of *Harper's Magazine* with "I don't use the word 'pet.' I think it's speciesist language. I prefer 'companion animal.' We would no longer allow . . . pet shops . . . Eventually companion animals would be phased out." Is this extremism thought? If so, it is from the co-founder and head of PETA, one of the nation's largest animal rights organizations.

> ***"Even painless research is fascism, supremacist, because the act of confinement is traumatizing in itself."***
>
> Ingrid Newkirk, PETA (*Washingtonian Magazine*, August 1986)

MAN'S INHUMANITY TO MAN

More than 30 years ago, one learned professor lived in a country known as Biafra. The country no longer exists (it was previously located in the eastern region of Nigeria). There, as a young child, he saw what his grandfather had described to him occurring in South Africa two generations before him, what his grandmother described to him occurring in Nazi Germany, and what his other grandfather described to him that occurred to American Indians just a century earlier. He witnessed the beating, the torture, and the brutal murder of men, women, and children. These individuals were murdered because they dared to be born to a different tribe (Ibos instead of Hausas). Never did he see such treatment against animals. In fact, if a person were to harm a vulture (a protected animal in Biafra and Nigeria) he or she would be severely punished. It was acceptable to kill for food, but never to torture. It was overlooked (and thus permitted), however, to torture a member of another tribe.

Fifty years ago, the Nazis adopted laws protecting animals. Hitler, himself a vegetarian, felt it was inhumane to experiment on animals, to kill them for food, etc. Instead, Hitler and others who worked for him assigned a lesser value to groups of human beings who were thus experimented on, tortured, and brutally murdered, eleven million people in all (see Figures 12–1 and 12–2). Animal life was much more valued than the lives of Jews, gypsies, blacks, homosexuals, those who were crippled, and others he deemed "undesirable."

What have we learned since the time of the American Indians' exploitation, Nazi Germany, and Biafra? How about what happened more recently in Cambodia, Bosnia, Rwanda, and Kosovo and the violent and sometimes racist crime in America?

Ingrid Newkirk said, "People complain about six million Jews being killed in the Holocaust when six billion chickens are slaughtered every year in the United States . . . homelessness drives me crazy! I have more sympathy for animals because they don't deserve anything that happens to them."

How can anyone compare a homeless child to a homeless dog or cat, or the murder of six million human beings to that of a chicken? And yet, the founder of one of the largest animal rights groups in America does so. We need to care about our homeless pets. That is why many people have

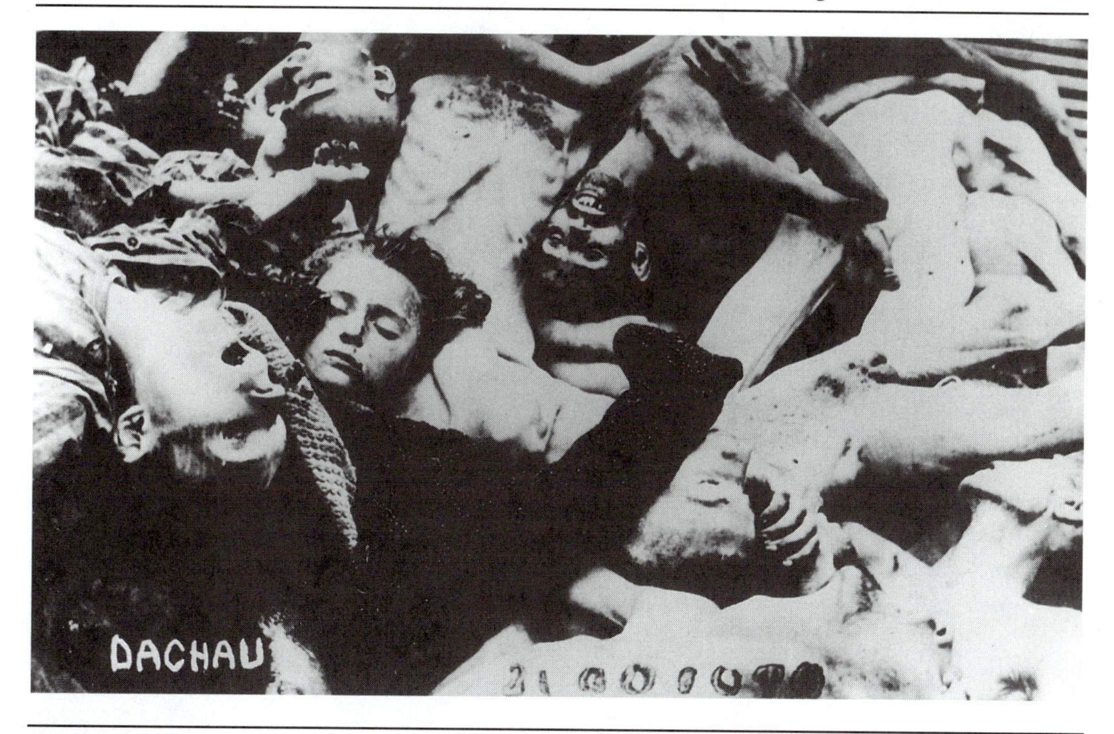

Figure 12–1 Dachau. *Courtesy of the Simon Wiesenthal Center Library and Archives*

suggested taking a large part of the $50 million that the HSUS and PETA collects every year and putting that money toward helping these "unwanted animals." At the same time, most Americans find it morally abhorrent to assign more worth to this unwanted pet than to a homeless child or a victim of the Nazi's final solution. Others, including PETA cofounder Newkirk, feel that "humans have grown like a cancer. We're the biggest blight on the face of the Earth." (Reader's Digest, June 1990) Why are we *more* concerned about homeless animals than homeless children (see Figure 12–3)? The children did not choose to live in the elements, to eat trash, and to lie in filth. Why do we not hear an outcry for the children as we do for our companion animals and other animals? This question is a moral question that must be asked and should be discussed in all ethics courses.

TERRORISM —DAMAGE TO RESEARCH AND SOCIETY

The highest price resulting from animal research sabotage is not the dollar loss, but rather the millions of people and animals suffering from diseases and ailments for which there are no cures or treatments.

Figure 12–2 "We have not forgotten." *1939-1945 (The Holocaust)*

Figure 12–3 "Stray dog at garbage dump."*Courtesy of PETA*

From 1982 to 1990, more than 100 criminal acts were committed against animal-research labs. Twenty-three percent of these crimes involved vandalism, seventeen percent were bombings or other threats, and ten percent were cases involving actual or attempted arson, attempted murder, bombing, or firebombing. Credit for some of these acts has been claimed by the ALF. In addition, there have been thousands of threatening letters and phone calls sent to researchers.

In 1985, as part of a post-graduate assembly in surgery at the University of California, San Diego, Dr. Abdool Moossa, MD, scheduled a course on the use of surgical staplers. Surgeons who wanted a practical course in the use of the stapler would practice stapling techniques on eighty dogs. This procedure would speed operating room procedures and allow doctors to see more patients, thus saving money for the patient and perhaps saving lives. Dr. Moossa received obscene calls, letters, and threats, including: "Either Dr. Moossa stops the course, or I will shoot him in the head." Dr. Moossa has four children and a wife and did not want to risk their welfare; he thus canceled the course.

In April 1989, activists stole 1,231 animals from a laboratory at the University of Arizona at Tucson. Activists hindered some twenty research projects, smashed expensive equipment, set fire to two buildings, and caused an estimated $250,000 in damage.

Another laboratory raid at Texas Tech University in Lubbock caused $70,000 in damage and held up research on *Sudden Infant Death Syndrome* (SIDS) for eight months.

At the University of California-Davis, arsonists set fire to a laboratory that was under construction, causing more than $4 million in damage. This laboratory was a diagnostic laboratory (not research). The individual was later identified as a Davis resident and a member of the ALF.

One of the more frightening incidents happened in England. Animal rights extremists bombed a car belonging to a London medical researcher. A 13-month-old baby was severely injured in the explosion. Alex Pacheco, co-founder of PETA, said that "arson, property destruction, burglary, or theft are acceptable crimes when they directly alleviate the pain and suffering of an animal." Pacheco told the *New York Times* that "we feel that animals have the same rights as a retarded human child, because they are equal mentally in terms of dependence on others." Pacheco defended the use of violence when used for the "animal cause." Once again, the value of animals was placed higher than the life of the innocent 13-month-old child. Is this situation morally abhorrent?

ANIMAL EXTREMISM—CRUELTY TO BOTH MAN AND ANIMAL

When free discussion stops, when only the bomb, the gunshots, and the fires decide what we do, what we say, and how and when we say it, then both man and animal have lost. What have we lost? We have lost our freedom to choose to live. Someone else has decided for us which diseases, if any, we will prevent and treat, and which surgeries we will or will not perform. America was built on freedom of choice. We agree that we disagree. We do the disagreeing peacefully and politely. When violent extremism dictates our choices, we all lose.

WHAT IF ANIMAL RESEARCH STOPS?

According to animal rights philosopher Tom Regan, if the scrapping of animal-based research "means there are some things we cannot learn, then so be it. We have no basic right not to be harmed by those natural diseases we are heir to."

The National Research Council spent three years and $315,000 studying whether animal research was necessary, because animals could now be replaced by artificial constructs and computer-generated methodology.

The council concluded that "animal experiments are still critically important to further improvements in medicine and biomedical science."

The report describes contributions that such experiments have made to the treatment of patients with polio, AIDS, cardiovascular and kidney diseases, and psychological illnesses, as well as those who require surgery of any kind, particularly organ transplantation. Because research with human subjects must be limited for obvious reasons, animals are needed for experiments in biomedical and behavioral research, teaching, and drug and product testing. For example, dogs are important to cardiology research, cats to research on blindness and deafness, cows to research on artificial heart valves, sheep for fetal development and birth defects research, and pigs for skin disease and burn research. Primates, especially chimpanzees, are critical to the understanding and treatment of AIDS.

If animal research were halted today, new surgical procedures to repair congenital heart defects will have to be abandoned or tried for the first time on children. A cure for diabetes will be beyond reach, there will be no hope for finding a safe and effective vaccine against AIDS, and development of techniques that may help restore function to paralyzed victims of spinal cord injuries will not continue. The 30,000 young Americans with cystic fibrosis will have little hope of a normal lifespan, the 250,000 people with multiple sclerosis will lose the promise of new treatments for the symptoms of this degenerative disease, and thousands of schizophrenics will be institutionalized because of lack of understanding of the disease and its treatment. Methods to prevent many cancers will never be found, because theories about genetic and environmental causative factors cannot be tested in humans, and the improvement of hearing through electronic simulation of the inner ear might never benefit any of the 17 million hearing-impaired Americans. Testing breakthrough products such as artificial blood, which shows promise for saving the lives of critically injured accident victims will be too dangerous. Researchers will be unable to clarify the cause of Alzheimer's disease (without that knowledge, the prognosis for the three million Alzheimer's victims in America will remain bleak). The development of urgently needed, new drugs to treat heart disease, cancer, and a host of other diseases will also be severely curtailed.

The objective of the animal rights movement is to legislate laws and regulations that will ban all experiments and testing of animals. The movement questions whether research with animals is necessary for scientific and medical progress and whether all of the experiments and the animals used are justified and required. Animal rights groups question whether basic biomedical research has substantial clinical value. Many in the movement have advocated the abandonment of a search for an animal model for AIDS because there are so many patients upon which to experiment. They have also suggested using prisoners, the elderly, or the brain dead for research investigation. Those who have conducted any research, know from a scientific standpoint why this situation can never happen (because the researcher needs to eliminate too many variables when using unrelated human volunteers, as compared to genetically similar mice). Secondly (or primarily), we must consider the moral dilemma of using humans as research models (similar to the acts in Nazi Germany and Tuskeegee).

To accept the philosophical and moral viewpoints of the animal rights movement, we would have to enact a total ban on the use of animals in biomedical research. The consequences of such an act would certainly slow the pace of health-related research and impede any new procedures and technologies from reaching clinical medicine. Among other benefits, research on animals has led to vaccines to protect cats, cattle, dogs, horses, pigs, poultry, and sheep against a spectrum of diseases ranging from anthrax to Venezuelan equine encephalomyelitis. The knowledge transfer works both ways. Animal researchers found that the drug that kills parasitic worms in livestock and heartworms in dogs also attacks the tiny worms that cause river blindness among millions of Africans. Because of that discovery, Merck & Co., a pharmaceutical company headquartered in New Jersey, has donated its personnel and products to greatly reduce the suffering of millions of humans in the jungles of Africa.

Although the AMA supports regulatory policies that protect animals from unnecessary pain or inappropriate use, pain and suffering in both animals and humans will increase if policies advocated by animal rights groups are adopted. Therefore, the AMA and AVMA believe that we must balance the cost of such protection in terms of animal and human health.

In 1990, the incurably ill For Animal Research (iiFAR) staged a protest in front of PETA headquarters in Maryland. Television coverage showed iiFAR signs saying, "Animal rights is human hate," "Animal Research Saves Lives," and "Thanks to animal research, they'll be able to protest 20.8 years longer." Of course, this and other pro-research group are small in comparison to the 400-plus animal rights organizations.

The ultimate question that Americans must answer was framed by Harvard physiologist Walter B. Cannon in 1908, during the first of several periodic revolts against the welfare of the sick:

> ***"Shall men suffer and die to save the lives of the experimental animal, or shall the experimental animal die to mitigate pain and wide calamity among men?"***

Recommended Readings

1. *Activists Call for Direct Action Against Research*. AMPEF.
2. *Animals' Rights: What's at Stake*. AMPEF.

3. *Animal Rights Extremists: Impact on Public Health.* NABR.
4. Is This Really What You Believe?
5. Loeb, J. et al. *Human vs. Animal Rights. In Defense of Animal Research. JAMA* Vol. 262(19):2716-2720. 1989.
6. Morrison, A.R. "Understanding (and Misunderstanding) the Animal Right Movement in the United States." *Animal and Human Experimentation.* 1993.
7. PETA factsheets (eleven of them are found in Section 2)
8. "Public Health and the Role of Animal Testing." USPHS. 1992.
9. Review iiFARsighted Reports (1988, 1990).
10. *Some Call It Protecting Animal Rights. The Fact Is It's Terrorism.* AMPEF.
11. *The Animal Rights Movement...In the Words of Its Leaders.* AMPEF.
12. *21 Things You May Not Know About the Animal Rights Movement.* AMPEF.
13. *We're going to expose the naked truth about PETA tomorrow.* CBRA.
14. World Medical Association Statement on Animal Use in Biomedical Research. 1989.

Reprint Articles and Miscellaneous Reference Material from All Sides

American Medical Association
Physicians dedicated to the health of America

Use of Animals in Biomedical Research

The Challenge and Response

An American Medical Association White Paper

1989
1992 (revised)

Preface

Since the late 1800s, the American Medical Association (AMA) has consistently supported the humane and responsible use of animals for biomedical research. Research involving animals is essential to improving the health and well-being of the American people, and the AMA actively opposes any legislation, regulation, or social action that inappropriately limits such research. Most Americans support the use of animals in research but want assurance that animals are treated humanely and used only when necessary. Over the years, animal activists have exploited this appropriate concern for animal welfare and have attempted to impede or stop biomedical research with animals. As a result, some Americans believe that the abuse of laboratory animals is common. Recently, the animal "rights" movement has been making substantial inroads in obtaining philosophic and financial support for legislative and regulatory changes that would compromise the future of biomedical research.

To protect this future and to clarify the issue of animals in research for the American public, we have prepared the accompanying White Paper: Use of Animals in Biomedical Research—The Challenge and Response. In this document, we describe the importance of animals to past, present and future biomedical research and medical practice. The history and philosophies of the animal rights movement are considered together with arguments raised by animal welfare groups and responses by the scientific community. We hope that our approach, written dispassionately and devoid of the rhetoric usually associated with this topic, will help the reader understand the complexity of the issues and the need for thoughtful choices by individuals and society.

For further information concerning the use of animals in biomedical research, contact:

Group on Science and Technology
American Medical Association
515 North State Street
Chicago, IL 60610
1 800 621-8335

I. Executive Summary

The use of animals in biomedical research has been contested on various grounds for decades. With the recent advent of the animal "rights" movement, attempts to limit or abolish experimentation with animals have been increasing. The moral and ethical arguments raised by animal rights proponents have placed scientists in a defensive position. Public education initiatives by scientists to explain the use of animals in biomedical research and to demonstrate that research animals are treated humanely have been sporadic and generally ineffective, and animal activist groups have made substantial progress in winning public opinion and financial support for their cause.

Scientists believe that both basic (ie. laboratory) and applied (ie. clinical) research require the use of animals for continued medical progress. Animals are used in experiments for biomedical and behavioral research, for teaching, and for drug and product testing. One major objection raised to these experiments is that animals are used as surrogates for human beings. Scientists argue that this use is necessary because research with human subjects must be limited for obvious ethical reasons.

In general, the animal cause movement can be divided into two broad groups: 1) individuals concerned with animal welfare are not necessarily opposed to biomedical research using animals but want assurance that the animals are treated as humanely as possible, that the number of animals used is the minimal required and that animals are used only when necessary; and 2) individuals concerned with animal rights who insist that animals have moral rights equal to those of humans and who are totally opposed to biomedical research using animals. People for the Ethical Treatment of Animals (PETA) and the Animal Liberation Front (ALF) have been the most vocal and visible groups associated with the animal rights movement. The Physicians Committee for Responsible Medicine (PCRM), a group that has often been associated with PETA, is also a prominent opponent of biomedical research using animals. PCRM was denounced by the House of Delegates of the American Medical Association in 1990 for deliberately misrepresenting the critical role animals play in medical research.

Most of the arguments used by animal rights and animal welfare groups today have historic counterparts. A resurgence of the antivivisection movement in the last 20-25 years has been fueled, at least in part, by publication of a series of books that raise moral arguments about the quality of all forms of life. In these books, the concepts of animal rights and speciesism are placed into a contemporary framework to serve as a polemical call to action of animal advocacy groups world-wide. The number and size of these groups has been growing rapidly, and current estimates suggest that more than 1,000 groups exist. According to The Nonprofit Almanac 1992-93, 0.9% of the US charity dollar, or $1.6 billion, goes to animal-related causes. In contrast, funds for the defense of animal research are minimal. Animal rights activists have been successful on multiple fronts using tactics that include publicity, lobbying of elected and appointed officials, public demonstrations and petitions, threatening and terrorizing researchers and administrators and their families, and causing physical damage to property and research data. For example, a lab under construction at the University of California at Davis was burned in 1987, with damage estimated in excess of $3 million. In 1989, three buildings at the University of Arizona (Tucson) were vandalized and heavily damaged by a series of fires set by a radical animal rights group and more than 1200 animals were stolen from the university.

The objective of the animal rights movement is to achieve laws and regulations banning all experiments and tests using animals. Bills to prevent the use of animals in biomedical research have been introduced in several state legislatures. Other techniques have included attempts to divert federal monies allocated for biomedical research to development of so-called alternatives to such experiments, gain legal sanction to act on behalf of animals, and secure the right to examine all research proposals to determine whether, in their view, the research is necessary or might cause pain to animals.

Animal rights activists question whether research with animals is necessary for scientific and medical progress, and if all of the experiments and the animals used are justified and required. Scientists feel that the contributions made by animals to human and animal health are overwhelming; 60 of 82 Nobel prizes awarded in Physiology or Medicine since 1901 have been made for discoveries involving the use of animals. These discoveries include important breakthroughs in diagnosis, treatment and prevention of many conditions: heart disease, cancer, poliomyelitis, diabetes, measles, smallpox and massive burns. Many procedures, unthinkable 50 years ago, are today commonplace: open heart surgery, blood transfusions, organ transplantation, laser neurosurgery, etc. Each of these technologies was developed entirely, or at least in part, by experiments with animals. Animal activists claim that these advances are relatively insignificant, because they believe most major advances in life-span and health are attributable almost exclusively to improvements in nutrition, public health and sanitation.

Animal rights groups question whether basic biomedical research has substantial clinical value. Scientists answer that while all research might not have immediate value, most clinical breakthroughs are based upon multiple fundamental studies. Moreover, the process of serendipity (making a discovery that was unanticipated) has proven time and again to be crucial in clinical medicine. Animal rights activists also argue against research designed to "prove the obvious", and against experiments that repeat studies already conducted. Scientists counter that the scientific method necessitates experimental proof for all new findings, and that experiments are not judged valid until results are confirmed by others.

Although the exact number of animals used in biomedical research is not known, the best estimates are that 12-15 million animals are used each year, with rodents contributing 75-90% of the total. Scientists agree that the number of animals used in biomedical research should be reduced wherever possible, and there is evidence that this is occurring. For example, substantial decreases have been reported in the number of animals used in drug and product testing. However, complete abolition of animal use for this purpose must be considered in light of the need to protect the public from dangerous and toxic products.

Many animal rights advocates are especially concerned with perceived abuses of primates by medical researchers. While an occasional abuse has been documented, there is substantial evidence that essentially all primate research is conducted ethically and that federal, state, local, and institutional guidelines are being followed. Research involving primates is critical to understanding and treating one of the most vexing health problems in centuries—AIDS. Recently, one prominent AIDS researcher noted that vaccine development was being hampered by the inability to obtain chimpanzees—a fact he attributed directly to efforts of animal rights activists. Both the primate research community and Congress have been sensitive to concerns raised by the animal rights community. The Animal Welfare Act passed in 1985 has resulted in alterations in physical facilities requested by animal activists. Scientists agree that these changes have improved primate facilities, although at a substantial financial cost to institutions.

The concept of "alternatives" to animal use is a major focus of animal rights advocates. Specifically, replacement of animals with computer simulations and isolated cell cultures has received considerable support from these groups. Scientists avoid the use of animals wherever possible both because scientists do not wish to inflict suffering on animals, and because animals are expensive to purchase and maintain. Cell cultures and computer simulations do allow, for the most part, a reduction in the number of animals used, however, they are not useful alternatives to animal use but rather, serve as adjuncts to their use. Efforts have been made by animal rights activists to divert funds earmarked for research into efforts to develop alternatives to such research. Federal bills to prohibit research with animals have been rejected by Congress; however, the Health Research Extension Act of 1985 requires the National Institutes of Health to promote methods for conducting, evaluating and disseminating research that does not require animal use.

Both animal welfare and animal rights groups have repeatedly sought the adoption of laws and regulations that establish standards for the care and treatment of laboratory animals, and scientists have consistently supported such efforts. Many new requirements are now in place, and most scientists believe that these changes have improved the care and environment for lab animals. More stringent new rules are constantly promoted by animal activists that scientists feel would have a negative impact upon future research. For example, proposals to ban the use of animals obtained from public animal welfare shelters (pounds) will significantly increase the cost of research by requiring specific purpose-bred animals for research. Ironically, these proposals would result in the death of two animals—the research animal and the unclaimed pound animal. In public referenda, voters have consistently supported the release of pound animals to research labs. Hence, scientists question whether implementation of these laws represent what the American people want. However, such laws have now been passed in a number of states, and additional bills are presently under consideration.

Conclusion

The activities and arguments of animal rights groups raise difficult moral and ethical questions. To accept the philosophical and moral viewpoint of the animal rights movement would require a total ban on the use of animals in biomedical research. The consequences of such a step would certainly slow the pace of health-related research and impede new procedures and technologies from reaching clinical medicine. Laws and regulations directed toward improving the conditions under which biomedical experiments that use animals are conducted have had a measurable impact. The AMA supports regulatory policies enacted to protect animals from unnecessary pain or inappropriate use—however, pain and suffering in human beings will occur if policies advocated by animal rights groups are adopted. Therefore, the AMA believes that what must be recognized and weighed in the balance is the cost of such protection and comfort in terms of human health.

II. Introduction

During the past 15-20 years, a major controversy over the use of animals in biomedical and behavioral research has arisen. At the leading edge of the controversy are groups of animal "rights" activists whose purpose is to put an end to all experimentation and testing with animals. Their methods range from the use of publicity, misinformation and the filing of lawsuits, to raiding laboratories and threatening and intimidating scientists and their families.

Traditional animal welfare and humane societies also have been active. Through legislation, regulation, negotiation and intimidation via publicity, they generally have sought to limit and control experiments conducted with animals, to reduce the number of animals being used, or to exempt certain animals from use in biomedical research.

Representatives of these animal cause groups appeal to the emotions and raise ethical or moral questions. They question human's right to use animals in a number of contexts, including biomedical research; contest the value of animal research and the need to conduct certain types of experiments; challenge and generate concern over the amount of pain and suffering endured by animals during experiments; and promote the use of alternative methods of experimentation in place of the use of animals.

The activities of these organizations have grown in scope and intensity during the past 10 years and have had a number of important consequences. They have created confusion and doubt in the minds of some regarding the need for animal experiments, prompted the passage of federal and state laws regulating or restricting the use of animals in research, and led to the destruction or termination of many research projects.

The impact that these organizations have had is a matter of great concern to scientists. Many scientists perceive that, while most Americans continue to support the use of animals in research, they do not fully appreciate the role and importance of such research in scientific and medical progress. They also believe that most persons do not understand the implications of the legislation and regulations that have been passed or proposed to resist and regulate biomedical research with animals. There also is a concern that many Americans have been misinformed or misled by the arguments that animal rights activists have advanced against research using animals.

Scientists believe that the issue is not animal rights or the welfare of individual animals used in biomedical research. They believe instead that the controversy should focus on the human and animal welfare that depends on continuing medical and scientific progress. This welfare requires that researchers be able to conduct experiments, including those utilizing animals, when necessary. Scientists are convinced that restrictions placed on that ability, no matter how well intended, will impede such progress.

The arguments and activities of the animal rights organizations and the concerns they have aroused among those engaged in scientific research ultimately must be resolved by the American people, either directly at the polls through referenda, through their elected and appointed representatives or through the courts. The ways in which the issues are resolved will affect the future development of medical knowledge; hence, the AMA believes that there is a crucial need for the American people to understand the challenge to research using animals and to become fully informed on the various questions and issues.

This White Paper describes the role research with animals plays in medical science, identifies the nature of the challenges that have been raised, and examines the various issues that have been presented by the animal rights and animal welfare advocates. It concludes by presenting the choices that face the American people, and the consequences of decisions among those choices.

III. Use of Animals for Scientific Research

Animals have been used in experiments for at least 2,000 years, with the first reference made in the third century BC in Alexandria, Egypt when the philosopher and scientist, Erisistratus, used animals to study body functions.

Five centuries later, the Roman physician Galen used apes and pigs to prove his theory that veins carry blood rather than air. In succeeding centuries, animals were employed to discover how the body functions or to confirm or disprove theories developed through observation. Advances in knowledge made through these experiments included Harvey's demonstration of the circulation of blood in 1622, the effect of anesthesia on the body in 1846, and the relationship between bacteria and disease in 1878.

A. Types of experiments that require the use of animals

Today, animals are used in experiments for three general purposes: (1) biomedical and behavioral research, (2) education, (3) drug product testing.

Biomedical research increases our understanding of how biological systems function and advances medical knowledge. Biomedical experiments are conducted in accordance with the principles of the scientific method developed by the French physiologist, Claude Bernard, in 1865. This method established two requirements for the conduct of a valid experiment: (1) control of all variables so that only one factor or set of factors is changed or at a time, and (2) the replication of results by other laboratories. Unless these requirements are met, an experiment is not considered scientifically valid. Behavioral research is a type of biomedical research that is directed toward determining the factors that affect behavior and how various organisms and organs respond to different stimuli. Much behavioral research is environmental in nature but some involves the study of responses to physical stimuli or manipulation of biological systems or organs, such as the brain.

Educational experiments are conducted to educate and train students in medicine, veterinary medicine, physiology, and general science. In many instances, these experiments are conducted with dead animals.

Animals are also employed to determine the safety and efficacy of new drugs or the toxicity of chemicals to which humans or animals may be exposed. Most of these experiments are conducted by commercial firms to fulfill government requirements.

As with all scientific research, biomedical research may be subdivided into two types: basic and applied. Basic biomedical research is conducted to increase the base knowledge and understanding of the physical, chemical, and functional mechanisms of life processes and disease. The aim of applied research is to attain specific, targeted objectives, such as the development of a new drug, therapy, or surgical procedure, and usually involves the application of existing knowledge, much of which is obtained by basic research. Applied research can be either experimental—that is, conducted with animals or some non-animal adjunctive method—or clinical—that is, conducted with human beings. Clinical trials are the last step in the biomedical research process and are performed only after the procedure, drug, or device being investigated has been thoroughly tested in experimental research conducted under guidelines designed to protect both patients and volunteers.

B. Use of animals rather than humans

A basic assumption of all types of research is that humans should strive to relieve human and animal suffering. One objection to the use of animals in biomedical research is that the animals are used as surrogates for human beings. This objection presumes the equality of all forms of life; animal rights advocates argue that if the tests are for the benefit of humans, then humans should serve as the subject of the experiments. There are limitations, however, to the use of human subjects both ethically, such as in the testing of a potentially toxic drug or chemical, and in terms of what can be learned. The process of aging, for instance, can best be observed through experiments with rats, which live an average of two to three years, or with some types of monkeys, which live 15 to 20 years. Some experiments require numerous subjects of the same weight or genetic makeup or require special diets or physical environments; these conditions make the use of human subjects difficult or impossible. By using animals in such tests, researchers can observe subjects of uniform age and background in sufficient numbers to determine if findings are consistent and applicable to a large population.

Animals are important in research precisely because they have complex body systems that react and interact with stimuli much as humans do. The more true this is with a particular animal, the more valuable that animal is for a particular type of research. One important property to a researcher is discrimination—the extent to which an animal exhibits the particular quality to be investigated. The greater the degree of discrimination, the greater the reliability and predictability of the information gathered from the experiment.

For example, dogs have been invaluable in biomedical research because of the relative size of their organs compared to humans. The first successful kidney transplant was performed in a dog and the techniques used to save the lives of "blue babies," and babies with structural defects in their hearts, were developed with dogs. Open heart surgical techniques, coronary bypass surgery and heart transplantation all were developed using dogs.

Another important factor is the amount of information available about a particular animal. Mice and rats play an extensive role in research and testing, in part because repeated experiments and controlled breeding have created a pool of data to which the findings from a new experiment can be related and given meaning. In fact, nearly 90% of animals used in medical research are rodents bred specifically for that purpose. Their rapid rate of reproduction also has made them important in studies of genetics and other experiments that require observation over a number of generations. Moreover, humans cannot be bred to produce "inbred strains" as can be done with animals; therefore, humans cannot be substituted for animals in studies where an inbred strain is essential.

Scientists argue repeatedly that research is necessary to reduce human and animal suffering and disease. Biomedical advances depend on research with animals, and not using them would be unethical because it would deprive humans and animals of the benefits of research.

IV. Challenge to Animal Research

Controversy over the use of animals in biomedical research is not new. Such use has been opposed on philosophic, scientific, and emotional grounds in Europe for at least 300 years and in the US for more than 100 years. Many current activities and arguments have historic precedents or counterparts.

The animal cause movement in the US has always been complex and diverse; the organizations involved continue to span a broad spectrum of attitudes, concerns, activities, relationships, and tactics. Historically, however, most groups within the movement have tended to be identified primarily as anticruelty or antivivisectionist in orientation. Today, the distinction is between "animal rights" and "animal welfare."

A. Animal welfare movement

Most of those concerned with animal welfare do not propose to ban the use of research animals. The primary efforts of animal welfare organizations in the area of animal research in the past 10 years have been in the passage of laws to ban experimentation with animals obtained from pounds and to improve the care and housing of animals kept in research institutions. They have been joined in those efforts by some animal rights organizations and they have had success in both areas.

B. Animal rights movement.

1. History. The animal rights movement is more complex and embraces a broader range of attitudes and activities. This movement had its genesis in three books published in the early 1970s. These books created the atmosphere and introduced the terms, concepts, and rationale for what has become the animal "rights" philosophy and movement.

The first of these books, Animals, Men and Morals by Godlovitch, Godlovitch, and Harris, was published in 1971 and was an anthology that revived and presented to a new generation many long-dormant thoughts and views regarding the relationship between humans and animals. One of these was the view that animals have rights, an idea first proposed in 1894 that had had little impact in its own time or in the intervening years. Animals, Men and Morals renewed interest within the intellectual community of the relationship between human beings and animals and led to the publication in 1975 of the two books that form the basis of the animal rights philosophy as it exists today. One book was Victims of Science, written by Richard Ryder. Ryder's contribution was the introduction of the concept of "speciesism" (a concept he equated with fascism as described later). The other book was Animal Liberation: A New Ethic for Our Treatment of Animals by the Australian philosopher, Peter Singer. It is this book that is generally considered to be the progenitor of the animal rights movement.

Singer placed the concepts of animal rights and speciesism in a philosophical framework and questioned human use of animals in a number of contexts, including as food and clothing. In so doing, he revived a debate that raged between Cartesian and Utilitarian philosophers from the 16th to 18th centuries and made it an issue of the 1970s and 1980s.

This debate was opened by the French philosopher, Rene Descartes, who defended the use of animals in experiments by insisting that because animals could respond to stimuli in only one way—"according to the arrangement of their organs"—they lacked the ability to reason and think. This, he said, made them similar in nature and function to a machine. Humans, on the other hand, through their ability to think and talk, were capable of responding to stimuli in a variety of ways. These differences, Descartes argued, made animals inferior to humans and, given their machine-like nature, humans might use them as they would use a machine, including as experimental objects. He also argued that animals can only learn by experience and not by "teaching-learning" that man uses; ie. humans do not have to experience everything (eg. touching a hot stove) to know whether it is harmful or not. On the other hand, animals do.

A response to this thesis was developed in the next century by the Utilitarian philosopher, Jeremy Benthem of England. "The question", said Benthem in An Introduction to the Principles of Morals and Legislation, "is not, can they reason? nor, can they talk? but can they suffer?" Answered in the affirmative and placed in an emotional context, that question has become the focus of antivivisectionists and animal welfare organizers since the 19th century. It also has served as the thesis of animal rights. Singer defined as a "right" any claim that a being can make to have its "interests" considered equally with the interests of others. Since, in Utilitarian terms, man and animals are linked by their common ability to suffer, animals have a right to have their interests in not suffering (or dying) considered equally with man's. This concept of prevention of suffering can be extrapolated to bees, insects, fish, worms, flies, etc.

Singer's book was not just a philosophical treatise; it was also a polemical call to action. Invoking Ryder's concept of speciesism, he deplored the historic attitude of humans toward "non-humans" as a "form of prejudice no less objectionable than racism or sexism" and urged that the "liberation" of animals become the next great cause after civil rights and the women's movement.

Singer's book has enjoyed wide popularity and was reprinted in a second edition in 1990. The book has produced two important effects. One, it reintroduced to the cause of antivivisectionism an intellectual basis, a philosophical orientation, and a moral focus. These gave the issue a new attraction to those who had been generally indifferent to or repelled by the essentially emotional appeal based on love of animals that antivivisectionism had been waging for the past century. This legitimacy and support among some intellectuals has given the animal rights movement credibility over the past decade.

The second important effect of Singer's book was to attract to the animal rights cause a host of new activists, particularly in England, Canada, and on the European continent. In the US, some activists were veterans of the civil rights, women's rights and antiwar movements of the 1960s and 1970s. They brought to the cause of animal rights the same commitment, zeal, and tactics that they employed in support of those movements. These activists established many new organizations that have become the leading edge in a revitalized and reoriented animal cause movement in the United States.

2. Growth and organization. It is not possible to state with any authority how many animal rights activists or organizations there are nationwide but they exist at three levels—national, state and local—under a variety of names. Some estimate the total number of animal welfare and animal rights organizations in the country at more than 1,000. Many local animal rights organizations are associated with one of several national or state organizations. Local organizations have recruited thousands of new members who spend much time conducting demonstrations, lobbying legislators, circulating petitions, raising funds and working with the media. Recent estimates suggest that as many as 10 million Americans annually provide some form of financial support to one or more animal activist organizations. According to The Nonprofit Almanac 1992-93, 0.9% of the US charity dollar, or $1.6 billion goes to animal-related causes. The large animal welfare and animal rights organizations have considerable financial resources that are being increased constantly by new contributors. In contrast, much less money is available to defend animal research.

In general, animal rights activists and organizations can be classified as moderate, militant, or terrorist.

The moderates are similar in spirit and motivation to animal welfare activists and simply wish to put an end to what they see as an abuse of animals. They pursue this objective through education, legislation, and/or negotiation. Moderates seem to play only a small role in the animal rights movement and certainly are not among the leadership.

The militants are more diverse in motivation, more aggressive in method, and more vocal. Some have a particular interest or goal, such as vegetarians who want to put an end to the eating of meat. Others see the movement as one part of a larger social revolution intended to remold western society by changing its values, institutions, and laws. For most militants, the purpose and ultimate goal of the movement is that proposed by Singer: to alter all human relationships with animals and to put an end to all forms of exploitation, whether for research, business, pleasure, fur, sport, transportation, or food. The present concentration of effort on animal experimentation and factory farms, another target, was a strategic decision based on the relief that these would be the easiest causes for which to generate public support.

The most active and visible militant organization nationally is the **People for the Ethical Treatment of Animals** (PETA). It is believed that PETA has at least 350,000 dues paying members in the US and an annual budget of about $10 million. PETA has been especially aggressive in promoting the animal rights concept and in attempting to radicalize the animal cause movement. Its basic philosophy is that all mammals are created equal—a belief that has been expressed by a PETA director in the aphorism, "A rat is a pig is a dog is a boy"—and that the human treatment of animals represents "superracism" and "fascism." Its activities have included the prompting of a police raid on a laboratory in Maryland where it had planted a worker undercover, filing a lawsuit to obtain possession of the animals seized in the raid and the distribution of edited videotapes and other material taken by terrorist organizations in raids on laboratories. In 1982, members of PETA and another animal rights organization attempted a hostile takeover of the New England Antivivisectionist Society, one of the oldest, largest, and richest animal cause organizations in the country. Although this attempt failed, a renewed takeover was successful in 1987. Animal rights activists also have gained control of the Torono Humane Society in Canada.

The **Physicians Committee for Responsible Medicine** (PCRM) was founded in 1984 by a Washington, DC psychiatrist. PCRM has been a vocal opponent of virtually all biomedical research that uses animals. PCRM claims a membership of some 3,000 physicians and an overall membership of about 60,000. Even assuming 3,000 physician members is accurate, that number represents less than 0.005% of the total US physician population. However, PCRM has already helped bring to a halt important scientific work in several critical areas, including head trauma. PCRM falsely claims that animal testing helps delay the recalling of consumer products found to be carcinogens; that prevention and nutrition alone can eliminate disease (PCRM advocates total vegetarianism); that alternatives exist that can eliminate the use of animals in research; that animal testing is never directly relatable to possible human response; and that money spent on research should be diverted to promoting prevention rather than finding cures even with diseases that currently cannot be prevented. PCRM was denounced by the AMA House of Delegates in 1990 for implying that physicians who support the use of animals in biomedical research are irresponsible, for misrepresenting the critical role animals play in research and teaching, and for obscuring the overwhelming support for such research that exists among practicing physicians in the United States.

PCRM members use the tactics of propaganda to promote their views and distort the truth. They have manipulated facts, oversimplified complex issues and complex research, taken words out of context, omitted key facts, and misrepresented the goals of specific research projects. The overall effect of these activities has been to mislead. This is particularly true because of the apparent "scientific" credibility of PCRM.

The leading radical organization has been the **Animal Liberation Front** (ALF). ALF has been identified by Scotland Yard as an international underground terrorist organization that is active in the United Kingdom, France, Canada, and the US. Unlike some terrorist groups, which have restricted their activities to entering laboratories on "raids of liberation" to free animals, ALF has destroyed equipment, records, and facilities, including the burning of a laboratory under construction in California. It has been declared one of the most dangerous terrorist organizations in the state by the Attorney General of California. In 1989, the Animal Liberation Front openly took credit for vandalism and arson at the University of Arizona in

Tucson. Three buildings were heavily damaged and more than 1200 laboratory animals were stolen in this raid, with damage estimates exceeding $250,000. Some of the stolen mice were infected with cryptosporidium, an organism not dangerous to mice but extremely dangerous to infants or those with immune deficiency syndromes. Similarly in 1989, the ALF vandalized a laboratory at Texas Tech School of Medicine where research directed toward understanding basic mechanisms in Sudden Infant Death Syndrome (SIDS) was being carried out. In 1990, ALF raided a laboratory at the University of Pennsylvania and, more recently in 1992, a laboratory at Michigan State University was destroyed by ALF vandals.

Of greater concern to scientists have been efforts of animal rights activists to interfere with the conduct of research. The most direct threat has been the attacks on laboratories and the harassment, including death threats, of scientists and their families. At least 25 research laboratories in the US have been vandalized during the past three years by various animal rights terrorist groups. These "raids of liberation" have caused millions of dollars in damage to laboratories, the theft of more than 2500 laboratory animals, and the destruction of numerous records. The burning of a laboratory under construction at the University of California at Davis School of Veterinary Medicine in 1987 has been estimated to result in a loss in excess of $3 million.

These break-ins have achieved their purpose. They have led to the interruption, delay, and even abandonment of research, and they have increased the costs of conducting research by requiring more security at laboratories, the replacement of stolen animals, and the repetition of experiments. The loss of valuable data and records has hindered efforts to treat human diseases. The impact of these raids has been considerable in terms of costs and morale and they have created great concern within the scientific community. These activities also have had a significant impact upon recruitment of young people into careers in science.

Although there has been some attempt to unify the movement through coordinating committees, sharing of information and activities and some interorganizational support, until recently the animal cause movement has been more notable for its disunity than its unity, not only between the two wings but among organizations within the animal rights groups in the movement. However, within the past several years, the animal rights movement seems to be striving to forge a united animal cause organization. These efforts have included an attempt to eliminate attacks by activists on the position of animal welfare groups and attempts to achieve a consensus on goals and tactics. Such a consensus would require a shifting to the left by the animal welfare groups and an acceptance of some of the goals and tactics of the animal rights movement. Many believe that this is occurring.

Unification would greatly increase the resources available to the animal rights movement. Much of the money in the movement —an estimated $200 million annually—is controlled by the leading animal welfare organizations, most of it donated by those who are primarily motivated by their love of animals. Animal rights groups are attempting to merge with those animal welfare organizations that have large financial resources and radicalize their policies.

3. Strategic objectives and activities. The ultimate objective of the animal rights movement with respect to biomedical research, is to obtain the passage of laws that would ban the performance of any experiments or tests using animals. Optimally, such laws would recognize the right of animals not to be used in experiments and would be part of a package of laws banning any exploitation of animals by humans.

Short of banning the use of animals, it is the intent of many within the animal rights movement to impede, curtail, or control the conduct of such research by a combination of means. These include attacking laboratories; raising the costs of conducting experiments by increasing the regulations controlling the use of animals; diverting funds from research projects; establishing the right of human beings to act as guardians of animals or as their agents in the enforcement of animal welfare and other laws; and gaining the right to examine all animal research protocols and to prohibit any that they consider to be either cruel or unnecessary.

Bills to completely ban the use of animals in biomedical research and to establish the belief that animals have rights have been introduced in several state legislatures. During 1987, more than 85 separate bills were introduced in state legislatures, covering the availability of pound animals, regulation of animal facilities, status of animal cruelty statutes and the use of animals in testing and education. In addition, Massachusetts passed a law several years ago in which the use of pound animals for research and testing is banned with the Commonwealth. In New York in 1987, the legislature passed and the Governor signed a law that prohibits the release of any dog or cat from any type of shelter except for adoption or return to its owner.

Elements within the animal cause movement have pursued a variety of tactics to reduce or eliminate the conduct of animal research. The most successful of these has been the effort to ban the use of animals obtained from pounds for laboratory experiments, an effort that has succeeded in a number of states.

Other attempts have been less successful. One was the attempt to divert 30% to 50% of the money allocated by the National Institutes of Health from research utilizing animals into the development of alternatives to such experiments. A bill to require the diversion was introduced in several sessions of Congress but has to date not received serious consideration.

Some animal rights organizations are seeking legal sanctions to act on behalf of animals. These efforts, which are being pursued both through courts and legislatures, seek to establish either "standing to sue" rights on behalf of animals, or a private right of action to compel enforcement of a law, such as an anticruelty statute. Lawsuits have been filed in both federal and state courts. In a precedent setting action, a Federal District Court, the US Court of Appeals, and the US Supreme Court denied PETA and other animal rights groups the right to assume guardianship over primates taken from a Maryland laboratory in a police raid prompted by PETA in 1981. To grant guardianship, said the Court of Appeals, "might unleash a spate of private lawsuits that would impede advances by medical science in the alleviation of suffering."

Massachusetts, in conjunction with its repeal of the pound law in 1983, directed the Department of Public Health to establish regulations for licensure of research institutions using dogs or cats for research or education. Under these regulations the state's Society for the Prevention of Cruelty to Animals and the Animal Rescue League have the authority to act as agents of the Department in unannounced inspections of research facilities at least four times annually. Furthermore, under the Massachusetts law amended in 1983, the New England Antivivisection Society attained special police powers enabling them, as agents of the Department, to arrest and detain any person violating any law for the prevention of cruelty to animals.

Through passage of federal, state, and local regulations, animal rights activists also are attempting to gain the right to examine all research proposals to determine the experiment, in their view, is unnecessary or would cause the animal(s) pain. If so, they hope to have the right to prohibit the experiment.

Animal rights activists support these activities through publicity campaigns designed to influence public and legislative opinion; arguments and assertions of activists regarding the validity and performance of animal experiments have heightened public awareness on the issue and created doubt and concerns about the necessity for research with animals and about specific aspects of its conduct. Only rarely do animal rights brochures or audiovisuals note that 75-90% of lab animals used in research are not dogs or cats, but rodents.

V. Response by the Research Community to Arguments Raised by Animal Rights Activists

The arguments advanced by animal rights activists in opposing the use of animals in biomedical research follow the historic pattern and are scientific, emotional, and philosophic. In debate, categorical distinctions are rarely drawn and the arguments tend to become commingled as part of a broad general attack on the use of animals in research. However, each presents a separate set of issues and questions that can best be understood by examining them separately.

A. Validity and conduct of animal research

The scientific challenge raised by animal rights activists goes to the heart of the issue by asking whether animal experiments are necessary for scientific and medical progress and whether all the experiments being performed and all the animals being used are justified and required. Scientists insist that they are; animal rights activists insist that they are not.

1. **Biomedical research.** Scientists justify use of animals in biomedical research on two grounds: the contribution that the information makes to human and animal health and welfare, and the lack of any alternative way to gain the information and knowledge. Animal rights activists contest experiments that utilize animals on both these grounds and assert that this practice no longer is necessary because alternative methods of experimentation exist for obtaining the same information.

In an appearance on the Today show in 1985, Ingrid Newkirk, representing PETA stated: "If it were such a valuable way to gain knowledge, we should have eternal life by now." This statement is similar in spirit to one made in 1900 by an antivivisectionist who stated that, given the number of experiments on the brain done up to then, the insane asylums of Washington, DC should be empty. More recently when queried about animal research, Ms. Newkirk stated "...even if it produced a cure for AIDS we'd be against it...."

Scientists believe that such assertions miss the point. The issue is not what has not been accomplished by animal use in biomedical research, but what has been accomplished. A longer life span has been achieved, decreased infant mortality has occurred, effective treatments have been developed for many diseases, and the quality of life has been enhanced for mankind in general.

One demonstration of the critical role that animals play in medical and scientific advances is that 60 of 82 Nobel Prizes awarded in physiology or medicine since 1901 have been for discoveries and advances made through the use of experimental animals. Among these have been the Prize awarded in 1985 for the studies (using dogs) that documented the relationship between cholesterol and heart disease; the 1966 Prize for the studies (using chickens) that linked viruses and cancer; and the 1960 Prize for studies (using cattle, mice, and chicken embryos) that established that a body can be taught to accept tissue from different donors if it is inoculated with different types of tissue prior to birth or during the first year of life, a finding expected to help simplify and advance organ transplants in the future. Studies using animals also resulted in successful culture of the poliomyelitis virus; a Nobel Prize was awarded for this work in 1954. The discovery of insulin and treatment of diabetes, achieved through experiments using dogs, also earned the Prize in 1923.

In fact, virtually every advance in medical science in the 20th century, from antibiotics and vaccines to antidepressant drugs and organ transplants, has been achieved either directly or indirectly through the use of animals in laboratory experiments. The result of these experiments has been the elimination or control of many infections diseases—smallpox, poliomyelitis, measles—and the development of numerous life-saving techniques—blood transfusions, burn therapy, open-heart and brain surgery. This has meant a longer, healthier, better life with much less pain and suffering. For many, it has meant life itself. Often forgotten in the rhetoric is the fact that humans do participate in biomedical research in the form of clinical trials. They experience pain and are injured and in fact, some of them die from this participation. Hence, scientists are not asking animals to be "guinea pigs" alone for the glory of science. Some medical breakthroughs accomplished through research with animals are described in Table 1.

Scientists believe it is essential for the public to understand that had scientific research been restrained in the first decade of the 20th century as antivivisectionists and activists were then, and are today urging, many millions of Americans alive and healthy today would never have been born or would have suffered a premature death. Their parents or grandparents would have died from diphtheria, scarlet fever, tuberculosis, diabetes, appendicitis, and countless other diseases and disorders.

Selected medical advances made possible through the use of animals

Pre-1900	Treatment of rabies, anthrax, beriberi (thiamine deficiency), and smallpox
	Principles of infection control and pain relief
	Management of heart failure
Early 1900s	Treatment of histamine shock, pellagra (niacin deficiency) and rickets (Vitamin D deficiency)
	Electrocardiography and cardiac catheterization
1920s	Discovery of thyroxin
	Intravenous feeding
	Discovery of insulin — diabetes control
1930s	Therapeutic use of sulfa drugs
	Prevention of tetanus
	Development of anticoagulants, modern anesthesia and neuromuscular blocking agents
1940s	Treatment of rheumatoid arthritis and whooping cough
	Therapeutic use of antibiotics, such as penicillin, aureomycin and streptomycin
	Discovery of Rh factor
	Treatment of leprosy
	Prevention of diptheria
1950s	Prevention of poliomyelitis
	Development of cancer chemotherapy
	Open heart surgery and cardiac pacemaker
1960s	Prevention of rubella
	Corneal transplant and coronary bypass surgery
	Therapeutic use of cortisone
	Development of radioimmunoassay for the measurement of minute quantities of antibodies, hormones and other substances in the body
1970s	Prevention of measles
	Modern treatment of coronary insufficiency
	Heart transplant
	Development of non-addictive pain killers
1980s	Use of cyclosporin and other anti-rejection drugs
	Artificial heart transplantation
	Identification of psychophysiological factors in depression, anxiety and phobias
	Development of monoclonal antibodies for treating disease
	Discovery of HIV as causative agent for AIDS
1990s	Pancreas and liver transplantation
	Thrombolytic therapy for acute myocardial infarction
	Human gene therapy

Table 1

Animal rights activists attribute advances in longevity and health to public health measures and better nutrition. Scientists agree that for a number of infectious diseases such as typhoid fever, influenza and tuberculosis, such measures were important; however, for most infectious diseases, improved public health and nutrition have played only a minor role. This is clear when one considers the marked reduction in the incidence of infectious diseases such as whooping cough, rubella, measles and poliomyelitis. Despite advances in public health and nutrition, eradication or control of these and most other infectious diseases was not achieved until the development of vaccines and drugs through research using animals. Further proof is provided by the widespread prevalence of chicken pox. Virtually every child in America still suffers from chicken pox because no vaccine is yet available; continued experiments and testing with animals has recently resulted in the development of a vaccine now undergoing clinical trials. Many physicians have become concerned recently over pertussis immunization resulting in some cases of brain damage. Only through additional studies of pertussis vaccines will scientists be able to eliminate these unpredictable events, and animals are imperative for these studies. Similarly, the development of a vaccine against AIDS is dependent upon continued studies conducted in animals.

Animal use in biomedical research also has directly benefitted animals themselves. The most obvious benefits are the advances made in veterinary medicine and surgery, such as the vaccines for rabies and distemper and the recent development of a pill to protect against heartworm, a painful and ultimately fatal affliction in dogs. In farm animals, the control of hog cholera and tuberculosis and brucellosis in cattle was accomplished through research with animals. Some breakthroughs in veterinary medicine attributable to research with animals are shown in Table 2.

Selected veterinary advances made possible through the use of animals

- Immunization against distemper, rabies, parvo virus, infectious hepatitis, anthrax and tetanus
- Treatment for animal parasites
- Orthopedic surgery for horses
- Surgery to correct hip dysplasia in dogs
- Experimental radiation techniques and immunotherapy for cancer in dogs
- Identification and prevention of brucellosis and tuberculosis in cattle
- Treatment of feline leukemia
- Improved nutrition for pets

Table 2

2. **Behavioral research.** Of potentially even greater importance to animals are the lessons learned through behavioral research. Behavioral research with animals in a laboratory setting has been a particular target of animal rights activists. They claim it subjects animals to stress without producing meaningful results. Behavioral scientists believe that discoveries made through such research may not only save the lives of individual animals but provide information necessary to save entire species from extinction.

Most valuable have been the experiments that may help to save species. For example, information gained through experiments begun 50 years ago on "imprinting"—the tendency of an animal to identify with and relate to the first species it came into contact with—has been used to train captive-born animals to relate to members of their own species. Recently, this has been important in helping condors to survive and propagate in the wild. Animal studies are helping to save the musk ox in Alaska from extinction. Research done on the sexual behavior of animals threatened with extinction has led to successful reproduction in captivity, a necessary step in restoring their numbers in the wild. Some of this research is being conducted in the wild but much of it can only be conducted in the laboratory.

Behavioral research has also been of immense benefit to humans. For example, fundamental information on how people learn was discovered by experiments on animals in laboratories; the learning principles and behavior modification therapies discovered or developed through such experiments are today being used to treat conditions such as anuresis (bed-wetting), addictive behaviors (tobacco, alcohol and other drugs), and compulsive behaviors such as anorexia nervosa.

Biofeedback techniques that have become a major means of treatment for a number of conditions were developed through behavioral research with animals. The use of biofeedback enables people to control what are normally automatic body functions, such as blood pressure, heart rate, and muscle tension. Biofeedback helps cardiac patients reduce the risk of heart attack by controlling their blood pressure, assists persons paralyzed with spinal injuries to raise their blood pressure and permit them to sit up, and relieves the discomfort of migraine headaches, insomnia, and low back pain. Many of these afflictions had no effective treatment before biofeedback, which was developed through studies of the nervous system of the rat.

Experiments on cats have enhanced the understanding of the corpus callosum, a band of fibers that connects the left and right sides of the brain needed for transfer of information from one side to another. This finding, for which a Nobel Prize was awarded in 1981, led directly to the development of new treatments for patients with strokes, language disorders, brain damage, intractable epilepsy, and other neurologic conditions.

One objection of animal rights advocates to behavioral research is their belief that many tests are conducted merely to confirm or prove long-accepted or obvious concepts, such as that a child will suffer when deprived of love or a parental figure. However, what appears to be an obvious truth often proves to be false when subjected to close scrutiny in experiments. This includes the idea that all animals suffer when separated from a parent. In tests conducted over a number of years with rhesus monkeys, scientists discovered that, whereas some infants became withdrawn and anxious, others grew stronger and showed fewer stress-related symptoms when exposed to new or threatening situations. This has prompted new research into possible genetic and physiologic reasons why people react differently to stress and which types of persons are more likely to develop conditions such as depression.

Other behavioral research utilizing animals is leading to a deeper understanding of links between the mind and the body that may have important ramifications for the prevention and treatment of disease. Studies conducted with mice and monkeys have helped to establish and explore the relationship between stress and conditions such as heart disease, hypertension, and breakdowns in the immune system that leave individuals vulnerable to disease. Such studies may lead to an understanding of the nature of psychosomatic illnesses in humans.

B. Questions regarding the conduct of research

Challenges regarding the necessity for certain types of experiments and the way they are conducted represents another point of attack by animal rights activists. The following answers by scientists are given to questions posed by animal rights groups:

1. Does basic biomedical research have clinical value? The purpose of basic biomedical research with animals is the same today as it was for the early scientists: to increase the understanding of how the body functions in health and disease. Thus, while such research may not always have a direct clinical purpose, it does play a critical role in the development of medical knowledge by increasing the knowledge bare essential for the conduct of experi-

ments that do. Basic research is the foundation for most important medical advances. Progress in clinical medicine would rapidly diminish without basic research using animals. Although it is true that basic research experiments do not always produce positive results, it is not possible to predict which ones will be productive. If it were, scientists would not perform them.

One important aspect of both basic and applied research is the discovery of the unexpected. Often, a research project will produce a finding that has no relevance to the original purpose of the experiment but will contribute to another line of inquiry or even lead to a new discovery. This factor, known as serendipity, occurs with both basic and applied research and it has been an important element in many medical advances, such as the discovery of penicillin. Another example is a natural antibiotic from the skin of frogs that was discovered during an experiment involving the removal of eggs from the frog through an incision in the skin; the investigators noted that the skin of the frog was not infected with bacteria under conditions highly favorable for bacterial growth and surmised that some antibacterial substance must be present in the frog skin.

2. Are experiments needlessly duplicated? Criticism by animal rights adherents is directed at three types of experiments: those that are conducted for educational purposes; those that are conducted merely to "prove the obvious"; and those that repeat experiments already conducted.

Fewer animals are used for education than for any other purpose. They are used either to instruct students in courses such as physiology or to teach techniques, such as surgery. The use of animals in education varies from school to school and from program to program, even within medical schools. Educators generally agree, however, that students are better trained and patients better served when the students are given "hands-on" experience with living tissue, especially for training in surgery. Even the British, who otherwise ban the use of animals for educational purposes, accept the necessity of using animals under anesthesia to teach the techniques of microsurgery.

Animal rights activists also object to the duplication of experiments including those conducted for educational purposes and those that have failed elsewhere. The performance of experiments when the result is known in advance is a sound educational technique and is employed in many fields that do not involve animals, such as mathematics, physics, and chemistry. By utilizing an experiment in which the result is known in advance, it can be determined whether the experiment is performed correctly or incorrectly.

The criticism relating to the repetition of failed experiments has some merit, scientists agree, and emphasizes the need to improve communication among scientists in both the research and testing communities. Commercial firms that perform many of the testing experiments, such as drug and chemical companies have taken steps both directly and through their trade associations to respond to this need by establishing mechanisms for sharing of data and results among firms. The creation of a comprehensive data bank for all research and testing is a far more complex undertaking, however, and an attempt by a private firm to establish a limited system failed earlier this decade.

A third reason for duplication of experiments is the requirement of the scientific method for a new finding to be verified by scientists in other laboratories before the finding can be considered valid. Such replication is necessary and quite often uncovers a mistake in technique or design or some other flaw in the original experiments that will render it invalid.

Both economic pressure and the peer review process used to evaluate research proposals make the conduct of unnecessary experiments unlikely. Research today involves intense competition for funding; for example, less than 25% of studies proposed to, and approved by, federal agencies each year are actually funded.

Therefore, scientists on research evaluation committees are not likely to approve redundant or unnecessary experiments. Also, given the competition for funds, scientists are unlikely to waste valuable time and resources conducting unnecessary or duplicative experiments.

3. Are more animals used than needed? The number of animals being used in biomedical research is not known. Animal rights activists place the figure as high as 150 million, but such estimates have no basis in any known data. An authoritative estimate was made by the Office of Technology Assessment (OTA) in its report for Congress: Alternatives to Animal Use in Research Testing and Education. The OTA examined data from both public and private sources and estimated that, for 1982 and 1983, the number of animals used in laboratory experiments in the United States was between 17 and 22 million. The overwhelming majority of these animals were rats and mice especially bred for that purpose. In its report, the OTA estimated the number of rats and mice used in 1982 and 1983 was 12.2 to 15.2 million each year, or about 75-90% of all animals used. The figures for other species were: 2.5 to 4 million fish; 100,000 to 500,000 amphibians; 100,000 to 500,000 birds; 500,000 to 550,000 rabbits; 500,000 to 520,000 guinea pigs; approximately 450,000 hamsters; 182,000 to 195,000 dogs; 55,000 to 60,000 cats; and 54,000 to 59,000 primates.

The OTA drew no conclusion on the trend in animal use. Surveys conducted by the Institute of Laboratory Animal Resources of the National Research Council indicate that the number of animals being used may be decreasing. In its 1978 survey, the Institute estimated the total was 20 million, a 40% reduction from the number noted in its 1968 survey. The National Academy of Sciences, with the assistance of the National Institutes of Health, has conducted an in-depth study on the use of animals in research.

Efforts are underway to reduce the number of animals being used and to ensure that those used represent the minimum possible. Congress, in passing the Health Research Extension Act in 1985, instructed the National Institutes of Health to assure that scientists are trained in methods to reduce the number of animals used in experiments. The OTA identified a number of approaches, including the sharing of animals or their tissues; the use of common or historical "controls" (the animals to which experimental animals are compared); improved designs of experiments; and wider use of existing data bases.

Efforts also are being made to reduce the number of animals used in, and perhaps eventually to abolish, the lethal dose test (LD_{50}) that long has been used to determine the toxic levels of chemicals. In the test, a series of doses of a chemical is administered to 30 to 100 animals, usually rats or mice, and the LD_{50} point is reached when one-half of the animals die. Although never specifically required by the government, the test has become the standard measuring tool for submissions for Food and Drug Administrations (FDA) approval of drugs and for meeting certain requirements of the Environmental Protection Agency (EPA) and the Consumer Product Safety Commission. The FDA has declared that it does not require that data be based on the LD_{50} test for approval and the EPA has established circumstances in which the test can be replaced by a "limit" test that used far fewer animals (4 to 10) to test for toxicity. Reduced use of the LD_{50} test has spared many animals from such testing.

C. Emotional issues

The emotional appeals made by animal rights activists are often directed toward the use of apes and monkeys in experiments, the pain and death associated with animal experimentation, and what they assert is the abuse that animals are subjected to by scientists.

1. Use of primates. Although animal rights activists object to the use of all animals in experiments, they make a special plea and have shown a particular interest on behalf of the

nonhuman primates. Many of the sit-ins, demonstrations, and raids on laboratories have been directed at primate research centers. The argument they advance is that, because apes and monkeys are so much like us, they experience suffering much as humans do and this should exempt them from use in experiments, just as it does humans. Even keeping them in cages for long periods or isolating them from others of their kind is both cruel and destructive to psychological well-being, the activists argue.

Scientists argue that it is this very factor—their similarity to humans—that makes apes and monkeys so valuable to research and their use in some experiments indispensable. Apes and monkeys have both strong physiologic and strong behavioral similarities to humans. They are susceptible to many of the same diseases and have similar immune systems. They also possess intellectual, cognitive, and social organizational skills far above those of other animals, and these characteristics have made them invaluable in research related to language, perception, and visual and spatial skills. Primates are used in experiments in relatively few numbers—approximately one-half of 1% of all animals used. However, their contributions to both biological and behavioral sciences have been numerous, significant, and in some cases crucial, as with pilomyelitis and hepatitis.

Primates played three different roles in the development of the poliomyelitis vaccines, all of them essential. Although many studies on poliomyelitis in humans were conducted in the late 19th century, the cause of the disease remained unknown until scientists were able to transmit the virus to monkeys in 1908. There followed many years of research with primates until scientists were able, in the early 1950s, to grow the virus in human cell cultures and development of a vaccine became possible. At that point, to ensure the safety and effectiveness of the vaccines, tests were conducted with monkeys. To produce the vaccines in pure form in great quantities, it was necessary to use kidney tissue taken from monkeys. Today, the use of the monkey kidney tissue is no longer necessary because vaccines now are produced through self–propagating cells—an alternative to the use of animals developed through appropriate research.

In current biomedical research, monkeys are useful in studies to examine the causes of high blood pressure. This is because the natural hormones that control blood pressure in man and other primates are identical, in contrast to other animals in which they are dissimilar. Researchers also are using the monkeys to examine the genetic transmission of high blood pressure. Research with rhesus monkeys and chimpanzees led to the development of a vaccine, derived from human plasma, to combat hepatitis B infection. Infection with hepatitis B is estimated to cost, directly and indirectly, more than $1 million a day in this country. Green monkeys served as the animal model for the successful development of the drug, acyclovir, that has proved to be effective in treating genital herpes.

Rhesus monkeys may also hold the key to developing an understanding of Parkinson's disease, a neurologic disorder that destroys brain cells and afflicts older persons with palsy and rigid muscles. Monkeys treated with a drug that destroys specific cells in the same area of the brain affected by Parkinson's disease not only exhibited all the symptoms of the disease but also responded well to doses of levodopa, a drug used to treat the disease. Scientists are hopeful that this new line of research will provide the information necessary to understand why and how Parkinson's disease occurs. Recent experiments with monkeys have resulted in the development of a new surgical procedure that shows great promise in treating patients with Parkinson's disease.

The chimpanzees may play a critical role in developing a therapy for AIDS. Medical researchers note that, in contemporary medicine, AIDS is the first infectious disease that is virtually 100% fatal. Chimpanzees are the only animal that scientists have been able to infect with the AIDS virus. To date, none have developed the disease, leading some animal rights activists and even some scientists to question its value in the research. However, the fact that chimpanzee do not become ill may in itself provide a clue to combating the disease if the reason can be discovered through research. Primates have already been vitally important in the development and testing of potential vaccines for AIDS. Recently a prominent AIDS researcher at the National Institutes of Health commented that the difficulty of procuring animals for experiments with the AIDS virus may significantly slow the development of effective drugs and vaccines. He attributes these problems directly to the efforts of animal rights groups.

Chimpanzees and other primates also have been important in behavioral research. Recent studies of how chimpanzees learn language are helping therapists to teach severely retarded children and adults to communicate through the use of a language based on geometric symbols in which each symbol represents a word and is printed by pressing keys on a video display keyboard.

Both the primate research community and Congress have been sensitive to the concerns expressed by animal rights activists and others over the psychological well-being of primates kept in laboratories and have taken steps to respond to them. The amendments to the Animal Welfare Act passed in 1985 required the Department of Agriculture, the federal agency responsible for enforcing the law, to establish standards to create a physical environment adequate to promote the psychological well-being of primates. Even prior to passage of the law, a number of primate centers and research laboratories took action on their own to create such an environment. At some facilities this has included the creation of outdoor areas with play equipment for young apes and the pairing of animals in larger indoor cages for companionship and social interaction. At the Yerkes Regional Primate Research Center in Atlanta, Georgia, indoor-outdoor enclosures are provided for small groups of primates and large outdoor enclosures are provided for large social groups of animals. These social enclosures contain equipment on which animals may swing, play or just sit. In addition, young primates that need nursing care usually live in groups and have cloth swings, toys and other devices to manipulate.

2. **Animal pain and abuse.** Animal rights advocates assert that most biomedical research conducted with animals causes severe pain and that many animals used in experiments are abused either deliberately or through indifference. In fact, scientists respond, most experiments today do not involve pain, most animals used in experiments do not suffer pain, and the degree of pain that is inflicted during some experiments has been greatly reduced through the establishment of rules for the humane conduct of experiments and the development of new types of instruments and techniques that sometimes can preclude the need for animal experiments. These include noninvasive imaging techniques, microinstrumentation, fiber optics, and the laser, among others. Many of these techniques were developed in part through experiments with animals.

The fact that most experiments do not expose animals to pain was confirmed by a report issued by the Department of Agriculture in 1984. The Department found that 61% of animals were used in experiments that did not expose them to painful procedures. Another 31% were used in experiments that would subject them to pain but only after they received anesthesia or pain-relieving drugs. The remaining 8% were not protected from pain because this would have interfered with the purpose of the experiment. It is important to note that the perpetrators of a recent break-in at the University of Oregon stated in a letter to the press that no Department of Agriculture violations were found.

In any experiment in which an animal is likely to suffer pain, federal law now requires that a veterinarian be consulted in the planning of the experiment and that anesthesia, tranquilizers, and anelgesics be used except when they would interfere with the purpose of the experiment. In experiments in which the animals are not protected from pain, it is often because pain itself is the subject under study. These experiments are conducted to understand the basis of pain and to develop methods to negate or control pain. Chronic pain is one of the most costly health problems in America; in addition to the suffering it causes, it costs an estimated $50 billion a year in direct medical expenses, lost productivity, and income.

Recent research has disclosed specific pathways in the brain that strongly resist pain, and that the brain produces its own painkillers. With this knowledge, scientists are trying to develop nonaddictive painkillers. After experiments with animals had shown that use of stimulating electrodes could reduce or inhibit pain, electrodes are now being implanted in specific parts of the brain of persons suffering from chronic pain. By activating the electrodes with radio transmitters, a number of patients have experienced long periods of relief. This understanding of the brain could never have been achieved without the use of animals, for such experiments could never have been performed initially on human beings. Most such experiments cause little pain because direct brain manipulation is not felt by the animal.

There is no question that some experiments cause pain to the animals used. Experiments, such as those on head trauma, may leave animals injured or disabled in some fashion, usually because such experiments are designed to observe the effects of these injuries over a period of time and the adjustment of the animal to injury. Scientists feel these experiments are critical to develop treatments for nerve or muscle damage and brain injury. Animal rights activists often cite such experiments as examples of abuse of animals by researchers and have displayed photographs of animals used in such experiments as evidence. Scientists insist that a distinction must be drawn between pain or impairment that is a necessary part of an experiment and that which constitutes abuse. True abuse results from neglect or malicious mistreatment of animals and is rare. Most scientists do not mistreat animals and would prefer not to use animals at all if that were possible. In fact, mistreated or stressed animals do not make good research subjects, particularly in behavioral research.

More than 300,000 Americans suffer severe traumatic injuries each year from car accidents and other causes that leave them permanently disabled. If experiments seeking a better understanding of these injuries and ways to treat them are banned because the pain they inflict on animals is considered to be abusive, those and future victims will have to live without hope of improvement in their condition, for there is no other way to gain the knowledge and understanding necessary to help them.

D. Alternatives to use of animals in research and testing

Among all the issues raised by animal rights activists, that of "alternatives" has probably caused the most confusion. The reason for that, say scientists, is the misrepresentation and misunderstanding that has been created regarding what is possible. The concept of alternatives was first introduced in 1959 by two British scientists and was defined in terms of what has become known as the Three R's—refinement, reduction and replacement. In their book, The Principles of Humane Experimental Techniques, W.M.S. Russell and R.L. Burch contended that humane research techniques require scientists to work toward refinement of techniques to reduce potential suffering, toward reduction in the number of animals needed, and, where possible, toward replacement of animals by non-animal techniques.

In the present debate, the focus is on the last of these—replacement—as animal rights activists have insisted that certain experimental methods that investigators have developed can be used in lieu of animal experiments. Thus, as used in the debate, the word "alternatives" has become virtually synonymous with the word "substitutes" and animal rights activists have tended to focus on two: in vitro research (cell, tissue, and organ cultures) and computer simulation of biological systems in the form of mathematical models.

Both of these methods play important roles in biomedical research and have allowed the performance of experiments not possible with animals. In addition, both have avoided the need to use animals in some stages of research. However, they cannot serve, either individually or in combination with any other research method, as total replacement for use of live animals in experiments, for they cannot reproduce exactly the intact biological system provided by live animals. Each method suffers from at least some inherent deficiency.

The technology and use of cell cultures has grown dramatically during the past 20 years, making possible the performance of experiments that were previously impossible. Cells in isolation, however, do not act or react the same as cells in an intact system. As the Congressional Office of Technology Assessment (OTA) noted in a report it prepared for Congress, Alternatives to Animal Use in Research, Testing, and Education: "... isolated systems give isolated results that may bear little relation to results obtained from the integrated systems of whole animals."

The same is true of tissues and organs placed in cultures. In addition, scientists point out, tissue and organs are difficult to nourish and maintain and tend to disintegrate automatically or lose their ability to function when maintained in cultures for long periods.

Computer simulations, often promoted as the great hope of the future by animal rights activists, have been invaluable in developing or suggesting new lines of scientific inquiry and in developing new mechanisms or techniques. However, both computers and computer simulations have inherent limitations that make it unlikely that they will ever totally replace animals in experiments. One of these limitations is the nature of simulation. The validity of any model depends on how closely it resembles the original in every respect. Much about the body and the various biological systems of humans and animals is not known. For example, how the body breaks down each chemical or drug or the manner in which brain cells transmit the sensory signals that create vision are not known; therefore, they cannot be programmed into any model. Until full knowledge of a particular biological system is developed, no model can be constructed that will in every case predict or accurately represent the reaction of the system to a given stimulus.

As the OTA report notes with respect to behavioral research: "To construct a computer simulation that would fully replace the use of a live organism in behavioral research should require knowing that everything about the behavior in question, which in turn would eliminate the need for computer simulation for the research proposed." That observation is equally valid for all biomedical research.

No other method of study can exactly reproduce the characteristics and qualities of a living intact biological system or organism. Therefore, in order to understand how such a system or organism functions in a particular set of circumstances or how it will react to a given stimulus, it becomes necessary at some point to conduct an experiment or test to find out. There simply is no alternative to this approach and therefore no alternative to using animals for most types of health related research.

In their zeal to promote the concept of alternatives, some animals rights organizations introduced legislation in several sessions of Congress that would have required that 30% to 50% of the funds allocated for biomedical research by the National Institutes of Health be spent on the development of alternatives. Scientists feel that this is the wrong way to develop alternative methods of research. The best way to develop alternatives is through the conduct of basic biological research. Again, as the OTA report noted: "Basic biomedical research at all levels, some of it involving live animals, will continue to provide the new knowledge required to improve existing simulations and develop models where no satisfactory one exists." To divert funds away from basic research to fund special programs for the development of alternatives could have the ironic effect of delaying that development.

Congress did not pass the bill to divert funds. However, it did make a commitment to the search for alternatives in the Health Research Extension Act passed in 1985. This bill requires the National Institutes of Health to establish a plan for conducting, evaluating, and disseminating research to train scientists in methods of biomedical and behavioral experimentation that do not require the use of animals, methods that reduce the number of animals, and methods that produce less pain and distress than those currently in use.

One testing procedure for which a concerted effort to develop an alternative has been made is the Draize test in which chemicals and other substances are sprayed into the eyes of rabbits to determine potential toxic effects. This test, which can cause pain and blinding, has been the special target of one particular animal rights group and through its efforts several million dollars have been contributed by cosmetics companies, foundations, and animal welfare organizations to seek substitute tests. Although attempts are being made by scientists to limit the use of this test and some alternatives seem promising, no adequate replacements have yet been devised that would in all situations provide accurate information to protect the public from toxic new drugs, household products, and other chemicals. Work is continuing in this area.

Improvements in techniques or the development of alternatives that have reduced the need for research animals resulted in large part from studies using animals. Scientists note that it was research scientists themselves who first developed both the concept and models being promoted by animal rights activists, and the development and refinement of such models continues today; they agree that this is sound public policy. They also recognize that no animal research program can be considered humane unless it includes a commitment to employ alternative methods whenever possible. What scientists are concerned about and wish to avoid are policies or efforts that are built on misunderstanding or misrepresentation as to what is possible. Unfortunately, that appears to be occurring. Popular expectations for the use of alternatives outdistance actual performance. This can be attributed, at least in part, to the expectations and false hopes raised by animal rights activists.

VI. Legislative and Regulatory Activities of Animal Cause Organizations

Moving in the wake of, and using the agitation and publicity generated by the militant and terrorist animal rights activists, animal welfare and animal rights organizations have been able to convince Congress to adopt new legislation. These laws and regulations establish standards for, and govern the treatment of research animals. These same groups have persuaded a number of state, county, and municipal legislatures to pass laws to prohibit the release of animals (primarily dogs and cats) from public pounds to researchers.

A. Animal care laws

Animal welfare and animal rights activists have sought the adoption of federal laws and regulations to establish standards for the care and treatment of animals used in experiments. They have been largely successful in these efforts. Scientists strongly support setting standards for the proper care and responsible treatment of laboratory animals. They point out that it was the research community itself that first developed the concept and principles of humane treatment at the turn of the century and established the first formal guidelines for the humane use of research animals through the National Institutes of Health in 1963. Their concern is with setting standards that are unnecessarily rigid and prohibitively expensive to implement and with establishing a monitoring system that diverts significant time and money from the conduct of research. All of these, they fear, are happening today.

Historically, the concept of humane treatment was directed toward the pain endured by animals only during the performance of an experiment or as a consequence of it. However, in recent years, the concepts have been expanded to embrace all the care and treatment given to an animal before, during, and after an experiment or test. It is this expanded concept that has formed the basis of most of the federal legislation and regulations adopted during the past 20 years and particularly within the past five years.

The first federal legislation concerning the use of animals in medical research was the Animal Welfare Act, originally passed in 1966. Beginning in 1985, both the National Institutes of Health and Congress established specific and stricter requirements for the care and treatment of all animals used in federally funded research. The NIH guidelines must be used by all researchers who receive funding from NIH and a number of other federal agencies.

These regulations require the appropriate use of pain-relieving drugs, veterinary care, and euthanasia on all animals used in experiments and set detailed and specific requirements regarding cage sizes, feeding and watering, lighting, and sanitation for animals kept in laboratories. To ensure that the guidelines are adhered to, each research institution is required to establish an animal care oversight committee to supervise all the procedures involving animals, including regular inspections of animal housing areas, and to review all protocols to ensure compliance. This committee must be composed of at least one veterinarian, one non-scientist, and one person not affiliated with the institution.

In 1985, Congress passed the Health Research Extension Act, which is actually an amendment to the Public Health Service Act of 1978. This act made the NIH voluntary guidelines law. In the same year, Congress passed amendments to the Animal Welfare Act that required the establishment of a physical environment to promote the psychological well-being of primates, provided for the physical exercising of dogs, and specified that there be research on alternative methods of experimentation.

Scientists have supported the intent and purpose of these laws but have expresed concern about their impact on the conduct of research and the manner of their implementation. Many institutions already meet the requirements or have taken steps to conform in every detail; however, to bring all laboratories in the country into strict conformance with rigid standards set by the laws and guidelines require an estimated expenditure of close to half a billion dollars. In a time of budget constraint, those funds could only come from diversion from the monies allocated to research itself.

Scientists also are concerned about the potential impact on research of the newer statutes. One would require that the animal care committees enforce the act by regularly inspecting the laboratory of each researcher in the institution for compliance with the law. Animal care committees do not have the manpower or administrative or financial resources to carry out this directive.

Another major concern expressed by scientists is the lack of allowance for the differences in the size of research institutions, the kinds of research being conducted, or the types of animals being used. There is also concern about the increasing effort of animal rights activists to become members of institutional animal care committees and to convince local governmental bodies to form animal research oversight committees. The fear among scientists is that animal rights activists would disrupt the work of institutional animal care committees and/or gain control of local animal research oversight committees and, given their philosophy, would deny most proposed experiments.

B. Pound laws

Pound laws are used in experiments for both scientific and ecomomic reasons. Many experiments require the use of what are known as "purpose-bred" animals—those bred for a particular trait or whose genealogy or physiology must be known in order for the experimental results to be valid. Such animals are bred and sold by professional dealers. Other experiments can be conducted with what are known as "random source" animals whose ancestry and physiologic history are not known. Such animals are preferable in some experiments because their unknown and varied backgrounds approximate that found among a human population. One source of such animals has been pounds.

During the past 10 years, the passage of laws to ban the use of animals obtained from public pounds or private shelters in laboratory experiments has been the issue of highest priority to many animal welfare organizations. For these organizations, the passage of such laws has been an end in itself and is inspired by a concern that lost, abandoned, or surrendered pets might become experimental subjects in a research laboratory. In recent years, they have been joined by animal rights organizations who have seen the issue as a device for uniting the movement and as a tactic to reduce or restrict the performance of experiments.

All 50 states and most communities have some form of law requiring the seizure, detention, and humane destruction of stray or unclaimed animals. The stated purpose of these laws is to protect human health and safety. Under these laws, each year between 10 and 15 million animals are taken in by pounds in the United States. Most of these are animals turned in by their owners or are abandoned strays. If unclaimed within a specified time, usually several days, the animals are killed by the pound. This delay has been built into the laws to protect any lost pet that gets picked up. Few animals are released to a researcher or put to death unless this time has elapsed. The number of animals released to researchers represents less than 2% of the number put to death in pounds each year.

State, county, and municipal laws vary in their provisions for releasing animals to research laboratories. Among the states, many have no specific provision at all and, unless prohibited by local statute, pound animals in these states are generally released to laboratories. A number of other states have specific provisions requiring the release of animals to laboratories. These provisions were generally adopted in the 1950s when scientific research began to burgeon and a ready supply of animals was needed.

In response to the activities of the animal welfare and animal groups rights that began in the late 1970s, a number of states, particularly those in the East, have passed or have considered laws to restrict or ban the use of animals taken from pounds or shelters in laboratory experiments. Some states prohibit the use of pound animals obtained within the state but permit their importation from other states. The most comprehensive pound law adopted by Massachusetts, bans the use of pound animals irrespective of course.

The use of pound animals reduces the expense of conducting experiments. Animals obtained from commercial sources cost from two to five times more than animals obtained from pounds. With its ban on the use of all pound animals, it is estimated that the additional cost of conducting experiments in Massachusetts will be $6 million a year. If funds to cover the increased expense are not forthcoming from some source, many experiments that should be conducted will be abandoned. Even in those states that permit the importation of pound animals from other states, scientists point out that the effect of the laws is to increase the costs of conducting experiments because of the expense of purchase and transportation. A ban on the use of pound animals nationally would add millions each year to the cost of conducting research.

Another effect of these laws, scientists note, is to cause additional animals to lose their lives. When animals from pounds are not available, other animals must be used. Since animals may be put to death at the completion of an experiment, this means that two animals die instead of one—the experimental animal and the unclaimed pound animal.

Scientists also question whether these laws really represent what the people want. No evidence exists that the majority of pet owners giving their animals to pounds object to the use of these animals for research. Although many state and local legislative bodies, under strong pressure from animal cause groups have restricted or banned the use of pound animals, in communities or counties where the issue has been put on the ballot for decision by referendum, the voters have chosen not to adopt such a ban.

VII. Philosophical and Moral Issues

The goal of improved animal welfare, scientists say, may represent only a part of the agenda of the animal rights movement. In the view of some scientists, published statements and public positions used by animal rights activists are motivated by an anti-science and anti-technology philosophy. Disaffection with modern society and the rapid changes associated with the growth and development of technology appear to be factors which underlie these attitudes.

The philosophical issue posed by the animal rights movement is derived largely from the philosophy of Peter Singer: whether animals have rights. Much time and many words have been spent arguing this question during the past 10-15 years. As with all philosophical debate, the issue can only be resolved on the basis of assumptions. If one accepts, as Singer and other animal right advocates do, that all forms of animal life possess qualities in common that endow them with equivalent moral status to humans, then it can be argued logically that all have equivalent rights to humans which must be recognized and enforced by law. If, on the other hand, one believes that only forms of life that possess a moral sense or some other attribute that only humans possess can have rights, as some have argued, then animals cannot be said to have rights.

Some within the animal rights movement, notably American philosopher, Tom Regan, have shifted away from the concept of rights (as has Singer himself to some degree) and advanced the argument that it is precisely because animals are inferior to humans that they are entitled to and should be given the same consideration as "moral patients" (ie, that given to human babies and the mentally incompetent). Humans, as the superior and rational being, have the moral obligation to protect them and not to abuse them or to use them against their own interests, say these animal rights advocates.

This may shift the question from one of rights to be claimed by animals to one of responsibilities to be met by humans, but it does not avert the central issue: whether humans can rightfully use animals against their own individual interests for our own benefit. As with most moral issues, the question has another side. Do we in light of the consequences, have the right NOT to use animals in experiments?

How humane or moral is it, to deprive the human victims of Alzheimer's disease, stroke, disabling injury, heart disease, cancer, and countless other diseases or disorders, of hope, knowing that the solution to these problems that afflict a large and growing proportion of the population can only be solved or lessened through the conduct of biomedical research with animals? To many scientists, that is the real moral issue. Sometimes it is necessary to inflict pain or to sacrifice the life of an animal in the laboratory to avoid or eliminate suffering and pain—or even save a greater number of human and animal lives.

Many primates suffered and died in the process of developing the vaccines for poliomyelitis, but because they did, the number of poliomyelitis cases in the United States alone declined from 58,000 in 1952 to four in 1984. Results on a world wide basis are equally impressive. That means that in each of the intervening years and for every year in the future, thousands of human beings have not, do not, and will not spend their lives in wheelchairs or walking in braces or lie trapped and suffering in iron lungs. Given these results, scientists and other defenders of animal use in biomedical research might ask whether the use of those primates served a moral purpose? The answer given by most people would be a resounding yes.

VIII. Conclusion

The activities and arguments of animal rights and animal welfare activists and organizations present the American people with some fundamental decisions that must be made regarding the use of animals in biomedical research.

The fundamental issue raised by the philosophy of the animal rights movement is whether humans have the right to use animals in a way that causes them to suffer and die. To accept the philosophical and moral viewpoint of the animal rights movement would require a total ban on the use of animals in any scientific research and testing. The consequences of such a step were set forth by the OTA in its report to Congress: "Implementation of this option would effectively arrest most basic biomedical and behavioral research and toxicological testing in the US." The economic and public health consequences of that, the OTA warned Congress, "are so unpredictable and speculative that this course of action should be considered dangerous."

No nation and no jurisdiction within the US has yet adopted such a ban. Although, as noted earlier, laws to ban the use of animals in biomedical research have been introduced into a number of state legislatures, neither a majority of the American people nor their elected representatives have ever supported these bills.

Another aspect of the use of animals in biomedical research that has received little consideration is the economic consequences of regulatory change. Clearly, other nations are not curtailing the use of animals to any significant degree. Some of these, like Japan, are major competitors of the US in biomedical research. Given the economic climate in the US, our massive trade imbalance and our loss of leadership in many areas, can the US afford not to keep a leading industry, ie, biomedical science, developing as rapidly as possible? Many nations are in positions to assume leadership roles, and the long term economic impact on our citizens could be profound. This economic impact would be expressed in many ways, not the least of which would certainly be a reduction in the quality and number of health services available for people who need them.

Through polls and by other means, the American people have indicated that they support the use of animals in research and testing. At the same time they have expressed a strong wish that the animals be protected against any unnecessary pain and suffering. The true question, therefore, is how to achieve this without interfering with the performance of necessary research. Scientists already comply with a host of federal, state, municipal, and institutional guidelines and laws. However, in this era of cost containment, they fear that overregulation will become so costly that research progress will suffer. Scientists emphasize that a reasonable balance must be achieved between increased restrictions and increased cost.

What must be recognized, say scientists, is that it is not possible to protect all animals against all pain and still conduct meaningful research. No legislation and no standard of humane care can eliminate this necessity. The only alternative is either to eliminate the research, as animal rights adherents urge, and forego the knowledge and the benefits of health-related research that would result, or to inflict the pain and suffering on human beings by using them as research subjects.

The desire by animal welfare proponents to ensure maximum comfort and minimal pain to research animals is understandable and appeals to scientists, the public, and to legislators. But what also must be recognized and weighed in the balance is the price paid in terms of human pain and suffering if overly protective measures are adopted that impede or prevent the use of animals in biomedical research.

In short, the American people should not be misled by emotional appeals and philosophic rhetoric on this issue. Biomedical research using animals is essential to continued progress in clinical medicine. Animal research holds the key for solutions to AIDS, cancer, heart disease, aging and congenital defects. In discussing legislation concerning animal experimentation, the prominent physician and physiologist, Dr. Walter B. Cannon, stated in 1896 that "... the antivivisectionists are the second of the two types Theodore Roosevelt described when he said, 'Common sense without conscience may lead to crime, but conscience without common sense may lead to folly, which is the handmaiden of crime.'"

The AMA has been an outspoken proponent of biomedical research for over 100 years, and that tradition continues today. The AMA believes that research involving animals is absolutely essential to maintaining and improving the health of the American people. The AMA is opposed to any legislation or regulation that would inappropriately limit such research, and actively supports all legislative efforts to ensure the responsible use of animals in research, while providng for their humane treatment.

IX. Bibliography

Nearly 200 papers, books, monographs, and articles were consulted in the preparation of this document. The following short list will give the reader perspectives from the viewpoint of the scientist and the animal rights advocate.

Godlovitch S, Godlovitch R, Harris J (eds): Animals, Men and Morals. New York, Taplinger, Inc, 1971.

Ryder R: Victims of Science. London, Davis-Poynter, 1975.

Singer P: Animal Liberation: A New Ethic for Our Treatment of Animals. New York, Random House, 1975.

Rowan AN, Rollin BE: Animal research—for and against: A philosophical, social, and historical perspective. Perspect Biol Med 27:1-17, 1983.

US Congress, Office of Technology Assessment: Alternatives to Animal Use in Research, Testing, and Education. Washington, DC, US Government Printing Office, OTA-BA-273, February 1986.

Trull FL, Kalikow NA: Animal rights movement: A threat to biomedical research: Cancer Invest 2:479-482, 1984.

McCabe K: Who will live, who will die? The Washingtonian. 21:112-156, 1986.

The Biomedical Investigator's Handbook. Foundation for Biomedical Research, Washington DC, 1987.

Feeney DM: Human rights and animal welfare. Am Psychologist 42:593-599, 1987.

Leader RW, Stark O: The importance of animals in biomedical research. Perspect. Biol. Med. 30:470-485, 1987.

Krasney JA: Some thoughts on the value of life. Buffalo Physician 18:5-13, 1984.

Smith SJ, Evans RM, Sullivan-Fowler M, et al: Use of animals in biomedical research: Historical role of the American Medical Association and the American physician. Arch Int Med 148:1849-1853, 1988.

Smith SJ, Hendee WR: Commentary on animals in research. JAMA 259:2007-2008, 1988.

Jasper JM, Nelkin D. The Animal Rights Crusade. The Growth of a Moral Protest. New York, New York, The Free Press, 1992.

Loeb JM, Hendee WR, Smith SJ, Schwarz MR: Human vs Animal Rights. JAMA 262:2716-2720, 1989.

Council on Scientific Affairs: Animals in Research. JAMA 261:3602-3606, 1989.

Council on Scientific Affairs: Use of Animals in Medical Education. JAMA 266:836-837, 1991.

Americans for Medical Progress Educational Foundation
*Helping the Public Understand Animal Research in Medicine*SM

421 King Street • Suite 401 • Alexandria, VA 22314-3121
Phone (703) 836-9595 • Fax (703) 836-9594 • Email AMPEF@aol.com • http://www.ampef.org

Animal Rights '97 National Convention

Activists Call for Direct Action Against Research

An AMD Intelligence Report, July 1997

Discerning the current state of the animal rights movement, which held a national convention June 26-30 in the Washington, D.C. area, calls into play the same skills once held by Kremlinologists. Those experts could tell us much about the internal politics of the former Soviet Union by the relative positions of the leaders on a parade reviewing stand, who introduced whom, or who was missing from an event. So at the National Animal Rights Convention '97, insight into the leadership, evolution and future direction of the movement came not only from what was said, but by who was doing the talking, who was listening and who sat with whom on the dais.

The Scene: Straight-Edgers and Vegan Merchandisers

The convention, which ran from Thursday evening, June 26 and continued through a day of lobbying on Capitol Hill Monday, June 30, attracted between 400 and 500 participants. Its theme was "Charting Our Course Into the Next Century..." The convention is usually held by the National Alliance for Animals, but according to an activist from Vegan Outreach, this year FARM (Farm Animal Reform Movement) took over leadership because the Alliance was "burnt out" from last year's activities. You will remember that last year Americans for Medical Progress, working with AIDS research advocates, generated much negative publicity for the convention and the animal rights movement as a whole, by focusing on the threat animal rights activism posed to AIDS and other disease research.

It was estimated that last year's events, including a march on Capitol Hill, drew 3,000 animal activists, up to that point an all-time low for such gatherings. This year, organizers did not court media attention for the convention, nor did they seem to actively attempt to recruit participants beyond a narrow circle of activists.

Perhaps a third of the participants in this year's gathering fit the traditional image of an animal activist—mostly female, middle income, aged 30-50, with a primary interest in local animal rights organizing. However, by far, the majority at the convention were younger activists, in their 20s, mostly female. There was a noticeable contingent of "Straight Edge" males, marked by extremely short hair and body piercing. A faction of the "Straight Edge" movement is believed to be involved with recent criminal actions of the Animal Liberation Front. While at the convention, this observer did not see a single non-caucasian participant.

In the third floor exhibit space, after the vegan and New Age merchandisers, PETA was the overwhelming presence, with separate tables devoted to its Boys Town and Premarin campaigns. There was no reference in PETA's materials to its recent undercover operation at

Huntingdon Life Sciences, most likely due to the court-imposed gag order against it. In Defense of Animals' table was predictably focused on its campaigns against The Rockefeller University, The Coulston Foundation, and Oregon Health Sciences University. The Humane Society of the U.S., Physicians Committee for Responsible Medicine, Farm Animal Reform Movement, American Anti-Vivisection Society, World Society for the Protection of Animals and the Animal Protection Institute all had tables filled with their grassroots literature, merchandise and sign-up sheets. Of special interest was a table by the relatively new group SAEN—Stop Animal Exploitation Now. Founder Michael Burkie was offering boilerplate FOIA forms and providing individual factsheets about animal research being conducted in over a dozen universities. In a half hour conversation, he even tried to recruit this observer into starting a SAEN chapter.

Younger Leaders Head Drive for Direct Action

We have all sensed over the past few years that the animal rights movement is in transition, with leadership roles that had belonged to a founding generation of activists—including Cleveland Amory, Ingrid Newkirk, Kim Stallwood, Henry Spira, Don Barnes, Elliott Katz—being conferred on a new group of younger activists—JP Goodwin, Cres Velluci, Larry Carter-Long, Wayne Pacelle, Heidi Prescott, Lisa Lange and others. This was borne out in the programming of the convention. To be sure, the old guard was present (with some notable exceptions such as Amory and PETA co-founder Alex Pacheco) and fully participated in the workshops and panel discussions. The Friday night banquet paid tribute to movement pioneers. But the energy and focus on future tactics was provided by the upcoming generation of movement leaders.

This dynamic was clearly seen in one of the opening workshops, which featured PETA co-founder Ingrid Newkirk, movement veteran Kim Stallwood and JP Goodwin of the Committee to Abolish the Fur Trade. The session, entitled Gaining Public Attention, was moderated by Stallwood, editor of *The Animal's Agenda*. Over 200 activists were in attendance.

PETA co-founder Ingrid Newkirk did tell the audience "I wish we all would get up and go into the labs and take the animals out or burn them down," but in general her approach was more to the moderate members of the audience who might be uncomfortable with direct action. "There are all sorts of quick things people can do," she encouraged, mentioning talking to store managers, calling talk shows, writing letters to the editor and putting vegan stickers on mirrors in restaurant rest rooms. "Our opposition is very 9-5, thank God. They are very punch time-clock kind of people and if they are not being paid to oppose us, they are usually not bothering to oppose us, whereas our strength is in our volunteerism—is in the fact we care." "Use your mouth, use your voice, use your library, use your body, use your wallet, use your life," she cajoled.

She addressed the criticism within the movement of PETA's tactics. "I never thought there was one true way to accomplish animal rights. We are all very different people and we all gave very different things that move us, so somebody's style is not somebody else's style...Just thank goodness that anyone is doing anything at all."

Newkirk seemed concerned about the growing split in the movement between those supporting the Animal Liberation Front and those less sanguine about illegal acts being committed in the name of animal rights. "Some people are uncomfortable about the ALF, some people are uncomfortable about wearing costumes.... Animal rights is not about comfort. The movement is not a walk in the park, it is not a tea party, it is a revolution," Newkirk said.

Goodwin, on the other hand, clearly was speaking strictly to those who supported animal "liberations" and other direct actions. "We have found that civil disobedience and direct action has been powerful in generating massive attention in our communities and in energizing the local activists and giving them a newfound momentum, and has been very effective in traumatizing our targets." He spoke of hunger strikes, and "some pretty intense anti-vivisection actions" at UC Davis and Yerkes. "There is a whole new upsurge of direct action grassroots activism that has occurred in the past few years. Some of the hot spots are Syracuse, Dallas, Atlanta and Minnesota—we are all friends, we are all connected." "The animal rights movement is a serious revolutionary movement—because this is revolutionary—a serious movement for social change and a threat to anyone who profits from the exploitation of animals."

Most of the other workshops featured the same paradigm—a veteran of the movement would moderate and provide context, but it was the emerging leaders, honed by recent campaigns attacking research laboratories, wildlife conservation and fur farms, who carried the discussion with encouragements for greater involvement and tougher tactics.

The workshops were divided into 3 types: issues (such as The Tragedy of Animal Agriculture, Abuse of Companion and Exhibit Animals), organizing (Group Relationships, Conflict Resolution, Fundraising), and outreach (Reaching School Children, Reaching Out With Computers).

New Initiatives Against Animal Research

The animal research issue was given close scrutiny with three separate sessions on Friday afternoon: The Tragedy of Vivisection, Engaging Vivisectors, and Laboratory Animal Tactics.

The big news coming out of the panels was a sense that the animal activists, pleased with the outcome of ballot initiatives on the hunting issue last fall, will begin to use the same tactic on animal research. This was predicted by AMP in the wake of the initiatives battles last fall.

"A lot of people think that the next stage of ballot initiatives might be relating to vivisection, dissection, and so forth," Dean Smith, Director of Research for the American Anti-Vivisection Society told the convention. "This is something definitely to pursue. There are good potentials and good possibilities with ballot initiatives especially regarding this issue...Regarding product testing and some things like that there are some real possibilities, so take a look at that avenue as well." His comments were echoed by Elliott Katz, founder of In Defense of Animals, who was moderating the session.

Katz spoke of his frustration with the legislative process. "The biomedical community is a tremendous lobbying force," he said. He told of working with Sen. Barbara Boxer of California over a specific animal research project to have been conducted by the U.S. Army. IDA had the experiment stopped. "But there was such a tirade of letters and calls to Boxer from literally every university and medical school that now she will only look at whale and dolphin (issues)." "To get a piece of legislation through, especially now in the case of a Republican congress, is virtually nil," unless it is attached to an appropriations bill, Katz cautioned.

Martin Stephens, Director of Laboratory Animal Issues for the Humane Society of the United States agreed. "The tactic of trying to get things attached to larger bills can be quite effective. You can usually do that without too many people noticing and get the funding cut for something. We were almost successful for getting the funding cut for NASA's Bion space project last year. Sometimes you can do things quietly like that." (NB: Bion's funding was cut in the House, but restored in the Senate. AMP was the only pro-research organization to testify in favor of Bion and oppose the HSUS during the funding battle.)

On two separate occasions, Don Barnes of the National Anti-Vivisection Society talked about reigniting the fires he claimed were burning in the early 1980s during the time of the Silver Springs Monkeys case. He urged the audience to use the occasion of Dr. Edward Taub's recent honor from the American Physiological Society. "We lost a great deal of media interest when the Taub issue went away...This may be an opportunity for those of us who were involved in those days to scream in outrage...Let's go back on the debate circuit. If we can resurrect the pictures and images from the Taub trial we may be able to do a great deal more damage than we did the first time around. We need your help to get the vivisection issue back into the headlines again."

Neil Barnard of PCRM spoke both of his three-year-old, 80 city campaign to stop charities from funding animal research and of his attempts to stop dissection in education. He is claiming partial victory among medical schools, now that fully 50 percent no longer have dog labs, and of the remaining schools, all dog labs are optional, with the exception of the military medical school.

"These are just areas that were waiting for us to pluck them and win," he said. "Some of them you win in a minute, some may take 5 years or 10 years. We can outlast them, we are clearly right and sooner or later we will win."

Larry Carter-Long, now Coordinator for Science and Research Issues for the Animal Protection Institute in Sacramento, offered advice on "engaging vivisectors" in the media, debates, and other public fora:

1. Appeal to emotion.

2. Talk about the human costs; "the people not treated because of the time and resources taken up by animal experimentation."

3. Talk about the non-human costs, "the waste of non-human animal lives."

"Show the pictures! Show the videos! Make people aware of what is actually going on and say 'if it disturbs you, folks, it should. It is disturbing to see. It is disturbing to think about.' Our role here is to get people to think about it," he urged. "You're not talking to the vivisector—you are talking to the audience. The question is how are you going to get them?"

Both Carter-Long and Dean Smith of AAVS urged the activists not to argue the science of research in a debate. "If you are talking to a scientist who knows their stuff, 99 per cent of the time they will be able to make you look like a fool. You know ethics, and 99 per cent of the time you can make the scientist look like a fool when talking about the ethics of the issue. So stick to that; stick to the ethics and really bring that up because they can't respond."

Smith also talked about how the AAVS identifies its campaign targets. He says there are three criteria for an ideal research target:

"1. A target that gets an emotional reaction from the public. Animals that people know a lot about: cats, dogs, primates. This also plays the function of making the animal experimenter lose credibility. It makes them look callous when they are using these types of animals—these are people's pets.

2. Performance—where they are performing unnecessary procedures...(such as) animal labs where students are given an option not to do it. If they are given an option, then why are we doing it at all?

3. The use of animals where there are alternatives available. This breaks down the notion that the researchers really care about the animals."

Smith urged activists to develop a long term strategy for their campaigns against research. "A protest is only effective for a day or so. We have to develop a long-range plan with increasing pressures, pressures that will build over a year or two. Correspond with the institution or government officials first. Then move on to something a little heavier—advertising, leafletting—so then they start getting reaction from the public. Then move on to protests."

Martin Stephens of the HSUS talked in terms of pressure points on the research community. "We tend usually just to think of the uses of animals as pressure points, but you can also think of the whole cycle of animal research, starting out with the procurement of animals. Where do these animals come from? Under what conditions are they captured or bred? How are they transported to the facility? How are they cared for and housed? All of these things are pressure points. Of course, how are they used, but even afterwards, how are they 'disposed of?' All of these things are opportunities for targeting efforts."

Stephens, Smith and Katz all urged the activists to work with the national organizations and offered the resources of their respective groups. "But you guys'd be the ones out there," cautioned Katz. We'd be in the background because many communities don't want to be told by outsiders what needs to be done."

"Don't feel incapacitated because the end of vivisection is not in sight," championed Stephens of the HSUS. "There is plenty of work to be done in the meantime."

Making Their Voices Heard

While the activists didn't solicit media coverage for their convention, the movement received national news coverage on Sunday, when over 200 protesters marched from the convention hotel to a McDonalds restaurant that had the misfortune of being located across the street. Among the signs the activists carried were those containing the slogans "Meat is Murder" and "Dairy is Rape."

Several activists entered the restaurant and began harassing customers and throwing chairs, trays and condiments. Over 60 local and state police officers responded in full riot gear, and were forced to use pepper gas to disperse the demonstrators. The protesters spit at the police and threw rocks at them. The busy highway in front of the McDonalds was shut down for nearly two hours. At least 18 activists, including several juveniles, were arrested. News reports termed the disturbance a riot.

Earlier in the weekend, demonstrators from the convention attacked Miller's Furs in downtown Washington, D.C., shattering every window and burning an effigy at the front door of the store. There was an incident at a circus being held at RFK Stadium in which several rocks were thrown. Activists also gathered in front of the Health and Human Services building in Washington on Friday for an early morning protest calling for an end to animal research.

Into the Next Century

It should not be overlooked that the convention, which had the theme "Charting Our Course Into the Next Century..." was cosponsored not only by PETA, FARM, In Defense of Animals, the New England Anti-Vivisection Society and the International Society for Animal Rights, but also by the Humane Society of the United States. Responding to recent AMP materials demonstrating the links between the HSUS and PETA, HSUS officials steadfastly maintain that they are not part of the extreme animal rights movement. Yet six members of the upper management of the HSUS spoke at the convention and Vice President Wayne Pacelle served as coordinator of the convention's Lobby Day activities. For HSUS spin doctors to insist the organization is not part of the movement's leadership is disingeneous, at best.

Every indication from the convention—the rhetoric of the speakers, the responsiveness of the attendees, the emotionalism of the literature and the militancy of the protest—suggests the course into the next century will be marked by an escalation in confrontational tactics, violence, and personal attacks against researchers as a new generation of leaders solidifies its control over the animal rights movement.

Americans for Medical Progress Educational Foundation
Helping the Public Understand Animal Research in Medicine℠

421 King Street • Suite 401 • Alexandria, VA 22314-3121
Phone (703) 836-9595 • Fax (703) 836-9594 • Email AMPEF@aol.com • http://www.ampef.org

Animal Rights: What's at Stake

Americans have a life expectancy that's more than 25 years longer that it was at the turn of the century, **because of scientists' work with animals in developing medical cures and treatments.**

Polio, smallpox, cholera, diphtheria, rabies, whooping cough and measles are no longer the threats to public health they once were. Effective treatments for killers such as diabetes, pneumonia, and leukemia have been found. Sophisticated diagnostic tests mean early treatment of cancer. Complicated surgical techniques have opened the way for coronary bypasses, joint replacements and organ transplants. **None of these advances would have been possible without responsible animal research.**

Now, continuing their work with laboratory animals, scientists are on the brink of discovering ways to combat Alzheimer's, Parkinson's, AIDS, cystic fibrosis, spinal cord injuries, and other ills. **Research with laboratory animals is the vital link between ideas on the drawing board and human application.**

Incredibly, "animal rights" extremists oppose any use of animals in medical research. As one of their leaders, PETA's Ingrid Newkirk, has stated, even if animal research resulted in a cure for AIDS, "we'd be against it."

Animal activist groups raise over $200 million each year and spend it on campaigns of misinformation, intimidation, even violence.

"Animal rights" lobbyists have had great success stopping research with legislative red tape and legislation. These activists are responsible for a blizzard of bills before the U.S. Congress and state legislatures each year, designed to drive up the costs of medical research without contributing to the actual welfare of the animals involved. They have helped shift billions of dollars in medical research funds to compliance with excessive regulations that simply waste scarce resources.

The more extreme elements are resorting to harsher tactics—**firebombing, vandalism, threats, slander, burglary, physical harassment and even attempted murder**—in their efforts to hobble the advance of medical science and the saving of human lives. The U.S. Justice Department has documented over 300 such incidents. Security experts believe that number is just a page out of the actual catalog of violence. The same Justice Department report estimates security costs for animal facilities has risen anywhere from 10 to 20 percent because of violent extremist activity—**an increase reflected in higher health care costs for society.**

Lies, intimidation and violence by the animal activists have delayed the benefits of vital research from reaching the public, driven scientists out of certain areas of inquiry, and dissuaded promising students from careers in medical research.

Americans for Medical Progress works to ensure that scientists and doctors have the freedom and resources necessary to pursue their life-saving and life-enhancing research. Through newspaper and magazine articles, broadcast debates and public education materials, AMP exposes the dangers of the "animal rights" philosophy and serves as the premiere public advocate for medical research involving animals.

Americans for Medical Progress Educational Foundation
Helping the Public Understand Animal Research in Medicine℠

421 King Street • Suite 401 • Alexandria, VA 22314-3121
Phone (703) 836-9595 • Fax (703) 836-9594 • Email AMPEF@aol.com • http://www.ampef.org

Testimony to the Value of Animal Research

"Animal experimentation has been essential to the development of all cardiac surgery, transplantation surgery, joint replacements and all vaccinations."

—Joseph E. Murray, M.D.
Professor Emeritus, Harvard Medical School
1990 Nobel Laureate

"There is no alternative to the use of animals for analyzing the complexity of immunity. Progress in all areas of medicine is enormously enhanced by the new gene 'knock-out' and transgenic mouse technologies."

—Peter C. Doherty, Ph.D.
Chair, Department of Immunology
St. Jude Children's Research Hospital
1996 Noble Laureate

"Now, more than ever, research with laboratory animals is required to bring the benefits of advances in molecular genetics, neuroscience, and other highly productive fields to clinical application through the study of intact organisms."

—Harold E. Varmus, M.D.
Director, National Institutes of Health
1989 Nobel Laureate

"Without the use of animals and human beings, it would have been impossible to acquire the important knowledge needed to prevent much suffering and premature death not only among humans, but also among animals."

—Albert B. Sabin, M.D.
Developer of Polio Vaccine

"Animal experiments have contributed greatly to the understanding, prevention and control of human illness. Animal experiments will continue to contribute greatly to the understanding and control of human illness."

—Alan Goldberg, Ph.D.
Director, Center for the Development of
Alternatives to Animal Testing
John Hopkins University

"Animal research—followed by human clinical study—is absolutely necessary to find the causes and cures for so many deadly threats from AIDS to cancer."

—Daniel H. Johnson, Jr., M.D.
President, American Medical Association

Americans for Medical Progress Educational Foundation
*Helping the Public Understand Animal Research in Medicine*SM

421 King Street • Suite 401 • Alexandria, VA 22314-3121
Phone (703) 836-9595 • Fax (703) 836-9594 • Email AMPEF@aol.com • http://www.ampef.org

21 Things You May Not Know About the Animal Rights Movement

...But Probably Should

1. The animal rights movement believes that a rat's life has the same value as a child's.
2. Animal rights leaders believe the following activities are morally wrong: owning a pet, going to the zoo or circus, swatting a mosquito, wearing leather, suede and silk, eating meat, drinking milk, riding a horse and fishing.
3. PETA's co-founder, Ingrid Newkirk, compared killing broiler chickens with killing Jews in Nazi concentration camps.
4. The animal rights industry raises and spends over $300 million annually.
5. The U.S. Department of Justice reports that animal activists have shown "an increasing willingness to engage in more militant and costly activities."
6. PETA is a media mouthpiece for the Animal Liberation Front, which the FBI calls a domestic terrorist organization.
7. PETA's co-founder, Alex Pacheco, says arson, property destruction, burglary and theft are 'acceptable crimes' when used for the animals' cause.
8. In 1995, PETA sent $45,200 to convicted animal rights terrorist Rodney Coronado.
9. Animal rights violence has increased the cost of medical research by as much as 20% according to the U.S. Department of Justice.
10. PETA killed 32 animals it had "rescued" because it "ran out of space to house them."
11. The Humane Society of the United States (HSUS) paid its two top executives more than $440,000 combined in salary last year.
12. The HSUS is now as extreme as PETA. Michael Fox, a HSUS vice president, wrote, "The life of an ant and that of my child should be granted equal consideration."
13. The HSUS took $1 million from its Canadian affiliate's bank account and was ordered by the Canadian courts to pay it back. The judge derided the HSUS' "overweening arrogance" in taking the funds.
14. The HSUS raises more than $40 million annually, yet does not operate a single shelter.
15. Approximately 11 million animals are destroyed in shelters, yet PETA spent less than one-half of one percent of its $10.9 million in 1996 on shelters in the U.S.
16. PETA co-founder Newkirk says even if animal research resulted in a care for AIDS, "we'd be against it."
17. Animal activists blocked for two years research aimed at stopping transmission of HIV from mother to child. That research ultimately demonstrated how AZT can prevent babies from getting AIDS.
18. Animal activists oppose breast cancer research involving mice, rats and fruit flies.
19. PETA celebrity spokesman Alec Baldwin says we don't need animal research because there are "a lot of human subjects who would be more than willing to become live experiments."
20. PETA attacks children's charities such as the Make-A-Wish Foundation, Boys Town and the March of Dimes.
21. Animal rights = no more cures for diseases.

Full-page ad in *Roll Call*, Capitol Hill's newspaper, June 30, 1997

Some Call It Protecting Animal Rights

Photo courtesy of UC Davis News Service archives. Jim West Photographer.

The Fact Is It's Terrorism.

The animal rights movement employs a two-pronged strategy. Its covert wing firebombs buildings and destroys medical research, while its political wing sharpens its soundbites and heads for Capitol Hill.

The animal rights pitch is polished. Animal rights leaders masterfully hoodwink animal lovers into believing their mission is to protect suffering animals. But their true agenda calls for a radical redefinition of our relationship with animals. And these activists will stop at nothing to see it through—lies, intimidation and even outright violence.

People for the Ethical Treatment of Animals (PETA), one of the largest animal rights groups, considered by many to be "mainstream," serves as a media mouthpiece for the Animal Liberation Front (ALF). The ALF is a shadowy group which claims responsibility for most animal rights terrorism and destruction, including the 1987 University of California, Davis attack pictured above.

Arson, property destruction, burglary, and theft are "acceptable crimes" when used for the animals' cause.

Alex Pacheco, Co-founder, Director People for the Ethical Treatment of Animals (PETA)
The Associated Press, January 15, 1989

In 1995, PETA even sent $45,200 of its donors' money to terrorist Rodney Coronado who was convicted of firebombing a Michigan State University research facility—while donating less than $5,000 to shelters nationwide.

The U.S. Department of Justice estimates animal rights violence has increased the cost of medical research by as much as 20%. Millions of taxpayer dollars have gone up in smoke due to animal rights terrorism. Most importantly, the search for cures and treatments has been obstructed.

Just this spring, FBI Director Louis Freeh told Congress the activities of the Animal Liberation Front and Coronado are prime examples of domestic terrorism that must be stopped. We agree.

When animal rights lobbyists knock on your door, tell them you have a zero-tolerance policy for terrorists. And send them on their way. That's the way your voters would want it.

For more information on the animal rights threat to health, contact:

Americans for Medical Progress Educational Foundation
Helping the Public Understand Animal Research in Medicine℠
421 King Street • Suite 401 • Alexandria, VA 22314 • (703) 836-9595
E-mail: AMProgress@aol.com • Website: http://www.ampef.org

Americans for Medical Progress Educational Foundation
Helping the Public Understand Animal Research in Medicine℠

421 King Street • Suite 401 • Alexandria, VA 22314-3121
Phone (703) 836-9595 • Fax (703) 836-9594 • Email AMPEF@aol.com • http://www.ampef.org

The Animal Rights Movement...
...In the Words of Its Leaders

Ingrid Newkirk, Founder, People for the Ethical Treatment of Animals (PETA)

- "Animal liberationists do not separate out the human animal, so there is no rational basis for saying that a human being has special rights. A rat is a pig is a dog is a boy. They're all mammals." (*Vogue*, September, 1989)
- "Six million Jews died in concentration camps, but six billion broiler chickens will die this year in slaughterhouses." (*Washington Post*, 1983)
- "Humans have grown like a cancer. We're the biggest blight on the face of the earth." (*Reader's Digest*, June, 1990)
- Even if animal research resulted in a cure for AIDS, "we'd be against it." (*Vogue*, September, 1989)
- "It (animal research) is immoral even if it's essential." (*Washington Post*, May 30, 1989)
- "Probably everything we do is a publicity stunt...We are not here to gather members, to please, to placate, to make friends. We're here to hold the radical line." (*USA Today*, September 3, 1991)
- "I am not a morose person, but I would rather not be here. I don't have any reverence for life, only for the entities themselves. I would rather see a blank space where I am. This will sound like fruitcake stuff again but at least I wouldn't be harming anything." (*Washington Post*, November 13, 1983)
- "You don't have to own squirrels and starlings to get enjoyment from them...One day, we would like an end to pet shops and the breeding of animals. [Dogs] would pursue their natural lives in the wild...They would have full lives, not waiting at home for someone to come home in the evening and pet them and then sit there and watch TV." (*Chicago Daily Herald*, March 1, 1990)
- "Eventually companion animals would be phased out." (*Harper's Magazine*, August, 1988)

Alex Pacheco, Chairman, People for the Ethical Treatment of Animals (PETA)

- "We feel that animals have the same rights as a retarded human child because they are equal mentally in terms of dependence on others." (*New York Times*, January 14, 1989)
- "Arson, property destruction, burglary, and theft are 'acceptable crimes' when used for the animal cause." (*Gazette Mail*, Charleston, WV, January 15, 1989)

The Animal Liberation Front

- "Killing any sentient being is morally wrong but, bearing in mind the number of animal lives it would save and the amount of suffering it would relieve, I would not weep for the individual vivisector." (Robin Webb, *The Times Saturday Review*, November 7, 1992)

- "Animal liberation is...a fierce struggle that demands total commitment...There will be injuries and possibly deaths on both sides... this is sad but certain." (Ronnie Lee, *The Times Saturday Review*, November 7, 1992)

- "I would be overjoyed when the first scientist is killed by a liberation activist." (*USA TODAY*, September 3, 1991)

The Animal Liberation Action Foundation (a letter to Dr. John Orem, Texas Tech University)

- "Unless you concede to the demands of mainstream anti-vivisection groups, my organization will engage in a campaign of destruction and bloodletting upon your facilities and person more devastating than you can possibly conceive of."

- "Let me emphasize that the law is powerless to protect you. There are too many hours in too many days (and nights), and my colleagues and I are too well structured and committed."

- Dr. Orem's Texas Tech University laboratory was destroyed on July 3, 1989.

Dr. Neal Barnard, President, Physicians Committee for Responsible Medicine

- "You are going to have to worry...If you plan to do animal experiments you are vulnerable...Yes you are." (Texas Tech University, 1989)

Researchers have "set themselves up for it." (Barnard on why animal rights activists attack researchers, Texas Tech University, 1989)

Kevin Beedy, Political Scientist

- "Terrorism carries no moral or ethical connotations. It is simply the definition of a particular type of coercion...It is up to the animal rights spokespersons either to dismiss the terrorist label as propaganda or make it a badge to be proud of wearing." (*Animals' Agenda*, March, 1990)

- "Ten years from now, the animal rights movement could still be grappling with questions of dying human babies being saved because of techniques learned through animal experimentation. This argument must be rendered politically useless." (*The Animals' Agenda*, March, 1990)

Americans for Medical Progress Educational Foundation
Helping the Public Understand Animal Research in Medicine℠

421 King Street • Suite 401 • Alexandria, VA 22314-3121
Phone (703) 836-9595 • Fax (703) 836-9594 • Email AMPEF@aol.com • http://www.ampef.org

Lives Saved Thanks to Animal Research

Biomedical research using animals has contributed to an increased life expectancy of about 25 years in the U.S. since 1900. The health and well being of millions of people and animals depends on animal research.

Without Animal Research...

• Polio would kill or cripple thousands of unvaccinated children and adults this year.

• 7,500 newborns who develop jaundice each year would develop cerebral palsy, now preventable through photo therapy.

• Most of the nation's one million insulin-dependent people with diabetes wouldn't be insulin dependent—they would be dead.

• Many of the 392,111 individuals who benefited from coronary bypass surgery in 1990 would never have seen 1991.

• 1.5 million Americans would contract rubella—more than 400 times the current annual incidence of the disease.

• 60 million Americans would risk death from heart attack, stroke or kidney failure from lack of medication to control their high blood pressure.

• The more than 100,000 people with arthritis who each year receive hip replacements would walk only with great pain and difficulty or be confined to wheelchairs.

• More than one million Americans would lose vision in at least one eye this year because cataract surgery would be impossible.

• There would be no kidney dialysis to extend the lives of thousands of victims of end-stage renal disease.

• Physicians would have no chemotherapy to save the 70 percent of children who now survive acute lymphocytic leukemia.

• Hundreds of thousands of people disabled by stroke or head and spinal cord injury would not benefit from rehabilitation techniques.

• Hundreds of thousands of dogs, cats and other pets who would have died from anthrax, distemper, canine parvovirus, feline leukemia, rabies and more than 200 other diseases now preventable thanks to animal research.

Working With Animals in Research is Necessary

Scientists need to test medical treatments for efficacy and test new drugs for safety or toxicity before beginning human testing. Small animals, usually rats, are used to determine the possible side-effects of new drugs including infertility, miscarriage, birth defects, liver damage and cancer-causing potential. After animal tests have proven the safety of new drugs, patients asked to participate in further studies can be assured that they may fare better—and will not do worse—than if they were given standard treatment or no treatment.

New surgical techniques first must be carefully developed and tested in living, breathing, whole organ systems with pulmonary and circulatory systems much like ours. The physicians and physicians-in-training who perform today's delicate cardiac, ear, eye, pulmonary and brain surgeries must develop the necessary skills before patients' lives are entrusted to their care. Neither computer models, cell cultures, nor artificial substances can simulate flesh, muscle, blood, bone and organs working together in the living system.

Animals provide needed models for the study of diseases of humans and animals. For example, with biliary tracts similar to humans, prairie dogs make fine models for studying gallstone disease.

Basic biomedical research forms the cornerstone of our understanding of how the body functions in health and disease. Basic research increases the knowledge of biological and chemical processes that is essential to understanding and treating human disease.

There is No Alternative to Animal Research

Living systems are complex. The nervous system, blood and brain chemistry, gland and organ secretions, and immunological responses are all interrelated. ***It is impossible to explore, explain or predict the course of many diseases or the effects of many treatments without observing and testing the entire living system.***

Cell and tissue cultures, often suggested as "alternatives" to using animals, have been supplements to biomedical research for many years. But these are isolated tests. And isolated tests will yield isolated results, which may bear little relation to a whole living system. Scientists do not know enough about living systems or diseases to replicate one on a computer. Further, information required to build such a database will be based on animal studies.

Animals Involved in Biomedical Research Are Mostly Rodents

The number of animals used in research has declined dramatically in recent years—down 40 percent since 1968. ***The vast majority (90 to 95 percent) of animals involved in U.S. research are rats, mice and other rodents* specifically bred for research.** Only 1 to 1.5 percent of research animals are dogs and cats; only 0.5 percent are non-human primates.

Fewer than one dog or cat is involved in research for every 50 destroyed by animal pounds and humane societies engaged in animal control. Between 10.1 and 16.7 million dogs and cats are put to death in pounds and shelters annually (American Humane Society data).

Americans for Medical Progress Educational Foundation
*Helping the Public Understand Animal Research in Medicine*SM

421 King Street • Suite 401 • Alexandria, VA 22314-3121
Phone (703) 836-9595 • Fax (703) 836-9594 • Email AMPEF@aol.com • http://www.ampef.org

Federal Regulation of Animal Research

Medical research is one of the most heavily regulated activities in this country, and within the research community, animal research is most heavily regulated. In fact, up to an estimated $2 billion—about 20 percent of the medical research budget—is spent annually complying with new regulations created in response to activist pressure, and in increased security for research labs and personnel.

Fact: The Animal Welfare Act has regulated the animal care for research involving animals except rats, birds and mice since 1966 and has been amended as recently as 1991. It requires appropriate veterinary care, housing, feeding, handling, sanitation, ventilation and sheltering.

Fact: The United States Department of Agriculture (USDA) requires institutions to report the number of animals used in research and the number of animals that experience not only pain, but distress, along with an explanation of why the research had to be performed in this manner. A veterinarian must also be consulted for such research. Ninety-three percent of the experiments reported to the USDA do not involve any pain or distress.

Fact: The Animal Welfare Act requires the USDA's law enforcement arm, the Animal and Plant Health Inspection Service (APHIS), inspect all registered facilities at least once a year. The latest USDA Annual Report showed that the USDA is making an average of two visits per institution per year. The NIH also conducts unannounced visits to ensure compliance with their regulations.

Fact: The vast majority of research facilities voluntarily seek accreditation from the American Association for the Accreditation of Laboratory Animal Care (AAALAC), an organization with still higher standards for laboratory animal care.

Fact: More than half of all medical research in the U.S. is funded by the National Institutes of Health (NIH), and is therefore governed by the Public Health Service's (PHS's) *Policy on Humane Care and Use of Laboratory Animals* and the NIH's *Guide for the Care and Use of Laboratory Animals*. The *Guide* applies to all animals in a facility including rats, birds and mice and provides standards for housing care and use that must be met to retain funding.

Fact: The Animal Welfare Act and the NIH *Guide* require research institutions to establish animal care committees. The committee must contain one or more persons who are unaffiliated with the institution and one veterinarian. The committee is responsible for reviewing care and treatment of animals, inspecting the facilities and reporting any violations to the NIH. Researchers must justify to the committee in writing the number of animals they plan to use, why a certain species is necessary, and what steps will be taken to see that the animals do not suffer unnecessarily. The committee has the power to reject any research proposal.

Americans for Medical Progress Educational Foundation
Helping the Public Understand Animal Research in Medicine℠

421 King Street • Suite 401 • Alexandria, VA 22314-3121
Phone (703) 836-9595 • Fax (703) 836-9594 • Email AMPEF@aol.com • http://www.ampef.org

The Vital Role of Animals in AIDS Research

"With animals, we may have a cure for AIDS in 10 years. Without animals we will never cure AIDS in our lifetime."

—Robert Gallo, M.D.
Co-Discoverer of HIV

"We would be in absolute, utter darkness about AIDS if we hadn't done decades of basic research in animal retroviruses."

—C. Everett Koop, M.D.
Former U.S. Surgeon General

"The use of animals in biomedical research and testing has been, and will continue to be, absolutely crucial to the progress against AIDS and a wide range of other applications in both humans and animals."

—Jocelyn Elders, M.D.
Former U. S. Surgeon General

"The importance of animal models in biomedical reserach is indisputable. Without these models many years of valuable time would be lost in the discovery and perfection of treatments and vaccines for a multitude of illnesses, including AIDS."

—Kristine Gebbie, R.N.
Former Director, Office of National AIDS Policy

"Americans must decide whether they support animal research or 'animal rights.'...Once again the animal activists are wrong and we can't let a potential treatment for AIDS fall victim to their specious rhetoric."

—Joseph E. Murray, M.D.
Nobel Laureate; performed the first human kidney transplant

"Animal rights groups are the thorn in the side of progressive AIDS research."

—Golden Gate ACT-UP news release
December, 1995

"Many of the cures for diseases that are now long gone and out of the way came from animal research. If animal rights activists had their way 50 years ago, we'd be sitting around talking about all the hundreds of thousands of people dying from polio, not AIDS."

—Jeff Getty
Recipient of baboon bone marrow in AIDS research experiment

Appeared as a full-page ad in *The Hollywood Reporter* March 24, 1997, the day of the Academy Awards ceremony.

Celebrity Supporters of PETA:

Alec Baldwin
Candace Bergen
Sandra Bernhard
Ellen DeGeneres
David Duchovny

Jennie Garth
Richard Gere
Jonathan Silverman
Oliver Stone
Lily Tomlin

Is This *Really* What You Believe?

This Oscar Night Message Brought to You by:

Americans for Medical Progress Educational Foundation
*Helping the Public Understand Animal Research in Medicine*SM
421 King Street • Suite 401 • Alexandria, VA 22314 • (703) 836-9595
E-mail: AMProgress@aol.com • Website: http://www.ampef.org

We're going to expose the naked truth about PETA tomorrow.

Who are we? We are people with HIV/AIDS. Breast cancer survivors. Spinal cord injury patients. Children with birth defects. Parents of babies lost to SIDS. Teens crippled by muscular dystrophy. Women suffering from heart disease.

We know that PETA has attacked medical research that's aimed at curing every one of these ills. And now PETA dares to come to Hollywood and ask you to show your support by attending its Awards Gala tomorrow night.

Over the years, the entertainment community has been generous in lending its time and talent to raise funds for medical research. Countless lives have been saved through these efforts. And it's not just human lives. Animals, too, are benefiting from advances in biomedical research.

That is why we believe that when you discover how PETA is working against research leading to new treatments and cures for human and animal diseases, you will understand **that a choice has to be made**. You can't support both medical progress and PETA.

We urge you to stay home tomorrow night, and not attend the Gala. Send PETA's leaders a message that you cannot support their cause unless they drop their campaign against doctors and scientists.

Or better yet, join us outside the Gala at Paramount Studios tomorrow night as we expose the naked truth about PETA. Come to show the world that you've made your choice—for medical progress.

Medical Progress or PETA?
It's Your Choice.

Americans for Medical Progress Educational Foundation
Helping the Public Understand Animal Research in Medicine[SM]

421 King Street • Suite 401 • Alexandria, VA 22314-3121
Phone (703) 836-9595 • Fax (703) 836-9594 • Email AMPEF@aol.com • http://www.ampef.org

This Message Supported By:
American Heart Association[SM]
Fighting Heart Disease and Stroke
California Affiliate

California Biomedical Research Association

iiFAR
incurably ill For Animal Research
Building hope for tomorrow's cures through support of research today

Appeared as a full-page ad in *Daily Variety* March 25, 1996, the day of the Academy Awards ceremony.

The Red Ribbon You'll Wear Tonight Tells the World You've Made a Choice:

A Cure for AIDS Over Animal Rights

As you get ready for the Oscars ceremony tonight, you'll certainly reach for that one accessory that has become *derigueur* – a slip of red ribbon. As you wear it, you'll show your care and compassion for people with HIV/AIDS, and your support for finding a cure.

When a cure for AIDS is found, it will come through research with animals. Opposition by animal rights groups to such vital research threatens new treatments and potential cures for people living with HIV/AIDS.

For more information on the critical role of animal research in finding a cure for AIDS, call 1 800 4 AMP USA (1-800-426-7872)

You can't be for AIDS research AND Animal Rights

Americans for Medical Progress Educational Foundation

Helping the Public Understand Animal Research in Medicine℠

421 King Street Suite 401
Alexandria, VA 22314
email: AMProgress@aol.com
http://www.ampef.org

Americans for Medical Progress Educational Foundation
Helping the Public Understand Animal Research in Medicine℠

421 King Street • Suite 401 • Alexandria, VA 22314-3121
Phone (703) 836-9595 • Fax (703) 836-9594 • Email AMPEF@aol.com • http://www.ampef.org

HSUS—PETA Connections

The executive offices of The Humane Society of the United States are filled with officers with strong ties to People for the Ethical Treatment of Animals and its leaders:

- Wayne Pacelle, former national director of the Fund for Animals and friend of PETA founder Alex Pacheco, is running both the Media and Governmental Affairs Sections of the HSUS. Pacelle is a rabid animal rights proponent and is close to PETA's chair, Ingrid Newkirk. Pacelle also remains close to Heidi Prescott, who took his job at the Fund for Animals. Cleveland Amory, whom many credit with mentoring Pacheco and Pacelle, is still close to them and to Newkirk. Amory was instrumental in PETA's takeover of the New England Anti-Vivisection Society several years ago.

- Rick Swain, former Managing Director of PETA, is now vice president of Investigations for HSUS. A 25 year veteran of the Montgomery County (Maryland) Police Department, Swain first became associated with Ingrid Newkirk and Alex Pacheco in 1981 as part of the landmark "Silver Spring Monkeys" case and joined PETA's paid staff immediately after his retirement from the police force.

- John Kulberg, former president of the American Society for the Prevention of Cruelty to Animals, now heads the Wildlife Lands Trust of the HSUS. He is friends with Ingrid Newkirk and other PETA leaders. Kulberg was an outspoken advocate of hardline animal rights while he was with the ASPCA.

- Martin Stephens, a self-proclaimed antivivisectionist who is married to animal rights attorney Jo Shosmith, heads the Lab Animals Section. Working with him is Jonathan Balcombe, a former PETA employee and animal rights activist.

Back in the ranks, placed so they can be part of the up-and-coming pool of new directors and section leaders, are a virtual horde of former PETA employees:

- Companion Animals Section: Rachel Lamb and Leslie Isom, both former PETA employees.

- Investigations: Virginia Bollinger, PETA's longtime chief investigator, and Cristobel Block, who trained under Phil Hirschkoff, PETA's legal brain trust, before coming to HSUS.

- Peggy Carlson, an emergency room doctor who worked with Neal Barnard at the PETA-created and incongruously named Physicians' Committee for Responsible Medicine, is now a research scientist at HSUS.

- International Section: Both Kim Roberts and Leslie Gerstenfeld came from PETA.

- Howard Edelstein is the HSUS's computer programmer and troubleshooter, with full access and capability to travel in and out of the HSUS computer network. Edelstein is yet another former PETA employee and he was Virginia Bollinger's roommate.

PETA's influence extends into the regional offices of the HSUS. For example, in California Michael Winikoff, a former PETA investigator, works out of the Sacramento office.

Some People Just See A Rat. We See A Cure For Cancer.

Hopefully, sooner or later there will be a cure for cancer, a vaccine against AIDS and an effective treatment for Alzheimer's.

And when these breakthroughs occur it will be thanks to the rats and other laboratory animals that are so vital to medical research.

Because, historically, no cure, no vaccine, no revolution in surgery was ever discovered without animal research.

The war against disease is far from over. There are still arthritis, hepatitis, lupus and dozens of other terrifying and debilitating and life-sapping illnesses for which cures must be found. And thanks to dedicated scientists, the research continues, and new cures are being worked on.

But all of this life-promising work is being threatened by "animal rights" activists.

They oppose the use of animals in medical research and they oppose the use of animals to rescue human lives.

This so-called "animal rights" movement believes that animals and humans are equal. They say, even if animal research resulted in a cure for AIDS, "we'd be against it."

Through a well-financed disinformation campaign -- often directed at impressionable school children -- medical science is losing its allure and the future of medicine is being undermined.

The notion that a rat and a child are equal is irrational to most Americans. The belief that research which can save the lives of millions of humans (and animals) should be paralyzed, is an outrage.

Americans for Medical Progress is spearheading the critical effort to educate American students, citizens and opinion leaders about the need for animal research if new cures are to be found.

You are invited to join Americans for Medical Progress. Your membership will show strong support for those working to ease suffering and save lives.

For more information and to receive your free "Research Saves Lives" bumper sticker, please call 800-4-AMP-USA or write to Americans for Medical Progress, 421 King Street, Suite 401, Alexandria, VA 22314.

Using Animals in Research is Necessary

Prepared by the Saving Lives Coalition March 1993

Scientists need to test medical treatments for efficacy and test new drugs for safety or toxicity before beginning human testing.

Small animals, usually rats, are used to determine the possible side-effects of new drugs including infertility, miscarriage, birth defects, liver damage and cancer-causing potential. After animal tests have proven the safety of new drugs, patients asked to participate in further studies can be assured that they may fare better—and will not do worse—than if they were given standard treatment or no treatment.

New surgical techniques first must be carefully developed and tested in living, breathing, whole organ systems with pulmonary and circulatory systems much like ours.

The physicians and physicians-in-training who perform today's delicate cardiac, ear, eye, pulmonary and brain surgeries must develop the necessary skills before patients' lives are entrusted to their care. Neither computer models, cell culture nor artificial substances can simulate flesh, muscle, blood, bone and organs working together in the living system.

Animals provide needed models for the study of diseases of human and animals.

For example, with biliary tracts similar to humans, prairie dogs make fine models for studying gallstone disease; the natural aging process in polar bears appears to resemble Alzheimer's disease in humans.

Basic biomedical research forms the cornerstone of our understanding of how the body functions in health and disease.

Basic research increases knowledge of biological and chemical processes that is essential to understanding and treating human disease.

Computer and Laboratory Tests Can't Replace Animal Research

Living systems are complex. The nervous system, blood and brain chemistry, gland and organ secretions, and immunological responses are all interelated. It is impossible to explore, explain or predict the course of many diseases or the effects of many treatments without observing and testing the entire living system.

Cell and tissue cultures, often suggested as "alternatives" to using animals, have been adjuncts of biomedical research for many years.

Animals Used in Biomedical Research

Number of Animal Models Used for Biomedical Research

On the average, 20 million animals are used in research annually. This number has declined dramatically in recent years—down 40 percent since 1968. The vast majority (90 to 95 percent) of animals used in U.S. research are rats, mice and other rodents specifically bred for research. Only 1 to 1.5 percent of research animals are dogs and cats; only 0.5 percent are non-human primates.

Fewer than one dog or cat is used for research purposes for every 50 destroyed by animal pounds and humane societies engaged in animal control. Between 10.1 and 16.7 million dogs and cats are put to death in pounds and shelters annually (American Humane Society data).

Sources of Animals for Biomedical Research

The majority of animals used in research are bred specifically for that purpose by suppliers licensed by the U.S. Department of Agriculture. Ungulates (swine, cattle and sheep) are supplied by agricultural sources. Non-human primates come from scientific breeding centers. Of the 200,000 dogs and cats abandoned weekly by pet owners (10.4 million annually), fewer than three percent are used in research.

Medical Advances Resulting from Animal Research

Major Discoveries and Developments

DNA, viruses and retroviruses, pump-oxygenator, electrocardiograph, electroencephalograph, angiogram, cardiac catheters, radiation therapy, kidney dialysis machines, iron lung, blood pressure measures, cardiac pacemaker, artificial limbs and joints, x-rays, monoclonal antibodies, surgical dressings, ultrasound, CT and PET scans, magnetic resonance imaging (MRI), extracorporeal membrane oxygenation (ECMO), cardiopulmonary resuscitation (CPR), treatment of injury victims to prevent paralysis

Treatable Conditions

Gallstones, tooth and gum disease, anemia, hay fever, schizophrenia, Rh disease, PKU (phenylketonuria), cataracts, corneal defects, ear infections, vitamin deficiency, bone fractures, herpes, depression, pneumonia, bronchitis, acne, allergies, peptic ulcers, premature birth, chlamydia, cancer, adult and juvenile diabetes, cystic fibrosis, hypertension, epilepsy, muscular dystrophy, emphysema, hemophilia, leukemia, coronary disease, infertility

Protective Vaccines

Anthrax, cholera, diphtheria, hepatitis, influenza B, rabies, measles, mumps, polio, rubella, smallpox, tetanus, typhoid, whooping cough, yellow fever

Drugs

Insulin, anticonvulsives, antibiotics, anti-inflammatory drugs, birth control pills, sulfa drugs, tranquilizers, anesthetics, pain killers, antihistamines, antimalarial drugs, interferon, cortisone, fertility drugs, anticoagulants, chemotherapy

Surgery

Blood transfusions, coronary bypass, heart replacements, abdominal and reconstructive surgery, skin grafts (burn therapy), neurosurgery, microsurgery, mastectomy, organ and bone marrow transplants

Other Benefits

Stroke rehabilitation, intravenous feeding and medication, artificial limbs, identification of occupational and environmental hazards, improved nutrition, insights into effects of smoking, alcohol, drugs and other "lifestyle choices" on health

Veterinary Advances Resulting from Animal Research

Protective Vaccines

Distemper, rabies, parvovirus, infectious hepatitis, anthrax, tetanus and feline leukemia for domestic and wild animals, including endangered species

Treatable Conditions

Parasites, tuberculosis and brucellosis in cattle, heartworm infection and arthritis in dogs, leukemia and other cancers

Surgery

Hip dysplasia in dogs, orthopedic surgery and rehabilitation for horses

Lives Saved Because of Animal Research

Biomedical research using animals has contributed to an increased life expectancy of about 25 years in the United States since 1900. Many peoples' daily lives and health depend on the products of animal research.

Without animal research...

- Polio would kill or cripple thousands of unvaccinated children and adults this year.
- 7,500 newborns who develop jaundice each year would develop cerebral palsy, now preventable through phototherapy.

- Most of the nation's one million insulin-dependent people with diabetes wouldn't be insulin dependent—they would be dead.
- Many of the 392,111 individuals who benefitted from coronary bypass surgery in 1990 would never have seen 1991.
- 1.5 million Americans would contract rubella—more than 400 times the current annual incidence of the disease.
- 60 million Americans would risk death from heart attack, stroke or kidney failure from lack of medication to control their high blood pressure.
- The more than 100,000 people with arthritis who each year receive hip replacements would walk only with great pain and difficulty or be confined to wheelchairs.
- More than one million Americans would lose their vision in at least one eye this year because cataract surgery would be impossible.
- There would be no kidney dialysis to extend the lives of thousands of victims of end-stage renal disease.
- Physicians would have no chemotherapy to save the 70 percent of children who now survive acute lymphocytic leukemia.
- Hundreds of thousands of people disabled by stroke or head and spinal cord injury would not benefit from rehabilitation techniques.

Ongoing Research

The following are examples of continuing research that requires animal studies.

Cancer

Research investigates drugs and other treatments that may be effective in controlling or killing cancer cells, mechanisms for early detection, and a better understanding of factors that might reduce the risk of developing the disease. One cancer-fighting technique that shows promise is gene therapy. Researchers have had success with this technique using rodents; they hope to apply it to humans in the future.

Diabetes

Research continues on transplantation of the pancreas and of islet (insulin-producing) cells.

Alzheimer's Disease

Alzheimer's afflicts four million Americans and one in twelve over the age of 65. The cause of the disease, which is both debilitating and extraordinarily costly (from $15 to $25 billion per year) is still a mystery. Several drug treatments, which appear to alleviate symptoms in some patients, have been tested for safety in animals and for effectiveness in human clinical studies.

Infectious Diseases

More than 800 million people worldwide suffer from the debilitating and often fatal diseases schistosomiasis, malaria, leprosy, leishmaniasis and trypanosomiasis. Animals allow researchers to measure the host response to disease-causing agents, the effectiveness of agents designed to prevent infection, and their short- and long-term side effects.

AIDS

Hopes for eventual development of a vaccine to prevent infection with HIV, the virus that causes AIDS, are increasing, based on promising results of experiments with mice and specific types of primates. Drug therapies for AIDS patients, all have been safety tested on animals before they are evaluated in experimental human clinical studies.

Cystic Fibrosis

Cystic fibrosis (CF) provides an excellent example of the "payoff" from and importance of basic biological studies. Scientists are now experimenting with gene therapy in rats and believe they have taken major steps toward curing CF, the most common genetic killer of American children.

Also

New hope for effective treatments or cures are being sought for arthritis, asthma, cholera, Down's syndrome, genetic disorders, Huntington's disease, infertility, meningitis, migraine headaches, multiple sclerosis, Parkinson's disease, shock, strole, sickle cell anemia and sudden infant death syndrome.

PETA
Factsheet—Animal Experiments

Companies That Don't Test on Animals

People for the Ethical Treatment of Animals
501 Front Street
Norfolk, VA 23510
Tel: 757-622-PETA

Frequently Asked Questions

What Products Does This List Cover?

The list only includes manufacturers of cosmetics, toiletries, and household products. It does not include strictly pharmaceutical companies, because pharmaceuticals are still required by law to be tested on animals, nor does it include manufacturers of other types of products, such as garden and automotive chemicals and food items, which do not fall under the same regulatory guidelines as cosmetics.

How Does a Company Get on the List?

Companies listed have completed PETA's Statement of Assurance certifying that neither they nor their subsidiary companies conduct or commission any animal tests on ingredients, formulations, or finished products and won't do so in the future.

We are in the process of putting into place an even more stringent cruelty-free standard developed by a coalition of animal protection groups to create a unified policy and make it easier for consumers and companies to identify which products meet ethical standards. In addition to meeting our current requirements, the Corporate Standard of Compassion for Animals (CSCA) will require companies to obtain statements of assurance from all their suppliers to the effect that no animal tests were performed on ingredients supplied to them. We are currently collecting new data from companies, and the new standard should be reflected in this list by 1998.

How Do I Know That These Companies Really Don't Test on Animals?

To a degree, the Statement of Assurance is a matter of trust. However, companies are putting their integrity on the line when they respond to consumers. A company that has publicly announced an end to its animal tests and states in writing that it doesn't test on animals would face a public relations disaster (and potential lawsuits) if it were caught misrepresenting its policies. Companies are well aware that consumers are serious about the issue of animal testing, and they know that they would ruin the public's confidence in their products if it were discovered that they were being dishonest about their animal-testing policies.

How Often Are PETA's Product Lists Updated?

PETA's "Do Test" and "Don't Test" factsheet is updated every two months to reflect additions (e.g., if we are informed of a new company's non-animal-testing policy), deletions (e.g., if a non-animal-testing company is purchased by an animal-testing company or if a company goes out of business), changes in contact information, etc. Please contact PETA if you have any questions about the status of listed companies or if you know the address of a company that is not listed. PETA reserves the right to choose which companies will be included, based on company policy.

PETA also publishes product lists in two other formats: a pocket-sized "Cruelty-Free Shopping Guide," updated twice yearly, and the *Shopping Guide for Caring Consumers*, an annually updated directory of companies that don't test on animals, which includes lists of their products, coupons, and contact information. To order either, contact PETA.

The following companies manufacture products that <u>**ARE NOT**</u> tested on animals. Those marked with an asterisk (*) manufacture strictly vegan products—made without animal ingredients, such as milk and egg byproducts, slaughterhouse byproducts, sheep lanolin, honey, or beeswax. Companies without an asterisk may offer some vegan products. Listed in parentheses are either examples of products manufactured by that company or, if applicable, its parent or subsidiary company. Please enclose a self-addressed, stamped envelope when writing the companies for more information.

***ABBA Products, Inc.**, 2010 Main St., #1000, Irvine, CA 92714

ABEnterprises, 145 Cortlandt St., Staten Island, NY 10302-2048

Abercrombie & Fitch (The Limited), 4 Limited Pkwy. E., Reynoldsburg, OH 43068

Abkit, Inc. (CamoCare), 207 E. 94th St., Suite 201, New York, NY 10128

***Abracadabra, Inc.**, P.O. Box 1040, Guerneville, CA 95446

Adrien Arpel, Inc., 720 Fifth Ave., New York, NY 10019

***Advanage Wonder Cleaner**, 16615 S. Halsted St., Harvey, IL 60426

African Bio-Botanica, Inc., 602 N.W. Ninth Ave., Gainesville, FL 32601

***Ahimsa Natural Care**, 99 Dolomite Dr., Ontario M3J 2N1 Canada

Alba Botanica Cosmetics, P.O. Box 40339, Santa Barbara, CA 93140

Alexandra Avery Purely Natural Body Care, 4717 S.E. Belmont, Portland, OR 97215

Alexandra de Markoff (Parlux Fragrances), 3725 S.W. 30th Ave., Ft. Lauderdale, FL 33312

***Alexia Alexander Corp.**, 24937 West Ave. Tibbits, Valencia, CA 91355

***Allens Naturally**, P.O. Box 514, Dept. M, Farmington, MI 48332-0514

Almay (Revlon), 625 Madison Ave., New York, NY 10022

Aloegen Natural Cosmetics (Levlad), 9200 Mason Ave., Chatsworth, CA 91311

***Aloe Gold (Green Mountain)**, 2755 Highway 55, St. Paul, MN 55121

Aloette Cosmetics, 1301 Wright's La., West Chester, PA 19380

Aloe Up, Inc., P.O. Box 2913, Harlingen, TX 78551

Aloe Vera of America, Inc., 9660 Dilworth, Dallas, TX 75243

Alvin Last, 19 Babcock Pl., Yonkers, NY 10701-2714

***Amazon Products**, 275 N.E. 59th St., Miami, FL 33137

***American Formulating & Manufacturing**, 350 W. Ash St., #700, San Diego, CA 92101

American International, 2220 Gaspar Ave., Los Angeles, CA 90040

American Safety Razor (Personna, Flicker, Bump Fighter), P.O. Box 500, Staunton, VA 24401

***America's Finest Products Corp.**, 1639 Ninth St., Santa Monica, CA 90404

Amitée Cosmetics, Inc., 151 Kalmus Dr., Suite H3, Costa Mesa, CA 92626

Amoresse Labs, 4121 Buchanan St., Riverside, CA 92503

Amway, 7575 E. Fulton St, Ada, MI 49355-0001

***The Ananda Collection**, 14618 Tyler Foote Rd., Nevada City, CA 95959

Ancient Formulas, Inc., P.O. Box 1313, Wichita, KS 67206

Andrea International Industries, 2220 Gaspar Ave., Los Angeles, CA 90040

An-Tech, 201 N. Figueroa, Los Angeles, CA 90012

Aramis, Inc., 767 Fifth Ave., New York, NY 10153

Arbonne International, Inc., 15 Argonaut St., Aliso Viejo, CA 92656

Ardell International, 2220 Gaspar Ave., Los Angeles, CA 90040

Arizona Natural Resources, 2525 E. Beardsley Rd., Phoenix, AZ 85027

***Aromaland, Inc.**, Rt. 20, Box 29 AL, Santa Fe, NM 87501

Aroma Vera Co., 5901 Rodeo Rd., Los Angeles, CA 90016-4312

***Astonish Industries**, 423 Commerce La., Unit 2, Berlin, NJ 08091

***ATMOSA Brand Aromatherapy Products**, 1420 Fifth Ave., 22nd Fl., Seattle, WA 98101-2378

Atta Lavi, 443 Oakhurst Drive S., #305, Beverly Hills, CA 90212

Aubrey Organics, 4419 N. Manhattan Ave., Tampa, FL 33614

Aunt Bee's Skin Care, P.O. Box 21127, Albuquerque, NM 87154

***Aura Cacia**, P.O. Box 399, 716 Main St., Weaverville, CA 96093

***Auroma International (Ayurherbal Corp.)**, P.O. Box 1008, Silver Lake, WI 53170

***Auromère Ayurvedic Imports**, 2621 W. Hwy. 12, Lodi, CA 91768

The Australasian College of Herbal Studies, P.O. Box 57, Lake Oswego, OR 97035

Autumn-Harp, Inc., P.O. Box 267, Bristol, VT 05443

***Avanza Corp. (Nature Cosmetics)**, 11818 San Marino St., Rancho Cucamonga, CA 91730

Aveda, 4000 Pheasant Ridge Rd., Blaine, MN 55449

***Avigal Henna**, 45-49 Davis St., Long Island City, NY 11101

Avon, 9 W. 57th St., New York, NY 10019

***Ayurherbal Corp. (Auroma International)**, P.O. Box 1008, Silver Lake, WI 53170

***Ayurveda Holistic Center**, 82A Bayville Ave., Bayville, NY 11709

***Ayus/Oshadhi**, 15 Monarch Bay Plaza, Suite 346, Monarch Beach, CA 92629

***Baby Massage**, P.O. Box 51867, Bowling Green, KY 42101

Barbizon International, Inc., 1900 Glades Rd., Suite 300, Boca Raton, FL 33431

***Bare Escentuals**, 600 Townsend St., Suite 329-E, San Francisco, CA 94103

***Basically Natural**, 109 E. G St., Brunswick, MD 21716

***Basic Elements Hair Care System, Inc.**, 505 S. Beverly Dr., Suite 1292, Beverly Hills, CA 90212

Basis (Beiersdorf), BDF Plaza, 360 Martin Luther King Dr., Norwalk, CT 06856

Bath & Body Works (The Limited), 97 W. Main St., New Albany, OH 43054

Bath Island, Inc., 469 Amsterdam Ave., New York, NY 10024

Baudelaire, Inc., Forest Rd., Marlow, NH 03456

BeautiControl Cosmetics, 2121 Midway Rd., Carrollton, TX 75006

Beauty Naturally, P.O. Box 4905, 859 Cowan Rd., Burlingame, CA 94010

***Beauty Without Cruelty**, P.O. Box 750428, Petaluma, CA 94975-0425

Beehive Botanicals, Inc., Rt. 8, Box 8257, Hayward, WI 54843

Beiersdorf, Inc. (Nivea, Eucerin, Basis, La Prairie), BDF Plaza, 360 Martin Luther King Dr., Norwalk, CT 06856

Bella's Secret Garden, 1601 Emerson Ave., Channel Islands, CA 94975-0428

Belle Star, Inc., 23151 Alcalde, #C11, Laguna Hills, CA 92653

Berol (Sanford Corp.), 2711 Washington Blvd., Bellwood, IL 60104

Beverly Hills Cold Wax, P.O. Box 600476, San Diego, CA 92160

Beverly Hills Cosmetic Group, 289 S. Robertson Blvd., #461, Beverly Hills, CA 90211

Bill Blass (Revlon), 625 Madison Ave., New York, NY 10022

***BioFilm, Inc.**, 3121 Scott St., Vista, CA 92083

***Biogime**, 1665 Townhurst, #100, Houston, TX 77043

***Bi-O-Kleen, Inc.**, P.O. Box 82066, Portland, OR 97282-0066

Biokosma (Caswell-Massey), 121 Field Crest Ave., Edison, NJ 08818

***Bio-Pac**, RR #1, Box 407, Union, ME 04862

Bio-Tec Cosmetics, Inc., 92 Sherwood Ave., Toronto, Ontario M4P 2A7 Canada

Biotone, 4564 Alvarado Canyon Rd., #1, San Diego, CA 92120

Black Pearl Gardens, 425 S. Main St., Franklin, OH 45005

Bo-Chem Co., Little Harbor, Marblehead, MA 01945

Body Encounters, 604 Manor Rd., Cinnamonson, NJ 08077

Bodyography, 10250 Santa Monica Blvd., Suite 305, Los Angeles, CA 90067

The Body Shop, Inc., P.O. Box 1409, Wake Forest, NC 27588

***Body Suite**, 874 Center St., San Luis Obispo, CA 93401

Body Time, 1341 Seventh St., Berkeley, CA 94710

Bon Ami/Faultless Starch, 510 Walnut St., Kansas City, MO 64106-1209

Bonne Bell, 18579 Detroit Ave., Lakewood, OH 44107

Börlind of Germany, P.O. Box 130, New London, NH 03257

***Botan Corporation**, 4012 Dupont Cir., Suite 401, Louisville, KY 40207

***Botanics Skin Care**, 3001 South State, #29, Ukiah, CA 95482

Botanicus Retail, 7610 T. Rickenbacker Dr., Gaithersburg, MD 20879

***Brocato International**, 7650 Currell Blvd., Suite 240, Woodburg, MN 55125

Bronson Pharmaceuticals, 1945 Craig Rd., St. Louis, MO 63146

***Brookside Soap Company**, P.O. Box 55638, Seattle, WA 98155

CAERAN, 25 Penny La., Brantford, Ontario N3R 5Y5 Canada

California Skin Therapy, 1100 Glendon Ave., Suite 1250, Los Angeles, CA 90024

CamoCare Camomile Skin Care Products (Abkit, Inc.), 207 E. 94th St., Suite 201, New York, NY 10128

***Candy Kisses Natural Lip Balm**, 16 E. 40th St., 12th Fl., New York, NY 10016

Carina Supply, Inc., 464 Granville St., Vancouver, B.C. V6C 1V4 Canada

The Caring Catalog, 7678 Sagewood Dr., Huntington Beach, CA 92648

Carlson Laboratories, 15 College Dr., Arlington Heights, IL 60004

Carma Laboratories, 5801 W. Airways Ave., Franklin, WI 53132

Carter's Naturals, 3A Hamilton Business Park, Dover, NJ 07801

Caswell-Massey, 121 Field Crest Ave., Edison, NJ 08837

Celestial Body, 21298 Pleasant Hill Rd., Boonville, MO 65233

Chanel, Inc., 9 W. 57th St., New York, NY 10019

Charles of the Ritz (Revlon), 625 Madison Ave., New York, NY 10022

Chatoyant Pearl Cosmetics, P.O. Box 526, Townsend, WA 98368

Chempoint Products, P.O. Box 2597, Danbury, CT 06813-2597

Chica Bella, Inc., Interlink 580, P.O. Box 02-5635, Miami, FL 33152

***CHIP Distribution Co.**, 8321 Croydon Ave., Los Angeles, CA 90045

Christian Dior, 9 W. 57th St., New York, NY 10019

Christine Valmy, Inc., 285 Change Bridge Rd., Pine Brook, NJ 07058

Chuckles, Inc. (Farmavita USA), 59 March Ave., Manchester, NH 03103

CiCi Cosmetics, 9500 Jefferson Blvd., Culver City, CA 90232

Cinema Secrets, Inc., 4400 Riverside Dr., Burbank, CA 91505

***Citius USA, Inc.**, 120 Interstate N. Pkwy. E., Suite 106, Atlanta, GA 30339

Citré Shine, 151 Kalmus Dr., Suite H3, Costa Mesa, CA 92626

Clarins of Paris, 135 E. 57th St., New York, NY 10022

***Clearly Natural Products**, Box 750024, Petaluma, CA 94975

***Clear Vue Products, Inc.**, P.O. Box 567, 417 Canal St., Lawrence, MA 01842

Cleopatra's Secret, 130 W. 25th St., 10th Fl., New York, NY 10001

Clientele, 5207 N.W. 163rd St., Miami, FL 33014

Clinique Laboratories, 767 Fifth Ave., New York, NY 10153

Color & Herbal Co., P.O. Box 5370, Newport Beach, CA 92662

***Colorations**, 6500 McDonough Dr., #D4, Norcross, GA 30093

Color Me Beautiful, 14000 Thunderbolt Place, Suite E, Chantilly, VA 22021

Color My Image, 5025B Backlick Rd., Annandale, VA 22003

Columbia Cosmetics Mfg., 1661 Timothy Dr., San Leandro, CA 94577

Come To Your Senses, 321 Cedar Ave. S., Minneapolis, MN 55454

Comfort Mfg. Co., 1056 W. Van Buren St., Chicago, IL 60607

Common Scents, 134 Main St., Port Jefferson, NY 11777

Compar, Inc., 70 East 55th St., New York, NY 10022

The Compassionate Consumer, P.O. Box 27, Jericho, NY 11753

Compassionate Cosmetics, P.O. Box 3534, Glendale, CA 91201

Compassion Matters, P.O. Box 3614, Jamestown, NY 14702-3614

Conair (Jheri Redding), 1 Cummings Point Rd., Stamford, CT 06904

Concept Now Cos. (CNC), 10200 Pioneer Blvd., #100, Santa Fe Springs, CA 90670

Cosmair (L'Oréal, Lancôme, Maybelline), 575 Fifth Ave., New York, NY 10017[1]

Cosmyl, Inc., 1 Cosmyl Place, Columbus, GA 31907

***Cot'N Wash, Inc.**, 502 The Times Bldg., Ardmore, PA 19003

Country Comfort, 28537 Nuevo Valley Dr., Nuevo, CA 92567

***Country Save Corp.**, 3410 Smith Ave., Everett, WA 98201

***Countryside Fragrances**, Pacific First Centre, 1420 Fifth Ave., 22nd Fl., Seattle, WA 98101-2378

Crabtree & Evelyn (KLK), Peake Brook Rd., Box 167, Woodstock, CT 06281

Crebel International, 4401 Ponce de Leon Blvd, Coral Gables, FL 33146

Creighton's Naturally, 11243-4 St. Johns Ind. Pkwy. S., Jacksonville, FL 32246

Crème de la Terre, 30 Cook Rd., Stamford, CT 06902

***Crown Royale, Ltd.**, P.O. Box 5238, 99 Broad St., Phillipsburg, NJ 08865

***CYA Products, Inc.**, 211 Robbins La., Syosset, NY 11791

***Davidoff Fragrances**, 745 Fifth Ave., 10th Fl., New York, NY 10151

Decleor USA, Inc., 500 West Ave., Stamford, CT 06902

***Deodorant Stones of America**, 9420 E. Doubletree Ranch Rd., Unit 101, Scottsdale, AZ 85258

Dep Corporation (Agree, Hälsa, Lilt, Lavoris), 2101 E. Via Arado, Rancho Dominguez, CA 90220-6189

Derma-E, 9400 Lurline Ave., # C-1, Chatsworth, CA 91311

Dermalogica (International Dermal Institute, Dermal Products), 1001 Knox St., Torrance, CA 90502

Dermatologic Cosmetic Labs, 360 Sackett Point Rd., North Haven, CT 06473

Desert Essence, 9510 Vassar Ave., Unit A, Chatsworth, CA 91311

Desert Naturels, 83-612 Ave. 45, Suite 5, Indio, CA 92201

DeSoto, Inc. (Keystone Consolidated Industries), 900 E. Washington St., P.O. Box 609, Joliet, IL 60434

Diamond Brands, Inc., 1660 S. Highway 100, Suite 340, Minneapolis, MN 55416

Dr. A.C. Daniels, Inc., 109 Worcester Rd., Webster, MA 01570

***Dr. Bronner's "All-One" Products**, P.O. Box 28, Escondido, CA 92025

Dr. Hauschka Cosmetics USA, Inc., 59C North St., Hatfield, MA 01038

***D.R.P.C. (AmerAgain)**, 567-1 S. Leonard St., Waterbury, CT 06708

Earth Friendly Products, P.O. Box 607, Wood Dale, IL 60191

***Earthly Matters**, 2719 Phillips Hwy., Jacksonville, FL 32207

Earth Science, 23705 Via Del Rio, Yorba Linda, CA 92687-2717

***Earth Solutions, Inc.**, 427 Moreland Ave., #100, Atlanta, GA 30307

Eberhard Faber (Sanford Corp.), 2711 Washington Blvd., Bellwood, IL 60104

E. Burnham Cos., 7117 N. Austin Ave., Niles, IL 60714

Ecco Bella Botanicals, 1133 Route 23, Wayne, NJ 07470

EcoDent, 3130 Spring St., Redwood City, CA 94063

Eco Design Company, 1365 Rufina Cir., Santa Fe, NM 87501

Ecover Products, Carpenter Rd., P.O. Box SS, Philmont, NY 12565

Edward & Sons Trading Co., P.O. Box 1326, Carpinteria, CA 93014

Elizabeth Grady Face First, 200 Boston Ave., Suite 3500, Medford, MA 02155

Elizabeth Van Buren Aromatherapy, Box 7542, 303 Potrero St., #33, Santa Cruz, CA 95061

Elvira's Halloween Cosmetics, P.O. Box 38246, Hollywood, CA 90038

Enfasi Hair Care, 2937 S. Alameda St., Los Angeles, CA 90058

English Ideas, 3 Cavalier, Laguna Beach, CA 92677

Epilady International, Inc., 39 Cindy La., Ocean, NJ 07712-7249

***Espial Corp.**, 7045 S. Fulton St., #200, Englewood, CO 80112-3700

***Essential Aromatics**, 205 N. Signal St., Ojai, CA 93023

The Essential Oil Company, P.O. Box 206, Lake Oswego, OR 97034

***Essential Products of America**, 5018 N. Hubert Ave., Tampa, FL 33614

Estée Lauder (Clinique, Origins), 767 Fifth Ave., New York, NY 10153

Eucerin (Beiersdorf), BDF Plaza, 360 Martin Luther King Dr., Norwalk, CT 06856

European Gold, 33 S.E. 11th St., Grand Rapids, MN 55744

***EuroZen**, 10 S. Franklin Tpk., #201, Ramsey, NJ 07446

Eva Jon Cosmetics, 1016 E. California St., Gainesville, TX 76240

Evans International, 14 E. 15th St., Richmond, VA 23224-0189

Every Body, Ltd. (Mountain Ocean), 1175 Walnut St., Boulder, CO 80302

The Face Food Shoppe, 185 Bruce Hill Rd., Cumberland, ME 04021

Faces by Gustavo, 1200 N. Veitch St., Suite 1243, Arlington, VA 22201

Facets/Crystalline Cosmetics, 8436 N. 80th Pl., Scottsdale, AZ 85258

***Faith in Nature**, Unit 5, Kay St., Bury, Lancashire BL9 6BU England

Faith Products, Ltd., Unit 5, Kay St., Bury, Lancashire BL9 6BU England

Farmavita USA (Chuckles, Inc.), P.O. Box 5126, Manchester, NH 03109

Faultless Starch (Bon Ami), 510 Walnut St., Kansas City, MO 64106-1209

Fernand Aubry, Paris, 14, rue Alexandre Parodi, 75010 Paris, France

Finelle Cosmetics, 137 Marston St., Lawrence, MA 01841-2297

Fleabusters/Rx for Fleas, Inc., 6555 N.W. Ninth Ave., Suite 412, Ft. Lauderdale, FL 33309

Flex (Revlon), 625 Madison Ave., New York, NY 10022

***Flower Essences of Running Fox Farm**, P.O. Box 381, Worthington, MA 01098

Focus 21 International, 2755 Dos Aarons Way, Vista, CA 92803

Food Lion (*house brand products only*), P.O. Box 1330, Salisbury, NC 28145-1330

Forest Essentials, 385 Bel Marin Keys Blvd., Suite H, Novato, CA 94947

Forever Living Products, P.O. Box 29041, Phoenix, AZ 85038

***Forever New International**, 4791 N. Fourth Ave., Sioux Falls, SD 57104-0403

For Pet's Sake Enterprises, Inc., 3780 Eastway Rd., Suite 10A, S. Euclid, OH 44118

***Fort Howard Corp.**, P.O. Box 19130, Green Bay, WI 54307-9130

***IV Trail Products**, P.O. Box 1033, Sykesville, MD 21784

Fragrance Impressions, Ltd., 116 Knowlton St., Bridgeport, CT 06608

Framesi, USA, Inc., 400 Chess St., Coraopolis, PA 15108

***Frank T. Ross (Nature Clean)**, 6550 Lawrence Ave. E., Scarborough, Ontario M1E 4R5 Canada

Freeda Vitamins, Inc., 36 E. 41st St., New York, NY 10017

Freeman Cosmetics Corp., 10000 Santa Monica Blvd., #400, Los Angeles, CA 90067

***Free Spirit Enterprises**, 2064 Dennis La., Santa Rosa, CA 95403

Frontier Cooperative Herbs, 3021 78th St., P.O. Box 299, Norway, IA 52318

Fruit of the Earth, P.O. Box 152044, Irving, TX 75015-2044

Garden Botanika, 8624 154th Ave. N.E., Redmond, WA 98052

Garnier (L'Oréal), 575 Fifth Ave., New York, NY 10017[1]

Gigi Laboratories, 2220 Gaspar Ave., Los Angeles, CA 90040

Giovanni Cosmetics, 5415 Tweedy Blvd., Southgate, CA 90280

***Golden Lotus**, P.O. Box 51867, Bowling Green, KY 42101

Golden Pride/Rawleigh, 1501 Northpoint Pkwy., W. Palm Beach, FL 33407

Goldwell Cosmetics (USA), 9050 Junction Dr., Annapolis Junction, MD 20701

***Green Ban**, P.O. Box 146, Norway, IA 52318

Green Earth Office Supply, P.O. Box 719, Redwood Estates, CA 95044

***Green Mountain (Baby Massage)**, P.O. Box 51867, Bowling Green, KY 42101

Greentree Laboratories, Inc., P.O. Box 425, Tustin, CA 92681

***Greenway Products**, P.O. Box 183, Port Townsend, WA 98368

Gryphon Development (The Limited), 767 Fifth Ave., New York, NY 10153

Gucci Parfums (Wella), 15 Executive Blvd., Orange, CT 06477

Halo Purely for Pets, 3438 E. Lake Rd., #14, Palm Harbor, FL 34685

Halston Borghese, Inc. (Princess Marcella Borghese), 767 Fifth Ave., 49th Fl., New York, NY 10153-0002

Hard Candy, Inc., 110 N. Doheny Dr., Beverly Hills, CA 90211

***Hargen Distributing, Inc.**, 3422 W. Wiltshire Rd., Suite 13, Phoenix, AZ 85034

***Harvey Universal**, 1805 W. 208th St., Torrance, CA 90501

***Healthy Times**, 461 Vernon Way, El Cajon, CA 92020

Helen Lee Skin Care & Cos., 205 E. 60th St., New York, NY 10022

Henri Bendel (The Limited), 712 Fifth Ave., New York, NY 10019

Herbal Products & Development, Box 1084, Aptos, CA 95001

***The Herb Garden**, P.O. Box 773-P, Pilot Mountain, NC 27041

***h.e.r.c., Inc.**, 3622 N. 34th Ave., Phoenix, AZ 85017

Heritage Store, P.O. Box 444, Virginia Beach, VA 23458

Hewitt Soap Company, 333 Linden Ave., Dayton, OH 45403

Hobé Laboratories, Inc., 4032 E. Broadway, Phoenix, AZ 85040

Homebody (Perfumoils, Inc.), 143A Main St., Brattleboro, VT 05301

Home Health Products, P.O. Box 8425, Virginia Beach, VA 23450

***Home Service Products Co.**, P.O. Box 245, Pittstown, NJ 08867

House of Cheriss, 13475 Holiday Dr., Saratoga, CA 95070

H20 Plus, Inc., 676 N. Michigan Ave., Suite 3900, Chicago, IL 60611

Huish Detergents, Inc., 3540 W. 1987 South, Salt Lake City, UT 84104

Ida Grae (Nature's Colors Cosmetics), 424 La Verne Ave., Mill Valley, CA 94941

Il-Makiage, 107 E. 60th St., New York City, NY 10022

Ilona, 3201 E. Second Ave., Denver, CO 80206

Image Laboratories, 2340 Eastman Ave., Oxnard, CA 93030

i natural cosmetics (cosmetic source), 3202 Queens Blvd., Long Island City, NY 11101

***Innovative Formulations, Inc.**, 1810 S. Sixth Ave., Tucson, AZ 85713

***International Rotex**, Box 20697, Reno, NV 89515

International Vitamin Corp., 209 40th St., Irvington, NJ 07111

Internatural, P.O. Box 1008, Silver Lake, WI 53170

***Invalco d/b/a Rudolfo Alosio**, 3251 Corte Malpaso, Suite 504, Camarillo, CA 93021

IQ Products Company, 16212 State Hwy. 249, Houston, TX 77086

Jacki's Magic Lotion, 258 A St., #7A, Ashland, OR 97520

James Austin Company, Box 827, 115 Downieville Rd., Mars, PA 16046-0827

Jason Natural Cosmetics, 8468 Warner Dr., Culver City, CA 90232

J.C. Garet, Inc., 2471 Coral St., Vista, CA 92083

Jean Naté, 625 Madison Ave., New York, NY 10022

Jeanne Rose Aromatherapy, 219 Carl St., San Francisco, CA 94117

Jessica McClintock, Inc., 1400 16th St., San Francisco, CA 94103-5181

Jheri Redding (Conair), 1 Cummings Point Rd., Stamford, CT 06904

Joe Blasco Cosmetics, 7340 Greenbriar Pkwy., Orlando, FL 32819

John Amico Expressive Hair Care, 7327 W. 90th St., Bridgeview, IL 60455

***John Paul Mitchell Systems**, 26455 Golden Valley Rd., Santa Clarita, CA 91350

***JOICO International**, P.O. Box 42308, Los Angeles, CA 90042-0308

Jolen Creme Bleach, 25 Walls Dr., P.O. Box 458, Fairfield, CT 06430

***J.R. Liggett, Ltd.**, RR2, Box 911, Cornish, NH 03745

Jurlique Cosmetics, 1411 Dresden Dr., Atlanta, GA 30319

Kallima International, 1802 Tobin Trail, Garland, TX 75043

Katonah Scentral, 51 Katonah Ave., Katonah, NY 10536

K.B. Products, 20 N. Railroad Ave., San Mateo, CA 94401

Kenic Pet Products, Inc., 109 S. Main St., Lawrenceburg, KY 40342

***Ken Lange No-Thio Perm. Waves**, 7112 N. 15th Pl., Suite 1, Phoenix, AZ 85020

Kenra Laboratories, 6501 Julian Ave., Indianapolis, IN 46219

Kimberly Sayer, 125 W. 81st St., #2A, New York, NY 10024

Kiss My Face, P.O. Box 224, 144 Main St., Gardiner, NY 12525

Kleen Brite Laboratories, Box 20408, Rochester, NY 14602

KMS Research, 4712 Mountain Lakes Blvd., Redding, CA 96003

***KSA Jojoba**, 19025 Parthenia St., #200, Northridge, CA 91324

La Costa Products, 2875 Loker Ave. E., Carlsbad, CA 92008

***LaCrista**, P.O. Box 240, Davidsonville, MD 21035

***LaNatura**, 425 N. Bedford Dr., Beverly Hills, CA 90210

Lancaster Group, 745 Fifth Ave., New York, NY 10151

Lancôme (Cosmair), 575 Fifth Ave., New York, NY 10017[1]

Lander Company, Inc., 106 Grande Ave., Englewood, NJ 07631

***L'anza Research International**, 935 W. Eighth St., Azusa, CA 91702

La Prairie, Inc. (Beiersdorf), 31 W. 52nd St., New York, NY 10019

Lee Pharmaceuticals, 1444 Santa Anita Ave. S., El Monte, CA 91733

***Levlad/Nature's Gate**, 9200 Mason Ave., Chatsworth, CA 91311

***Liberty Natural Products, Inc.**, 8120 S.E. Stock St., Portland, OR 97215

Life Dynamics, Inc., 21640 N. 19th Ave., Suite C101, Phoenix, AZ 85027

***Life Tree Products (Sierra Dawn)**, P.O. Box 1203, Sebastopol, CA 95472

Lightning Products, 10100 N. Executive Hills Blvd., Suite 105, Kansas City, MO 64153

Lily of Colorado, P.O. Box 12471, Denver, CO 80212

Lime-O-Sol Company (The Works), 101 S. Parker Dr., Ashley, IN 46705

Lissée Cosmetics, 2937 S. Alameda St., Los Angeles, CA 90058

Liz Claiborne Cosmetics, Inc., 1441 Broadway, New York, NY 10018

***Lobob Laboratories**, 1440 Atteberry La., San Jose, CA 95131-1410

Logona USA, Inc., 554-E Riverside Dr., Asheville, NC 28801

L'Oréal (Cosmair, Maybelline, Lancôme), 575 Fifth Ave., New York, NY 10017[1]

Lotus Light, P.O. Box 1008, Silver Lake, WI 53170

Louise Bianco Skin Care, Inc., 13655 Chandler Blvd., Sherman Oaks, CA 91401

M.A.C. Cosmetics, 233 Carlton St., 2nd Fl., Toronto, Ontario M5A 2L2 Canada

Magick Botanicals, 3412-K W. MacArthur Blvd., Santa Ana, CA 92704

The Magic of Aloe, 7300 N. Crescent Blvd., Pennsauken, NJ 08110

Mallory Pet Supplies, 118 Atrisco Pl. S.W., Albuquerque, NM 87105

***Marcal Paper Mills, Inc.**, 1 Market St., Elmwood Park, NY 07407

Marché Image Corp., Box 1010, Bronxville, NY 10708

Marilyn Miglin Institute, 112 E. Oak St., Chicago, IL 60611

***Martin Von Myering**, 422 Jay St., Pittsburgh, PA 15212

***Masada**, P.O. Box 4767, N. Hollywood, CA 91617-0767

Mastey de Paris, Inc., 25413 Rye Canyon Rd., Valencia, CA 91355

Maybelline, Inc. (Yardley, L'Oréal), P.O. Box 372, Memphis, TN 38101-0372[1]

Mehron, Inc., 100 Red Schoolhouse Rd., Chestnut Ridge, NY 10977

Melaleuca, Inc., 3910 S. Yellowstone Hwy., Idaho Falls, ID 83402-6003[2]

***Mère Cie, Inc.**, 1100 Soscol Rd., #3, Napa, CA 94558

Merle Norman Cos., 9130 Bellanca Ave., Los Angeles, CA 90069

***Mia Rose Products**, 177-F Riverside Ave., Newport Beach, CA 92663

Michael's Health Products, 6820 Alamo Downs Pkwy., San Antonio, TX 78238

Michelle Lazar Cosmetics, Inc., 755 S. Lugo Ave., San Bernardino, CA 92408

***Micro Balanced Products**, 25 Aladdin Ave., Dumont, NJ 07628

Mira Linder Spa in the City, 29935 Northwestern Hwy., Southfield, MI 48034

Montagne Jeunesse, The Old Grain Store, 4 Denne Rd., Horsham, W. Sussex RH12 1JE England

Monteil Paris, 745 Fifth Ave., New York, NY 10151

***Mother's Little Miracle**, 930 Indian Peak Rd., Rolling Hills Estates, CA 90274

Mountain Ocean (Every Body Ltd.), P.O. Box 951, Boulder, CO 80306

Mr. Christal's, 1100 Glendon Ave., Suite 1250, Los Angeles, CA 90024

Nadina's Cremes, 3600 Clipper Mill, Suite 140, Baltimore, MD 21211

***Nala Barry Labs**, P.O. Box 151, Palm Desert, CA 92261

Narwhale of High Tor, Ltd., 591 S. Mountain Rd., New City, NY 10956

***Natracare**, 191 University Blvd., Suite 129, Denver, CO 80206

Naturade Cosmetics, 7110 E. Jackson St., Paramount, CA 90723

Natural Animal Health Products, Inc., 7000 U.S. 1 N., St. Augustine, FL 32095

***Natural Bodycare, Inc.**, 511 Calle San Pablo, Camarillo, CA 93010

Natural Chemistry, Inc., 244 Elm St., New Canaan, CT 06840

***Naturally Free, The Herbal Alternative**, Rt. 2, Box 248C, Lexington, VA 24450

***Naturally Yours, Alex**, P.O. Box 3398, Holiday, FL 34690-0398

Natural Organics, 548 Broadhollow Rd., Melville, NY 11747

***Natural Products Co.**, 7782 Newburg Rd., Newburg, PA 17240-9601

***Natural Research People, Inc.**, South Route, Box 12, Lavina, MT 59046

***Natural Science**, 41 Madison Ave., 4th Fl., New York, NY 10010

Natural (Surrey), 13110 Trails End Rd., Leander, TX 78641

***Natural Therapeutics Centre**, 2500 Side Cove, Austin, TX 78704

Natural Touch, P.O. Box 2894, Kirkland, WA 98083-2894

Natural Wonder, 625 Madison Ave., New York, NY 10022

Natural World, Inc., 88 Danbury Rd., Wilton, CT 06897

***Nature Cosmetics (Avanza Corp.)**, 11818 San Marino St., Rancho Cucamonga, CA 91730

***Nature de France**, 100 Rose Ave., Hempstead, NY 11550

Nature's Acres, 8984 E. Weinke Rd., New Freedom, WI 53951

***Nature's Best (Natural Research People)**, South Route, Box 12, Lavina, MT 59046

***Nature's Country Pet**, 1765 Garnet Avenue, Suite 12, San Diego, CA 92109

Nature's Gate (Levlad), 9200 Mason Ave., Chatsworth, CA 91311

Nature's Plus, 548 Broadhollow Rd., Melville, NY 11747

Nectarine, 1200 Fifth St., Berkeley, CA 94710

Nemesis, Inc., 4525 Hiawatha Ave., Minneapolis, MN 55406

***Neocare Labs**, 3333 W. Pacific Coast Hwy., 4th Fl., Newport Beach, CA 92663

***New Age Products**, 16200 N. Highway 101, Willits, CA 95490-9710

***Neway**, Little Harbor, Marblehead, MA 01945

Neways, Inc., 150 E. 400 N., Salem, UT 84653

***New Moon Extracts, Inc.**, 99 Main St., Brattleboro, VT 05301

Nexxus, Box 1274, Santa Barbara, CA 93116

***Nirvana**, P.O. Box 18413, Minneapolis, MN 55418

Nivea (Beirsdorf), BDF Plaza, 360 Martin Luther King Dr., Norwalk, CT 06856

No Common Scents, King's Yard, 220 Xenia Ave., Yellow Springs, OH 45387

Nordstrom Cosmetics, 865 Market St., San Francisco, CA 94103

***Norelco**, 1010 Washington Blvd., Stamford, CT 06912

North Country Soap, 7888 County Rd., #6, Maple Plain, MN 55359

N/R Laboratories, Inc., 900 E. Franklin St., Centerville, OH 45459

NuSkin International, One NuSkin Plaza, 75 West Center, Provo, UT 84601

NutriBiotic, 865 Parallel Dr., Lakeport, CA 95453

***Nutri-Cell, Inc.**, 1038 N. Tustin, Suite 309, Orange, CA 92667-5958

Nutri-Metics International USA, Inc., 12723 E. 166th St., Cerritos, CA 90703

Nutrina Company, Inc., 1117 Foothill Blvd., La Canada, CA 91011

***Oasis Biocompatible**, 1020 Veronica Springs Rd., Santa Barbara, CA 93105

The Ohio Hempery, Inc., 7002 S.R. 329, Guysville, OH 45735

***Oil of Orchid**, P.O. Box 1040, Guerneville, CA 95446

***Oliva Ltd.**, P.O. Box 4387, Reading, PA 19606

OPI Products, 13034 Saticoy St., N. Hollywood, CA 91605

***Orange-Mate**, P.O. Box 883, Waldport, OR 97394

Organic Aid, 8439 White Oak Ave., #106, Cucamonga, CA 91730

Organic Moods (KMS Research), 4712 Mountain Lakes Blvd., Redding, CA 96003

Oriflame Corp., 76 Treble Cove Rd., N. Billerica, MA 01862

Origins Natural Resources (Estée Lauder), 767 Fifth Ave., New York, NY 10153

Orjene Natural Cosmetics, 5-43 48th Ave., Long Island City, NY 11101

Orlane, 555 Madison Ave., New York, NY 10022

Orly International, 9309 Deering Ave., Chatsworth, CA 91311

Otto Basics–Beauty 2 Go!, P.O. Box 9023, Rancho Santa Fe, CA 92067

***Oxyfresh U.S.A.**, 12928 E. Indiana Ave., Spokane, WA 99220

***Pacific Scents, Inc.**, P.O. Box 8205, Calabasas, CA 91375-8205

Parfums Houbigant Paris (Alyssa Ashley), 1135 Pleasant View Terr. W., Ridgefield, NJ 07657

***Park-Rand Enterprises**, 12896 Bradley Ave. #F, Sylmar, CA 91342

***Parlux Fragrances, Inc. (Perry Ellis, Todd Oldham)**, 3725 S.W. 30th Ave., Ft. Lauderdale, FL 33312

Pathmark Stores, Inc. (*house brand products only*), 301 Blair Rd., Woodbridge, NJ 07095

Patricia Allison Natural Beauty, 4470 Monahan Rd., La Mesa, CA 91941

***Paul Mazzotta, Inc.**, P.O. Box 96, Reading, PA 19607

Paul Penders USA, 1340 Commerce St., Petaluma, CA 94954

The Peaceable Kingdom, 1902 W. Sixth St., Wilmington, DE 19805

Perfect Balance Cosmetics, Inc., 2 Ridgewood Rd., Malvern, PA 19355-9629

The Pet Connection, P.O. Box 391806, Mountain View, CA 94039

PetGuard, Inc., 165 Industrial Loop S., Unit 5, Orange Park, FL 32073

***Pets 'N People**, 930 Indian Peak Rd., Suite 215, Rolling Hills Estates, CA 90274

Pharmagel Corporation, P.O. Box 50531, Santa Barbara, CA 93150

***Pilot Corporation of America**, 60 Commerce Dr., Trumbull, CT 06611

***Planet, Inc.**, 10114 McDonald Park Rd., C-16 RR3, Sidney, B.C. V8L 3X9 Canada

PlantEssence, P.O. Box 14743, Portland, OR 97293-0743

Potions & Lotions/Body & Soul, 10201 N. 21st Ave., #8, Phoenix, AZ 85021

Prescriptions Plus, 25028 Kearney Ave., Valencia, CA 91355

Prescriptives, 767 Fifth Ave., New York, NY 10153

Prestige Cosmetics, 1330 W. Newport Center Dr., Deerfield Beach, FL 33442

Prestige Fragrances, 625 Madison Ave., New York, NY 10022

Princess Marcella Borghese (Halston Borghese), 767 Fifth Ave., 49th Fl., New York, NY 10153-0002

The Principal Secret, 41-550 Ecclectic St., Suite 200, Palm Desert, CA 92260

Professional Choice Hair Care, 2937 S. Alameda St., Los Angeles, CA 90058

***Professional Pet Products**, 1873 N.W. 97th Ave., Miami, FL 33172

Pro-Ma Systems, 477 Commerce Way, #113, Longwood, FL 32750

***Pro-Tan (Green Mountain)**, P.O. Box 51867, Bowling Green, KY 42101

Pro-Tec Pet Health, 2395-A Monument Blvd., Concord, CA 94520

P.S.I. Industries, 1619 Shenandoah Ave., Roanoke, VA 24017

***Pulse Products**, 2021 Ocean Ave., #105, Santa Monica, CA 90405

***Pure & Basic Products**, 20600 Belshaw Ave., Carson, CA 90746

***Pure Touch Therapeutic Body Care**, P.O. Box 1281, Nevada City, CA 95959

***Quan Yin Essentials**, P.O. Box 2092, Healdsburg, CA 95448

Queen Helene, 100 Rose Ave., Hempstead, NY 11550

Rachel Perry, 9111 Mason Ave., Chatsworth, CA 91311

Rainbow Research, 170 Wilbur Pl., Bohemia, NY 11716

The Rainforest Company, 3830 Washington Blvd., St. Louis, MO 63108

Ralph Lauren Fragrances (Cosmair), 575 Fifth Ave., New York, NY 10017[1]

***Ranir/DCP Corporation**, 4701 E. Paris, Grand Rapids, MI 49512

***Real Animal Friends**, 101 Albany Ave., Freeport, NY 11520

Redken Laboratories (Cosmair), 575 Fifth Ave., New York, NY 10017[1]

Redmond Products, Inc. (Aussie), 18930 W. 78th St., Chanhassen, MN 55317

***Rely Enterprises Corp.**, 7 Stonebridge Ct., Manalapan, NJ 07726

Reviva Labs, 705 Hopkins Rd., Haddonfield, NJ 08033

Revlon (Almay, Jean Naté), 625 Madison Ave., New York, NY 10022

***Royal Laboratories**, 2849 Dundee Rd., Suite 112, Northbrook, IL 60062

***Royal Labs Natural Cosmetics**, Box 22434, Charleston, SC 29413

Rusk, Inc., 1 Cummings Point Rd., Stamford, CT 06904

Safeway, Inc. (***house brand products only***), Fourth & Jackson sts., Oakland, CA 94660

***Sagami, Inc.**, 825 N. Cass Ave., Suite 101, Westmont, IL 60559

Sanford Corp. (Berol, Eberhard Faber), 2711 Washington Blvd., Bellwood, IL 60104

San Francisco Soap Company, P.O. Box 750428, Petaluma, CA 94975-0428

***Santa Fe Fragrance, Inc.**, P.O. Box 282, Santa Fe, NM 87504

***Santa Fe Soap Company**, 369 Montezuma, #167, Santa Fe, NM 87501

***Sappo Hill Soap Works**, 654 Tollman Creek Rd., Ashland, OR 97520

Schiff Products, Inc., 1911 S. 3850 W., Salt Lake City, UT 84104

Scruples Salon Products, 8231 214th St., W. Lakeville, MN 55044

Sebastian International (Wella), 6109 DeSoto Ave., Woodland Hills, CA 91367

***SerVaas Laboratories**, P.O. Box 7008, 1200 Waterway Blvd, Indianapolis, IN 46207

***Seventh Generation**, One Mill St., Burlington, VT 05401-1530

***The Shahin Soap Co.**, 427 Van Dyke Ave., Haledon, NJ 07538

Shaklee Worldwide, Shaklee Terraces, 444 Market St., San Francisco, CA 94111

Shené Cosmetics, 22761 Pacific Coast Hwy., Suite 264, Malibu, CA 90265

Shikai (Trans-India Products), P.O. Box 2866, Santa Rosa, CA 95405

Shirley Price Aromatherapy, 462 62nd St., Brooklyn, NY 11220

Shivani Ayurvedic Cosmetics, P.O. Box 377, Lancaster, MA 01523

***Simplers Botanical Co.**, Box 39, Forestville, CA 95436

Simple Wisdom, 775 S. Graham, Memphis, TN 38111

Sinclair & Valentine, 480 Airport Blvd., Watsonville, CA 95076

***Sirena Tropical Soap Co.**, P.O. Box 797217, Dallas, TX 75379

Smith & Vandiver, Inc., 480 Airport Blvd., Watsonville, CA 95076

SoapBerry Shop Company, 50 Galaxy Blvd., Unit 12, Rexdale, Ontario M9W 4Y5 Canada

Sojourner Farms Natural Pet Products, P.O. Box 8062, Ann Arbor, MI 48107

Solgar Vitamin Co., 500 Willow Tree Rd., Leonia, NJ 07605

Sombra Cosmetics, 5600 G McLeod N.E., Albuquerque, NM 87109

Song of Life, Inc., 152 Fayette St., Buckhannon, WV 26201

SoRik International, 278 Taileyand Ave., Jacksonville, FL 32202

Soya Systems, Inc., 1572 Page Industrial Dr., St. Louis, MO 63132

Spa Natural Beauty, 1201 16th St., #212, Denver, CO 80202

***The Spanish Bath**, P.O. Box 750428, Petaluma, CA 94975-0428

Staedtler, Ltd., Cowbridge Rd., Pontyclym, Mid Glamorgan, Wales

Stanley Home Products, 50 Payson Ave., Easthampton, MA 01027-2262

***Stature Field Corp.**, 1143 Rockingham Dr., Suite 106, Richardson, TX 75080

Steps in Health, P.O. Box 1409, Lake Grove, NY 11755

***Stevens Research Salon Products**, 19009 61st Ave., N.E., Unit 1, Arlington, WA 98223

Studio Magic, 1417-3 Del Prado Blvd., Suite 480, Cape Coral, FL 33990

Sukesha (Chuckles, Inc.), P.O. Box 5126, Manchester, NH 03108

***Sumeru**, P.O. Box 2110, Freedom, CA 95019

***SunFeather Herbal Soap Co.**, HCR 84, Box 60A, Potsdam, NY 13676

Sunrider International, 1625 Abalone Ave., Torrance, CA 90501

Sunrise Lane Products, 780 Greenwich St., Dept. PT, New York, NY 10014

***Sunshine Natural Products**, Route 5P, Renick, WV 24966

***Sunshine Products Group**, 2545-D Prairie Rd., Eugene, OR 97402

Supreme Beauty Products Co., 820 S. Michigan, Chicago, IL 60605

Surrey, Inc., 13110 Trails End Rd., Leander, TX 78641

Tammy Taylor Nails, 1800E Skypark Cir., Irvine, CA 92714

Taut by Leonard Engelman, 9428 Eton, #M, Chatsworth, CA 91311

TerraNova, 1200 Fifth St., Berkeley, CA 94710

***Terressentials**, 2650 Old National Pike, Middletown, MD 21769-8817

Thursday Plantation Pty., Ltd., P.O. Box 5613, Montecito, CA 93150-5613

***Tisserand Aromatherapy**, P.O. Box 750428, Petaluma, CA 94975-0428

Tom's of Maine, Lafayette Center, Box 710, Kennebunk, ME 04043

Tova Corporation, 192 N. Canon Dr., Beverly Hills, CA 90210

Trader Joe's Company, P.O. Box 3270, 538 Mission St., S. Pasadena, CA 91030

Travel Mates America, 1760 Lakeview Rd., Cleveland, OH 44112

Tressa, Inc., P.O. Box 75320, Cincinnati, OH 45275

TRI Hair Care Products, 1850 Redondo Ave., Long Beach, CA 90804

Trophy Animal Health Care, 2796 Helen St., Pensacola, FL 32504

***Tropical Botanicals, Inc.**, P.O. Box 1354, 15920 Via del Alba, Rancho Santa Fe, CA 92067

***Tropix Suncare Products**, 217 S. Seventh St., Suite 104, Brainerd, MN 56401

Truly Moist (Desert Naturels), 74-940 Hwy. 111, Suite 437, Indian Wells, CA 92210

Tyra Skin Care, 9019 Oso Ave., Suite A, Chatsworth, CA 91311

***The Ultimate Life**, P.O. Box 31154, Santa Barbara, CA 93130

Ultima II, 625 Madison Ave., New York, NY 10022

***Ultra Glow Cosmetics**, P.O. Box 1469, Station A, Vancouver, B.C. V6C 1P7 Canada

Unicure, 4437 Park Dr., Norcross, GA 30093

Upper Canada Soap & Candle Makers, 1510 Caterpillar Rd., Mississauga L4X 2W9 Canada

***USA King's Crossing, Inc.** (Total Shaving Solution), P.O. Box 832074, Richardson, TX 75083

***U.S. Sales Service (Crystal Orchid)**, 1414 E. Libra Dr., Tempe, AZ 85283

Vapor Products, P.O. Box 568395, Orlando, FL 32856-8395

***Vegelatum**, P.O. Box 51867, Bowling Green, KY 42101

Vermont Soapworks, Rt. 7, Brandon, VT 05733

***Veterinarian's Best**, P.O. Box 4459, Santa Barbara, CA 93103

Victoria's Secret (The Limited), 4 Limited Pkwy., Reynoldsburg, OH 43068

Virginia Soap, Ltd., Group 60, Box 20, RR#1, Anola, Manitoba R0E 0A0 Canada

V'tae Parfum & Body Care, 576 Searls Ave., Nevada City, CA 95959

Wachters' Organic Sea Products, 360 Shaw Rd., S. San Francisco, CA 94080

***Wala-Heilmittel**, P.O. Box 407, Wyoming, RI 02898

***Warm Earth Cosmetics**, 1155 Stanley Ave., Chico, CA 95928-6944

***The WARM Store**, 31 Mill Hill Rd., Woodstock, NY 12498

Weleda, P.O. Box 249, Congers, NY 10920

The Wella Corporation (Gucci, Sebastian), 524 Grand Ave., Englewood, NJ 07631

***Whip-It Products, Inc.**, P.O. Box 30128, Pensacola, FL 32503

Wind River Herbs, P.O. Box 3876, Jackson, WY 83001

***Winter White**, P.O. Box 51867, Bowling Green, KY 42101

Wisdom Toothbrush Co., 151 Pfingsten Rd., Deerfield, IL 60015

WiseWays Herbals, 99 Harvey Rd., Worthington, MA 01098

Womankind, P.O. Box 1775, Sebastopol, CA 95473

Woods of Windsor, Ltd., 125 Mineola Ave., Suite 304, Roslyn Heights, NY 11577

Wysong, 1880 N. Eastman Rd., Midland, MI 48640

Yardley (Maybelline, L'Oréal), P.O. Box 372, Memphis, TN 38101-0372[1]

Zia Cosmetics, 410 Townsend St., 2nd Fl., San Francisco, CA 94107-1524

[1]L'Oréal (Cosmair, Lancôme, Maybelline) signed PETA's Statement of Assurance and told PETA in face-to-face meetings that the company no longer conducts animal tests of any kind. However, later, when L'Oréal executives were asked to guarantee that they will not test ingredients on animals, they refused.

[2]In 1996, in the course of litigation, Melaleuca, Inc., commissioned an LD50 test on a competitor's product, resulting in the suffering and deaths of 10 rats. Melaleuca has provided PETA with a letter expressing regret for this action and renewing its commitment to a non-animal-testing policy.

Companies That Test on Animals

People for the Ethical Treatment of Animals
501 Front Street
Norfolk, VA 23510
Tel: 757-622-PETA

The following companies manufacture products that **ARE** tested on animals. Those marked with an asterisk (*) are presently observing a moratorium on animal testing. Please encourage them to announce a permanent ban. Listed in parentheses are either examples of products manufactured by that company or, if applicable, its parent company. Companies on this list may manufacture individual lines of products without animal testing (e.g., Del Laboratories claims its Naturistics and Natural Glow lines are not animal tested). They have not, however, eliminated animal testing on their entire line of cosmetics and household products.

Alberto-Culver Co. (Tresemmé, Sally Beauty Supply, Alberto V05, TCB Naturals) 2525 Armitage Ave., Melrose Park, IL 60160, 708-450-3000

Allergan, Inc. 2525 Dupont Dr., P.O. Box 19534, Irvine, CA 92713, 714-752-4500, 800-347-4500

Arm & Hammer (Church & Dwight) 469 N. Harrison St., Princeton, NJ 08543, 609-683-5900, 800-524-1328

Aziza (Chesebrough-Ponds) 33 Benedict Pl., Greenwich, CT 06830, 203-661-2000, 800-243-5804

Bausch & Lomb (Curél, Soft Sense, Clear Choice) P.O. Box 450, Rochester, NY 14692-0450, 716-338-5386, 800-344-8815

Benckiser (Coty, Lancaster, Jovan) 237 Park Ave., New York, NY 10017-3142, 212-850-2300

Bic Corporation 500 Bic Dr., Milford, CT 06460, 203-783-2000

Block Drug Co., Inc. (Polident, Sensodyne, Tegrin, Lava, Carpet Fresh) 257 Cornelison Ave., Jersey City, NJ 07302, 201-434-3000, 800-365-6500

Boyle-Midway (Reckitt & Colman) Box 7, Station "U," 2 Wickman Rd., Toronto, Ontario 5M5 M82 Canada, 416-255-2300

***Braun (Gillette Company)** 66 Broadway, Rte. 1, Lynnfield, MA 01940, 800-272-8611

Bristol-Myers Squibb Co. (Clairol, Ban Roll-On, Keri, Final Net) 345 Park Ave., New York, NY 10154, 212-546-4000

Carter-Wallace (Arrid, Lady's Choice, Nair, Pearl Drops) 1345 Ave. of the Americas, New York, NY 10105, 212-339-5000

Chesebrough-Ponds (Fabergé, Cutex, Vaseline) 33 Benedict Pl., Greenwich, CT 06830, 203-661-2000, 800-243-5804

Church & Dwight (Arm & Hammer) 469 N. Harrison St., Princeton, NJ 08543, 609-683-5900, 800-524-1328

Clairol, Inc. (Bristol-Myers Squibb) 345 Park Ave., New York, NY 10154, 212-546-5000, 800-223-5800

Clorox (Pine-Sol, S.O.S., Tilex, ArmorAll) 1221 Broadway, Oakland, CA 94612, 510-271-7000, 800-227-1860

Colgate-Palmolive Co. (Palmolive, Ajax, Fab, Speed Stick, Mennen, SoftSoap) 300 Park Ave., New York, NY 10022, 212-310-2000, 800-221-4607

Coty (Benckiser) 237 Park Ave., New York, NY 10017-3142, 212-850-2300

Dana Perfumes (Alyssa Ashley) 635 Madison Ave., New York, NY 10022-1009, 212-751-3700

Del Laboratories (Flame Glow, Commerce Drug, Sally Hansen) 565 Broad Hollow Rd., Farmingdale, NY 11735, 516-293-7070, 800-645-9888

***Dial Corporation (Purex, Renuzit)** 1850 North Central, Phoenix, AZ 85004, 602-207-7100, 800-528-0849

DowBrands (Glass Plus, Fantastik, Vivid) P.O. Box 68511, Indianapolis, IN 46268, 317-873-7000

Drackett Products Co. (S.C. Johnson & Son) 1525 Howe St., Racine, WI 53403, 414-631-2000, 800-558-5252

EcoLab, Inc. Ecolab Center, St. Paul, MN 55102, 612-293-2233, 800-352-5326

Erno Laszlo 200 First Stamford Pl., Stamford, CT 06902-6759, 203-363-5461

***Gillette Co. (Liquid Paper, Flair, Jafra, Braun, Duracell)** Prudential Tower Bldg., Boston, MA 02199, 617-421-7000, 800-872-7202

Givaudan-Roure 1775 Windsor Rd., Teaneck, NJ 07666, 201-833-2300

Harris Research, Inc. (Chem-Dry) 1530 N. 1000 W., Logan, UT 84321, 801-755-0099, 800-CHEMDRY

Helene Curtis Industries (Finesse, Unilever, Suave) 325 N. Wells St., Chicago, IL 60610-4713, 312-661-0222

Jhirmack (Playtex) 215 College Rd., P.O. Box 728, Paramus, NJ 07653, 201-295-8000, 800-222-0453

Johnson & Johnson (Neutrogena) 1 Johnson & Johnson Plaza, New Brunswick, NJ 08933, 908-524-0400

S.C. Johnson & Son (Pledge, Drano, Windex, Glade) 1525 Howe St., Racine, WI 53403, 414-631-2000, 800-558-5252

Kimberly-Clark Corp. (Kleenex, Scott Paper, Huggies) P.O. Box 2020, Neenah, WI 54957-2020, 414-721-2000, 800-544-1847

Kroger (*house brand products only*) 1014 Vine St., Cincinnati, OH 45202, 513-762-1578

Lamaur P.O. Box 1221, Minneapolis, MN 55440-1221, 612-571-1234

L & F Products One Philips Pkwy., Montvale, NJ 07645-1810, 201-573-5700

***Mary Kay Cosmetics** 16251 Dallas Pkwy., Dallas, TX 75248, 214-687-6300, 800-201-1362

Mead Courthouse Plaza N.E., Dayton, OH 45463, 513-222-6323

Mennen Co. (Colgate-Palmolive) E. Hanover Ave., Morristown, NJ 07962, 201-631-9000

Murphy-Phoenix Co. (Colgate-Palmolive) P.O. Box 39670, Solon, OH 44139, 800-486-7627

Naturelle 325 N. Wells St., Chicago, IL 60610-4713, 312-661-0222

Neoteric Cosmetics 4880 Havana St., Denver, CO 80239-0019, 303-373-4860

Neutron Industries, Inc. 7107 N. Black Canyon Hwy., Phoenix, AZ 85021, 602-864-0090

Noxell (Procter & Gamble) 11050 York Rd., Hunt Valley, MD 21030-2098, 410-785-7300, 800-572-3232

Olay Co./Oil of Olay (Procter & Gamble) P.O. Box 599, Cincinnati, OH 45201, 800-543-1745

Oral-B (Gillette Comapny) 1 Lagoon Dr., Redwood City, CA 94065-1561, 415-598-5000

Pantene (Procter & Gamble) Procter & Gamble Plaza, Cincinnati, OH 45202, 800-945-7768

Parfums International (White Shoulders) 1345 Ave. of the Americas, New York, NY 10105, 212-261-1000

Parker Pens (Gillette Company) P.O. Box 5100, Janesville, WI 53547-5100, 608-755-7000

Pennex 1 Pennex Dr., Verona, PA 15147, 412-828-2900, 800-245-6110

Perrigo 117 Water St., Allegan, MI 49010, 616-673-8451, 800-253-3606

Pfizer, Inc. (Bain de Soleil, Plax, Visine, Desitin, BenGay) 235 E. 42nd St., New York, NY 10017, 212-573-2323

Physicians Formula Cosmetics (Pierre Fabré) 230 S. Ninth Ave., City of Industry, CA 91749, 818-968-3855

Playtex Products, Inc. (Banana Boat, Woolite) 215 College Rd., P.O. Box 728, Paramus, NJ 07653, 201-265-8000

Procter & Gamble Co. (Crest, Tide, Cover Girl, Max Factor, Giorgio) P.O. Box 599, Cincinnati, OH 45201, 513-983-1100, 800-543-1745

Publix Super Markets (*house brand products only*) P.O. Box 407, Lakeland, FL 33802-0407, 813-688-1188

Reckitt & Colman (Lysol, Mop & Glo) 1655 Valley Rd., Wayne, NJ 07474-0945, 201-633-6700, 800-232-9665

Richardson-Vicks (Procter & Gamble) P.O. Box 599, Cincinnati, OH 45201, 513-983-1100, 800-543-1745

Sally Hansen (Del Laboratories) 565 Broad Hollow Rd., Farmingdale, NY 11735, 516-293-7070, 800-645-9888

Sanofi (Yves Saint Laurent) 90 Park Ave., 24th Fl., New York, NY 10016, 212-907-2000

Schering-Plough (Coppertone) 2000 Galloping Hill Rd., Kenilworth, NJ 07033, 908-298-4000, 800-842-4090

Schick (Warner-Lambert) 201 Tabor Rd., Morris Plains, NJ 07950, 201-540-2000, 800-323-5379

SmithKline Beecham 100 Beecham Dr., Pittsburgh, PA 15230, 412-928-1000, 800-456-6670

SoftSoap Enterprises (Colgate-Palmolive) 1107 Hazeltine Blvd., Suite 370, Chaska, MN 55318, 612-448-1118

Sunshine Makers (Simple Green) P.O. Box 2708, Huntington Beach, CA 92649, 714-840-1319, 800-228-0709

Sun Star 600 Eagle Dr., Bensenville, IL 60106-1977, 708-595-1660, 800-821-5455

3M (Scotch, Post-It) Center Bldg., 220-2E-02, St. Paul, MN 55144-1000, 612-733-1110, 800-364-3577

Unilever (Lever Bros., Calvin Klein, Elizabeth Arden, Helene Curtis) 390 Park Ave., New York, NY 10022, 212-888-1260, 800-745-9696

Vidal Sassoon (Procter & Gamble) P.O. Box 599, Cincinnati, OH 45201, 800-543-7270

Warner-Lambert (Lubriderm, Listerine, Schick) 201 Tabor Rd., Morris Plains, NJ 07950, 201-540-2000, 800-323-5379

Westwood Pharmaceuticals 100 Forest Ave., Buffalo, NY 14213, 716-887-3400, 800-333-0950

Whitehall Laboratories (American Home Products) 5 Giralda Farms, Madison, NJ 07940, 201-660-5000

All PETA product lists are updated every two months. Please call if you have any questions about the status of any companies or organizations listed.

PETA
Factsheet
Miscellaneous #12
People for the Ethical Treatment of Animals
501 Front Street, Norfolk, VA 23510
Tel: 757-622-PETA
http://envirolink.org/arrs/peta

Animals in the Classroom: Lessons in Disrespect

Rabbits, mice, guinea pigs, frogs, parakeets, rats, snakes, fish, turtles, and countless other animals suffer abuse and neglect in school classrooms every year as teaching "tools" and classroom "pets." Many teachers bring animals into the classroom with good intentions—to interest or amuse students, to teach responsibility, or to convey information about the animals themselves. But animals suffer because of those practices. Ironically, students can and do learn responsibility, as well as animal behavior and hands-on science, without the presence of animals in their classrooms. There are far more constructive ways to learn about living beings than by holding animals captive in school, where they are vulnerable to hazards and neglect.

Paying the Price

Poked with pencils, accidentally dropped, given the wrong food, ignored, or kept in far too small an enclosure, animals often pay the price for educators' lack of ingenuity in devising appropriate lessons.

Those who seek to keep animals in schools are hindering an encouraging cultural trend toward greater sensitivity to animals' well-being. Opponents of efforts to protect animals overstate the educational benefits of using animals in the classroom while underestimating the animals' suffering and the difficulty of protecting them. In defending such practices, they consistently deny the animals' interests.

Some reports PETA has received of animal abuse in U.S. schools include:

- Tarantulas, turtles, and lizards died because a high school science department failed to provide adequate food or maintain proper temperatures.(1)
- An elementary-school girl, pecked by a chick she was holding, threw the chick against the wall.(2)
- A guinea pig taken home for winter break suffered a broken back.(3)
- A parakeet left in a school building over winter break died when the temperature dropped too low.(4)
- Five rabbits, two goats, a sheep, and a pig were stabbed or crushed to death despite being kept inside a school's padlocked barn.(5)
- Thirty-five to 40 gerbils who were kept as classroom "pets" were intentionally stomped on by children at the end of the school year.(6)

Many "textbook" uses of animals are intrinsically inhumane. In chick-hatching programs, popular in elementary schools, teachers often fail to turn the eggs on schedule, or a custodian unplugs the incubator thinking it's an appliance inadvertently left running. The result: dead and crippled hatchlings, as well as distraught children.

One school's science project calls for students to place fish in warm water, then cold water, to demonstrate the effects of the change on respiratory rates. One student's parent was told that some of the students had made the warm water too hot; some of the fish leapt out of the water and died. When students asked that the project be stopped, the teacher refused.

Some school districts inject male chickens with estrogen and females with testosterone to show "opposite-sex mating behavior." One caller reported that a chicken died immediately upon injection. Many chicks die in shipment for this and other school activities.

Even classroom "pets" can suffer from neglect and abuse. Once animals are brought to school, important aspects of their nature are ignored altogether. Mice and most other small mammals are nocturnal, yet they are kept in brightly lit classrooms and removed from their cages during the day. Snakes and other non-domestic animals demonstrate "predation" to children who laugh, scream, or turn away as live mice or rats are fed to the animals.

School's Out - Now What?

When the school year ends, these once-valuable "pets" suddenly pose a problem: how to dispose of them? All too often, the animals end up the responsibility of already-overworked shelters or are given to students who can provide "good homes." Unfortunately, important screening procedures, such as home checks and interviews with the entire family, are often inconvenient or overlooked by a hurried and overworked teacher at the end of the school year. Even if a student has behaved responsibly toward an animal in a classroom, siblings might be abusive or reckless. Sadly, many children become bored with animals once they take them home. Water bottles may be left dry, food dishes empty, and cages dirty. Animals also can be ignored, or deprived of human contact or appropriate companions of the animals' own species.

What Lesson Learned?

Keeping animals in the classroom teaches the wrong lessons about animals. Students might learn personal responsibility from cleaning a rat's cage or filling a hamster's water bottle on time, but this kind of basic discipline is easily taught without animals. Rather than teaching the broader responsibility for animals' total well-being, allowing animals to be used as "learning tools" inevitably lowers their status in the minds of students. Young people generalize from the "do's" and "don'ts" of authority figures, and human beings of all ages can contrive an endless variety of animal "uses" once they are taught that animals are tools.

Teaching the responsibility involved in caring for captive animals ignores the question of whether animals belong in cages at all. Instead, teachers who want their classes to learn about animals can lead discussions of these ethical issues rather than dictate by their actions that the position most detrimental to animals is correct.

School hatching projects should be replaced with modern teaching programs. Films, videos, state-of-the-art computer programs, and plastic models can demonstrate the major stages of animal development—even of chicks still inside the egg. Such programs are already in use in other areas of biology education and can easily be adapted to fit classroom needs.

What You Can Do

A more respectful understanding of animals can be encouraged by quietly observing animals in their natural surroundings. If your school is planning a hatching project, urge the science curriculum coordinator, the classroom teacher, or whomever is responsible to use an alternative project, such as visiting your local animal shelter or wildlife rehabilitation center.

If your school is keeping animals—either as "pets" or as teaching tools—protest to the teacher, the administrator, and, if necessary. the school board. Ask your school board to forbid the use of animals in classrooms.

(1) Cantor, David, "Animals Don't Belong in School," *The American School Board Journal,* October 1992. pp.39–40

(2) Ibid.

(3) Ibid.

(4) Ibid.

(5) Ibid.

(6) "School Probes Report That Kids Stomped on Gerbils as Year Ended," *Grand Rapids Press,* June 17, 1991.

PETA
Factsheet
People for the Ethical Treatment of Animals
501 Front Street, Norfolk, VA 23510
Tel: 757-622-PETA (7382)

Animal Experimentation: Sadistic Scandal

Animal Experiments #1

Vivisection, the practice of experimenting on animals, began because of religious prohibitions against the dissection of human corpses. When religious leaders finally lifted these prohibitions, it was too late—vivisection was already entrenched in medical and educational institutions.

Estimates of the number of animals tortured and killed annually in U.S. laboratories diverge widely - from 17 to 70 million animals.(1) The Animal Welfare Act requires laboratories to report the number of animals used in experiments, but the Act does not cover mice, rats, and birds (used in some 80 to 90 percent of all experiments).(2) Because these animals are not covered by the Act, they remain uncounted and we can only guess at how many actually suffer and die each year.

The largest breeding company in the United States is Charles River Breeding Laboratories (CRBL) headquartered in Massachusetts and owned by Bausch and Lomb. It commands 40-50 percent of the market for mice, rats, guinea pigs, hamsters, gerbils, rhesus monkeys, imported primates, and miniature swine. (3)

Since mice and rats are not protected under Animal Welfare Act regulations, the United States Department of Agriculture (USDA) does not require that commercial breeders of these rodents be registered or that the USDA's Animal and Plant Health Inspection Service (APHIS) inspect such establishments. (4)

Dogs and cats are also used in experiments. They come from breeders like CRBL, some animal shelters and pounds, and organized "bunchers" who pick up strays, purchase litters from unsuspecting people who allow their companion animals to become pregnant, obtain animals from "Free to a Good Home" advertisements, or trap and steal the animals. Birds, frogs, pigs, sheep, cattle, and many naturally free-roaming animals (e.g., prairie dogs and owls) are also common victims of experimentation. At this writing, animals traditionally raised for food are covered by Animal Welfare Act regulations only minimally, and on a temporary basis, when used in, for example, heart transplant experiments; but they are not covered at all when used in agriculture studies. Unfortunately, vivisectors are using more and more animals whom they consider less "cute," because, although they know these animals suffer just as much, they believe people won't object as strenuously to the torture of a pig or a rat as they will to that of a dog or a rabbit.

Paying for Pain

The National Institutes of Health (NIH) in the United States is the world's largest funder of animal experiments. It dispenses seven billion tax dollars in grants annually, of which about $5 billion goes toward studies involving animals.(5) The Department of Defense spent about $180 million on experiments using 553,000 animals in 1993. Although this figure represents a 36% increase in the number of animals used over the past decade, the military offered no detailed rationale in its own reports or at Congressional hearings.(6) Examples of torturous taxpayer-funded experiments at military facilities include wound experiments, radiation experiments, studies on the effects of chemical warfare, and other deadly and maiming procedures.

Private institutions and companies also invest in the vivisection industry. Many household product and cosmetics companies still pump their products into animals' stomachs, rub them onto their shaved, abraded skin, squirt them into their eyes, and force them to inhale aerosol products. Charities, such as the American Cancer Society and the March of Dimes, use donations from private citizens to fund experiments on animals.

Agricultural experiments are carried out on cattle, sheep, pigs, chickens, and turkeys to find ways in which to make cows produce more milk, sheep produce more wool, and all animals produce more offspring and grow "meatier."

Bad Science

There are many reasons to oppose vivisection. For example, enormous physiological variations exist among rats, rabbits, dogs, pigs, and human beings. A 1989 study to determine the carcinogenicity of fluoride illustrated this fact. Approximately 520 rats and 520 mice were given daily doses of the mineral for two years. Not one mouse was adversely affected by the fluoride, but the rats experienced health problems including cancer of the mouth and bone. As test data cannot accurately be extrapolated from a mouse to a rat, it can't be argued that data can accurately be extrapolated from either species to a human.

In many cases, animal studies do not just hurt animals and waste money; they harm and kill people, too. The drugs thalidomide, Zomax, and DES were all tested on animals and judged safe but had devastating consequences for the humans who used them. A General Accounting Office report, released in May 1990, found that more than half of the prescription drugs approved by the Food and Drug Administration between 1976 and 1985 caused side effects that were serious enough to cause the drugs to be withdrawn from the market or relabeled. All of these drugs had been tested on animals.

Animal experimentation also misleads researchers in their studies. Dr. Albert Sabin, who developed the oral polio vaccine, cited in testimony at a congressional hearing this example of the dangers of animal-based research: "[p]aralytic polio could be dealt with only by preventing the irreversible destruction of the large number of motor nerve cells, and the work on prevention was delayed by an erroneous conception of the nature of the human disease based on misleading experimental models of the disease in monkeys." (7)

Healing Without Hurting

The Physicians Committee for Responsible Medicine reports that sophisticated non-animal research methods are more accurate, less expensive, and less time-consuming than traditional animal-based research methods. Patients waiting for helpful drugs and treatments could be spared years of suffering if companies and government agencies would implement the efficient alternatives to animal studies. Fewer accidental deaths caused by drugs and treatments would occur if stubborn bureaucrats and wealthy vivisectors would use the more accurate alternatives. And tax dollars would be better spent preventing human suffering in the first place through education programs and medical assistance programs for low-income individuals—helping the more than 30 million U.S. citizens who cannot afford health insurance—rather than making animals sick. Most killer diseases in this country (heart disease, cancer, and stroke) can be prevented by eating a low-fat, vegetarian diet, refraining from smoking and alcohol abuse, and exercising regularly. These simple lifestyle changes can also help prevent arthritis, adult-onset diabetes, ulcers, and a long list of other illnesses.

It is not surprising that those who make money experimenting on animals or supplying vivisectors with cages, restraining devices, food for caged animals (like the Lab Chow made by Purina Mills), and tiny guillotines to destroy animals whose lives are no longer considered useful insist that nearly every medical advance has been made through the use of animals. Although every drug and procedure must now be tested on animals before hitting the market, this does not mean that animal studies are invaluable, irreplaceable, or even of minor importance or that alternative methods could not have been used.

Dr. Charles Mayo, founder of the Mayo Clinic, explains, "I abhor vivisection. It should at least be curbed. Better, it should be abolished. I know of no achievement through vivisection, no scientific discovery, that could not have been obtained without such barbarism and cruelty. The whole thing is evil." (8)

Dr. Edward Kass, of the Harvard Medical School, said in a speech he gave to the Infectious Disease Society of America: "[I]t was not medical research that had stamped out tuberculosis, diphtheria, pneumonia and puerperal sepsis; the primary credit for those monumental accomplishments must go to public health, sanitation and the general improvement in the standard of living brought about by industrialization."(9)

Changing the System

Write to your legislators today to express your concern for animals used in experiments. Urge them to do everything in their power to push researchers out of the Dark Ages where animals are still butchered in the name of science.

Write to People for the Ethical Treatment of Animals for a free factsheet detailing the many humane alternatives to vivisection.

References

1. Orlans, F. Barbara, "Data on Animal Experimentation in the United States: What They Do and Do Not Show," *Perspectives in Biology and Medicine*, 37, 2. Winter 1994.
2. Ibid.
3. Reddy, Kal, THETA Corporation, Research Animal Markets Report, No. 982, September 1989.
4. Soos, Troy, "Charles River Breeding Labs," *The Animals' Agenda*, Dec. 1986, p. 10.
5. Stoller, Kenneth, M.D., "Animal Testing: Why a Doctor Opposes It," *The Orlando Sentinel*, June 25, 1990.
6. Krizmanic, Judy, "Military Increases Animal Experiments," *Vegetarian Times*, August 1994.
7. Stoller, op. cit.
8. Quoted by William H. Hendrix, *New York Daily News*, Mar. 13, 1961.
9. Prouix, Lawrence, "A History of Progress," *Washington Post*, Feb. 21, 1995.

For more information on animal testing, E-mail your request along with your name and mailing address.

PETA
Factsheet
People for the Ethical Treatment of Animals
501 Front Street, Norfolk, VA 23510
Tel: 757-622-PETA
http://envirolink.org/arrs/peta

Factory Farming: Mechanized Madness

Vegetarianism #3

Life on "Old MacDonald's Farm" isn't what it used to be. The green pastures and idyllic barnyard scenes portrayed in children's books are quickly being replaced by windowless metal sheds, wire cages, "iron maidens," and other confinement systems integral to what is now known as "factory farming."

Deprivation and Disease

Simply put, the factory farming system of modern agriculture strives to produce the most meat, milk, and eggs as quickly and cheaply as possible, and in the smallest amount of space possible. Cows, calves, pigs, chickens, turkeys, ducks, geese, rabbits, and other animals are kept in small cages or stalls, often unable to turn around. They are deprived of exercise so that all of their bodies' energy goes toward producing flesh, eggs, or milk for human consumption. They are fed growth hormones to fatten them faster and are genetically altered to grow larger or to produce more milk or eggs than nature originally intended.

Because crowding creates a prime atmosphere for disease, animals on factory farms are fed and sprayed with huge amounts of pesticides and antibiotics, which remain in their bodies and are passed on to the people who eat them, creating serious human health hazards.

Chickens are divided into two groups: layers and broilers. Five to six laying hens are kept in a 14-inch-square mesh cage, and cages are often stacked in many tiers. Conveyor belts bring in food and water and carry away eggs and excrement. Because the hens are severely crowded, they are kept in semi-darkness and their beaks are cut off with hot irons (without anesthetics) to keep them from pecking each other to death. The wire mesh of the cages rubs their feathers off, chafes their skin, and cripples their feet.

Approximately 20 percent of the hens raised under these conditions die of stress or disease.(1) At the age of one to two years, their overworked bodies decline in egg production and they are slaughtered (chickens would normally live 15-20 years).(2) Ninety percent of all commercially sold eggs come from chickens raised on factory farms.(3)

More than six billion "broiler" chickens are raised in sheds each year.(4) Lighting is manipulated to keep the birds eating as often as possible, and they are killed after only nine weeks. Despite the heavy use of pesticides and antibiotics, up to 60 percent of chickens sold at the supermarket are infected with live salmonella bacteria.(5)

Genetic selection to keep up with demand and also reduce production costs, causes extremely painful joint and bone conditions, making any movement difficult. PETA's 1994 undercover investigation into the "broiler" chicken industry also revealed birds suffering from dehydration, respiratory diseases, bacterial infections, heart attacks, crippled legs, and other serious ailments.

Cattle raised for beef are usually born in one state, fattened in another, and slaughtered in yet another. They are fed an unnatural diet of high-bulk grains and other "fillers" (including sawdust) until they weigh 1,000 pounds. They are castrated, de-horned, and branded without anesthetics. During transportation, cattle are crowded into metal trucks where they suffer from fear, injury, temperature extremes, and lack of food, water, and veterinary care.

Calves raised for veal—the male offspring of dairy cows—are the most cruelly confined and deprived animals on factory farms. Taken from their mothers only a few days after birth, they are chained in stalls only 22 inches wide with slatted floors that cause severe leg and joint pain. Since their mothers' milk is usurped for human consumption, they are fed a milk substitute laced with hormones but deprived of iron: anemia keeps their flesh pale and tender but makes the calves very weak. When they are slaughtered at the age of about 16 weeks, they are often too sick or crippled to walk. One out of every 10 calves dies in confinement.(6)

Ninety percent of all pigs are closely confined at some point in their lives, and 70 percent are kept constantly confined.(7) Sows are kept pregnant or nursing constantly and are squeezed into narrow metal "iron maiden" stalls, unable to turn around. Although pigs are naturally peaceful and social animals, they resort to cannibalism and tailbiting when packed into crowded pens and develop neurotic behaviors when kept isolated and confined. Pork producers lose $187 million a year due to dysentery, cholera, trichinosis, and other diseases fostered by factory farming.(8) Approximately 30 percent of all pork products are contaminated with toxoplasmosis.(9)

Laws and Lifestyles

Factory farming is an extremely cruel method of raising animals, but its profitability makes it popular. One way to stop the abuses of factory farming is to support legislation that abolishes battery cages, veal crates, and intensive-confinement systems. But the best way to save animals from the misery of factory farming is to stop buying and eating meat, milk, and eggs. Vegetarianism and veganism mean eating for life: yours and theirs.

References

1. "Factory Farming," United Animal Defenders, Inc., p. 3.
2. Mason, Jim and Peter Singer, *Animal Factories*, p. 5.
3. *Poultry Digest*, July 1978, p. 363.
4. *Animal Factories*, op.cit., pp. 6–8.
5. Burros, Marian, "Clinton Plan Would Move Meat and Poultry Inspections to F.D.A.," *The New York Times*, Sept. 13, 1993.
6. "Factory Farming," p. 2.
7. *Animal Factories*, op.cit., p. 8.
8. Ibid, p. 76.
9. Dubey, J.P., "Toxoplasmosis," *Journal of the American Veterinary Medical Association*, Vol. 189, No. 2, 1986, p. 168.

PETA
Factsheet
People for the Ethical Treatment of Animals
501 Front Street, Norfolk, VA 23510
Tel: 757-622-PETA
http://envirolink.org/arrs/peta

Veal: A Cruel Meal

Vegetarianism #4

The veal calf industry is one of the most reprehensible of all the kinds of intensive animal agriculture. Veal calves are a by-product of the dairy industry; they are "manufactured" by "milk machines"—dairy cows. Female calves are raised to be dairy cows: They are confined and fed synthetic hormones to increase growth and production and antibiotics to keep them alive in their unhealthy, unnatural environments. They are artificially inseminated and, after giving birth, are milked for several years until their production levels drop, then they are slaughtered.

Male calves are taken from their mothers shortly after birth. Some are slaughtered soon after birth for "bob veal." Others are raised in "open pens," a kind of minimum security prison, and even then they are sometimes chained. Most are destined for the veal crate.

Solitary Confinement

The veal crate is a wooden restraining device that is the veal calf's permanent home. It is so small (22" x 54") that the calves cannot turn around or even lie down and stretch and is the ultimate in high-profit, confinement animal agriculture.(1) Designed to prevent movement (exercise), the crate does its job of atrophying the calves' muscles, thus producing tender "gourmet" veal.

"Feeding" Time

The calves are generally fed a milk substitute intentionally lacking in iron and other essential nutrients. This diet keeps the animals anemic and creates the pale pink or white color desired in the finished product. Craving iron, the calves lick urine-saturated slats and any metallic parts of their stalls. Farmers also withhold water from the animals, who, always thirsty, are driven to drink a large quantity of the high-fat liquid feed.

Because of such extremely unhealthy living conditions and restricted diets, calves are susceptible to a long list of diseases, including chronic pneumonia and "scours," or constant diarrhea. Consequently, they must be given massive doses of antibiotics and other drugs just to keep them alive. (The antibiotics are passed on to consumers in the meat.) The calves often suffer from wounds caused by the constant rubbing against the crates.

A Fate Worse Than Death

About 14 weeks after their birth, the calves are slaughtered. The quality of this "food," laden with chemicals, lacking in fiber and other nutrients, diseased and processed, is another matter. The real issue is the calves' experience. During their brief lives, they never see the sun or touch the Earth. They never see or taste the grass. Their anemic bodies crave proper sustenance. Their muscles ache for freedom and exercise. They long for maternal care. They are kept in darkness except to be fed two to three times a day for 20 minutes. The calves have committed no crime, yet have been sentenced to a fate comparable to any Nazi concentration camp.

Reflecting on the fate of a calf raised for veal, Peter Lovenheim writes, "I don't believe that the human animal is inherently cruel. But over the centuries we have lost contact with, and compassion for, the rest of nature. This process has allowed us to make countless errors along the way. Human warfare, pollution, racism, sexism, and other 'isms' are largely a result of the 'me first' attitude that began with the subjugation of animals. If we are to survive as a species and part of a living ecosystem, we are faced with no options other than adoption of a new attitude toward nature and our role in the system. A logical and ethical place to start is to eliminate unnecessary exploitation and suffering such as that of the veal calf."(2)

What You Can Do

To help stop veal calf abuses, don't buy or eat veal, and tell friends, relatives, and neighbors why. Tell restaurant managers about veal cruelties and ask them to remove veal from their menus. Also, don't buy or eat dairy products, because of the dairy industry's role in veal production. Ask your state legislators to sponsor bills that would prohibit the use of veal crates.

References

1. Singer, Peter, *Animal Liberation,* 1975, p. 123.
2. Lovenheim, Peter, "Veal: The Great White Hoax?," *New Yorker,* November 5, 1979, p. 66.

PETA
Factsheet
People for the Ethical Treatment of Animals
501 Front Street, Norfolk, VA 23510
Tel: 757-622-PETA (7382)

Companion Animals: Pets or Prisoners?

Companion Animals #19

In a perfect world, animals would be free to live their lives to the fullest: raising their young, enjoying their native environments, and following their natural instincts. However, domesticated dogs and cats cannot survive "free" in our concrete jungles, so we must take as good care of them as possible. People with the time, money, love, and patience to make a lifetime commitment to an animal can make an enormous difference by adopting from shelters or rescuing animals from a perilous life on the street. But it is also important to stop manufacturing "pets," thereby perpetuating a class of animals forced to rely on humans to survive.

The sad truth is, not everyone loves animals. Ask any animal control officer about the animals found bruised, bloodied, and emaciated; the litters of puppies and kittens rescued from taped-up boxes alongside highways, or from sealed plastic garbage bags thrown into lakes and rivers; the animals abandoned because they "bark too much," or because they are aging, or because the family is moving.

Even people who care about animals are often unable to recognize or meet animals' many needs. Domesticated animals are in a catch-22 situation—they can no longer survive on their own, yet they retain many of their basic instincts and drives. Usually, they are isolated from their natural packs. Their bodies and souls yearn to roam—but, for safety's sake, they are confined to a house or yard, always dependent on their guardians, even for a drink of water, food to eat, or social contact.

As long as people treat animals as toys, possessions, and commodities, rather than as individuals with feelings, families, and friendships, widespread neglect and abuse is destined to continue.

Breeding's Sad Legacy

Approximately 2,500 kittens and puppies are born each hour in the U.S.—70,000 each day. One unspayed dog can lead to 28,244 puppies in nine years. One unspayed cat can lead to 14 million kittens in nine years!

Because the number of animals far exceeds the demand for them, millions of homeless cats and dogs suffer from abandonment, abuse, starvation, disease, freezing, highway death, or procurement for laboratories.

More than 70 percent of people who acquire animals end up giving them away, abandoning them, or taking them to shelters(1), which receive about 27 million animals annually. More than half—about 17 million—must be destroyed for lack of homes.(2) Most are under 18 months of age, and 90 percent are healthy and adoptable.

In light of these tragic statistics, no breeding can be considered "responsible." Those who breed animals for profit and individuals who let their dog or cat have "just one litter," however well-intentioned they may be, contribute to the severe dog and cat overpopulation crisis. Every newborn puppy or kitten means one less home for a dog or cat desperately waiting in a shelter or roaming the streets.

Problems With Purebreds

Purebred breeding (breeding animals to have certain appearances or traits) has caused a wide range of health defects in animals. For example, "flat-faced" dogs, like bulldogs or Boston terriers, experience respiratory difficulties due to shorter breathing passages; bloodhounds and Shar Peis are prone to skin infections from excessively wrinkled skin(3); other dogs suffer from epileptic seizures, hip dysplasia, painful back problems—the list goes on—as a result of human manipulation.(4)

Sadly, while breeders "custom-design" millions of dogs and cats each year, millions of equally deserving dogs and cats languish in shelters. About 25 percent of animals euthanized by shelters are purebreds.(5)

Certain dogs bred originally for fighting, like the "pit bull" breeds, can have additional problems. Pit bull terriers were originally bred to fight chained bulls and bears. Today, they are frequently used by drug dealers to guard drugs and money, and in inner-city fighting rings where they often die very violent deaths.

Few good homes are open to dogs perceived as overly aggressive. Breed-specific legislation (with a "grandfather clause" for those dogs already in existence) can be an important tool in ending the tragic exploitation of these breeds.

Obedience Training

PETA wholeheartedly supports humane, interactive training: It gives dogs more freedom and understanding of our world. Untrained dogs must be constantly restrained from running off into a street or punished for anti-social behavior. Dogs should be trained only by those they live with; turning a dog over to someone else to train not only allows for unseen abuse, it also prevents guardians from learning how to communicate effectively with their animal companions.

Compassion, clarity, and consistency are the most important elements of dog training. Training should not include any activity that endangers animals or puts undue stress on them. Good books on the subject include: *Mother Knows Best: The Natural Way to Train Your Dog,* by Carol Lea Benjamin, *Communicating With Your Dog: A Humane Approach to Dog Training,* by Ted Baer, and *Dog Talk,* by John Ross.

Working Dogs

Relationships of mutual respect and benefit are truly wonderful. However, working dogs are often used as a substitute for innovative programs that intelligently address human needs. Sometimes they are used in situations considered too dangerous for a human being and, therefore, too dangerous for the animal. They may even be treated cruelly in preparation for, and during, their lives of servitude. Some people with working dogs love them, and some don't, so working dogs cannot always count on having homes where they are well-treated. Also, some working dog programs contribute to dog overpopulation by breeding their dogs (with the notable exception of programs for the deaf that rescue dogs from shelters).

When working dogs become too old to work, they may be separated from their human companions and either "retired" with another family (always wondering, no doubt, what they did wrong or where their lifelong human companion went), returned to the training center, or even killed. Optimally, human services for the disabled should be improved rather than relying on the breeding and exploitation of animals.

Horse Woes

PETA favors horse rescue and opposes horse breeding. Horses are often acquired frivolously and then are neglected and cruelly treated. Many horses are either lonely and bored or lonely and overworked. Because many horses are boarded and spend much of their lives away from their owners' watchful eyes, there is an enormous opportunity for abuse by uncaring or untrained stable personnel. Horse breeding has caused the same overpopulation problem that plagues dogs and cats, and many horses who end up at slaughterhouses are former "pets." Once adopted, horses should never be sold or given away; it is virtually impossible to know where they will ultimately end up.

What You Can Do

- Spay or neuter dogs and cats.
- Adopt from shelters—and don't forget adult animals, who are often overlooked by people looking for a puppy or kitten.
- Take strays to humanely run shelters.
- Work within your community to legislate mandatory spaying and neutering. (See PETA's "Spaying and Neutering: A Solution for Suffering" factsheet for more information.)
- Speak up if someone is planning to breed an animal. Urge people who desire the companionship of animals to adopt from animal shelters.
- Point out neglect—talk to the animal's guardian, send an anonymous letter, or contact the humane society. Be persistent!
- Walk and play daily with your companion animals.
- If possible, adopt two animals. Animals need both human and animal companionship. Having an animal friend can help alleviate the boredom and loneliness of long hours spent waiting for you to come home.
- Read *Save the Animals! 101 Easy Things You Can Do*, by Ingrid Newkirk, for more tips on how you can help animals.

References

1. Whitemore, Hank, "Pet Owners: Do the Right Thing," *Parade Magazine*, Feb. 19, 1995.
2. Moulton, Carol, "Animal Shelters: Changing Roles," *The Animals' Agenda*, May 1988, p. 14-15.
3. "Breeder's Integrity a Factor in Health," *Washington Times*, March 23, 1995.
4. Shook, Larry, "Bad Dogs," *New York Times*, Aug. 8, 1992.
5. Associated Press, "Breeding Industry Swells Glut of Dogs," *The Morning Call*, Aug. 1, 1993.

PETA
Factsheet
People for the Ethical Treatment of Animals
501 Front Street, Norfolk, VA 23510
Tel: 757-622-PETA (7382)

Euthanasia: The Compassionate Option

Companion Animals #9

Approximately 27 million animals are handled by animal shelters in the United States each year. Of them, about 10 million are reclaimed or adopted, leaving 17 million unwanted dogs and cats with nowhere to go.(1) Shelters cannot house and support all these animals until their natural deaths. Some shelters sell unclaimed animals to laboratories to be used in experiments and eventually killed.

Companion animals cannot survive on the streets. If they don't starve, freeze, get hit by a car, or die of disease, they may be tormented and possibly killed by bored juveniles or picked up by a dealer who obtains animals to sell to laboratories.

Good and Bad Solutions

With the number of unwanted companion animals so high and nowhere to place them, sometimes the most humane act a shelter worker can do is to give an animal a peaceful release from a world in which dogs and cats are often considered "surplus" and unwanted. Euthanasia should be done by intravenous injection of sodium pentobarbital. Inhalants, decompression, electrocution, shooting, and other methods are unacceptable because they usually cause suffering before death occurs.

Some noninhalant pharmacologic agents can cause discomfort if injected too quickly or at too large a dose; some, such as strychnine, can cause animals to experience violent convulsions, muscle contractions, or cardiac arrest. One noninhalant commonly used in shelters and sometimes by veterinarians and laboratory personnel, T-61, caused discomfort to animals if administered at too great a dosage. Because T-61 was not a controlled substance, it was easy for anyone to obtain it regardless of understanding of proper usage of the drug, so the company that produced it voluntarily withdrew it from the market.

Inhalants such as nitrous oxide, halothane, and carbon monoxide may be expensive and unreliable and can cause irritation or excitability in animals.

The physical methods used to kill animals in shelters include shooting, electrocution, and decompression. The obvious problem with shooting is the potential for extreme pain if the person handling the gun is not competent, if the animal is struggling too much for the bullet to be placed precisely, or if the bullet is deflected and the animal survives. Electrocution can be extremely painful and traumatic and doesn't always work.

Decompression chambers simulate an ascent to thousands of feet above sea level in a matter of minutes. At many shelters that use this method, decompression occurs at speeds measuring 15 times the recommended rate. At the greater speed, the gases in animals' sinuses, middle ears, and intestines expand quickly, causing considerable discomfort or severe pain.(2) Accidental recompression can occur due to malfunctioning equipment or to personnel error or when small animals become trapped in air pockets. They must then be put through the procedure all over again.

If your local pound or shelter is using any method other than an intravenous injection of sodium pentobarbital, protest to local authorities and demand the implementation of humane practices. Check state and local laws for prescribed methods of euthanasia and insist that your local shelter comply with these requirements. Euthanasia should always be performed by well-trained, caring staff members, and animals should never be euthanized in view of other animals.

Until people who have dogs and cats help control their population, some people must do society's dirty work and try responsibly and humanely to stop the immense suffering caused by overpopulation. Sadly, euthanasia is the most compassionate option.

Your Companion Animal

When companion animals become very sick and are suffering, with no hope of recovery, and if they seem incapable of truly enjoying life, it may be time for their peaceful death by euthanasia. This is a difficult decision to make. Many veterinarians feel it is unethical to recommend euthanasia before a client asks for their opinion. In a difficult situation, ask your veterinarian to talk frankly with you, and consider getting a second opinion if you are in doubt. Be sure you are not prolonging your friend's suffering because of your own pain at letting go. The tendency is to wait too long, at the expense of the animal you love.

If your companion animal is very nervous, you may want to obtain a dose of tranquilizer from your veterinarian and administer it two hours before the appointed time for euthanasia. The veterinarian will be able to give the injection more easily to a relaxed patient who is not moving about. You will also be calmer when your companion is at ease. It is important to try to be cheery in front of your friend until after he or she has died.

Some veterinarians will come to your home to administer the shot. Otherwise, go to the animal hospital, perhaps taking a member of your family or a friend for moral support and to drive you home. If necessary, have hospital personnel help you carry your companion animal inside. If you plan to bury the body, rather than leave it at the hospital, arrange beforehand to have the doctor come to the car to give the injection.

Many people wish to stay with their companion animals while they are gently "put to sleep" by injection of sodium pentobarbital into a vein in the leg; some prefer not to. This is a very personal decision that you must make for yourself. While your companion animal's brain will "go to sleep" immediately, his or her heart may beat a few minutes longer because circulation may be slowed from the tranquilizer and/or old age. A careful veterinarian will monitor the heart until its last beat. You will never doubt that your friend had a peaceful departure from this life if you are there to say goodbye.

Finally, remember it is normal to feel deep grief and a great sense of loss over the death of a true friend. Some hospital and private grief counseling services now recognize the need to help people adjust to the loss of close friends and family members who just happen not to be human. Take comfort in knowing that you did all you could for your companion in those last hours.

References

1. Moulton, Carol, "Animal Shelters: Changing Roles," *The Animals' Agenda*, May 1988, pp. 14–15.
2. "1986 Report of the AVMA Panel on Euthanasia," *The Journal of the American Veterinary Medical Association*, Vol. 188, No. 3, Feb. 1, 1986, pp. 265–66.

PETA
Factsheet
People for the Ethical Treatment of Animals
501 Front Street, Norfolk, VA 23510
Tel: 757-622-PETA (7382)

AIDS: Contagion and Confusion

Animal Experiments #2

Acquired Immune Deficiency Syndrome (AIDS) is a disease that results from a viral infection that damages the immune system. A damaged immune system cannot protect the body from other infections and cancers, and these secondary illnesses often result in death.

There are actually at least 10 strains of the AIDS virus, called the Human Immunodeficiency Virus (HIV), with different strains existing in different geographical areas. People may carry more than one strain in their bodies.

How AIDS is Transmitted

Non-animal clinical, epidemiological, and in vitro studies successfully isolated the virus that causes AIDS and demonstrated how the virus is transmitted in people. Clinical evidence shows that AIDS is transmitted only through blood, semen, and vaginal fluids. The virus dies quickly outside the body, so it cannot be transmitted through the air or through casual contact. Because AIDS can develop years after initial infection, it is impossible to predict how many people might get the disease.

The origin of the AIDS virus is unknown. The AIDS virus is similar to a virus found in African green monkeys, but there is no evidence that the simian virus caused the human virus. Other theories suggest that the AIDS virus was genetically engineered in a laboratory studying simian viruses, or that it resulted from live viral vaccines, particularly the smallpox vaccine, altering the immune system.

Infecting Primates

Animal experiments are neither necessary nor useful in studying how AIDS infects or affects humans. Even when injected with the AIDS virus, chimpanzees do not develop the disease. Experimenters continue to use them in AIDS studies because chimpanzees and humans share 99 percent of their genetic composition.

Experimenters have reported that two baboons infected with the H.I.V. virus beginning in 1988 have developed AIDS-like symptoms.(1) However the "achievement" of subjecting another species to this horrible disease has not produced any benefits to human patients and draws precious research funds away from the study of the disease in infected humans, who are already in abundant supply.

An increasing number of scientists are questioning the appropriateness of using endangered species, including chimpanzees, in such harmful tests. The National Academy of Sciences' Institute of Medicine issued a report stating that they are "gravely concerned that chimpanzees have been and might be used for experiments for which the rationale is not compelling in light of the scarcity and irreplaceable nature of these animals." (2)

Chimpanzees Isolated

Experimenters often keep AIDS-infected chimpanzees locked in small steel-and-glass isolation chambers in laboratories, where these highly social animals often become insane from stress and loneliness. The stress of confinement also suppresses the chimpanzees' immune systems, making accurate AIDS studies impossible. According to the Physicians Committee for Responsible Medicine, "chimpanzees and other animals have contributed nothing to progress in AIDS research that could not have been gained in other ways." (3)

Despite the repeated failures of the chimpanzee experiments, the National Institutes of Health spent more than $10 million in 1987 to fund them, and an additional $4.5 million has been allocated for a chimpanzee breeding program called the Chimpanzee Management Plan.

Education Is Needed

Although AIDS research first focused on ways to prevent infection, the difficulty of producing a vaccine for a virus that attacks the immune system has prompted scientists to shift their efforts toward finding drugs that slow the virus' progression.

Many scientists believe that the only way to stop the disease is through public education about the way it is transmitted. In a speech at Ohio Wesleyan University, former Surgeon General C. Everett Koop stressed the need to focus on AIDS education and prevention, saying, "There never will be a cure, and the likelihood of a vaccine is dim."(4)

Given the evidence, our limited resources would be better spent teaching people how to avoid AIDS, rather than attempting to spread the disease to other primates.

References

1. Altman, Lawrence K., "AIDS Drugs Fail to Curb Dementia and Nerve Damage," *New York Times*, Nov. 1, 1994.
2. Institute of Medicine, "Confronting AIDS," National Academy of Sciences, 1986, pp. 207–8.
3. Physicians Committee for Responsible Medicine, "AIDS Research: Problems With the Animal Model," *PCRM Update*, March-April 1987.
4. Koop, Dr. C. Everett, speech at Ohio Wesleyan University, quoted in *The Advocate*, Oct. 18, 1994.

PETA
Factsheet
People for the Ethical Treatment of Animals
501 Front Street, Norfolk, VA 23510
Tel: 757-622-PETA (7382)

Living in Harmony with Nature

Wildlife #7

We cause our wild animal neighbors far more trouble than they do us, as each day we invade thousands of acres of their territories and destroy their homes. Here are some ways to live in harmony with them.

Around the House

Cap your chimney. When birds sit atop chimneys for warmth they can inhale toxic fumes, and if the chimney is uncapped they can fall in and die.

Because we have destroyed so many den trees, many raccoons nest in chimneys. If you hear mouse-like squeals from above your fireplace damper, chances are they're coming from baby raccoons. Don't light a fire—you'll burn them alive. Just close the damper securely and do nothing until the babies grow older and the family leaves. When you're absolutely sure everyone's out, have your chimney professionally capped—raccoons can quickly get through amateur cappings. Also, a mother raccoon or squirrel will literally tear apart your roof if you cap one of her young inside your chimney.

If for some reason you must evict a raccoon family before they leave on their own, put a radio tuned to loud talk or rock music in the fireplace and hang a mechanic's trouble light down the chimney. (Animals like their homes dark and quiet.) Leave these in place for a few days, to give mom time to find a new home and move her children. You might also hang a thick rope down the chimney, secured at the top, in case your tenant is not a raccoon and can't climb up the slippery flue. If the animal still cannot get out, call your conservation department for the name of a state-licensed wildlife relocator. Don't entrust animals' lives to anyone else, especially "pest removal services," no matter what they tell you.

You can also use the light-radio-patience technique to evict animals from under the porch or in the attic. (Mothballs may also work in enclosed places like attics, although one family of raccoons painstakingly moved an entire box of mothballs outside, one by one.) Remember, when sealing up an animal's home, nocturnal animals, like opossums, mice, and raccoons, will be outside at night, while others, like squirrels, lizards, and birds, will be outside in the daytime.

If an animal has a nest of young in an unused part of your house and is doing no harm, don't evict them. Wait a few weeks or so, until the young are better able to cope. We owe displaced wildlife all the help we can give them.

Wild bird or bat in your house? If possible, wait until dark, then open a window and put a light outside it. Turn out all house lights. The bird should fly out to the light.

Uncovered window wells, pools, and ponds trap many animals, from salamanders to muskrats to kittens. To help them climb out, lean escape planks of rough lumber (to allow for footholds) from the bottom to the top of each uncovered window well, and place rocks in the shallow ends of ponds and pools to give animals who fall in a way to climb out. Also, a stick in the birdbath gives drowning insects a leg up.

Relocating animals by trapping them with a humane trap is often unsatisfactory; animals may travel far to get back home. Also, you may be separating an animal from loved ones and food and water sources. It is far better and easier to use one of the above methods to encourage animals to relocate themselves.

Bats consume more than 1,000 mosquitoes in an evening, so many people encourage them to settle in their yards by building bat houses. Contrary to myth, bats won't get tangled in your hair, and chances of their being rabid are miniscule. If one comes into your home, turn off all lights and open doors and windows. Bats are very sensitive to air currents. If the bat still doesn't leave, catch him or her very gently in a large jar or net. Always wear gloves if you attempt to handle a bat, and release him or her carefully outdoors. Then find and plug the entrance hole.

Leave moles alone. They are rarely numerous, and they help aerate lawns. They also eat the white grubs that damage grass and flowers. Gophers can be more numerous, but they, too, do a valuable service by aerating and mixing the soil and should usually be left alone.

Snakes are timid, and most are harmless. They control rodent populations and should be left alone. To keep snakes away from the house, stack wood or junk piles far from it, as snakes prefer this type of cover. Your library can tell you how to identify any poisonous snakes in your area; however, the vast majority are non-poisonous.

People unintentionally raise snake and rat populations by leaving companion animal food on the ground or keeping bird feeders. It is far better to plant bushes that will give birds a variety of seeds and berries than to keep a bird feeder.

Denying mice and rats access to food in your home will do the most to discourage them from taking up residence there. Do not leave dog and cat food out for long periods of time. Store dry foods such as rice and flour in glass, metal, or ceramic containers rather than paper or plastic bags. Seal small openings in your home. One PETA member drove mice from her cupboards by putting cotton balls soaked in oil of peppermint in them.

If you must trap an occasional rodent, use a humane live trap made for this purpose. If the trap is made of plastic, make sure it has air holes and check it often.

Be careful not to spill antifreeze which is highly toxic to animals, who like its sweet taste. Better, shop for Sierra antifreeze, which is non-toxic and biodegradable.

Garbage Dump Dangers

Many animals die tragically when they push their faces into discarded food containers to lick them clean and get their heads stuck inside. Recycle cans and jars. Rinse out each tin can, put the cover inside so no tongue will get sliced, and crush the open end of the can as flat as possible. Cut open one side of empty cardboard cup-like containers; inverted-pyramid yogurt cups have caused many squirrels' deaths. Also, cut apart all sections of plastic six-pack rings, including the inner diamonds. Choose paper bags at the grocery store, and use only biodegradable or photodegradable food storage bags, available from Co-op America (1612 K St., NW, Suite 600, Washington, DC 20006).

Be sure any garbage cans under trees are covered— baby opossums and others can fall in and not be able to climb out. If animals are tipping over your can, store it in a garage or make a wooden garbage can rack. Garbage can lids with clasps sometimes foil the animals. One homeowner solved the strewn garbage problem by placing a small bag of "goodies" beside his garbage can each night. Satisfied, the midnight raider left the garbage alone.

Dumpsters can be deadly—cats, raccoons, opossums and other animals climb into them and cannot climb out because of the slippery sides. Every dumpster should have a vertical branch in it so animals can escape. (Ask your local park district to put branches in park dumpsters.)

Orphaned or Sick Animals

Wild youngsters are appealing, but never try to make one your pet. It's unfair; they need to be with others of their kind. If you tame one, when the time comes for release, the animal will not know how to forage for food or be safe in the woods. Tame released animals normally follow the first humans they see, who often think, "Rabies!" and kill them. If you find a youngster who appears orphaned, wait quietly at a distance for a while to be certain the parents are nowhere nearby. If they are not, take the little one to a professional wildlife rehabilitation center for care and eventual release into a protected wild area.

An injured bird can be carried easily in a brown paper bag, loosely clothes-pinned at the top.

On very hot days, some animals come out of hiding. Foxes have been known to stretch out on patios. Normally nocturnal adult animals seen in daytime should be observed—if they run from you, chances are they are healthy. If sick, they may be lethargic, walk slowly, or stagger. Distemper is more often the culprit than rabies. (Distemper is not contagious to humans.) Call a wildlife expert.

Get names and telephone numbers of wildlife rehabilitators from your local humane society or park authority; keep them in your home and car at all times in case of an emergency.

Create a Backyard Habitat

Don't use pesticides on your yard and leave part of it natural (unmanicured). Dead wood is ecological gold—more than 150 species of birds and animals can live in dead trees and logs and feed off the insects there. The U.S. Forestry Department says saving dead wood is crucial to kicking our pesticide habit. Top off, rather than chop down, dead trees 12 inches or more in diameter. Save fat dead logs. Leave plenty of bushes for wildlife cover. Keep a birdbath filled with water, and a pan for small mammals, and use heating elements in them in the winter.

PETA
Factsheet
People for the Ethical Treatment of Animals
501 Front Street, Norfolk, VA 23510
Tel: 757-622-PETA (7382)

Xenografts: Frankenstein Science

Animal Experiments #14

Xenografts are surgical transplants in which donor and recipient are members of different species. The success rate for xenografts is zero.(1) Transplantation of vital organs and other body parts, such as bone marrow, taken from other-than-human animals has been attempted as part of experimental treatments for degenerative organ diseases and viral infections like hepatitis and AIDS. After several decades of research, xenografts have proved to be extremely costly failures that may pose serious health risks to the public.

A Bloody Trail

Since 1905, at least 34 pigs, chimpanzees, monkeys, and baboons have been made the unwilling "donors" of kidneys, hearts, livers, and bone marrow for transplantation into humans.(2) The misery inflicted on such experimental animals begins at birth, with delivery by surgical hysterectomy, after which they are placed in an "isolette" in an attempt to keep them free of infectious agents.(3) Animals are subjected to the sensory deprivation of a sterile laboratory environment and denied all social interaction with members of their own species. When the time comes for them to "donate" their organs, they are killed.

Every one of these experiments has failed, with most recipients dying within a few hours, days, or weeks.

Bad Science

The human immune system is designed to identify and reject foreign objects. Human-to-human transplants have relied on immunosuppressive drugs to control rejection of the transplanted organ. Genetic differences make transplants from other species particularly noticeable to the human immune system. Even chimpanzees, our closest relatives, are six times as different from us as we are from each other, and the risk of rejecting a baboon organ is 25 times greater than for an unmatched human organ.(4) Xenograft researchers have developed increasingly powerful immunosuppressive therapies to try to overcome this natural reaction. The drawback is that these treatments create an immune deficiency that leaves the recipient vulnerable to often fatal infections.

Hidden Dangers

In several recent xenograft experiments, researchers have cited the differences among species to try to justify the use of animal organs. In 1992, a team led by Dr. Thomas Starzl of the University of Pittsburgh transplanted a baboon liver into a 35-year-old man who was suffering from hepatitis B. The experimenters reasoned that since the hepatitis virus does not cause liver damage in baboons, a baboon liver would increase his chances of survival. Two months later, the patient died of a massive brain hemorrhage.(5)

In 1995, AIDS patient Jeff Getty received a transplant of bone marrow taken from a baboon. Baboons infected with the Human Immunodeficiency Virus (HIV), the virus presumed to cause AIDS, do not develop the life-threatening immune deficiency that characterizes human AIDS. Bone marrow is an important component of the immune system, and Getty's doctors hypothesized that by transferring this component to their patient they could create a "parallel" immune system that would fight the virus. But just weeks after the transplant, the doctors were forced to admit that the experiment had failed and no trace of the baboon cells could be found in the patient.(6)

Prior to approving the Getty experiment, the Food and Drug Administration held a conference with experts in immunology to discuss dangers and potential benefits. There was general agreement that the procedure was more likely to kill the patient than to help him.(7) In fact, many in the scientific community are calling for a moratorium on all xenografts because of the danger of unleashing new diseases into the human population. Many microbes that are completely harmless in one species cause disease in others. Baboons, for example, routinely carry infectious agents that are harmful or deadly to humans. Among them are *Yersinia pestis*, which causes bubonic plague, the Marburg virus, and the lethal Ebola virus and hantavirus.(8) In addition to known pathogens, animals may also harbor as-yet-unidentified viruses, bacteria, and parasites which could prove deadly to people. Many human epidemics, AIDS included, can probably be traced to microbes "jumping" from one species to another.

Informed Consent?

As with any hazardous medical procedure, xenograft recipients are required to sign an informed consent form before undergoing the procedure, stating that the patient understands the risks involved and the alternatives available. It is doubtful that desperately ill patients are given all the facts when considering xenograft procedures. Many believe their doctors are attempting to save their lives, rather than performing futile experiments on them. In 1984, doctors at Loma Linda University in California transplanted a baboon heart into an infant born with serious heart defects. "Baby Fae" died 20 days later. Afterwards, an independent review panel determined that there were at least three other options—all more promising than a xenograft—available to treat her condition. The baby's mother, who was alone and virtually destitute, was never informed of these options. Many medical ethicists condemned Dr. Leonard Bailey, who performed the experiment, for leading Baby Fae's mother to believe that the doomed experiment offered hope for her baby's survival.

Costly Failures

In addition to the toll in human and animal lives, xenografts divert precious resources away from truly life-saving efforts to treat disease. Each xenograft procedure costs between $250,000 and $300,000 to perform.(9) The University of Pittsburgh's experimental transplant program alone receives more than $8 million each year in funding, largely through federal grants from the National Institutes of Health. Meanwhile, many promising new treatments for AIDS and other life-threatening diseases go unexplored because of lack of funding. National organ donor procurement programs receive less than half a million dollars annually.

Even basic programs which have proved to save lives, like those that provide housing, primary care, and treatment to people with AIDS, have suffered cutbacks due to resource constraints. Most of the diseases for which xenografts have been proposed, including AIDS, hepatitis B, and other degenerative organ diseases, are preventable, yet prevention programs receive little to no public funding.

Become an Organ Donor

Advocates of cross-species transplants point to the scarcity of human organ donors to justify continued efforts in this field. Every year, thousands of Americans are buried with organs that are suitable for donation, far exceeding the 3,400 who die while on organ donor waiting lists.(10) European organ donor policies assume that every person is an organ donor unless otherwise specified. The burden rests with individuals (or their families) if they do not wish to donate their organs. Even within the current system, patients have a better chance of long-term survival by waiting for a last-minute human organ than by choosing a xenograft.(11)

In 1986, PETA initiated a campaign to save animal and human lives by encouraging people to become organ donors. To request a free organ donor card, write to PETA, 501 Front Street, Norfolk, Va., 23510 or call 757-622-PETA.

References

1. Associated Press Wire Report, "Doctors Call Use of Baboon Livers 'Bad Science,'" *Los Angeles Times*, Feb. 1, 1993.
2. Steele, David J.R., and Hugh Auchincloss, "The Application of Xenotransplantation in Human—Reasons for Delay," *ILAR Journal*, 37(1): 13–15, 1995.
3. Michaels, Marian, and Richard Simmons, "Xenotransplant-Associated Zoonoses," *Transplantation*, 57: 1–7, 1994.
4. Lowenstein, Jerold, "Fundamental Damage," *BBC Wildlife*, Aug. 1986.
5. Doris, Margaret, "The Animal Within: The Risks and Ethics of Trans-Species Transplants," *New York Perspectives*, Oct. 16, 1992.
6. Boudreau, John, "Baboon Cells Not Found in Patient," *Contra Costa Times*, Feb. 8, 1996.
7. Altman, Lawrence, "Doctors Treating AIDS Patient Turn to Baboon Marrow Cells," *The New York Times*, Dec. 15, 1995.
8. Thacher, Wendy, "Transplantation Information," Physicians Committee for Responsible Medicine News Release, Jul. 1, 1992.
9. Altman.
10. Colburn, Don, "Organ Donations Hinge on Survivors' Consent," *The Washington Post Health Section*, Aug. 4, 1995.
11. Steele and Auchincloss.

PCRM
Physicians Committee For Responsible Medicine
5100 Wisconsin Avenue NW
Suite 404
Washington, DC 20016
(202) 686-2210
FAX (202) 686-2216

Doctors and laypersons working together for compassionate and effective medical practice, research, and health promotion.

Prevention

PCRM promotes preventive medicine through innovative programs:

- The Cancer Prevention and Survival Fund has provided vital information to tens of thousands of people.
- The Gold Plan is a program for healthful eating for businesses, hospitals, and schools.
- The New Four Food Groups is PCRM's innovative proposal for a federal nutrition policy that puts a new priority on health. The program also includes an effective nutrition curriculum for schools.
- Our *Saving Lives* public service announcement series featured Dr. Henry Heimlich demonstrating the Heimlich Manuever, as well as messages on heart disease and cancer prevention.

Research

We also encourage higher standards for ethics and effectiveness in research:

* We oppose unethical human experiments. While great strides have been made in eliminating such experiments, problems remain. For example, children are still given synthetic growth hormone in experiments to make them taller, and both children and adults are exposed to unnecessary new drugs which have toxic effects.

- We promote alternatives to animal research. We have worked to put a stop to gruesome experiments, such as the military's cat-shooting studies, DEA narcotics experiments, and monkey self-mutilation projects. We also promote non-animal methods in medical education. Currently, 25% of U.S. medical schools have dropped their animal labs for medical students.

Medical Care

PCRM advocates broader access to medical services:

- We promote better medical care for disenfranchised groups, including minorities, women, persons with AIDS, and homeless persons.

Organization

Founded in 1985, PCRM is a non-profit organization supported by over 3,000 physicians and 60,000 laypersons. Supporters receive *Good Medicine* each quarter. PCRM programs combine the efforts of medical experts and grassroots individuals.

Leadership

PCRM president Neal D. Barnard, M.D., is a popular speaker and the author of *Eat Right—Live Longer, Food for Life*, and other books on preventive medicine. He provides psychiatric care at a shelter for the homeless in Washington, D.C. PCRM's advisory board includes 26 physicians from a broad range of specialties.

David DeGrazia
Kennedy Institute of Ethics Journal, Volume I, Number I, March 1991.

The Moral Status of Animals and Their Use in Research: A Philosophical Review

Introduction

Times have changed. Twenty years ago discussing the moral status of animals probably would have qualified one as a kook. Today no moral philosopher can evade the subject. But despite increased attention to ethical issues involving animals, nothing approaching a societal consensus on their moral status has emerged. Opinions currently range from the view that the lives and welfare of animals are as important as those of humans, to the view that animals have no moral status. Thus, while ethical discussions concerning *human* subjects of research, for example, are quite refined—resting on substantial agreement about matters such as the importance of informed consent—academic debates about animals are at a more rudimentary stage.

In this article I offer a philosophical review of (1) leading theories of the moral status of animals, (2) pivotal theoretical issues on which more progress needs to be made, and (3) applications to the setting of animal research. Such an examination demonstrates, I believe, that the practical implications of leading theories converge far more than might be expected. In addition, I hope this review helps to clarify particularly troubling issues that remain so they can be treated adequately.

General Characterization of the Debate

The philosphical debate concerning animals is anomalous for a variety of reasons. First, the ethical theories underpinning the dominant views are polarized to an unusual degree: two of the contributors most commonly cited—Peter Singer and R.G. Frey—are among the purest utilitarians in philosophy; the theory of the other—Tom Regan—features rights that are nearly absolute. Their positions therefore run counter to the current trend of trying to bridge the gap between utilitarianism and rights theories (see, e.g., Griffin (1986), Sumner (1987), and Beauchamp and Childress (1989)) or at least to modify a version of one to bring it normatively closer to the other.[1]

Also striking is the fact that there is no well-developed theory explicitly addressing the moral status of animals that supports such current practices as factory farming, animal research, and hunting. No philosopher who has developed his or her views to the point of publishing a book on the subject has vindicated the status quo. Michael A. Fox did write a book calling for only modest reforms in current animal research practices (Fox 1986), but his argumentation was severely criticized. Within a year, he recanted his views and joined those opposing the status quo (Fox 1987). Widely perceived to be a staunch opponent of the animal welfare movement, R.G. Frey is often invited to conferences as the sole opponent of Singer, Regan, and others considered radically proanimal. Yet while Frey vigorously opposes Regan's argumentation for animal rights, his own argumentation suggests he is almost an antivivisectionist (see, e.g., Frey (1987a)). This surprising clustering of the leading theorists on the side of animal welfare changes the meaning of "radical," "moderate," and "conservative" as one moves from society at large—

which generally accepts meat eating, for example—to the academic arena of animal ethics.

Some will no doubt argue that I note this convergence too quickly, that I have overlooked lesser known philosophical efforts that attempt to justify more conservative positions on these issues. They will most likely point to articles by Carl Cohen and H.J. McCloskey, whose positions I will briefly summarize later, indicating why I do not think they represent significant contributions.

Another distinctive feature of this debate is a relative dearth of rigorous, sustained philosophical exploration. Not enough is done in the way of conceptual analysis, moral epistemology, the philosophy of mind, the philosophy of science, and so on. (One understandable reason for this is a desire on the part of some writers to reach a much wider audience than academic philosophy.)[2] In my opinion only five authors have made a significant philosophical contribution to the endeavor of placing animals in ethical theory: Singer, Frey, Regan, Mary Midgley, and S.F. Sapontzis.[3]

The First Generation: Singer, Frey, and Regan

What I call "the first generation" of theories consists of the views of Singer, Frey, and Regan. Despite their differences, they are all, in an important sense, philosophically mainstream. First, working in the tradition of liberal individualism, their moral focus is on the individual, whether as rights-bearer or as bearer of interests to be counted in utility-maximization. Thus they are not very receptive to community-based approaches, for example, which ground obligations in social relations more than in characteristics (such as sentience) possessed by an individual. Second, this group has great confidence in reason as the major, if not sole, arbiter of ethical disputes. Thus they abstain from—and are even hostile to—appeals to emotion in ethics. (Indeed, Singer and Frey go further and part ways with most of our tradition by castigating even the most cautious employment of moral intuitions.)[4] In a related way, the first generation never seems to question the systematic approach to ethics, according to which once the right ethical theory is discovered, specific normative answers simply follow from it. Of course, some doubt the possibility of discovering such a comprehensive system because of significant objections to every major system in contention.

As we will see, these shared philosophical values have affected the shape of the debate. I would even argue that their orthodoxy causes them to underappreciate the contributions of what I call "the second generation." In any event, let us turn to the views of this first generation.

More than any other work, Peter Singer's *Animal Liberation* (1975) incited the philosophical debate about the moral status of animals. Singer presents his arguments with exceptional clarity. But the book, which was not written for a philosophical audience, lacks some depth; it races over or skips many philosophical issues treated in greater detail by other philosophers.

In *Animal Liberation*—and more explicitly in other writings—Singer argues from the perspective of act utilitarianism with a value theory consisting of preferences. Accordingly, everyone's good counts equally, and individual actions are justified by direct appeal to the standard of maximizing the good.[5] (By contrast, rule-utilitarians justify actions by appeal to rules such as "do not lie." These rules, in turn, are justified by their tendency to maximize the good.) As a preference-utilitarian (Singer 1979, p. 12), Singer argues that the satisfaction of actual preferences is what is to be maximized. Singer uses the more common term "interests" in *Animal Liberation*, but he views all interests as preferences.

Singer argues on the basis of a combination of behavioral, physiological, and evolutionary evidence that animals have interests—at the very least an interest in not suffering (Singer 1975, pp. 11–16). Indeed he identifies the capacity to suffer as the basic admission ticket to the moral arena:

> The capacity for suffering and enjoyment is a prerequisite for having interests at all, a condition that must be satisfied before we can speak of interests in a meaningful way. (Singer 1975, p. 9).

Thus beings that lack sentience (the capacity to suffer and experience enjoyment) need not morally be taken into consideration, except as their treatment affects sentient beings.

From all this it follows that sentient animals, which he thinks include at least all vertebrates (Singer 1975, pp. 185–88), must be given equal consideration in ethical decision making. This does not entail precisely equal treatment. Dogs have no interest in voting, so equal consideration does not necessitate granting them the right to vote if we grant such a right to humans. However, it does mean that if a human and a mouse suffer equally in intensity and duration, their suffering has the same moral weight. Consequently, trampling on animal interests—in activities such as fur trapping and factory farming—for relatively marginal increases in our interests cannot be justified.

Singer's views have several difficulties, including the following: (I) the vulnerability of act utilitarianism to various objections (especially concerning justice), which has led or kept most moral philosophers away from this theory; (2) its difficulty with "marginal cases" (the problem that some humans are similar to animals in what are taken to be the morally relevant features, e.g., intelligence and sentience), as discussed below; and (3) the poverty of its value theory, which deems nothing valuable unless it is the object of someone's actual preference. One implication of (3) is that animals who lack the concept of life and therefore do not prefer to continue living are in no way harmed if killed painlessly in their sleep, even if in the midst of a flourishing sentient life.

One of the great merits of R.G. Frey's provocative *Interests and Rights* (Frey 1980) is the very thorough argumentation that Frey gives in conceptual analysis, ethical theory, metaethics, and the philosophy of mind. I suggested above that most scholars in animal ethics shy away from the "nitty-gritty" of philosophical problems, especially problems ouside ethics. Frey is the most notable exception.

Like Singer, Frey is a preference act utilitarian.[6] Unlike Singer, Frey contends, at least in this book, that animals have no interests. Although his views have now changed, the argument he presented proceeds as follows. First, he defends the view that the only morally relevant interests are desires (preferences). Then he contends that one cannot have a desire (to own a book, for example) without a corresponding belief (that I lack the book or that the statement "I lack the book" is true). And he argues that a belief—which always amounts to a belief that a certain sentence is true—requires language. He concludes that since animals lack language, they lack the beliefs requisite for desires and therefore lack desires (Frey 1980, ch. 7). Thus animals have no interests.

It is more difficult to refute Frey's arguments than to disagree with his counter-intuitive conclusions. It is very difficult to believe that kicking a cat does not cause the cat to suffer and does not frustrate her interests. It is not surprising, then, to find considerable philosophical strain at the end of his book when he avoids the word "suffering," which he apparently links to "interests." Frey states instead that animals can experience "unpleasant sensations" and that wantonly causing unpleasant sensations is wrong (Frey 1980, pp. 170–71). It would seem difficult to argue that it is not in one's interests to avoid unpleasant sensations. It is also difficult to believe that beings who can have unpleasant sensations cannot suffer. Apparently, Frey agrees, because he now seems to allow that animals suffer (see, e.g., Frey 1987b), and that they have interests. (I base the latter attribution on numerous conversations with him.) But even if he has rejected some of his earlier conclusions, others might find the arguments compelling so it is worth confronting them on more than the intuitive level.

Most of the premises of his argument summarized above could be challenged, but the most vulnerable may be the claim that beliefs are always beliefs that a certain sentence is true. If this can be refuted,[7] then animals can be acknowledged to have beliefs, leaving no obstacle to attributing desires (preferences), and therefore interests, to them. Frey also faces objections to act utilitarianism and problems with marginal cases.

Seizing on these problems, Tom Regan argues for an alternative. His painstakingly thorough *The Case for Animal Rights* (Regan 1983), which all things considered may be the best book in the field to date, begins by rejecting utilitarianism. Regan argues that because it is committed to maximizing the good with no prior commitment to how the good is to be distributed, utilitarianism fails to respect what has come to be called the separateness of individuals.[8] If slavery is wrong, it is only because the practice fails to maximize the good, not because of the inviolability or dignity of persons. It is not even clear that the carefully concealed, painless killing of one unconsenting person to retrieve organs to save two other people is wrong based on utilitarian reasoning. Believing that our carefully considered moral intuitions have a significant place in ethical reasoning, Regan cites the counterintuitive implications and methodology of utilitarianism in rejecting that theory.

Regan goes on to propose that we regard individuals as possessing equal inherent value. Who are "individuals?" They are beings to whom the ascription of inherent value is meaningful because they have a welfare. The notion of a welfare involves that of faring well (or badly) over time. Thus animals who have beliefs, desires, and a psychophysical identity over time—so-called "subjects-of-a-life"—have inherent value. This includes at least normal adult mammals. Inherent value implies a basic Respect Principle, which in turn implies the Harm Principle, that "subjects" are not to be harmed. (It is significant that Regan includes both inflictions and deprivations as harms. Therefore, death, which deprives one of the opportunities that life holds, is a harm whether or not one has a concept of life.) A full examination of the Respect Principle leads to the thesis that "subjects" have rights that are not to be overridden in the name of the common good, except in rare, carefully circumscribed circumstances.

Regan's theory has its weaknesses; his strong rights view has failed to convince many philosophers. It would not allow killing a small number of rats in research in a promising attempt to fight a raging epidemic, for example. He takes this view even though failing to kill the rats might show disrespect for those who would eventually die from the epidemic. At the same time, the fact that Regan specifies situations in which rights can be overridden (e.g., when every rightholder on a lifeboat will drown if none is sacrificed) might be incompatible with his strong interpretation of the notion of equal inherent value. This point is well made in Jamieson (1990).

The Second Generation: Midgley and Sapontzis

The first generation of animal ethics philosophers proffered some carefully worked out theories of the moral status of animals. Subsequent work has typically used these theories as a point of departure. I think the contributions of the philosophers in the second generation, Mary Midgley and S.F. Sapontzis, have been underappreciated for two reasons: (1) the almost habitual scholarly focus on the first generation; and (2) the fact that the debate tends to be carried out in terms encouraged by the philosophical orthodoxy I described previously.

The less orthodox flavor of the second generation is most evident in the fact that theirs are mixed theories with no visible overarching framework. Neither is a utilitarian exactly, nor a committed rights theorist, nor for that matter a virtue theorist. Neither can formulate his or her theory in the form of a single principle or a neat set of principles with clear relations between them. They also have considerably less confidence in reason as the arbiter of ethical disputes, displaying a somewhat more pragmatic bent. And evincing great impatience with Frey-like skepticism about animal interests, the second generation invests somewhat less in issues concerning the philosophy of mind.

Much of Mary Midgley's *Animals and Why They Matter* (Midgley 1984) is devoted to discrediting the view that animals are morally unimportant. In this endeavor she profitably distinguishes "absolute dismissal" from "relative dismissal." The latter allows animal interests to count, but only after all human interests are satisfied. In arguing against absolute dismissal, she stops short of entirely rejecting the idea that the needs of those closer to us have moral priority over the needs of those less close. Indeed, in a qualified endorsement of such a perspective, she departs from the individualist mainstream and invokes social-bondedness as morally paramount (while revealing her antirationalist sentiments):

> The special interest which parents feel in their own children is not a prejudice, nor is the tendency which most of us would show to rescue, in a fire or other emergency, those closest to us. We are bond-forming creatures, not abstract intellects. (Midgley 1984, p. 102)

By way of analogy she goes on to argue that a preference for our own species is acceptable within limits, in no way justifying either complete dismissal or relative dismissal. (For example, she deplores the current practice of factory farming.) So instead of painting a picture of the moral concerns of family, kin, species, etc., as forming concentric circles with "me" in the middle, she portrays them as overlapping concerns.

Midgley's discussion of "rationalism"—a term she uses extremely broadly to refer to all traditions that emphasize reason and downgrade emotion—is of interest both in motivating her normative view and in its own right. In criticizing "rationalism," she apparently is attacking both (1) the view that reason is the basis for moral status, and (2) views that regard intellect, as opposed to feelings, as the proper guide in ethics. She argues, for example, that the notion of rights is so unclear as to be almost completely unhelpful. This point is well made in Russow (1985, p. 173).

Midgley's arguments bring theoretical fresh air into the debate and provide a novel account for very stubborn convictions concerning the strengths of our obligations to family, neighbors, members of the same species, etc. Still, I think her proposals have the following difficulties: (1) consideration of long-term utility may be equally capable of explaining such convictions, vitiating one of the strongest supports for her approach; (2) it is never successfully explained why racism could not be justified along the lines of her defense of giving priority to those "closer" to us; (3) the relation between reason and emotion is never clearly explicated (Russow 1985, p. 174); and (4) the normative suggestions are very vague and leave no clear guidelines for the treatment of animals, so that giving animals very little consideration does not seem to have been precluded (even if the book's tone suggests otherwise).

A somewhat clearer normative position is taken by S.F. Sapontzis in *Morals, Reason, and Animals* (Sapontzis 1987). Eschewing attempts to ground ethics in ahistorical, transcendental norms, Sapontzis regards ethics as a pragmatic endeavor rooted in cultural traditions but capable of development or progress within a tradition. He contends that while the Western tradition does not question our casual consumption of animals, "there are fundamental elements of that morality that point in the direction of animal liberation" (Sapontzis 1987, p. 110). In view of the lack of a clearly dominating ethical theory, and suspicious of relatively simple frameworks (like utilitarianism and Regan's rights view), Sapontzis treats three fundamental goals of our moral tradition as on a par in addressing ethical issues. He concludes:

> Liberating animals from our routine sacrifice of their interests for our benefit

> would seem to be the right thing to do in order better to pursue our moral goals of developing moral virtues, reducing suffering, and being fair. (Sapontzis 1987, p. 109)

Thus his theory is, in the end, an amalgam of considerations of virtue, utility, and rights.

Although Sapontzis argues in favor of extending rights to animals who have interests, he does not settle for simple solutions. Perhaps surprisingly, he rejects, in Midgley-like fashion, Singer's idea that we owe all animals with interests equal consideration:

> In common morality we are not under an obligation to give equal consideration to everyone. On the contrary, we are . . . even obligated to give priority to the interests of families, friends, colleagues, and compatriots. (Sapontzis 1987, p. 151)

In spite of this inegalitarian move, Sapontzis ends up condemning current animal-consuming practices, although (as I note below) he allows for very limited use of animals in research. Along the way, and in taking up further topics, there are many illuminating discussions that display both subtlety and originality.[9]

An excellent, underappreciated work, *Morals, Reason and Animals* nevertheless has its shortcomings. The argumentation is unevenly rigorous. Sapontzis's discussions of, for example, the replacement argument, the environment, and the disvalue of death are excellent. His discussion of the limits of reason, however, often seems irrational, while his treatment of the problem of marginal cases seems based on a logical misunderstanding. Also, while his disenchantment with monolithic ethical theories is understandable and his pluralistic approach attractive, the latter is no stronger than the basis he provides for it: our moral tradition. Why assume that our moral tradition is ethically in the right? If one turns to it for fear that there is no other plausible foundation, there remains this problem: every exhortation to "progress" invokes normative standards that either (1) are independent of that tradition, ensnaring the metaethic in self-contradiction, or (2) leave us no reason to respect them.

Some Outstanding Theoretical Issues

Progress on ethical issues involving animals could be facilitated by clarifying important theoretical issues that have thus far resisted adequate treatment. Let us turn, then, to three of them: equal consideration; the value of life; and marginal cases.

Equal Consideration. Assuming that the interests of animals are morally important, do they have as much moral weight as human interests (as both the utilitarians and Regan assume), or do they have less weight, that is, deserve less consideration (as Midgley, for example, has maintained)?[10] This issue, though rarely discussed explicitly, is of the greatest importance in determining our moral relationship to animals.

It has been argued by Singer and R. M. Hare that equal consideration is a formal requirement of morality, that the very concept of morality, or the logic of moral language, includes this requirement.[11] This is surely false. Would a putatively moral system that stipulated that everyone's interests counted equally, except for Buddha's, which counted twice as much as the others, not qualify as a moral system?

On the other hand, the principle of universalizability seems to establish a presumption in favor of equal consideration. This principle states that if one makes a certain judgment in a particular set of circumstances, one must make the same judgment in circumstances that are similar in relevant ways. The upshot is that anyone who claims that the interests of animals have less weight than human intersts must produce a relevant difference between them. What makes this issue so difficult is having to determine what differences are morally relevant.

One could try to meet this burden of proof in a number of ways. I discuss only two. As we have seen, one possibility is to cite degrees of social-bondedness, or special relationships, as underlying differences of how much consideration to give the interests of others. Midgley and Sapontzis would justify some greater degree of consideration for other humans based on the argument that family members, friends, and so on deserve different degrees of moral consideration. To make a plausible case, a proponent of this view must show at least how these arguments do not support what seem to be unjust forms of discrimination, such as racial discrimination. One must also say more to the utilitarian who vindicates giving varying degrees of moral attention to members of different groups, but only on the basis of long-term efficiency (since, for example, children will generally be better taken care of if we care for our own).

Perhaps the most troubling challenge to equal consideration is what we call the "sui generis view"—a view that is almost never taken seriously by philosophers yet seems to capture the convictions of many people. Justifications for what counts as a morally relevant characteristic, argument goes, must ultimately depend on an undefended assumption. Utilitarians typically conclude that having interests is the relevant trait on the assumption that the essential job of morality is to promote interest. But while they sometimes argue for this assumption, such arguments never amount to a rigorous proof. Many discussants invoke traits like sentience, self-consciousness, etc., but no logical demonstration is ever offered to species membership per se is not morally important. Intuitively, it seems relevant. Further, its relevance would explain the tendency to think our duties to humans are stronger than our duties to animals and would dissolve the problem of marginal cases, which I discuss later.

In spite of some attractions, the sui generis view seems vitiated by the fact that universalizability establishes a presumption in favor of equal consideration—so that a convincing argument, not just a stalemate is needed to defeat equal consideration.[12] (To my mind no such argument has been found.)

The value of life. Another issue on which a great deal turns concerns the value of life. What sort of being must one be for one's life per se as opposed to the quality of the experiences it contains, to have moral weight. Does the act of killing an animal painlessly in its sleep have any disvalue by itself? The U.S. Public Health Service, in its *Guide for the Care and Use of Laboratory Animals*, suggests not (Department of Health and Human Services, 1985). It regards the infliction of suffering—but not killing—as demanding justification.[13] How one answers affects how one is likely to judge hunting, which often entails little or no suffering, as well as raising meat in nonintensive family farms. Let me sketch three rough positions on this issue of killing.

One answer, implied by the brand of utilitarianism advocated by Singer and Frey, is simple: all and only beings who have a preference to continue living (which requires having a conception of oneself as living) have lives with moral weight.[14] Only the satisfaction of preferences is valuble; if one does not prefer X, removing X does not cause harm. Probably all animals show a tendency to struggle for survival, but if it is instinctual and unaccompanied by particular mental states, then it fails to reveal a preference to live. Perhaps then, the members of few species are harmed by death per se, although identifying them, using this criterion, poses a formidable challenge to the philosophy of mind and animal psychology.

Of course, if not all interests are thought to be actual preferences, this view is less attractive. Consider a second position: the lives of all and only morally considerable beings (usually beings with interests) have value, but these lives may differ greatly in value.[15] Moreover, the mentally more complex being generally has the more valuable life. Different theorists explain this judgment in different ways, but I do not think anyone has explained it in sufficient depth. Regan's account captures the spirit of the central point. In defending the sacrifice of a dog over a normal human if one of them must be thrown off a lifeboat to prevent all aboard from drowning, Regan argues that death is less of a harm for the dog. Death is a harm because it forecloses all opportunities one has for obtaining the satisfactions available to members of one's species, and more opportunities are closed off in the case of a human than in the case of a dog (Regan 1983, p. 324). Rachels (1983, p. 254) similarly argues that "when a mentally sophisticated being dies, there are more reasons why the death is a bad thing."[16] I have contended that human life generally "has more interests and more very important interests [than animal life has], all of which are thwarted by death" (DeGrazia 1989, p. 99).

A third view reflects the hesitation of some philosophers to value lives differently, and in particular, to assign superior worth to normal human lives. In what way are more opportunities—or more interests—thwarted by a human death than by an animal death? Why are there more reasons that human death is an evil? Pressed to explain, proponents of this view might say that while humans have the kinds of experience that animals have, they also have other kinds of experiences out of the reach of animals, for example, the pursuit of life projects and the cultivation of deep personal relationships. In response, Sapontzis notes that animals have a multitude of experiences that humans cannot appreciate:

> We cannot enjoy the life of a dog, a bird, a bat, or a dolphin. Consequently, we cannot appreciate the subtleties of smell, sight, sound, and touch that these animals can apparently appreciate. Here we are the boors. (Sapontzis 1987, p. 219)

A common rejoinder switches emphasis from quantity to quality: some of the riches of typical human lives that are not shared by animals have special weight or value, and therefore, human lives are generally more valuable than the lives of animals.[17] Sapontzis retorts that we cannot know the value of an experience unless we have had it, so we have no business arguing that our experiences, and therefore our lives, are morally weightier than those of other animals (Sapontzis 1987, p. 219). Suffice it to say that the issue is unresolved.

Marginal cases. Perhaps the most agonizing of the unresolved issues is the problem of marginal cases. This is a problem only for those who, unlike Regan, believe that we may sometimes harm animals for human benefit. Medical research probably offers the strongest case. If we may harm animals in this way, then either (1) we may likewise harm humans who are similar to animals in relevant ways, or (2) we must explain why no humans are similar in relevant ways to the animals we are justified in harming. The latter is difficult. If we refuse to conscript humans for involuntary harmful research, what do all humans have that animals lack? Obviously not all humans are rational, self-conscious, autonomous, or even potentially so. On the other hand, the former is repellent to all of us, even those who take that tack.

Frey is famous for biting the normative bullet and arguing for the extremely limited (involuntary) use of humans in research. Whatever standard is employed to justify using animals (e.g., quality of life below a certain threshold) must be used consistently, so that some humans will be equally eligible. He believes the demand for consistency makes the case for antivivesection very strong, but as a utilitarian, he cannot justify foregoing the benefits for all animal research.[18] His view cannot be accused of logical inconsistency or evasive fudging. Yet only a perfectly uncompromising anti-intuitionist can deny that justifying the conscription and harming of innocent humans is an extremely unattractive feature of a theory.

A possible way to avoid this result without rejecting all animal research is to appeal to side effects. Although it is frequently mentioned, I do not think this possibility has been carefully enough explored. Anyone who is aware of the uproar caused by the debate over whether to use the organs or anencephalics—who are not even sentient—will tremble at the prospect of a public policy of conscripting sentient humans for research. People read newspapers; dogs do not. Just imagine the riots, threats against researchers, widespread paranoia, and so on. The suggestion that education about the justification of using some humans could lessen such side effects (see, e.g., Frey 1987a, p. 97) seems unrealistic, as does the claim that the conscription and use could be kept secret. But arguments on both sides need further development.

Another possibility is to allow, in principle, the use of humans who fail the tests that research animals fail, while emphasizing that such use is not obligatory. We may, as it were, confer rights on such humans simply because we do not want to use them. My giving a gift to one person does not oblige me to give gifts to others, for no gift-giving (special circumstances aside) is owed in the first place. Of course, some will object to justifying the use of humans in principle. Another problem is that exempting humans may be unfair to animals remaining in the eligibility pool, because they thereby become more likely to be conscripted.

James Nelson has provided a different response to the problem of marginal cases (Nelson 1988). He argues that humans with greatly subnormal capacities have suffered a tragic harm in that they will never enjoy the sort of lives normal people experience. A normal rat, in contrast, does not merit special sympathy because of its limited capabilities. To those who would argue that we should feel remorse for the rat because it lacks traits of normal humans, Nelson has a response:

> The rat could not be the possessor of [such] traits . . . and still be the same rat, or perhaps even a rat at all; such a radical alteration of its genetic structure would constitute an essential . . . change in its identity. (Nelson 1988, p. 192)

For three reasons, this difference justifies sparing such humans from research or at least greatly strengthening the presumption against using them: (1) such unfortunate humans call forth our compassion; (2) justice in distributing burdens requires us to prefer using those not already harmed; and (3) not treating them with special compassion would have pernicious consequences in the form of lessened moral sensitivity and discernment. I think this intriguing argument deserves further consideration, but I am inclined to argue instead that humans of subnormal capacities who never had greater capacities have neither been harmed nor suffered a loss. One is harmed only when one's situation is changed; one can lose only what one has.[19]

Applications to the Research Setting

Having examined several theories of the moral status of animals and a number of unresolved, theoretical issues, let us turn to the research setting and to specific working principles that are derived from, or at least suggested by, the theories. It will become apparent that there is significant convergence among the principles that flow from the theories; I believe that recognition of this convergence can facilitate progress in animal ethics.

To bolster my argument for this convergence, I will briefly explain why the contributions of philosophers H.J. McCloskey and Carl Cohen, who defend conservative views, fail to vitiate the others' positions. In "The Moral Case for Experimentation on Animals," H.J. McCloskey bases the case for animal research on prima facie duties—obligations that can conflict, none of which always trumps the others—to maximize good over evil, and to respect persons as persons, meaning respect their rights. What is the argument for these prima facie duties? McCloskey answers that "appeal must ultimately be made to the self-evidence of the principles and of the possession of these rights by persons [only]" (McCloskey 1987, p. 65).

The weakness of his case should be evident. Many utilitarians (e.g., Singer and Frey) do not find it self-evident that persons have such rights. For many others it is not self-evident that only humans have rights.[20] In essence, McCloskey has made an undefended assumption on precisely the crucial issue: the relative moral status of humans and other animals. Later, he claims that to forbid animal research but permit research on volunteering human subjects is "morally outrageous" (McCloskey 1987, p. 70). This is debatable to say the least—since the humans are presumed to consent—and reveals the futility of his earlier appeal to the self-evidence of the principles on which he bases his claims.

Carl Cohen's "The Case for the Use of Animals in Biomedical Research" is equally question-begging. Against the claim that animals have rights, he offers an analysis of rights that nearly excludes animals by definition:

> they are . . . claims, or potential claims, within a community of moral agents. Rights arise, and can be intelligibly defended, only among beings who actually do, or can, make moral claims against one another. (Cohen 1986, p. 865)

Cohen offers no defense of his analysis whatsoever.

Against utilitarians who oppose animal research, he argues that, even granting equal consideration of interests, maximizing utility requires increasing animal research because of the benefits it procures (Cohen 1986, p. 868). But because benefits are always only hoped for (and often seem obtainable by alternatives), while the harms to animals are certain, Cohen's calculations seem highly implausible. Singer, at least, has marshalled considerable empirical evidence that points in the opposite direction (Singer, 1975).

Other shortcomings are noteworthy. First, Cohen fails to note the distinction between equal consideration and equal treatment—arguing that equal consideration entails either that animals and humans both lack rights or that they have precisely the same rights (Cohen 1986, p. 867). As I indicated above, Singer obviated this non sequitur in his early work; many philosophers find the mistake particularly egregious, especially in relation to the value of life. In addition, Cohen responds to the problem of marginal cases by claiming that subnormal humans are of the same kind as normal humans and therefore have the same moral status (Cohen 1986, p. 866). But he never explains why being of a particular kind, and not one's own characteristics, determines one's moral status. Moreover, he does not defend the metaphysical claim that we are of the human kind, as opposed to primate, mammal, sentient creature, animal, etc. Finally, Cohen concludes his article with ad hominem arguments against antivivisectionists who inconsistently engage in other animal-consuming practices, as if their moral character determined the moral status of animals (Cohen 1986, p. 869).

Cohen and McCloskey seem to represent the most prominent philosophical efforts to justify a strongly proresearch position, so the failure of their arguments supports the thesis of convergence towards very progressive views.

Let me then turn to Regan's position on animal research, the simplest of those I will go on to discuss. In stating what his theory implies for animal research, Regan notes the comfort it offers for subnormal humans:

> Those who accept the rights view . . . will not be satisfied with anything less than the total abolition of the harmful use of animals in science—in education, in toxicity testing, in basic research. But the rights view plays no favorites. No scientific practice that violates human rights, whether the humans be moral agents or moral patients [who are not moral agents], is acceptable. (Regan 1983, p. 393)[21]

In the present context I will treat Singer and Frey as having a single view because there seem to be no significant differences between them (at least since Frey's apparent acknowledgement that animals have interests). Their utilitarian position allows for harming animals only when the beneficial consequences of doing so are likely to offset the harms. The principle of equal consideration, which precludes weighting consequences differentially according to species, also justifies harming humans when doing so would maximize the good. (But see the discussion of marginal cases above.)

It is notoriously difficult to translate utilitarianism into specific rules for the research setting. Clearly, however, they would be extremely restrictive because the harms done to animals must always be measured against benefits that are only hoped for. The hoped-for benefits must be multiplied by the probability of achieving them and must outweigh the virtually certain harms. The probability is always less than one—usually much less. And the lower the probability, the greater the amount of benefit must be to meet the utilitarian standard. Rarely could a research protocol meet this standard. Still, unlike Regan, a utilitarian would probably justify the use of a small number of rodents if doing so were likely to lead to an important medical breakthrough, and if there were no less harmful known way of achieving this benefit.[22]

Although Sapontzis's general theory resists the simple formulation permitted by the Regan and Singer-Frey views, he extracts some clearly stated principles from his considerations of utility, fairness, and virtue. Notice that the first and second principles generalize from federal guidelines covering adults, who can give informed consent, and children, who (like animals) cannot. Experiments are to be allowed only

> (i) on those who freely and with understanding consent . . .
>
> (ii) when, in situations beyond the subjects' ability to understand . . ., a guardian determines that participating in the experiment would (likely) be either innocuous to or beneficial for the research subjects and freely and with understanding consents for them, or (iii) when conducting this research on these subjects is the only available way to attain a clear and present, massive, desperately needed good that greatly outweighs the sacrifice involved . . . and where such sacrifice is minimized and fairly distributed among those likely to benefit from the research. (Sapontzis 1987, p. 226)

Most animal research is neither innocuous nor performed for the benefit of animals, so (iii) is the key principle. Because it requires a "good that *greatly* outweighs the sacrifice . . ." (emphasis added), Sapontzis's view appears to be normatively between Regan's view and utilitarianism. He says the practical implication is that "virtually all of that research would have to be radically restructured or terminated" (Sapontzis 1987, p. 228).

Midgley's theory does not provide anything like clear guidelines for the use of research animals. However, it seems safe to conclude from the tenor of the entire book, and from her criticisms of the typical justifications researchers provide for their excesses (see, e.g., Midgley 1984, pp. 37–39, and Midgley 1981), that Midgley's view supports considerably tighter restrictions than those currently in place. At the same time, it is clear from her arguments supporting a limited degree of species favoritism that her restrictions would be looser than those of the other views we have examined.

This review of the policy implications for animal research of what I consider the five leading theories in the animal ethics literature, leads to a striking conclusion: all the theories are very progressive. They range from total abolition to an implicit call for considerably tighter restrictions.

Looking Ahead

Let us conclude with a look into the future. What should be on the agenda for animal ethics? One issue requiring more careful treatment is the question of which species of animals are worthy of moral consideration. In tackling this question, the first task is to arrive at a more uniform understanding of what is required, in principle, for moral status. If the many who identify interests as the moral sine qua non are correct, residual issues in analyzing this concept need to be cleared up. Perhaps most importantly, can one have an interest in something of which one has no conception? This issue is enormously important for determining the values of different animal lives—one of the major outstanding issues I discussed earlier. Whatever our answers to these questions, an even more difficult task is determining which animals satisfy the conceptual conditions for moral status. Although there is a near consensus that at least vertebrate animals can suffer,[23] beyond that range much is in dispute, indicating the need for extensive empirical research.

Further exploration of the other outstanding issues that I raised earlier is also needed. Are all beings with moral status owed equal consideration? Or is some degree of preferential treatment for our own species justified? While nonhuman animals with interests clearly deserve some consideration, and while the leading theoretical efforts all conclude that they deserve significant consideration, it is not enough to leave it at that. The difference between allowing some prohuman discrimination and requiring equal consideration is significant, and little that is penetrating has been written on this issue. Finally, if we permit some harmful use of animals for human benefit, how are we to regard those humans who are in no greater possession of morally relevant characteristics than the animals? Is there any way of exempting all nonvolunteering humans from harmful use—short of abolishing animal research?

Another item on the agenda of animal ethics—which overlaps with the first—is to achieve a deeper and more comprehensive understanding of animal interests or, less technically, animal welfare. To effectively promote something requires understanding it. To make a responsible estimate of how large a monkey's cage should be, we need to understand in what sense and to what degree confinement is harmful to the monkey. To decide reasonably how many monkeys should be placed in a cage, we need to understand the companionship needs of monkeys and how monkeys interact in groups. And to do all of this requires grasping a great deal about the nature of monkeys—biological, psychological, and social. As another example, when we give repeated electric shocks to rats, how bad, qualitatively, are their experiences? Do they experience only pain, or also highly emotional forms of suffering? Answering such questions is necessary for determining the extent of harm caused by particular cases of inflicted pain, suffering, or other unpleasant experiences. Of course, the absence of such precise information in no way justifies not promoting animal welfare as well as possible under the circumstances.

A further item is to maximize points of contact in theories of the moral status of animals. Less emphasis on the differences between, for example, Regan's and Frey's views, and more on the common ground of rights-based and utilitarian approaches should be made. Even better, the development of theories that form clear and coherent compromises between the two basic approaches would be welcome, especially since the uncompromised versions presented by the first generation seem open to serious objections and are likely to satisfy only small sects of devoted dogmatists. Sapontzis has taken a big step in this direction, although I think he leaves much to be desired at the foundations. Wayne Sumner has also taken such a step (see Sumner 1987 and Sumner 1988), though his main theoretical work has not focused on animals in particular.

A final item on the animal ethics agenda takes us onto a somewhat different stage. If my thesis about convergence is correct, policy changes will be in order. If in addition, further efforts to improve theories and resolve outstanding issues lead to greater specificity about what actual policies should look like, the question of how to effectuate such changes will become prominent. Hence the vexing issue of what strategy to adopt in the face of institutional and political realities; resistance to significant change will be enormous. This is an issue of both prudence and ethics; what works and what is permissible? Which is more effective, reform from within or more radical measures? Is civil disobedience or even violence ever justified in the name of combatting injustice? If so, how egregious does the injustice have to be? A detailed exploration of these issues will have to await another occasion.

Notes

1. See, for example, Brandt (1979), in which utilitarianism takes the form of a system of rules, and Dworkin (1978), in which rights (which usually trump considerations of utility) may be overridden in particular cases if the consequences of preserving them would be extremely pernicious.
2. This desire clearly affected the style of Bernard Rollin's *Animal Rights and Human Morality* (1981). The book is rich with good ideas and the beginnings of good discussions, but sails through issues in a superficial, if highly readable, manner.
3. It is worth mentioning Bernard Rollin's *The Unheeded Cry* (1989), and James Rachels's *Created from Animals: The Moral Implications of Darwinism* (1990). The former is a significant contribution to the philosophy of mind, but presents no ethical theory; the latter is of interest to numerous areas of philosophy and includes illuminating discussions of various points implicit in utilitarianism, but presents no novel theory of the moral status of animals.
4. See DeGrazia (1989, ch. 3), for an attempted refutation of such extreme antiintuitionism.
5. For an explanation of how he arrives at this particular moral theory, see Singer (1989, p. 11).
6. It is difficult not to notice the influence here of the famous British philosopher R.M. Hare. Singer was also his pupil.
7. For attempted refutations see Regan (1983, pp. 37–49) and DeGrazia (1989, pp. 239–47). Perhaps the best attempt is a section in Rollin (1989), "The Claim That Animals Lack Concepts."
8. This term was popularized with Rawls's eloquent arguments in Rawls (1971).
9. See, for example, his careful discussion of what has been called "the replacement argument" (ch. 10) and his argument that, in an important sense, many animals can act virtuously (ch. 3).
10. This issue is addressed in DeGrazia (1989, ch. 2).
11. This argument is given in Singer (1979, pp. 10–11).
12. I also find compelling James Rachels's arguments that, if a Martian came to Earth to teach in a boys school and behaved just as humans do, we would not think it justified to treat him with less consideration just because he is not homo sapiens (Rachels 1989, pp. 103–4).
13. The attitude is well reflected in his statement: "[Various] methods can be used for euthanasia [any killing] of anesthetized animals because the major criterion of humane treatment has been fulfilled" (Health and Human Services, Department of, 1985, p. 38).
14. I do not know that they unequivocally hold this position, but it is implied by their version of preference-utilitarianism.
15. Singer and Frey also allow for differences in the value of lives, but for simplicity, I ignored this feature of their position.
16. A similar argument is made, though more tentatively, in Jamieson (1983, pp. 145–46).
17. Mill in his classic *Utilitarianism* writes that it is "better to be a human being dissatisfied than a pig satisfied," arguing that certain kinds of pleasures that humans experience are qualitatively superior to other, baser pleasures. In keeping with Sapontzis's arguments, Edward Johnson (1983) attempts to discredit Mill's thesis.
18. He writes: "Clearly we have here a powerful argument for antivivisection: it is one sure way to bar human experiments. The reason I cannot endorse it is that it [ignores] the benefits to be derived from (some, by no means all) experimentation" (Frey 1987a, p. 98).
19. But see Nelson's responses to criticisms like mine (Nelson 1988). For a more detailed treatment of the problem of marginal cases than I have offered, see Pluhar (1988).
20. McCloskey might take comfort in my discussion of the sui generis view above, in which I said it was not logically impossible to argue

that being human per se confers special moral status. But again, a compelling argument, and not just a stalemate, is needed to meet the burden of proof set by the principle of universalizability. McCloskey would also need to explain why so many reasonable people fail to perceive what he takes to be self-evident.

21. This unequivocal abolitionist view, as well as Regan's general theory, were endorsed by speakers at the well-attended animal rights march in Washington, D.C., on June 10, 1990.

22. At the same time, Sumner (1988) notes considerable convergence between Regan's view and utilitarianism in the research setting, and forges a compromise between them.

23. For a particularly comprehensive article, see Rose and Adams (1989).

References

Beauchamp, Tom L., and Childress, James F. 1989. *Principles of Biomedical Ethics*. 3rd ed. New York: Oxford University Press.

Brandt, R. B. 1979. *A Theory of the Right and the Good*. Oxford: Clarendon Press.

Cohen, Carl. 1986. The Case for the Use of Animals in Research. *New England Journal of Medicine* 315:865–70.

DeGrazia, David. 1989. *Interests, Intuition, and Moral Status* (a Georgetown University dissertation).

Dworkin, Ronald. 1978. *Taking Rights Seriously*. Cambridge: Harvard University Press.

Fox, Michael Allen. 1986. *The Case for Animal Experimentation: An Evolutionary and Ethical Perspective*. Berkeley: University of California Press.

_____. 1987. Animal Experimentation: A Philosopher's Changing Views. *Between the Species* 3:55–60.

Frey, R.G. 1980. *Interests and Rights: The Case Against Animals*. Oxford: Clarendon Press.

_____. 1987a. Animal Parts, Human Wholes: On the Use of Animals as a Source of Organs for Human Transplants. In *Biomedical Ethics Reviews* 1987, eds. James M. Humber and Robert A. Almeder, pp. 89–107. Clifton, New Jersey: Humana Press.

_____. 1987b. The Significance of Agency and Marginal Cases. *Philosophica* 39:39–46.

Griffin, James. 1986. *Well-Being: Its Meaning, Measurement, and Moral Importance*. Oxford: Clarendon Press.

Health and Human Services, Department of. 1985. *Guide for the Care and Use of Laboratory Animals* U.S. Department of Health and Human Services, Public Health Service, National Institutes of Health (NIH Publication No. 85–23).

Jamieson, Dale. 1983. Killing Persons and Other Beings. In *Ethics and Animals*, eds. Harlan B. Miller and William H. Williams, pp. 135–62. Clifton, N.J.: Humana Press.

_____. 1990. Rights, Justice, and Duties to Provide Assistance: a Critique of Regan's Theory of Rights. *Ethics* 100:349–62.

Johnson, Edward. 1983. Life, Death, and Animals. *In Ethics and Animals*, eds. Harlan B. Miller and William H. Williams, pp. 123–33. Clifton. N.J.: Humana Press.

McCloskey, H.J. 1987. The Moral Case for Animal Experimentation. *The Monist* 70 6:64–82.

Midgley, Mary. 1981. Why Knowledge Matters. In *Animals in Research*, ed. E. Sperlinger, pp. 216–22. New York: John Wiley.

_____. 1984. *Animals and Why They Matter*. Athens, Georgia: University of Georgia Press.

Nelson, James Lindemann. 1988. Animals, Handicapped Children and the Tragedy of Marginal Cases. *Journal of Medical Ethics* 14:191–93.

Pluhar, Evelyn. 1988. Speciesism: A Form of Bigotry or a Justified View? *Between the Species* 4:83–96.

Rachels, James. 1983. Do Animals Have a Right to Life? In *Ethics and Animals*, eds. Harlan B. Miller and William H. Williams, pp. 275–84. Clifton, N.J.: Humana Press.

_____. 1989. Drawin, Species, and Mortality. In *Animal Rights and Human Obligations*, 2nd ed., eds. Tom Regan and Peter Singer, pp. 98–113. Englewood Cliffs, N.J.: Prentice Hall.

_____. 1990. *Created from Animals: The Moral Implications of Darwinism*. New York: Oxford University Press.

Rawls, John. 1971. *A Theory of Justice*. Cambridge; the Belknap Press of Harvard University Press.

Regan, Tom. 1983. The Case for Animal Rights. Berkeley: University of California Press.

Rollin, Bernard, 1981. *Animal Rights and Human Mortality*. New York: Prometheus Books.

_____. 1989. *The Unheeded Cry*. Oxford: Oxford University Press.

Rose, Margaret, and Adams, David. 1989. Evidence for Pain and Suffering in Other Animals. In *Animal Experimentation: The Consensus Changes*, ed. Gill Gangley, pp. 42–71. London: Macmillan.

Russow, Lilly-Marlene. 1985. Review of Midgley, *Animals and Why They Matter. Environmental Ethics* 7:171–75.

Sapontzis, S. F. 1987. *Morals, Reason, and Animals*. Philadelphia: Temple University Press.

Singer, Peter. 1975. *Animal Liberation: A New Ethics for Our Treatment of Animals*. New York: New York Review of Books.

_____. 1979. *Practical Ethics*. Cambridge: Cambridge University Press.

_____. 1989. The Significance of Animal Suffering. *Behavioral and Brain Sciences* 13: 9–12.

Sumner, L.W. 1987. *The Moral Foundation of Rights*. Oxford: Clarendon Press.

_____. 1988. Animal Welfare and Animal Rights. *Journal of Medicine and Philosophy* 13:159–75.

Human vs Animal Rights

In Defense of Animal Research

JAMA, November 17, 1989 – Vol 262, No. 19

Jerod M. Loeb, PhD; William R. Hendee, PhD; Steven J. Smith, PhD; M. Roy Schwarz, MD

From the Group on Science and Technology, American Medical Association, Chicago, Ill.

Reprint requests to the American Medical Association, 535 N. Dearborn St., Chicago, IL 60610 (Dr Hendee).

For centuries, opposition has been directed against the use of animals for the benefit of humans. For more than four centuries in Europe, and for more than a century in the United States, this opposition has targeted scientific research that involves animals. More recent movements in support of animal rights have arisen in an attempt to impede, if not prohibit, the use of animals in scientific experimentation. These movements employ various means that range from information and media campaigns to destruction of property and threats against investigators. The latter efforts have resulted in the identification of more militant animal rights bands as terrorist groups. The American Medical Association has long been a defender of humane research that employs animals, and it is very concerned about the efforts of animal rights and welfare groups to interfere with research. Recently, the Association prepared a detailed analysis of the controversy over the use of animals in research, and the consequences for research and clinical medicine if the philosophy of animal rights activists were to prevail in society. This article is a condensation of the Association's analysis.

(JAMA. 1989;262:2716-2720)

RESEARCH with animals is a highly controversial topic in our society. Animal rights groups that intend to stop all experimentation with animals are in the vanguard of this controversy. Their methods range from educational efforts directed in large measure to the young and uninformed, to promotion of restrictive legislation, filing lawsuits, and violence that includes raids on laboratories and death threats to investigators. Their rhetoric is emotionally charged and their information is frequently distorted and pejorative. Their tactics vary but have a single objective—to stop scientific research with animals.

The resources of the animal rights groups are extensive, in part because less militant organizations of animal activists, including some humane societies, have been infiltrated or taken over by animal rights groups to gain access to their fiscal and physical holdings. Through bizarre tactics, extravagant claims, and gruesome myths, animal rights groups have captured the attention of the media and a sizable segment of the public. Nevertheless, people invariably support the use of animals in research when they understand both sides of the issue and the contributions of animal research to relief of human suffering. However, all too often they do not understand both sides because information about the need for animal research is not presented. When this need is explained, the presentation often reveals an arrogance of the scientific community and an unwillingness to be accountable to public opinion.

The use of animals in research is fundamentally an ethical question: is it more ethical to ban all research with animals or to use a limited number of animals in research under humane conditions when no alternatives exist to achieve medical advances that reduce substantial human suffering and misery? This question has been addressed at length in a White Paper on

Animal Research prepared by the American Medical Association. This article is a condensation of the White Paper; the complete paper can be obtained from the Office of the Vice President for Science and Technology, American Medical Association, 535 N. Dearborn St, Chicago, IL 60610.

Animals in Scientific Research

Animals have been used in research for more than 2000 years. In the third century BC, the natural philosopher Erisistratus of Alexandria used animals to study bodily function. In all likelihood, Aristotle performed vivisection on animals. The Roman physician Galen used apes and pigs to prove his theory that veins carry blood rather than air. In succeeding centuries, animals were employed to confirm theories about physiology developed through observation. Advances in knowledge from these experiments include demonstration of the circulation of blood by Harvey in 1622, documentation of the effects of anesthesia on the body in 1846, and elucidation of the relationship between bacteria and disease in 1878.[1] In his book *An Introduction to the Study of Experimental Medicine* published in 1865, Bernard[2] described the importance of animal research to advances in knowledge about the human body and justified the continued use of animals for this purpose.

In this century, many medical advances have been achieved through research with animals.[3] Infectious diseases such as pertussis, rubella, measles, and poliomyelitis have been brought under control with vaccines developed in animals. The development of immunization techniques against today's infectious diseases, including human immunodeficiency virus disease, depends entirely on experiments in animals. Antibiotics that control infection are always tested in animals before use in humans. Physiological disorders such as diabetes and epilepsy are treatable today through knowledge and products gained by animal research. Surgical procedures such as coronary artery bypass grafts, cerebrospinal fluid shunts, and retinal reattachments have evolved from experiments with animals. Transplantation procedures for persons with failed liver, heart, lung, and kidney function are products of animal research.

Animals have been essential to the evolution of modern medicine and the conquest of many illnesses. However, many medical challenges remain to be solved. Cancer, heart diseease, cerebrovascular disease, dementia, depression, arthritis, and a variety of inherited disorders are yet to be understood and controlled. Until they are, human pain and suffering will endure, and society will continue to expend its emotional and fiscal resources in efforts to alleviate or at least reduce them.

Animal research has not only benefited humans. Procedures and products developed through this process have also helped animals.[4,5] Vaccines against rabies, distemper, and parvovirus in dogs are a spin-off of animal research, as are immunization techniques against cholera in hogs, encephalitis in horses, and brucellosis in cattle. Drugs to combat heartworm, intestinal parasites, and mastitis were developed in animals used for experimental purposes. Surgical procedures developed in animals help animals as well as humans.

Research with animals has yielded immeasurable benefits to both humans and animals. However, this research raises fundamental philosophical issues concerning the rights of humans to use animals to benefit humans and other animals. If these rights are granted (and many people are loath to do so), additional questions arise concerning the way that research should be performed, the accountability of researchers to public sentiment, the nature of an ethical code for animal research, and who should compose and approve the code. Today, some animal activists are asking whether humans have the right to exercise dominion over animals for any purpose, including research. Others suggest that because humans have dominion over other forms of life, they are obligated to protect and preserve animals and ensure that they are not exploited. Still others agree that animals can be used to help people, but only under circumstances that are so structured as to be unattainable by most researchers. These attitudes may all differ, but their consequences are similar. They all threaten to diminish or stop animal research.

Challenge to Animal Research

Challenges to the use of animals to benefit humans are not new—their origins can be traced back several centuries. With respect to animal research, opposition has been vocal in Europe for more than 400 years and in the United States for at least 100 years.[6]

Most of the current arguments against research with animals have historic precedents that must be grasped to understand the current debate. These precedents originated in the controversy between Cartesian and utilitarian philosophers that extended from the 16th to the 18th centuries.

The Cartesian-utilitarian debate was opened by the French philosopher Descartes, who defended the use of animals in experiments by insisting the animals respond to stimuli in only one way—"according to the arrangement of their organs."[7] He stated that animals lack the ability to reason and think and are, therefore, similar to a machine. Humans, on the other hand, can think, talk, and respond to stimuli in various ways. These differences, Descartes argued, make animals inferior to humans and justify their use as a machine, including as experimental subjects. He proposed that animals learn only by experience, whereas humans learn by "teaching-learning." Humans do not always have to experience something to know that it is true.

Descartes' arguments were countered by the utilitarian philosopher Bentham of England. "The question," said Bentham, "is not can they reason? nor, can they talk? but can they suffer?"[8] In utilitarian terms, humans and animals are linked by their common ability to suffer and their common right not to suffer and die at the hands of others. This utilitarian thesis has rippled through various groups opposed to research with animals for more than a century.

In the 1970s, the antivivisectionist movement was influenced by three books that clarified the issues and introduced the rationale for increased militancy against animal research. In 1971, the anthology *Animals, Men and Morals*, by Godlovitch et al,[9] raised the concept of animal rights and analyzed the relationships between humans and animals. Four years later, *Victims of Science*, by Rider,[10] introduced the concept of "speciesism" as equivalent to fascism. Also in 1975, Singer[11] published *Animal Liberation: A New Ethic for Our Treatment of Animals*. This book is generally considered the progenitor of the modern animal rights movement. Invoking Ryder's concept of speciesism, Singer deplored the historic attitude of humans toward nonhumans as a "form of prejudice no less objectionable than racism or sexism." He urged that the liberation of animals should become the next great cause after civil rights and the women's movement.

Singer's book not only was a philosophical treatise; it also was a call to action. It provided an intellectual foundation and a moral focus for the animal rights movement. These features attracted many who were indifferent to the emotional appeal based on a love of animals that had characterized antivivisectionist efforts for the past century. Singer's book swelled the ranks of the antivivisectionist movement and transformed it into a movement for animal rights. It also has been used to justify illegal activities intended to impede animal research and instill fear and intimidation in those engaged in it.

Animal Rights Activism

The animal rights movement is supported financially by a wide spectrum of individuals, most of whom are well-meaning persons who care about animals and wish to see them treated humanely. Many of these supporters do not appreciate the diverse philosophies and activities of different groups of animal activists, and they have not explored differences between animal welfare and animal rights in any depth. They believe that their financial contributions pay for animal shelters and efforts to find homes for stray animals. They do not realize that their contributions also support illegal activities that have been classified as terrorist actions by the US Federal Bureau of Investigation and the New Scotland Yard. Many of these illegal activities are conducted by a clandestine group called the Animal Liberation Front. Other groups alledged to be engaged in illegal activities include Earth First, Last Chance for Animals, People for the Ethical Treatment of Animals, Band of Mercy, and True Friends.

In the United States, illegal activities conducted by these groups since July 1988 include the following (F. Trull, personal communication):

- break-in, theft, and arson at the University of Arizona in Tucson, with more than 1000 animals stolen and arson damage of $250,000;
- break-in and theft at Duke University in Durham, NC;
- break-in and theft at the Veterans loose Administration Medical Center in Tucson (Arizona);
- bomb threat to the director of the lab animal facility, Stanford (Calif) University;
- attempted bombing, US Surgical Corporation, Norwalk, Conn;
- vandalism, University of California, Santa Cruz; and
- break-in and theft, Loma Linda (Calif) University.

Illegal actions have been pursued with even greater vigor in the United Kingdom.

Recent examples related to medical research in the United Kingdom include the following (M. Macleod, personal communication):

- home and car damage of two investigators at a Wellcome research facility;
- home and car damage of three investigators at St. George's Hospital, Tooting;
- bomb planted and warning given to director of construction firm building laboratory for Glaxo Corporation;
- five incendiary devices that caused $50,000 damage to company that supplied portable offices to Glaxo Corporation during laboratory construction;
- property damage at Bromley High School, where animals are used in dissection classes; and
- mailing of hundreds of incendiary devices, including one to Prime Minister Margaret Thatcher.

Other countries that experienced similar terrorist activities include Italy, Japan, New Zealand, Sweden, Holland, Belgium, Canada, and West Germany. Some officials believe that these activities are part of an international conspiracy operating under the rubric of animal rights but dedicated to general anarchy. They also feel that support for these efforts is derived principally from thousands of well-meaning but naive individuals who contribute to organizations that plead the cause of animal welfare but actually serve as fronts for terrorist activities.

Defense of Animal Research

The issue of animal research is fundamentally an issue of the dominion of humans over animals. This issue is rooted in the Judeo-Christian religion of western culture, including the ancient tradition of animal sacrifice described in the Old Testament and the practice of using animals as surrogates for suffering humans described in the New Testament. The sacredness of human life is a central theme of biblical morality, and the dominion of humans over other forms of life is a natural consequence of this theme.[12] The issue of dominion is not, however, unique to animal research. It is applicable to every situation where animals are subservient to humans. It applies to the use of animals for food and clothing; the application of animals as beasts of burden and transportation; the holding of animals in captivity such as in zoos and as household pets; the use of animals as entertainment, such as in sea parks and circuses; the exploitation of animals in sports that employ animals, including hunting, racing, and animal shows; and the eradication of pests such as rats and mice from homes and farms. Even provision of food and shelter to animals reflects an attitude of dominion of humans over animals. A person who truly does not believe in human dominance over animals would be forced to oppose all of these practices, including keeping animals as household pets or in any form of physical or psychological captivity. Such a posture would defy tradition evolved over the entire course of human existence.

Some animal advocates do not take issue with the right of humans to exercise dominion over animals. They agree that animals are inferior to humans because they do not possess attributes such as a moral sense and concepts of past and future. However, they also claim that it is precisely because of these differences that humans are obligated to protect animals and not exploit them for the selfish betterment of humans.[13] In their view, animals are like infants and the mentally incompetent, who must be nurtured and protected from exploitation. This view shifts the issues of dominion from one of rights claimed by animals to one of responsibilities exercised by humans.

Neither of these philosophical positions addresses the issue of animal research from the perspective of the immorality of not using animals in research. From this perspective, depriving humans (and animals) of advances in medicine that result from research with animals is inhumane and fundamentally unethical. Spokespersons for this perspective suggest that patients with dementia, stroke, disabling injuries, heart disease, and cancer deserve relief from suffering, and that depriving them of hope and relief by eliminating animal research is an immoral and unconscionable act. Defenders of animal research claim that animals sometimes must be sacrificed in the development of methods to relieve pain and suffering of humans (and animals) and to affect treatments and cures of a variety of human maladies.

The immeasurable benefits of animal research to humans are undeniable. One example is the development of a vaccine for poliomyelitis, with the result that the number of cases in poliomyelitis in the United States alone declined from 58,000 in 1952 to 4 in 1984. Benefits of this vaccine worldwide are even more impressive.

Every year, hundreds of thousands of humans are spared the braces, wheelchairs, and iron lungs required for the victims of poliomyelitis who survive this infectious disease. The research that led to a poliomyelitis vaccine required the sacrifice of hundreds of primates. Without this sacrifice, development of the vaccine would have been impossible, and in all likelihood the poliomyelitis epidemic would have continued unabated. Depriving humanity of this medical advance is unthinkable to almost all persons. Other diseases that are curable or treatable today as a result of animal research include diphtheria, scarlet fever, tuberculosis, diabetes, and appendicitis.[3] Human suffering would be much more stark today if these diseases, and many others as well, had not been amendable to treatment and cure through advances obtained by animal research.

Issues in Animal Research

Animal rights groups have several stock arguments against animal research. Some of these issues are described and refuted herein.

The Clinical Value of Basic Research

Persons opposed to research with animals often claim that basic biomedical research has no clinical value and therefore does not justify the use of animals. However, basic research is the foundation for most medical advances and consequently for progress in clinical medicine. Without basic research, including that with animals, chemotherapeutic advances against cancer (including childhood leukemia and breast malignancy), ß-blockers for cardiac patients, and electrolyte infusions for patients with dysfunctional metabolism would never have been achieved.

Duplication of Experiments

Opponents of animal research frequently claim that experiments are needlessly duplicated. However, the duplication of results is an essential part of the confirmation process in science. The generalization of results from one laboratory to another prevents anomalous results in one laboratory from being interpreted as scientific truth. The cost of research animals, the need to publish the results of experiments, and the desire to conduct meaningful research all function to reduce the likelihood of unnecessary experiments. Furthermore, the intense competition for research funds and the peer review process lessen the probability of obtaining funds for unnecessary research. Most scientists are unlikely to waste valuable time and resources conducting unnecessary experiments when opportunities for performing important research are so plentiful.

The Number of Animals Used in Research

Animal rights groups claim that as many as 150 million animals are used in research each year, most of them needlessly. However, the US Office of Technology Assessment has estimated that only 17 to 22 million animals were involved in experimental studies in 1983, including 12.2 to 15.2 million rats and mice bred especially for research. Also used were 2.5 to 4 million fish, 100,000 to 500,000 amphibians, 100,000 to 500,000 birds, 500,000 to 550,000 rabbits, 500,000 guinea pigs, 450,000 hamsters, 182,000 to 195,000 dogs, 55,000 to 60,000 cats, and 54,000 to 59,000 primates.[14]

Animal activists claim that research with animals is institutionalized and that investigators do not consider ways to reduce the number of animals involved in research. In contrast, evidence suggests that the number of research animals is decreasing each year, according to surveys by the National Research Council of the National Academy of Sciences.[15] In 1978, for example, the Council estimated that the total of 20 million research animals was 50% less than the number for 1968.

The number of animals used in research is limited by the cost of animals (especially in the present period of limited funds for research), the availability and expense of facilities to house them, and by the compassion of investigators to use no more animals than needed to perform meaningful research. It also is controlled by institutional animal-use committees that are empowered to monitor animal experimentation to ensure that experimental design is appropriate and animals are treated humanely and compassionately. These committees also are obligated to ensure that animals are properly housed and cared for. The performance of these committees is evaluated by various federal and voluntary agencies, including the American Association of Laboratory Animal Care.

The Use of Primates in Research

Animal activists often make a special plea on behalf of nonhuman primates, and many of the sit-ins, demonstrations, and break-ins have been directed at primate research centers. Efforts to justify these activities invoke the premise that primates are much like humans because they exhibit suffering and other emotions.

Keeping primates in cages and isolating them from others of their kind is considered by activists as cruel and destructive of their "psychological well-being." However, the opinion that animals that resemble humans most closely and deserve the most protection and care reflects an attitude of speciesism (ie, a hierarchical scheme of relative importance) that most activists purportedly abhor. This logical fallacy in the drive for special protection of primates apparently escapes most of its adherents.

Some scientific experiments require primates exactly because they simulate human physiology so closely. Primates are susceptible to many of the same diseases as humans and have similar immune systems. They also possess intellectual, cognitive, and social skills above those of other animals. These characteristics make primates invaluable in research related to language, perception, and visual and spatial skills.[14] Although primates constitute only 0.5% of all animals used in research, their contributions have been essential to the continued acquisition of knowledge in the biological and behavioral sciences.[15]

Do Animals Suffer Needless Pain and Abuse?

Animal activists frequently assert that research with animals causes severe pain and that many research animals are abused either deliberately or through indifference. Actually, experiments today involve pain only when relief from pain would interfere with the purpose of the experiments. In any experiment in which an animal might experience pain, federal law requires that a veterinarian must be consulted in planning the experiment, and anesthesia, tranquilizers, and analgesics must be used except when they would compromise the results of the experiment.[16]

In 1984, the Department of Agriculture reported that 61% of research animals were not subjected to painful procedures, and another 31% received anesthesia or pain-relieving drugs. The remaining 8% did experience pain, often because improved understanding and treatment of pain, including chronic pain, were the purpose of the experiment.[14] Chronic pain is a challenging health problem that costs the United States about $50 billion a year in direct medical expenses, lost productivity, and income.[15]

Alternatives to the Use of Animals

One of the most frequent objections to animal research is the claim that alternative research models obviate the need for research with animals. The concept of alternatives was first raised in 1959 by Russell and Burch[17] in their book, *The Principles of Humane Experimental Technique*. These authors exhorted scientists to reduce the pain of experimental animals, decrease the number of animals used in research, and replace animals with nonanimal models whenever possible.

However, more often than not, alternatives to research animals are not available. In certain research investigations, cell, tissue, and organ cultures and computer models can be used as adjuncts to experiments with animals, and occasionally as substitutes for animals, at least in preliminary phases of the investigations. However, in many experimental situations, culture techniques and computer models are wholly inadequate because they do not encompass the physiological complexity of the whole animal. Examples where animals are essential to research include development of a vaccine against human immunodeficiency virus, refinement of organ transplantation techniques, investigation of mechanical devices as replacements for and adjuncts to physiological organs, identification of target-specific pharmaceuticals for cancer diagnosis and treatment, restoration of infarcted myocardium in patients with cardiac disease, evolution of new diagnostic imaging technologies, improvement of methods to relieve mental stress and anxiety, and evaluation of approaches to define and treat chronic pain. These challenges can only be addressed by research with animals as an essential step in the evolution of knowledge that leads to solutions. Humans are the only alternatives to animals for this step. When faced with this alternative, most people prefer the use of animals as the research model.

Comment

Love of animals and concern for their welfare are admirable characteristics that distinguish humans from other species of animals. Most humans, scientists as well as laypersons, share these attributes. However, when the concern for animals impedes the development of methods to improve the welfare of humans through amelioration and elimination of pain and suffering, a fundamental choice must be made. This choice is present today in the conflict between animal rights activism and scientific research. The American Medical Association made this choice more than a century ago and continues to stand squarely in defense of the use of animals for scientific research. In this position, the Association is supported by opinion polls that reveal strong endorsement of the American public for the use of animals in research and testing.[18]

The philosophical position of animal rights activists would require a total ban on research with animals. The consequences of such a ban were described in a 1986 report to Congress by the US Office of Technology Assessment: "Implementation of this option would effectively arrest most basic biomedical and behavioral research and toxicological testing in the United States." The economic and public health consequences of that option, the US Office of Technology Assessment warned, "are so unpredictable and speculative" that this course of action should be considered dangerous.[18] Although laws to ban the use of animals in research have been introduced into a number of states' legislatures, neither a majority of the American people nor their elected representatives have supported them.[18]

Terrorist acts are a serious threat to biomedical research and to the safety and security of those involved in it. However, such desperate acts are unlikely to stop the use of animals in research. The greater threat to research is legislation that imposes restrictions on how animals are housed and cared for that cost more than the research community can afford. Existing federal and state laws, together with peer review and supervision by institutions and funding agencies, provide adequate protection against misuse and abuse of animals. Additional legislation does not offer increased protection. It only increases the expense of research and, in some cases, stops it entirely by imposing exorbitant costs and demands.

The American Medical Association believes that research involving animals is absolutely essential to maintaining and improving the health of people in America and worldwide.[6] Animal research is required to develop solutions to human tragedies such as human immunodeficiency virus disease, cancer, heart disease, dementia, stroke, and congenital and developmental abnormalities. The American Medical Association recognizes the moral obligation of investigators to use alternatives to animals whenever possible, and to conduct their research with animals as humanely as possible. However, it is convinced that depriving humans of medical advances by preventing research with animals is philosophically and morally a fundamentally indefensible position. Consequently, the American Medical Association is committed to the preservation of animal research and to the conduct of this research under the most humane conditions possible.[19,20]

References

1. Rowan AN, Rollin BE. Animal research—for and against: a philosophical, social, and historical perspective. *Perspect Biol Med.* 1983;27:1–17.
2. Bernard C; Green HC, trans. *An Introduction to the Study of Experimental Medicine.* New York, NY; Dover Publications Inc; 1957.
3. Council on Scientific Affairs. Animals in research. *JAMA.* 1989;261:3602-3606.
4. Leader RW, Stark D. The importance of animals in biomedical research. *Perspect Biol Med.* 1987;80:470–485.
5. Kransney JA. Some thoughts on the value of life. *Buffalo Physician.* 1984;18:6–18.
6. Smith SJ, Evans RM, Sullivan-Fowler M, Hendee WR. Use of animals in biomedical research: historical role of the American Medical Association and the American physician. *Arch Intern Med.* 1988;148:1849–1853.
7. Descartes R. *'Principles of Philosophy,' Descartes: Philosophical Writings.* Anascombe E, Geach PT, eds. London, England: Nelson & Sons; 1969.
8. Bentham J. *Introduction to the Principles of Morals and Legislation.* London, England: Athlone Press; 1970.
9. Godlovitch S, Godlovitch, Harris J. *Animals, Men and Morals.* New York, NY: Taplinger Publishing Co Inc; 1971.
10. Ryder R. *Victims of Science.* London, England: Davis-Poynter; 1975.
11. Singer P. *Animal Liberation: A New Ethic for Our Treatment of Animals.* New York, NY: Random House Inc; 1975.
12. Morowitz HJ. Jesus, Moses, Aristotle and laboratory animals. *Hosp Pract.* 1988;23:23–25.
13. Cohen C. The case for the use of animals in biomedical research. *N Engl J Med.* 1986; 315:865–870.

14. *Alternatives to Animal Use in Research, Testing, and Education.* Washington, DC: Office of Technology Assessment; 1986. Publication OTA-BA-273.
15. Committee on the Use of Laboratory Animals in Biomedical and Behavioral Research. *Use of Laboratory Animals in Biomedical and Behavioral Research.* Washington, DC: National Academy Press; 1988.
16. *Biomedical Investigator's Handbook.* Washington, DC: Foundation for Biomedical Research; 1987.
17. Russell WMS, Burch RL. *The Principles of Humane Experimental Technique.* Springfield, Ill: Charles C Thomas Publisher; 1959.
18. Harvey LK, Shubat SC. *AMA Survey of Physician and Public Opinion on Health Care Issues.* Chicago, Ill: American Medical Association; 1989.
19. Smith SJ, Hendee WR. Animals in research. *JAMA.* 1988;259:2007–2008.
20. Smith SJ, Loeb JM, Evans RM, Hendee WR. Animals in research and testing: who pays the price for medical progress? *Arch Ophthalmol.* 1988;106:1184–1187.

Council Report
JAMA, August 14, 1991 – Vol 266, No. 6

Use of Animals in Medical Education

Council on Scientific Affairs, American Medical Association

The use of animals in general medical education is essential. Although several adjuncts to the use of animals are available, none can completely replace the limited use of animals in the medical curriculum. Students should be made aware of an institution's policy on animal use in the curriculum before matriculation, and faculty should make clear to all students the learning objectives of any educational exercise that uses animals. The Council on Scientific Affairs recognizes the necessary for the responsible and humane treatment of animals and urges all medical school faculty members to discuss this moral and ethical imperative with their students.

(*JAMA*. 1991;266:836-837)

LIVE ANIMALS have been used as an integral component in the education of physicans for centuries. Although the use of animals for laboratory demonstrations and exercises has been most common in contemporary departments of physiology and surgery, many other units in medical schools (eg, pharmacology, microbiology, biochemistry, neurology, pediatrics, advanced trauma life support, and orthopedics) also have used animals to achieve specific educational goals. Animals have been used in exercises designed to reinforce important basic science concepts that underlie clinical medicine and in hands-on training programs designed to allow medical students, residents, or fellows the opportunity to develop expertise in a given technique or procedure before performing the procedure on a human being.

In 1989 the American Medical Association surveyed more than 1500 physicians to ascertain attitudes about animal use in medical research and teaching.[1] Nearly 90% of the physicians surveyed reported that they had used animals for educational purposes during their formal medical education, and 43% indicated that they had used animals for advanced educational training after completing medical school. Not surprisingly, a much greater proportion (68%) of physicians in surgical specialties (vs physicians in other specialties) had used animals for advanced educational training. Among all physicians, 91% thought that

From the Council on Scientific Affairs, American Medical Association, Chicago, Ill.

This report was presented at the 1990 Interim Meeting of the House of Delegates as Report A of the Council on Scientific Affairs.

This report is not intended to be construed or to serve as a standard of medical care. Standards of medical care are determined on the basis of all the facts and circumstances involved in an individual case and are subject to change as scientific knowledge and technology advance and patterns of practice evolve. This report reflects the views of the scientific literature as of November 1990.

Reprint requests to Group on Science and Technology, American Medical Association, 515 N. State St, Chicago, IL 60610 (Jerod M. Loeb, PhD)

Members of the 1990 Council on Scientific Affairs included the following: William C. Scott, MD, Tucson, Ariz, Chair; Yank D. Coble, Jr, MD, Jacksonville, Fla; A. Bradley Eisenbrey, MD, PhD, Royal Oak, Mich, Resident Representative; E. Harvey Estes, Jr. MD, Durham, NC, Vice Chair; Mitchell S. Karlan, MD, Beverly Hills, Calif; William R. Kennedy, MD, Minneapolis, Minn; Michael P. Moulton, San Antonio, Tex, Medical Student Representative; Patricia J. Numann, MD, Syracuse, NY; W. Douglas Skelton, MD, Macon, Ga; Richard M. Steinhilber, MD, Cleveland, Ohio; Jack P. Strong, MD, New Orleans, La; Henry N. Wagner, MD, Baltimore, Md; William R. Hendee, PhD, Secretary; William T. McGivney, PhD, Assistant Secretary; Jerod M. Loeb, PhD, staff author.

the use of animals had been important for their own training, and 93% expressed support for the continued use of animals in medical education.

Many professional organizations have developed specific policy statements that emphasize the importance of using animals in medical education. At the same time, these policy guidelines also stipulate the need for humane treatment of animals used in teaching programs. For example, *Statements on Animal Usage*, a brochure published by the American Physiological Society, Bethesda, Md, in October 1987, says: "The American Physiological Society believes the use of animals is important in the education of students in the biomedical sciences. The use of animals gives the student a direct understanding of how living systems work, an understanding that cannot be gained by reading a textbook, watching a video, or using a computer. To achieve the best biomedical education, students must have a complete learning experience including the use of laboratory animals."

At the AMA 1986 Interim Meeting, the House of Delegates adopted Substitute Resolution 109, which "supports continued efforts to defend and promote the use of animals in meaningful research, product safety testing, and teaching programs."

Activism against the use of animals in educational programs has been increasing in recent years. Animal "rights" activities, response to new regulations, security, and significantly higher prices for animals are costing US medical schools approxiamtely $17.3 million annually, according to survey results released in July 1990 by the Association of American Medical Colleges, Washington, DC (written communication). Although these figures focus primarily on research using animals, educational use has contributed to the overall costs. All 126 medical schools that were asked to participate responded to the survey. Fifty-four schools reported being contacted and told that their institution was the target of animal rights activities. In the past 5 years, 76 schools reported losing more than $4.5 million and 33,000 labor hours because of demonstrations, break-ins, vandalism, delays in construction, and other incidents. The schools also reported 3800 incidents of faculty and staff harassment, ranging from bomb and death threats to graffiti, picketing of homes, and threatening letters and telephone calls. Ninety-two schools reported using live animals in the medical curriculum. Of these 92 schools, the survey reported, 90% to 95% of their students voluntarily attend sessions that use live animals, and 61 schools offer alternative activities for students who object to the use of live animals. However, refusing to attend the sessions that used animals would affect a student's candidacy for admission or promotion in 22 schools. It is unclear on what basis students might decide that a given exercise that uses animals is pedagogically unjustified or unnecessary.

At the University of California at Berkeley, Students for Animal Liberation circulated a pamphlet that stated: "If you are asked to mutilate, cut apart, or psychologically abuse an animal for your education, ask respectfully for an alternate nonanimal model. Inertia, convention, and lack of imagination prevent their use." The American Medical Student Association House of Delegates in 1986 passed a resolution[2] that urged that "all medical school classes and labs involving the use of live animals . . . be optional for students who, for moral or pedagogical reasons, feel that such use is either unjustified or unnecessary."[2] In response to student and faculty concerns, the Association of American Medical Colleges, in May 1987, advised its members that all medical schools should have formal internal policies that delineate student participation requirements in exercises that use animals and the various options and consequences of noncompliance.[2]

Individual medical school policies regarding the use of animals in medical education vary widely. For example, Yale University policy[2] states that students must be notified by the first meeting of a course if the use of vertebrate animals will be required. If the student is morally opposed to participating in the laboratory exercise, it is the responsibility of the student to arrange for an alternative exercise that is satisfactory to the faculty member or to withdraw from the course. Stanford University[2] requires academic departments to "alert" prospective students about the use of animals in a given course. Instructors inform students of the use of animals or animal tissues during the first week of the class but are not obligated to alter course requirements that are consistent with university policies.

In early 1990, the curriculum committee of the University of Texas Medical Branch at Galveston adopted a statement on the use of animals in medical education that is meant to be provided to prospective students *before* they decide to matriculate at the medical school. This policy permits no exemptions from participation in laboratory exercises involving animals. The policy states that "a deep understanding of the mechanisms and functions of living mammalian systems is essential in the education of a modern physician. This understanding is initially gained from textbooks, lectures, computer simulations and group discussions, but cannot be complete unless students are exposed to the behavior and responses of living mammalian organisms. . . . It is inconceivable to us that a modern physician can be adequately prepared for the practice of medicine without exposure to studies of living, nonhuman mammals. We recognize that all persons must decide for themselves where they stand, morally and ethically, on the issue of the use of living animals in medical education, but believe that this stand must be taken prior to application and admission to the medical school, because every student will be expected to participate in the animal laboratory teaching sessions offered. These are considered an essential part of our medical education process and no exceptions will be made."

Some medical schools supplement the use of laboratory animal exercises with "alternatives" to live animal models. Although these alternatives have reduced the total number of animals used within the medical curriculum, a definite consensus about the educational utility of these approaches is lacking. A survey conducted by the Association of Chairmen of Departments of Physiology noted that most faculty believe that "alternative" techniques alone are insufficient educational tools to completely replace animals.[3] These alternative techniques include computer simulations, in vitro techniques, physical-chemical models, videotapes, and audiovisual aids. Medical schools also report that the number of live animals used in teaching programs has been reduced considerably by offering demonstrations to larger numbers of students allowing more students to share an animal, using single animals for multiple procedures, and, where appropriate, demonstrating physiologic variables using students. The physiology survey noted above delineated several disadvantages of replacing animals with alternative techniques. These include further distance of the ultimate subject (humans) from the teaching, the artificial nature of the system, inability to study interactions in the complex system, inability to study species-specific responses, loss of student experience of working with live subjects, and loss of student exposure to practicality and complications of gathering valid data.

The Association of American Medical Colleges recommends that medical school faculty members introduce any course that uses live animals with a clear statement about the learning objectives for the exercise and a discussion and demonstration of humane treatment of the animals.[2] Faculty members should be prepared adequately to respond to student concerns and should be prepared to do so either as part of the general class session or in separate one-on-one sessions. Similarly, students should be prepared to discuss issues associated with the use of animals in education and research with patients. Class discussions that explore and clarify students' beliefs and feelings would be of significant help. American Medical Association policy (1983 Interim Meeting Resolution 93) also encourages medical school faculty using animals in the education of students to continue instruction of students on the appropriate use and treatment of experimental animals. The Association of Chairmen of Departments of Physiology has prepared a brochure for medical students that explores the complexities associated with animal use in the medical school classroom, including the ethical considerations, practical considerations, and benefits of animal studies. This brochure, *Considerations for Medical Students Using Lab Animals*, is an excerpt from the Association of Chairman Departments of Physiology, 9650 Rockville Pike, Bethesda, MD 20814.

Comment

The Council on Scientific Affairs believes that the use of animals in the general education of physicians is essential. Although the council recognizes that several effective adjuncts to animal use in medical education have been developed, at present none of these can completely substitute for the limited use of animals. The council reaffirms the need for humane treatment of experimental animals used in medical education and urges all involved faculty members to discuss this moral and ethical imperative with their students.

The council recommends adoption of the following guidelines on the use of animals in medical school curricula, graduate medical education, and continuing medical education courses:

1. Medical school faculty should consider using nonanimal models in educational activities only when these nonanimal models would achieve the same educational goals as use of animal models; when animals are used in the curriculum, educational goals should be clearly stipulated.
2. Each medical school should disseminate a policy statement to students before matriculation that explains the students' participation in educational experiences involving animals.
3. All educational experiences involving animals should have the approval of the Institutional Animal Care and Use Committee.
4. Involved faculty members should discuss with students the learning objectives of any educational experience that uses animals, and faculty should remain available throughout the laboratory exercise to give advice and guidance on the conduct of the educational experience.
5. All educational experiences involving animals should be carried out in a humane manner that minimizes pain and uses anesthetic and analgesic drugs when procedures may cause more than momentary or slight pain.

References

1. American Medical Association; *Survey of Physicians' Attitudes Toward the Use of Animals in Biomedical Research*, Chicago, Ill: American Medical Association. 1989.
2. Hartman J, ed. *Saving Lives: Supporting Animal Research: A Resource Notebook for Institutional Leadership*. Washington, DC: Association of American Medical Colleges; 1989.
3. Greenwald, GS. Survey on use of animals in teaching physiology. *Physiologist*. 1985; 28: 478–480.

Council Report
JAMA. June 23/30, 1989 – Vol 261, No. 24

Animals in Research

Council on Scientific Affairs

WITHIN this century, spectacular advances have been made in the prevention, diagnosis, and treatment of many diseases. With expansion of our basic knowledge of processes involving biochemistry, physiology, and molecular biology, important new directions in medicine have been taken. Although medicine requires research for the development of new thought, the research process itself is less well understood. The brief glossary herein, which is organized to provide ready access to diseases and/or disciplines, cites some of the most important new discoveries in medicine and their bases. The list is not meant to be exhaustive, and research crucial to certain fields is omitted. Its purpose is to emphasize the research process and to point out the mechanisms by which answers often are obtained. The theme presented within the glossary is the indispensability of animals in research, often because animals and humans suffer from many of the same diseases. In fact, much of biomedical research using animals has resulted in significant benefits in veterinary medicine as well as human medicine.

Since the late 1800s, the American Medical Association has consistently supported the humane use of animals for biomedical research. Research that involves animals is essential to improving the health and well-being of the American people, and the American Medical Association actively opposes any legislation, regulation, or social action that inappropriately limits such research. Most Americans support the use of animals but want assurance that animals are treated humanely and used only when necessary. Over the years, animal rights activists have exploited this concern for animal welfare and have attempted to impede or stop biomedical research with animals. As a result, some Americans believe that abuse of laboratory animals is common. Recently, the animal rights movement has been making substantial inroads in obtaining philosophic and financial support for legislation and regulatory changes that could compromise the future of biomedical research.

The activities and arguments of animal rights groups present difficult ethical questions. To accept the philosophic and moral viewpoint of the

From the Council on Scientific Affairs, American Medical Association, Chicago, Ill.

This report was presented to the House of Delegates of the American Medical Association at the December 1988 Interim Meeting as an informational report of the Council on Scientific Affairs.

This report is not intended to be construed or to serve as a standard of medical care. Standards of medical care are determined on the basis of all of the facts and circumstances involved in an individual case and are subject to change as scientific knowledge and technology advance and patterns of practice evolve. This report reflects the views of scientific literature as of November 1988.

Reprint requests to Council on Scientific Affairs, American Medical Association, 535 N Dearborn St, Chicago, IL 60610 (William R. Hendee, PhD).

Members of the Council on Scientific Affairs include the following: George M. Bohigian, MD, St Louis, Mo, Chairman; Scott L. Bernstein, Champaign, Ill; Yank D. Coble, Jr, MD, Jacksonville, Fla; E. Harvey Estes, Jr, MD, Durham, NC; Ira R. Freidlander, MD, Minneapolis, Minn; Patricia J. Numann, MD; Syracuse, NY; William C. Scott, MD, Tucson, Ariz, Vice Chairman; Joseph H. Skormn, MD, Chicago, Ill; Richard M. Steinhilber, MD, Cleveland, Ohio; Jack P. Strong, MD, New Orleans, La; Henry N Wagner, Jr, MD, Baltimore, Md; William R Hendee, PhD, Secretary; William R. McGivney, PhD, Assistant Secretary; and Jerod M. Loeb, PhD, Staff Author.

animal rights movement requires a total ban on the use of animals in biomedical research. The consequences of such a step would slow the pace of health-related research and impede new procedures and technologies from reaching clinical medicine. Laws and regulations directed toward improving the conditions under which biomedical experiments that use animals are conducted have had a measurable impact. The American Medical Association supports regulatory policies enacted to protect animals from unnecessary pain or inappropriate use; however, pain and suffering in human beings will occur if policies advocated by animal rights groups are adopted. Therefore, the American Medical Association believes that what must be recognized and weighed in the balance is the cost of such protection and comfort in terms of human health.

The medical accomplishments delineated below were developed by research investigators using animals. The list reveals only a small part of the crucial role animals have played in contributing to major advances in reducing morbidity and mortality for all of society.

Aging

- Amyloidosis has been studied in dog kidneys because there are direct relationships between infections and immunologic disease and the deposition of amyloid. This has served well as an animal model for human amyloidosis.
- Since elderly dogs exhibit neuritic plaques, one of the pathological changes seen in Alzheimer's disease in humans, dogs serve as animal models for the study of this disease.
- Primate research has identified one of the important pathological features of Alzheimer's disease: the abundance of neuritic plaques or clusters of nerve endings surrounding a core of extracellular amyloid in the cerebral cortex.
- Inbred and randomly bred strains of rats show major, spontaneous, age-related lesions in the kidney and heart as well as in the nervous, skeletal, muscular, vascular, endocrine, and immune systems; thus, they have been used extensively in research on aging.
- Because the immune system in mice declines in functional ability with age in a manner similar to that in humans, the mouse has been used for immunologic studies of aging.
- The impact of diet on biologic mechanisms in aging has been delineated using primate models.

Acquired Immunodeficiency Syndrome

- Currently, chimpanzees are the only primate species capable of being infected with strains of human immunodeficiency virus, although none has yet developed an acquired immunodeficiency syndrome–like disease. This is not surprising since only a small number of chimpanzees have been infected for more than 4 years and the incubation period for acquired immunodeficiency syndrome may extend to 10 years or more in humans. Simian immunodeficiency viruses are closely related morphologically and antigenically to human immunodeficiency virus and other slow viruses. The original simian immunodeficiency virus was isolated from captive rhesus monkeys.

Anesthesia

- With the development of equipment for anesthesia and the maintenance of positive-pressure ventilation, it became possible to perform chest surgery on dogs. Subsequently, this procedure was adapted to use in humans. In the late 1930s, the first patient with lung cancer had a lung removed and survived the operation. The patient lived another 40 years and is a classic case in American surgery.
- The dog was one of the primary research subjects used in the development of hypothermia as a means of temporarily reducing brain metabolism and preventing shock and brain damage during prolonged neurosurgery.

Autoimmune Diseases

- Induced autoimmune disease, particularly allergic encephalomyelitis, thyroiditis, arthritis, myasthenia gravis, and renal disease, has been studied extensively in rats.

Basic Genetics

- Many aspects of the structure and function of the histocompatibility complex were delineated using rats.

- Comparative gene mapping between humans and experimental mammals has made it possible to identify with certainty animal models of many human genetic diseases.
- Virtually all basic information on genetic mechanisms obtained from the study of simpler organisms, such as yeast or *Drosophila*, has been tested for relevance to the biology of humans by first analyzing mechanisms in mice because many genetic and physiological traits are similar in mice and humans. Understanding of mRNA transcription, production of a polypeptide chain, folding of a polypeptide chain, glycosylation, assembly of the immunoglobulin molecule, cell secretion, immunoglobulin transport, and action at target sites have been made possible only through the availability of animal models.

Behavior

- Behavioral characteristics that are influenced by heredity, such as aggression, learning, shyness, capacity to adapt to the environment, and temperamental stability, have been studied in dogs. Behavioral patterns have been selected and preserved in many dog breeds by careful breeding. Relationships between behavior and cardiac activity (particularly the association of arrhythmias with fear and stress), anorexia nervosa developing from isolation or psychological trauma associated with food, and the hormonal control of behavior, learning, and dominance have been accomplished using the dog as a behavioral research subject.
- Evidence that the normal development of visually guided reaching behavior results from self-produced movement and the feedback associated with it was first obtained with cats.
- The biologic commonality in psychological functions in rats and men resolved the man-animal difference controversy and justified using animal models as a general approach in biomedical research.
- Rats were used to help prove mendelian inheritance in animals.
- The broad fields of educational and psychological measurement were developed using rats.
- Research on depression in young primates identified short- and long-term physiological changes in heart rate, immune response, sleep pattern, activation of the adrenocortical system, and neurotransmitter synthesis, uptake, and metabolism in the brain. These effects resemble the disordered physiology of human affective disorders.
- Studies on young monkeys have indicated that genetic factors and predispositions contribute to the development of behavioral problems such as chronic anxiety.
- Research on communicative abilities in primates has provided a practical benefit to human society: a new approach has been developed for teaching language to children who, because of severe mental retardation, cannot learn a language in the same way as normal children.
- Studies with cats, primates, and humans who had undergone commissurotomy for intractable epilepsy challenged the view that the right hemisphere of the brain is subordinate to a supposedly more highly evolved and intellectual left hemisphere; this work led to the observation that left-right hemisphere cognitive differences in man are subtle and qualitative.
- The behavioral effects identified after removal of precise areas of the brain in rats resulted in the development of many innovative neurosurgical procedures in humans.
- Experiments in rats provided the first clue that obesity might shorten the life span and that weight patterns might be determined by early nutrition.
- The mouse has been one of the principal research animals used to expand our understanding of genetically related aspects of mammalian behavior. The availability of inbred strains, mutants, congenics, and recombinant inbreds with differing neuroanatomy, biochemistry, and neurophysiology have provided scientists with the biologic means to study complex behavioral patterns.

Cancer

- One of the first studies of chemotherapy and cancer was carried out on dogs and resulted in the first demonstration that some forms of cancer might be caused by an infectious agent.
- Cats have been used to evaluate potential therapeutic techniques in the treatment of lymphosarcoma, adenocarcinoma, malignant

lymphoma, multiple myeloma, squamous cell carcinoma, acute lymphoblastic leukemia, aplastic anemia, mammary cancer, and hypercalcemia of malignancy.
- Because of its similarity to human breast cancer, feline mammary carcinoma continues to be an important model for the evaluation of new forms of therapy for breast cancer.
- Rats have served as excellent hosts for the study of transplanted tumors.
- The genetic control of the immune response to cancer has been studied extensively in virally induced cancers and in transplanted tumors in rats.
- The rat has been the major animal used in toxicology screening for carcinogenic compounds.
- Mice have contributed to our knowledge of oncology and the factors that contribute to the initiation, promotion, autonomy, and, sometimes, remission of cancer.
- Early work documented the role of genotype in modulating the frequency and type of neoplasia in the mouse; the H-2 complex, which specifies cell surface antigens and, in turn, histocompatibility, initially was found to affect tumor cell surface antigens in mice.
- The mouse model of human lymphoid cancer holds great promise for future research.
- The retroviruses or lentiviruses that contain RNA-dependent DNA polymerase (reverse transcriptase) were first described in the mouse.

Cardiovascular System

- Much of the information about the microbiology and physiology of bacterial endocarditis has been obtained from experiments in dogs; they are ideal animal models because endocarditis develops naturally in dogs.
- Many inherited cardiovascular defects occur in dogs, including patent ductus arteriosis, pulmonary stenosis, persistent right aortic arch, intraseptal defect, and tricuspid and mitral insufficiencies. All the surgical techniques currently used in humans for treatment of these diseases were tested initially in dogs.
- Surgical techniques for replacement of heart valves and segments of larger arteries using prosthetic devices also were tested initially in dogs.
- Drug therapy to reduce infarct size has been studied using cats and dogs.
- Cardic pacemaker technology was developed and tested using dogs.
- Many techniques and devices used to assist the failing heart (eg, the left ventricular assist device, the artificial heart, and the intra-aortic balloon pump) were evaluated using dog models.
- The most important model for the study of hypertension in humans has been the rat. Rats develop hypertension spontaneously, and it increases with age; the hypertension is more severe in males and leads to cerebral, myocardial, vascular, and renal lesions similar to those seen in humans. Blood pressure is responsive to control by use of antihypertensive agents in rats in a manner similar to that in humans.
- The rabbit has been used to show the relationship between genetics and blood pressure.
- Since cardiomyopathy in cats closely resembles that in humans, the cat has been used as a model to evaluate new therapeutic approaches.
- Rabbits are an important model for the study of stress-induced cardiomyopathy because there are many similarities between cardiomyopathy caused by stress in rabbits and in humans.
- Because dietary-induced atherosclerosis also occurs in cats, this species has been used as a model of atherosclerosis in humans.
- One of the first links between diet and atherosclerosis was established in studies with rabbits.
- The importance of dietary lipids and lipoproteins as risk factors in the development of atherosclerosis was demonstrated in primates.
- Studies have shown that diets devoid of, or low in, cholesterol can reverse atherosclerosis in monkeys.
- Hypercholesterolemia as a link between coronary heart disease and mortality was first documented in rabbits.
- Research with monkeys exercised on a treadmill indicates that exercise may help reduce cardiovascular disease and slow the accumulation of atherosclerotic plaque even when the diet is high in fat.
- All the techniques used in coronary artery bypass surgery and heart transplantation (including the heart-lung machine) were developed during experiments using dogs.

Childhood Diseases

- Using the monkey as a test subject in 1953, Jonas Salk developed the killed poliovirus vaccine used in many countries today. In 1959, Albert Sabin developed the live attenuated vaccine currently used in the United States.
- Therapies for and prevention of beriberi, smallpox, rubella, pertussis, pellagra, measles, mumps, and diphtheria all were pioneered with experiments that used mice, rats, chickens, and dogs.
- A vaccine for chickenpox, developed using animals, currently is undergoing clinical trials in the United States.

Cholera

- The usefulness of drug therapy for the treatment of patients with cholera was determined using dogs.

Convulsive Disorders

- Epilepsy in mice closely resembles that in humans, and biomedical studies in mice have indicated that epileptic brain foci have an increased sensitivity to neurotransmitters.

Diabetes

- Dogs were used in the classic experiments of Sir Frederick Banting and Charles Best that identified insulin as an important hormone in carbohydrate metabolism.
- Canine models of diabetes have led to advances in the treatment of ocular and vascular complications associated with diabetes.
- The technique of pancreatic transplantation was developed in dogs.
- The best model available for the study of spontaneous insulin-dependent (juvenile-onset) diabetes in humans has been the rat.

Gangliosidosis

- Disorders in lipid metabolism are common in cats, which have been used to evaluate therapy for diseases such as Tay-Sachs and Sandhoff in humans.

Gastrointestinal Tract Surgery

- The usefulness of a colostomy for the treatment of gastrointestinal tract cancer was tested in dogs.
- Rapid ambulation after gastrointestinal tract surgery as a means to reduce the formation of adhesions was demonstrated in dogs.
- The control of intestinal motility, the role of the liver and pancreas in digestion, and the production and elimination of gas were delineated in dogs and led to the development of surgical techniques to resect and rejoin the intestinal tract.
- Studies in dogs are under way currently that deal with intestinal immunity, gallstone formation, and improved treatment of fluid imbalance in various disease states.
- Virtually all new suture material currently is tested using dogs.
- Cats were used in the first visualization of the gastrointestinal tract as well as to define the relationship between autonomic neural function and gastrointestinal tract activity.
- The treatment of esophageal achalasia was developed in cats, which have a similar hereditary condition called "mega esophagus."

Hearing

- The pattern of hearing loss associated with age in mice closely parallels that in humans; therefore, the mouse has been an excellent experimental model for studying age-induced hearing loss in humans.
- The pattern of hearing loss due to noise exposure is parallel in mice and humans, and tests using mice have resulted in the establishment of rules and regulations to limit decibel levels in the environment.

Hemophilia

- Since hemophilia in dogs is nearly identical to that in humans, the dog has been a useful model to study the mechanisms of hemostasis and to determine treatment strategies.
- The first successful cure of a blood disorder by bone marrow transplantation occurred in a mouse; this experiment led to bone marrow transplantation in humans.

Hepatitis

- Because viral hepatitis occurs naturally in dogs, therapy for the disease was assessed in studies in dogs.
- The canine liver is subject to injury by a variety of chemicals, including ethanol; therefore,

many comparative studies on cirrhosis in dogs and humans have been performed.
- The world's first hepatitis B vaccine, which was dependent on research with chimpanzees and other primates (since hepatitis B cannot be transmitted to other laboratory animals) became available in 1981.

Infection
- Rats have been used extensively for the study of infectious disease caused by bacteria, fungi, *Mycoplasma, Rickettsia*, viruses, and parasites.

Malaria
- Primate studies on the pathogenesis of malaria have shown that all four species of human malaria pathogens have a fourth, or liver, stage that can remain asymptomatic indefinitely until the parasites are released into the bloodstream to produce relapses.
- Potential therapies for malaria, a disease that affects 200 million people worldwide, have been evaluated primarily in primates.

Muscular Dystrophy
- The mouse was the first animal model available to study genetic dystrophies, and early biochemical and physiological studies provide important information on Duchenne type muscular dystrophy.
- A model for muscular dystrophy exists in chickens and has been used for recent studies on therapeutic approaches to the treatment of muscular dystrophy.

Nutrition
- Effective therapy for beriberi resulted from a study of a similar niacin deficiency called "black tongue" that occurs in dogs.
- The dog has been used extensively in the study of rickets and calcium and vitamin D deficiency.
- Treatment of glycogen storage disease with portocaval shunt has been analyzed using dog models.

Ophthalmology
- Many common retinal diseases such as glaucoma, cataracts, and uveitis occur in dogs. Therefore, dogs have been used as animal models to determine effective therapies for these conditions.
- Studies on the development and function of the visual system in monkeys and cats revealed that these animals, and by implication humans, are not born with a fully developed visual cortex. The maturation of the connections between and within the cells depends on visual stimulation. If deprived of such stimulation, demonstrated when an eyelid is occluded experimentally in a monkey, changes occur in visual neurons in the brain and their connections.
- Research on cats with cataracts has contributed to knowledge about lens structure and function as well as corneal healing after surgery in humans.
- Mechanisms of visual development, plasticity of the visual cortex, and the processes that govern the balance between different visual stimuli that permit the cerebral cortex to integrate information were defined in primates.
- The understanding and treatment of children's visual disorders have been dependent on basic and applied research that uses young primates.

Organ Transplantation
- The dog has been used as the primary model for organ transplantation technology.
- The first successful kidney transplantations were performed in dogs in the late 1950s and have led to transplantation of the liver, heart, lung, and various endocrine organs in humans.
- The immunologic mechanisms involved in tissue rejection were determined in rats and have led to many advances in treatment of rejection.
- Rats have been used to investigate the nature of the immune response in sites such as the anterior chamber of the eye and the brain as well as to study the various strategies by which cells, passively administered antibodies, and cyclosporine treatment can enhance graft survival.

Parkinson's Disease
- In 1981, the first animal model for research on Parkinson's disease was developed through the long-term administration of the chemical MPTP (1-methyl-4-phenyl-1,2,3,6-tetrahydropyridine). This compound produces many of the clinical and physiological features of human

Parkinson's disease, including the loss of dopamine-producing brain cells.

Pulmonary

- The use of rats has been valuable in determining the effects of air emboli as well as the means to prevent and treat decompression sickness.
- The Blalock-Taussig operation, in which it was shown that the ductus arteriosis could be closed surgically and, thus, cyanosis prevented, was performed initially on dogs.
- The pump oxygenator used during heart and lung operations was developed in dogs.
- Dogs served as models in many of the physiological and biochemical studies necessary to determine the pathophysiological mechanisms involved in emphysema.
- Pulmonary edema and obstructive lung disease were studied in dogs to develop instruments for clinical measurements to evaluate similar phenomena in humans.
- The treatment of thermal burns of the lung, which often occur in association with chemical ingestion, fires, or automobile or aircraft accidents, was determined using dogs.
- Lung transplantation has been studied in dogs and has resulted in the recent development of surgical techniques for the treatment of end-stage pulmonary disease in man.
- Rabbits were used to prove that tuberculosis is transmissible by both inoculation and ingestion.
- Mucoid enteritis in rabbits seems to be an animal model of cystic fibrosis and is being used to develop new and innovative therapeutic regimens.

Rabies

- Louis Pasteur's original work on attenuated vaccines was done through passage in rabbits.

Radiobiology

- The properties of radiation, both as a hazard and as a therapy, have been defined using dogs.
- The earliest experiments on the influences of radiation on physiology were performed in mice.
- Early work using outbred and inbred strains of mice clearly demonstrates (1) that some animals are more radiosensitive than others; (2) that survival of cells in animals is dose dependent, but if the dose is divided over time an increased amount is required to achieve the same effect; and (3) that survival is related to the age at which exposure was first received.
- Experiments in mice have indicated that by reducing the molecular oxygen tension of cells, there is an increase in resistance to radiation, supporting the free-radical hypothesis.

Reproductive Biology

- The rat has been used in studies of reproductive endocrinology and of the morphology and control of the menstrual cycle. The physiological work led to much of our understanding of these processes in mammals.
- Recent studies in mice have documented the fact that there is a link between the pituitary gland and the initiation of menopause.
- Research on primate social behavior has delineated mechanisms responsible for female sexual maturation, reproductive strategies, infant care, and the influence of social, environmental, and biologic factors on reproduction. Results of this research have challenged the assumption that female primates are biologically emancipated from the endocrine influences that govern the reproductive behavior in the majority of mammalian species.
- Primate research also has delineated hormonal and other biologic mechanisms responsible for ovulation, steroid feedback, events governing gonadotropin secretion, adrenarche, puberty and adolescence, sperm maturation and transport, embryonic and fetal development, fetal exemption from the maternal immune response, and the onset of labor.
- Basic research studies carried out in the primate have led to the development of fertility control methods.
- Alternatives to in vitro fertilization have now been determined in primate research studies. These include low tubal transfer of eggs and surrogate embryo transfer.
- Pregnancy and fetal development have been studied extensively in primates, since they rarely can be investigated directly in humans.
- The identification of Rh (rhesus factor) was an early breakthrough in the area of the immunology of pregnancy and resulted from tests on primates.

Skeletal System, Fracture, and Related Studies

- The intramedullary pin for internal fixation of long bones and skeletal prostheses, including the artificial hip, were developed using dogs.
- The dog also has been used in the study of cartilage and tendon repair and surgical repair in osteoarthritis.
- Fusion of spinal vertebrae in chronic disk disease and surgery to provide relief of pain and herniated disks were procedures tested initially in dogs.
- Development of the skeleton and normal and abnormal growth patterns were first delineated using rabbits.

Spinal Cord Injury

- Long-term paralysis resulting from spinal cord injury has been reversed in cats by administering clonidine; preliminary studies in humans indicate that autonomic dysreflexia can be controlled and spasticity minimized using clonidine.

Toxoplasmosis

- Approximately 4500 infants are born with congenital toxoplasmosis annually in the United States. Virtually all research on this infection has been done using the cat as a research model.

Trauma and Shock

- The dog has been the primary research model used in the study of kidney function and shock as well as of the effects of fluid therapy and resuscitation. Most of the drugs used for cardiogenic shock were produced as a result of studies using dogs.
- Mechanical support measures to increase coronary artery flow in cardiogenic shock and to stimulate the development of intracoronary anastomosis were determined using dogs.
- The treatment of hepatic failure and abnormal metabolism associated with traumatic injury have been studied using dog models. Defense against infection (especially opportunistic infections) and organ failure also have been studied in dogs.

Yellow Fever

- Research with primates has clarified the differences between the urban and jungle types of yellow fever, and studies of yellow fever in monkeys have led to the production of the first yellow fever vaccine, which was developed specifically from research on the mouse.

Virology

- Understanding of viral diseases, including yellow fever, herpes simplex, rabies, encephalitis, and influenza, was accomplished using mice; sensitive diagnostic procedures to detect infection in humans were developed using mouse models.
- Mice have been used extensively in the study of viral immunity and have played a role in the development of vaccines against influenza, poliomyelitis, encephalomyelitis, eastern equine/western equine/Venezuelan encephalitis, and rabies.
- Chimpanzees were critical in the discovery that a slow virus is the infectious agent of kuru, a fatal brain disorder found in Southeast Asia.
- Primates currently are being used in research on the role of slow viruses in acquired immunodeficiency syndrome, Alzheimer's disease, and other degenerative disorders.

Conclusions

Regardless of the animal model, animals have proved to be invaluable in the pursuit of knowledge in the life sciences, and the knowledge gained often benefits both animals and humans. The director of People for the Ethical Treatment of Animals recently said, "If it were such a valuable way to gain knowledge, we should have eternal life by now." It should be clear, however, that the question should not be "What haven't we learned?" but rather "What have we learned?" Many of todays' most vexing health problems will be solved by research on animals. Acquired immunodeficiency syndrome, Alzheimer's disease, coronary heart disease, and cancer represent but a few of this nation's most troubling health problems. In scientific and medical journals, new pieces of information directed toward the conquest of these and other diseases continue to emerge. It is only with continued support for the research process (including funding, animal experimentation, and peer review) that American medicine can continue as a leader into the 21st century.

References

The following review articles reference more than 800 original scientific manuscripts that delineate medical advances that were dependent on experimental animals.

1. Gay WI. The dog as a research subject. *Physiologist.* 1984;27:133-141.

2. Warfield MS, Gay WI. The cat as a research subject. *Physiologist.* 1984;27:177-189.

3. Gill TJ. The rat in biomedical research. *Physiologist.* 1985;28:9-17.

4. Fox RR. The rabbit as a research subject. *Physiologist.* 1984;17:393-402.

5. Jonas AM. The mouse in biomedical research. *Physiologist.* 1984;27:330-346.

6. King FA, Yarbrough CJ. Medical and behavioral benefits from primate research. *Physiologist.* 1985;28:75-87.

7. King FA, Yarbrough CJ, Anderson DC, et al. Primates. *Science.* 1988;240:1475-1481.

8. Leader RW, Stark D. The importance of animals in biomedical research. *Perspect Biol Med.* 1987;30:471-485.

Foundation for Biomedical Research
818 Connecticut Avenue NW, Suite 303, Washington, DC 20006
Phone (202)457–0654 FAX (202)457–0659

FBR Facts
Vol. II, No. 4

Making Animal Tests the Scapegoat for Rare Side Effects

Animal rights activists claim that data from tests on animals cannot be extrapolated reliably to humans, and they use this claim to criticize the use of animal testing in the drug-approval process of the Federal Drug Administration (FDA). Animal rights literature implies that the FDA's reliance on animal testing leads it to approve drugs that turn out to be dangerous for humans. In reality, animal tests are conducted *in addition to*, not in place of, tests on humans, and rare, serious side effects that surface after drug approval are virtually impossible to prevent. The only way to eliminate all unforeseen side effects would be to prevent the marketing of any new drugs.

In attempts to bolster their claim, opponents of animal research repeatedly cite a 1990 report by the General Accounting Office (GAO), which examined the postapproval risks of drugs that had been approved by the FDA between 1976 and 1985. A typical interpretation is that of Neal Barnard of the Physicians Committee for Responsible Medicine (PCRM), who wrote, "Of the 198 new drugs for which data were available, 102 (51.5 percent) were more dangerous than premarket animal tests and limited human tests had indicated and had to be relabeled or withdrawn. Not an impressive track record."[1] The clear implication is that encouraging tests in animals led to reckless FDA approval, with the result that more than half of those drugs caused more harm than benefit to people.

A closer look at the GAO report shows that such a representation is very misleading. First, only six of the 198 drugs examined by GAO had been withdrawn; the rest were simply relabeled. Also, the GAO said the number of serious postapproval risks "is small when compared to the number of adverse reactions that had been identified at the time of approval."[2] Finally, the GAO report made it clear that it was not criticizing the FDA's drug approval process. ("We emphasize here that *we do not ascribe the serious postapproval risks to flaws in the drug development and approval process*; we have not yet analyzed the process."[3]) The recommendations made by GAO at the end of its report had to do with better ways to use information from clinical trials, not the animal tests that precede them.

Although every attempt is made to ensure a product's safety before it is released for general use by consumers, safety testing is done with a small number of animals and humans. When a product is used thousands, or even millions of times, it is possible that very rare side effects will occur that never could have been observed in smaller populations. This does not mean that safety testing was improper or inadequate, nor does it mean that the results of safety testing with animals are inapplicable to humans. It simply illustrates the mathematical reality that if a side effect occurs one time in a million, it is highly unlikely that it will be observed in only 1,000 or even 100,000 human tests. As the GAO wrote, "The preapproval human clinical trials for a drug involve testing with a relatively small sample of the potential user population under controlled conditions that *limit the extent of risk assessments*. However, when therapeutic benefits appear to outweigh the estimated potential risks, the new drug is approved as soon as possible for the benefit of those who can use it. After FDA approves the drug for marketing, it is then used by patients under conditions much less controlled than those that prevailed during testing."[4]

The animal rights tactic has been to attribute all unforeseen side effects to a reliance on animal testing, as if drugs found safe and effective in animals are rushed onto the market. Actually, animal tests are merely the first step in a meticulous, multi-layered drug-approval process. Animal studies are a principal source of information about a substance's biological and physiological effects before clinical trials in humans begin. Safety data obtained through animal testing influence not only the decision on whether to expose human subjects to the substance, but also the manner in which the clinical trials will be performed. Only after a drug has gone through clinical trials, which can take up to 10 years, will the FDA approve it for the public.

It is revealing that animal rights critics rarely, if ever, spell out an alternative drug-testing procedure, other than to make the general argument that animals should not be used. Relatively rare side-effects always will be experienced by humans, whether animals were tested or not. Removing the initial safety layer that animal tests provide would only make drug approval more dangerous for the people it is supposed to protect.

[1]Physicians Committee for Responsible Medicine, *Good Medicine*, Vol. II, No. 2, p. 7.

[2]GAO/PEMD-90-15 *FDA Drug Review: Postapproval Risks* 1976-85, p. 3.

[3]GAO, p. 124.

[4]GAO, p. 2.

Foundation for Biomedical Research
818 Connecticut Avenue NW, Suite 303, Washington, DC 20006
Phone (202)457–0654 FAX (202)457–0659

FBR Facts
Vol. II, No. 3

Historical Revisionism and Intellectual Dishonesty

Animal rights activists often take statements out of context to create the illusion that a particular scientist was opposed to animal research. In addition, medical history is many times grossly misrepresented to argue that animal research was not essential to a particular discovery. Most of these have been addressed before, but we at FBR thought it would be helpful to combine them for future reference while refuting animal rights claims.

Some Historical Myths

The discovery of insulin and its role was made without the aid of animal research. This argument is put forth often by Brandon Reines, D.V.M., who has even claimed that a book by Michael Bliss called *The Discovery of Insulin* supports this argument. In 1989, Bliss denied Reines' claim in the strongest possible terms, writing "Reines' interpretation of my work is thoroughly distorted, wrong-headed and silly. I informed him of this several years ago when I first read his mindless writing on the subject. I utterly repudiate his misunderstandings of my work. The discovery of insulin in the early 1920s stands as one of the outstanding examples in medical history of the successful use of animal experimentation to improve the human condition. Insulin would not have been isolated, at Toronto or anywhere else, without the sacrifice of thousands of dogs. These dogs made it possible for millions of humans to live."[1]

Research with the animal model of polio resulted in a misunderstanding of the mechanism of infection. Stephen Kaufman of the Medical Research Modernization Committee (MRMC) cites Dr. J.R. Paul's book, *The History of Polio*, as supporting this conclusion about animal research. But Dr. Paul's book refutes, rather than supports, Kaufman's claim.[2] Also, there is support for animal research from the very man credited with the oral polio vaccine—Dr. Albert Sabin. In a Sept. 13, 1991 letter, Dr. Sabin wrote: "My own experience of over 60 years in biomedical research amply demonstrated that without the use of animals and of human beings, it would have been impossible to acquire the important knowledge needed to prevent much suffering and premature death not only among humans but also among animals."[3]

Penicillin is very toxic to guinea pigs and hamsters. It is not harmful to hamsters when given in doses (relative to body weight) comparable to that given humans. The effect of penicillin on guinea pigs is similar, but it causes death indirectly. The British Research Defence Society addressed this in detail in the *RDS Newsletter*, June 1991.

Thalidomide was tested on animals, yet its potential for causing birth defects went undetected. Thalidomide was not tested on pregnant animals before being put on the market because such tests were not required then. Thalidomide was not available in this country because the animal test data were considered incomplete. When tested on pregnant animals, including rats, mice, rabbits, dogs and monkeys, thalidomide produced birth defects in a variety of species.

William Harvey formulated his theory of the circulation of blood without depending on

animal studies. This claim, made by Brandon Reines, was refuted by Adrian Morrison in the April 1993 issue of *The American Biology Teacher*. Morrison referred to Harvey's book, entitled *Exercitatio Anatomica de Motu Cordis et Sanguinis in Animalibus*, which in English translates to *Anatomical Studies on the Motion of the Heart and Blood in Animals*. As Morrison points out, Harvey states in Chapter 1 that although he experienced early difficulties in his experiments, "Finally, using greater care every day, with very frequent experimentation, observing a variety of animals and comparing many observations, I felt my way out of this labyrinth, and gained accurate information, which I desired, of the motions and functions of the heart and arteries."[4]

The commonly cited 1989 survey taken by the American Medical Association (AMA) showing overwhelming support for animal research among doctors is suspect because AMA members are not representative of all doctors in the country. Although the survey found that 99 percent of doctors agreed that animal experimentation had contributed to medical progress, animal rights activists try to discredit the study, implying that AMA members benefit financially from animal research and so hold a markedly different view of the practice. A little math shows how absurd this claim is: The population of the AMA survey was a random sample representative of all doctors in the U.S.; about half of those surveyed (52.1 percent) were AMA members and about half (47.9 percent) were not.[5] Since the sizes of the two groups were roughly equal, to end up with 99 percent overall supporting animal research, any difference between them would be minuscule, with both groups showing overwhelming support. Even if 100 percent of AMA members said animal research had contributed to medical progress, the non-AMA members agreeing with this would not be lower than 97.8 percent.

Eminent Scientists Supposedly Opposed to Animal Research

Charles Darwin: Darwin's name surfaces occasionally in antivivisectionist literature, although his support for animal research could not be more explicit. In a letter to a Swedish professor of physiology in 1881, Darwin wrote: "I know that physiology cannot possibly progress except by means of experiments on living animals, and I feel the deepest conviction that he who retards the progress of physiology commits a crime against mankind."[6]

Albert Schweitzer: In a letter to the New York Times, James A. Pittman, M.D., recalled visiting Schweitzer in 1957 in French Equitorial Africa; "At that time, I asked him specifically about his views on the use of laboratory animals for biomedical research. His response (as translated from the German) was: 'It is necessary for the advancement of medical understanding.' There was absolutely no equivocation in his statement."[7] For those who want Schweitzer's written words on animal research, they may consult *The Teaching of Reverence for Life* (Holt, Rinehart, Winston; 1965). Passages in the book show that the distinction made by Schweitzer is the same moral distinction made by the research community: while all life is meaningful, the goal of improving human and animal health requires the sacrifice of some life in order to preserve others.

[1]Letter from Michael Bliss to Charles S. Nicoll, & Sharon M. Russell, 1989.

[2]Miller, *Psychological Science*, Vol 2, No. 6, November 1991.

[3]Letter from Albert Sabin to Sharon M. Russell, Sept. 13, 1991.

[4]Harvey, W. (1928). *Exercitatio Anatomica de Motu Cordis et Sanguinis in Animalibus*, with an English translation and annotations by C.D. Leake. Springfield, IL: Charles C. Thomas, p. 26.

[5]American Medical Association, *Survey of Physicians' Attitudes Toward the Use of Animals in Biomedical Research*, 1989, p. 8.

[6]*The Life and Letters of Charles Darwin* (1959) Darwin, Francis, ed. New York: Basic Books, Inc., 382–383.

[7]Letter from James A. Pittman, M.D., Dean, University of Alabama School of Medicine, to the *New York Times*, May 26, 1990, p. 22.

The Payoff From Animal Research

A look at the Nobel Prizes for medicine awarded from 1901 to the present shows that animal research played a key role in these important discoveries. Animal research must continue for similar advances to occur in the future.

Year	Scientist(s)	Animal(s) Used	Contributions Made
1901	von Behring	Guinea pig	Development of diphtheria antiserum
1902	Ross	Pigeon	Understanding of malaria life cycle
1903	Pavlov	Dog	Animal responses to various stimuli
1905	Koch	Cow, sheep	Studies of pathogenesis of tuberculosis
1906	Golgi, Cajal	Dog, horse	Characterization of the central nervous system
1907	Laveran	Bird	Role of protozoa as cause of disease
1908	Metchnikov, Ehrlich	Bird, fish, guinea pig	Immune reactions and functions of phagocytes
1910	Kossel	Bird	Knowledge of cell chemistry through work on proteins including nuclear substances
1912	Carrel	Dog	Surgical advances in the suture and grafting of blood vessels
1913	Richet	Dog, rabbit	Mechanisms of anaphylaxis
1919	Bordet	Guinea pig, horse, rabbit	Mechanisms of immunity
1920	Krogh	Frog	Discovery of capillary motor regulating system
1922	Hill	Frog	Consumption of oxygen and lactic acid metabolism in muscle
1923	Banting, Macleod	Dog, rabbit, fish	Discovery of insulin and mechanism of diabetes
1924	Einthoven	Dog	Mechanism of the electrocardiograph
1928	Nicolle	Monkey, pig, rat, mouse	Pathogenesis of typhus
1929	Eijkman, Hopkins	Chicken	Discovery of antineuritic and growth stimulating vitamins
1932	Sherrington, Adrian	Dog, cat	Functions of neurons
1934	Whipple, Murphy, Minot	Dog	Liver therapy for anemia
1935	Spemann	Amphibian	Organizer effect in embryonic development
1936	Dale, Loewi	Cat, frog, bird, reptile	Chemical transmission of nerve impulses
1938	Heymans	Dog	Role of the sinus and aortic mechanisms in regulation of respiration
1939	Domagk	Mouse, rabbit	Antibacterial effects of prontosil

Year	Scientist(s)	Animal(s) Used	Contributions Made
1943	Dam, Doisy	Rat, dog, chick, mouse	Discovery of function of vitamin K
1944	Erlanger, Gasser	Cat	Specific functions of nerve cells
1945	Fleming, Chain, Florey	Mouse	Curative effect of penicillin in bacterial infections
1947	Carl Cori, Gerty Cori Houssay	Frog, toad, dog	Catalytic conversion glycogen; role of pituitary in sugar metabolism
1949	Hess, Moniz	Cat	Functional organization of the brain as a coordinator of internal organs
1950	Kendall, Hench, Reichstein	Cow	Antiarthritic role of adrenal hormones
1951	Theiler	Monkey, mouse	Development of yellow fever vaccine
1952	Waksman	Guinea pig	Discovery of streptomycin
1953	Krebs, Lipmann	Pigeon	Characterization of the acid cycle
1954	Enders, Weller, Robbins	Monkey, mouse	Culture of poliovirus that led to development of vaccine
1955	Theorell	Horse	Nature and mode of action of oxidative enzymes
1957	Bovet	Dog, rabbit	Production of synthetic curare and its action on vascular and smooth muscle
1960	Burnet, Medawar	Rabbit	Understanding of acquired immune tolerance
1961	von Bekesy	Guinea pig	Physical mechanism of simulation in the cochlea
1963	Eccles, Hodgkin, Huxley	Cat, frog, squid, crab	Ionic involvement in excitation and inhibition in peripheral and central portions of the nerve
1964	Block, Lynen	Rat	Regulation of cholesterol and fatty acid metabolism
1966	Rous, Huggins	Rat, rabbit, hen	Tumor-inducing viruses and hormonal treatment of cancer
1967	Harttline, Granit, Wald	Chicken, rabbit, fish, crab	Primary physiological and chemical processes of vision
1968	Holley, Khorana, Nirenberg	Rat	Interpretation of genetic code and its role in protein synthesis
1970	Katz, von Euler, Axelrod	Cat, rat	Mechanisms of storage and release of nerve transmitters
1971	Sutherland	Mammalian liver	Mechanism of the actions of hormones
1972	Edelman, Porter	Guinea pig, rabbit	Chemical structure of antibodies
1973	von Frisch, Lorenz, Tinbergen	Bee, bird	Organization of social and behavioral patterns in animals
1974	de Duve, Palade, Claude	Chicken, guinea pig, rat	Structural and functional organization of cells

Year	Scientist(s)	Animal(s) Used	Contributions Made
1975	Baltimore, Dulbecco, Temin	Monkey, horse, chicken, mouse	Interaction between tumor viruses and genetic material
1976	Blumberg, Gajdusek	Chimpanzee	Slow viruses, and new mechanisms for dissemination of diseases
1977	Guilemin, Schally, Yalow	Sheep, swine	Hypothalamic hormones
1979	Cormack, Hounsfield	Pig	Development of computer assisted tomography (CAT scan)
1980	Benacerraf, Dausset, Snell	Mouse, guinea pig	Identification of histocompatibility antigens and mechanism of action
1981	Sperry, Hubel, Wiesel	Cat, monkey	Processing of visual information by the brain
1982	Bergstrom, Samuelsson, Vane	Ram, rabbit, guinea pig	Discovery of prostaglandins
1984	Milstein, Kohler, Jerne	Mouse	Techniques of monoclonal antibody formation
1986	Levi-Montalcini, Cohen	Mouse, chick, snake	Nerve growth factor and epidermal growth factor
1987	Tonegawa	Mouse embryo	Basic principles of antibody synthesis
1989	Varmus, Bishop	Chicken	Cellular origin of retroviral oncogenes
1990	Murray, Thomas	Dog	Organ transplantation techniques
1991	Neher, Sakmann	Frog	Chemical communication between cells
1992	Fischer, Krebs	Rabbit	Regulatory mechanism in cells
1995	Lewis, Wieschaus	Fruit flies	Genetic control of early structural development

For further information contact:
Foundation for Biomedical Research
818 Connecticut Ave., NW, Suite 303
Washington, DC 20006
Phone: (202) 457-0654
Fax: (202) 457-0659
e-mail: NABR-FBR@access.digex.net
Web Site: http://www.fiesta.com/NABR/

Foundation for Biomedical Research
818 Connecticut Avenue NW, Suite 303, Washington, DC 20006
Phone (202)457–0654 FAX (202)457–0659

FBR Facts
Vol. II, No. 6

Making the Case for Drug-Addiction Research

Of all types of animal research, drug addiction studies probably come under the most intense criticism from animal rights activists. It is easy for activists to ask, for example, why the government is spending money having monkeys "smoke crack" to explore the effects of a condition that seems distinctly human. Since animals don't abuse drugs and can't verbalize how it makes them feel, why should they suffer for human depravity?

Such criticism has been common this year. In Defense of Animals (IDA), with the assistance of local animal rights groups, is identifying and targeting investigators around the country as part of what is called a national campaign. The usual allegations are not that the investigators have failed to comply with animal welfare laws. Rather, activists claim that addiction research is "cruel and useless" or "wasteful," and that federal research dollars would be better spent on the treatment of addicts. Although the question of balancing spending for research and treatment is always delicate, drug addiction is an area where additional research is clearly crucial. The medical community should be prepared to respond to such criticisms rather than hope they go unnoticed.

Activists misrepresent what we know about addiction and what the research is addressing. Although initial drug use is voluntary, addiction is not.

"We know that addiction is a brain disease," said Dr. Zach Hall, Director of NIH's National Institute of Neurological Disorders and Stroke. "As with any other brain disease, such as Alzheimer's or Parkinson's disease, research on animals is essential for understanding how addiction occurs, and how it can be prevented or treated."

According to Dr. Alan Leshner, Director of NIH's National Institute on Drug Abuse (NIDA), drug-addiction studies in animals are critical to understanding brain chemistry and the complex behaviors associated with addiction. This research usually addresses a drug's physiological affects on the body, the critical changes in the brain as a result of drug use, and the manner in which these brain changes can be reversed as effective treatments for addiction. This research also explores the biological mechanisms that make some individuals more susceptible to becoming addicted and have a harder time breaking their addictions once developed.

Addiction is more than a "habit" that people are too lazy to break. The very definition of addiction is that the addict continues using the substance even after recognizing its harmful effects. Just as research has productively revealed the biological bases and treatments for depression and schizophrenia, and allowed people to recognize them as diseases rather than "weaknesses," so too have drug studies found that addicted people have fundamental changes in their brain and are not just "weak-willed."

Money spent on addiction research is a drop in the health-care bucket. In 1995, the federal government appropriated $437.1 million to the National Institute for Drug Abuse (NIDA). In a country with roughly 250 million citizens, this comes to about $1.75 per person spent on drug-abuse research, with animal studies accounting for a fraction of that amount. By contrast, consider that in 1992 drug abuse had a cost to society of $165.5 billion. It is absurd to suggest that money spent on animal research is at the root of our health-care predicament.

In addition, hard-line animal activists tend to use arguments that contradict each other. On the one hand, they assert that humans are responsible for medical conditions brought on by their own unhealthy habits, such as smoking, eating fatty foods or doing drugs. (As PETA's Dan Mathews put it, "Don't get [diseases] in the first place, schmo.") At the same time, they complain that animal research diverts money away from treating these people.

Differences in economic class may play a role as well. As studies have shown, animal rights activists tend to be white and fairly affluent—and do not usually live in drug-ravaged neighborhoods. Those who *have* been exposed to repercussions of addiction would have an interesting perspective on the urgency of drug abuse research and the necessity of animal studies as a component.

nabr *ISSUE UPDATE*
Copyright 1995 NABR
National Association for Biomedical Research
818 Connecticut Avenue NW, Suite 303, Washington, DC 20006
Phone 202.857.0540 FAX 202.659.1902

The Humane Care & Treatment of Laboratory Animals

Summary

Virtually every major medical advance of the last century has depended upon research with animals. Data from experiments on humans are obviously the most scientifically reliable; however, in many cases human research is ethically unacceptable. In place of human models, researchers must use animals, the living systems most closely related to humans. Animals serve as surrogates in the investigation of human diseases and new ways to treat, cure or prevent them. The health of animals also has improved due to animal research.

Approximately 77% of the American public supports the necessary use of animals in biomedical research.[1] Yet, people are also justifiably concerned about the care and treatment of laboratory animals. They want assurance that animals are treated humanely, do not suffer, and are kept under conditions that allow them to be as healthy and comfortable as possible.

The scientific community recognizes its professional obligation to safeguard and improve the welfare of laboratory animals. In fact, individual researchers concerned about the care and treatment of laboratory animals were the first to set voluntary care standards at the turn of the century, long before federal laws and regulations were instituted. In 1909, the first voluntary procedures regarding lab animals were adopted and enforced in medical school laboratories. To care more effectively for research animals, veterinarians created a board-certified speciality in laboratory animal medicine in 1957. The scientific community founded a host of organizations to improve laboratory animal care, such as the American Association for Laboratory Animal Science (AALAS). A number of medical specialty socities and voluntary health organizations, including the Society for Neuroscience and the American Heart Association, have written standards for the care and treatment of laboratory animals. Researchers advocate high-quality animal care and treatment not only for reasons of conscience, but also for reasons of science. Good animal care is good science.

Unfortunately, most people are unaware that scientists are animal welfare advocates and that the research process is very complicated. This publication is designed to answer basic questions about the use of laboratory animals. Researchers are dedicated to the humane treatment of laboratory animals, to reducing the number of animals used whenever possible, and to refining experimental design to ensure no animal is used unnecessarily.

How Many Animals Are Used In Research?

In its most recent report, published in 1986, the U.S. Congress Office of Technology Assessment (OTA) estimated that 17 million to 22 million animals were used in research and testing in the United States; 85–90% of these animals were rats and mice.[2] OTA cited the U.S. Department of Agriculture's Animal and Plant Health Inspection Service (USDA/APHIS) as the best available data source. However, USDA does not include rats and mice in its numbers. Therefore, OTA estimated that the USDA data accounts for approximately 10% of the total animals.

According to the USDA, 1,842,420 animals—including dogs, cats, nonhuman primates, guinea pigs, hamsters, rabbits and farm animals—were used by research facilities in 1991.[3] The figures listed below are based on this data from OTA and USDA.

Rats and mice	85–90%
Hamsters, Rabbits & Guinea Pigs	Approximately 6.0%
Cats	Under .5%
Dogs	Under .5%
Primates	Under .5%

Do Laboratory Animals Experience Any Pain?

According to a 1991 report by the USDA, most research (94%) was not painful to the animals involved. In the majority of cases (61%), the animals were not exposed to or involved in any painful procedures. In approximately 33% of cases, animals were given anesthesia or pain-relieving drugs during procedures that could have involved some pain or distress to the animals. In about 6% of research projects, anesthetics or analgesics (pain-relieving drugs) were not used because they would have interfered with the end results. In rare cases, such as certain central nervous system studies, the research may have required that pain not be relieved.

What Professional Principles Do Animal Researchers Follow?

Good animal care is essential to good science. If a laboratory animal is unhealthy due to stress or disease, the researcher will be unable to collect reliable data. Animals that are treated well, on the other hand, provide the normal biological or behavioral responses that researchers need to examine. In protecting their lab animals, researchers are protecting the source of their scientific data.

Researchers are guided by the following four basic principles:

- Ensure all research animals receive good care and humane treatment.
- Use animal models only when nonanimal methods are inadequate or inappropriate.
- Use as few animals as possible.
- Design experiments so that all animal studies yield scientifically reliable results.

Numerous professional organizations comprising researchers and scientists have their own standards for lab animal care. The American Association for the Accreditation of Laboratory Animal Care (AAALAC) and the American Association for Laboratory Animal Science (AALAS) are two of the most prominent organizations. Leading veterinarians and researchers organized AAALAC in 1965 "to promote high standards of animal care, use and well-being and enhance life sciences research and education through the accreditation process." AAALAC conducts voluntary peer review evaluation of laboratory animal care facilities and programs which involve site visits, evaluation of site visit reports and recommendations concerning proposed accreditation status.

During the 1940s a group of professionals involved in animal research were concerned about the varying standards of lab animal care. In 1950, this group founded AALAS which today has more than 10,000 individual and institutional members ranging from veterinarians to lab technicians to university administrators. AALAS is dedicated to developing and maintaining the highest standards of animal care. The association serves as a forum for presenting and exchanging scientific information on all phases of laboratory animal welfare through its educational activities and certification programs.

Professional societies, such as the American Physiological Society, the American Psychological Association, and the Association for Research in Vision and Ophthalmology, have codes and policies governing animal research, which their members must follow. Voluntary health organizations, such as the American Heart Association, the American Cancer Society and the Juvenile Diabetes Foundation, have adopted official policies outlining acceptable standards for the care and use of lab animals. Research funded by these organizations must meet these criteria.

Are There Laws To Ensure Humane Care And Treatment Of Laboratory Animals?

In 1955 the National Institutes of Health (NIH) published the first federal lab animal guidelines, now entitled the *Guide for the Care and Use of Laboratory Animals*. In 1966, the U.S. Department of Agriculture (USDA) developed the first federal regulations under the Animal Welfare Act. (For detailed information, please refer to *NABR Issue Update; Regulation of Biomedical Research Using Animals*.)

Many people, however, are unaware of the extensive system of laws, guidelines, regulations and principles that ensure the welfare of laboratory animals in the U.S. Requirements address veterinary care (surgery, analgesics, anesthesia, and euthanasia methods) and housing conditions (food, water, sanitation, temperature, humidity, lighting, and drainage). All facilities must provide exercise for dogs and a physical environment adequate to promote the psychological well-being of nonhuman primates.

How Is A Researcher Granted Permission To Use Animal Models?

As required by the Animal Welfare Act, a researcher must submit a detailed animal care and use procedure plan to a review committee at the institution. The Institutional Animal Care and Use Committee (IACUC) is required by law, as is the animal care and use plan. According to federal regulations, a proposed plan must contain the following:

- Identification of the species and approximate number of animals to be used;
- A rationale for involving animals and the species and number to be used;
- A complete description of the proposed use of the animals;
- A description of procedures designed to assure that discomfort and pain to animals will be limited to that which is unavoidable for the conduct of scientifically valuable research, including the provision for the use of analgesic, anesthetic and tranquilizing drugs where indicated and appropriate to minimize discomfort and pain; and
- A description of any euthanasia method to be used.

The researcher also must provide the IACUC with written assurance that the plan does not unnecessarily duplicate previous research.

Finally, the researcher must consider alternatives to any procedure that may cause more than momentary or slight pain or distress to the animals and provide a written description of the methods and sources used to determine that nonanimal alternatives were not available.

After the researcher has provided the above information, the plan is reviewed by the IACUC. The IACUC may approve, reject, or ask for additional information about a plan. If the IACUC finds that the plan does not address each area of animal care sufficiently, the plan is rejected, and the researcher cannot begin the project. However, the researcher is given the opportunity to address the IACUC's concerns and may resubmit the denied plan with appropriate changes.

Once a plan is approved and under way, the IACUC has the authority to suspend any project. The IACUC must inspect the research institution's animal facilities at least once every six months. It should also be noted that all IACUCs must have no less then three members; one must be a veterinarian and one must not be affiliated with the institution in any way.

Who Takes Care Of Laboratory Animals?

Each animal research institution has an animal care staff. This staff works under the direct supervision of a veterinarian, generally a specialist in the practice of laboratory animal medicine or a related veterinary medical field such as comparative pathology. Like physician specialists who practice human medicine, these veterinarians undergo postgraduate and residency training to qualify for the rigorous certification examinations required for their specialties.

The animal care staff are laboratory animal technicians or technologists, occupations that combine traditional veterinary nursing skills with an understanding of research methods and requirements. Technicians check on each animal's health daily. They control the animal's environment, a responsibility that extends far beyond feeding and watering the animals and keeping them clean, dry and comfortable. Technicians continuously monitor external factors such as noise, light, heat and humidity, as well as the use of insecticides, detergents and disinfectants. Many animal technicians also are skilled veterinary nurses and medical assistants. They are trained to draw blood, take X-rays, give medications, administer fluid therapy, induce anesthesia, assist at surgery, give postoperative care and humanely, painlessly euthanize an animal.

Today, 59 degree-granting programs for animal technicians/technologists in 35 states are accredited by the American Veterinary Medical Association. Most are two-year, associate degree programs; some are four-year undergraduate programs offered by schools of agriculture or veterinary medicine. As part of its ongoing effort to set high quality animal care standards, AALAS instituted a national certification program for laboratory animal personnel. AALAS administers qualifying examinations and certifies successful candidates at three levels. Over 2,000 examinations are given annually. AALAS certification is highly encouraged for all animal care staff.

Why Is Research Sometimes Duplicated?

Duplication, or more accurately, replication of research is necessary to validate scientific findings. Each scientist's findings must be confirmed by others before they are considered valid. This requires some experimentation that may deviate in only minor ways from previous work and, therefore, may appear to be duplicative. Such replication, however, provides for rigorous testing for hypotheses and the formulation of conclusions that carry a higher degree of validity. Once validated, the data becomes biomedical knowledge that researchers throughout the world may draw upon for new investigations.

The scientific peer review process and keen competition for federal research funds prevents unnecessary duplication of research. Funds are not granted for projects that will not make significant contributions to the existing body of biomedical knowledge. The National Institutes of Health—the single major source of federal funding for biomedical research in this country—currently can support only about one-third of all worthy research proposals due to limited available funding. Certainly, scarce funds are not awarded for studies which will not add significantly to biomedical research.

Are There Alternatives To Using Animals In Research?

Scientists use a variety of methods in research. In some areas, the use of animals is neither necessary nor appropriate. In others, such as developing a fundamental understanding of how complex biological systems function, the use of animals has been and continues to be essential. In these cases there is no method to replace the use of animals.

Our knowledge of higher organisms is quite limited. Even though science has made remarkable progress, we cannot create an organ or even a cell. We cannot grow organs in culture dishes, as cells can be grown. With progressive knowledge, we hope that one day we will be able to grow groups of organs and actually make a whole organism from a few cells in a petri dish. That possibility is far into the future. Today, the replacement of whole animals with nonanimal models for advanced biomedical research is simply not possible.

As a byproduct of basic research, scientists have developed a number of valuable nonanimal research methods. Such methods are useful for some research, and in other cases they complement work in animal systems. Today, one of the widest uses of nonanimal tests is an initial screening of chemical substances for potentially toxic and harmful effects. With increasingly sophisticated computer technology and laboratory instrumentation, it is feasible to conduct many biological studies without using whole animals. A number of very important adjunct research methods are in use. Some examples of these nonanimal methods are PhysicoChemical techniques, computer and mathematical models, microbiological systems, and cell and tissue cultures.

The increased use of nonanimal adjunctive tests is reflected in the fact that there was a 40% drop in the number of animals used in research between 1968 and 1978.[4] However, no responsible researcher believes that the technology exists today, or in the foreseeable future, to replace the use of animals altogether in biomedical research.

[1]*The Washington Post*, Science Notebook—"Animal Research Favored," June 12, 1989.

[2]U.S. Congress, Office of Technology Assessment, *Alternatives to Animal Use in Research, Testing and Education* (Washington, DC: U.S. Government Printing Office, OTA-BA-273, February 1986), p. 5.

[3]U.S. Department of Agriculture, Animal and Plant Health Inspection Service, Regulatory Enforcement and Animal Care, *Animal Welfare Enforcement Fiscal Year 1991: Report of the Secretary of Agriculture to the President of the Senate and the Speaker of the House of Representatives*, Appendix, Table 2.

[4]Institute of Laboratory Animal Resources, Fiscal Year 1978, *National Survey of Laboratory Animal Facilities and Resources* (Washington, D.C.: U.S. Department of Health and Human Services, NIH Publication No. 80-2091, March 1980), p. 21.

nabr *ISSUE UPDATE*

National Association for Biomedical Research
818 Connecticut Avenue NW, Suite 303, Washington, DC 20006
Phone 202.857.0540 FAX 202.659.1902

The Use of Animals in Product Safety Testing

Summary

Scientists have reduced the number of animals used in product safety testing in recent times. Because this reduction has been possible, animal rights activists have led the public to believe that the use of laboratory animals can be eliminated in this field altogether. It is often claimed that valid alternatives to the use of animal tests already exist to evaluate product safety. This is simply not true.

The FDA states, "many procedures intended to replace animal tests are still in various stages of development . . . While the best means may begin with valuable adjunct tests, ultimately testing must progress to a whole intact, living system—an animal." Not using animal tests when necessary would subject humans and other animals to unreasonable risks.

- Manufacturers of food, drugs, household goods, cosmetic products, pesticides and other chemicals have both ethical and legal obligations to protect the safety of consumers.
- Federal statutes including the Food, Drug, and Cosmetic Act; Toxic Substance Control Act; Federal Insecticide, Fungicide, and Rodenticide Act; Clean Air and Water Act and the Consumer Product Safety Act mandate that the federal government be involved in assuring product safety and in protecting public health.
- Federal agencies such as the Food and Drug Administration, the Environmental Protection Agency, the Consumer Product Safety Commission and the Occupational Safety and Health Administration play a significant role in monitoring product safety.
- Federal agencies and professional groups agree that alternative methods are not available at this time to completely replace the use of animals in product safety testing.
- The vast majority of test animals are rodents (primarily mice and rats).
- The "classic" LD-50 test has been replaced, in almost all cases, by modified tests that require fewer animals. However, there are still rare instances when statistically precise acute-dose toxicity data are needed.
- The Draize test, developed in 1944, has changed considerably over the years to reduce or eliminate any pain test animals may experience. It is the FDA's position that "the Draize test is currently the most meaningful and reliable method for evaluating the hazard or safety of a substance introduced into or around the eyes."
- Companies that claim their products are not tested in animals—and, therefore, are "cruelty free"—mislead consumers since almost all products or the chemical compounds that comprise them were previously tested on animals. Such testing need not be repeated.

What is product safety testing?

The purpose of safety testing is to ensure that a product is safe when used as directed and, perhaps more importantly, to provide scientific data for poison control centers and emergency room personnel should a product be misused. Toxicology is the study of the harmful effects of substances on living systems. Toxicologists test prescription drugs, over-the-counter drugs, food additives, household products, pesticides, chemicals and cosmetics which include items like shaving cream, sunscreen and shampoo. A substance's relative safety is judged according to its effects under various conditions such as dose, route of administration, duration and frequency of exposure, as well as the actual chemical structure.

Do we really need to use animals for testing?

Because scientists have drastically reduced the number of animals needed for product safety testing, animal rights activists have led the public to believe that the use of laboratory animals can be eliminated from this field altogether. This is simply not true.

According to Dr. James Mason, former Assistant Secretary for Health at the Department of Health and Human Services, "Whole animals are essential in research and testing because they best reflect the dynamic interactions between the various cells, tissues and organs comprising the human body."[1]

The Food and Drug Administration (FDA) has stated that "many procedures intended to replace animal tests are still in various stages of development and that it would be unwise for us to urge manufacturers not to do any further testing . . . While the best means may begin with valuable adjunct tests, ultimately testing must progress to a whole intact, living system—an animal."[2]

While scientific societies, such as the Society of Toxicology, have issued strong statements in support of the development of nonanimal methodologies, they also caution that "at present, tests in intact animals are the only means of assessing the potential hazard from (dermal and ocular) exposure other than direct testing in man."[3]

Not performing adequate testing violates both our moral and legal obligation and places people as well as animals at risk.

Which federal agencies regulate product safety testing?

Federal laws require that the public be protected from hazardous commercial products, but specific animal tests are not statutorily mandated. Related federal regulations explicitly or implicitly call for animal testing. Four principal federal agencies have a significant role in animal testing for regulatory purposes. (See cover page.)

The FDA is responsible for administering statutes that regulate human and animal food and drugs, medical devices, biological products for human use, cosmetics, color additives and radiological products. FDA expressly requires that laboratory animal tests be conducted both for prescription drugs and over-the-counter drugs before these products can be tested further in humans. Antiperspirants and dandruff shampoos, for example, are considered to be over-the-counter drugs.

The Environmental Protection Agency (EPA) uses toxicity data derived from animal testing, as well as other data, to protect humans, animals and the environment from harmful effects of pesticides, industrial chemicals, air and water pollutants and hazardous wastes.

The Consumer Product Safety Commission (CPSC) relies on animal data in identifying and regulating risks to consumers of personal care and household products, while the Occupational Safety and Health Administration (OSHA) uses such data indirectly in requiring employers to maintain a safe workplace.

Types of Toxicology Tests

The nature and extent of testing may vary from one type of product to another. The intended use of the product, the ways in which people are likely to be exposed to it, the specific properties of the product and the dictates of federal law are all factors in determining which tests are needed and the extent of the evaluation.

Some of the common types of toxicology tests are acute lethality, and skin and eye irritancy (Draize) tests. There are also chronic and subchronic studies that examine the risks and extended exposure to new chemicals and drugs, such as long-term cancer treatment. Teratology and reproduction studies address the causes of birth defects, while mutagenicity studies assess a chemical's tendency to cause adverse changes in the genetic makeup of an organism.

What is the LD-50?

The LD-50 test is a measure of acute lethality. An LD-50 rating is calculated for the dose at which one-half of the test animals can be expected to die from ingestion of the test substance.

This "classic" LD-50 test, developed more than fifty years ago, has been discouraged by the EPA since 1984.[4] FDA has no requirements for LD-50 test data obtained by using the classic, statistically precise test.[5] CPSC also accepts estimated, rather than classic LD-50, acute lethality data.

Federal agencies now suggest using a tiered testing approach rather than the "classic" LD-50 test. First, data already available from existing or structurally related chemicals are reviewed in order to make preliminary safety evaluations. One useful data source for this purpose is the fourteen-year-old, industry-funded Cosmetic Ingredient Review (CIR). It brings together all available published and unpublished data on the safety of cosmetic ingredients.

The second step is the "limit" test which uses 10 to 20 animals, not 80 to 100 animals like the "classic" LD-50. In this test, animals are given a single dose of a product according to their body weight. The vast majority of test animals are rodents (mice and rats). Although the results of the "limit" test may not be as precise as the "classic" LD-50, this form of testing is almost always sufficient in estimating the toxicity of the substance in question.

Lastly, if necessary, multiple endpoint testing is done. Dosed animals are observed for abnormal behavior and are carefully watched during recovery. They are also autopsied to evaluate internal evidence of toxicity to various organ systems.[6]

While the "classic" LD-50 usually can be replaced by other tests that utilize fewer animals, there are still rare instances when precise acute dose toxicity data are needed. For example, the "classic" LD-50 test may be necessary when examining the potency of highly toxic drugs, such as new cytotoxic cancer drugs. Also, when determining the effective strengths of certain pesticides, precise acute toxicity data may be desired.[7]

Skin and Eye Irritancy Tests

Because many products are intended for topical use, or for use in and around the eyes, consideration for potential skin and eye irritancy is an important aspect of safety evaluation.

In some circumstances, it is not necessary to conduct any irritancy testing in animals. If, for example, initial chemical analysis shows that the product is a corrosive, the toxicologist assumes that the product will produce irritancy and that animal testing will yield no new information.

Similarly, testing for eye irratancy may not be needed if initial nonanimal or skin irritancy tests indicate that the product may injure eye tissue.

What is the Draize test?

The Draize eye and skin irritancy tests are used to determine whether substances that may come into contact with the eye or skin will cause irritation or injury.

In the eye test, drops of the test substance are placed into one eye in each of as few as three test animals. Rabbits are commonly used because their eyes are at least as sensitive as the human eye. Affected eyes are observed at intervals, and results are scored based on the degree of various reactions—corneal clouding, percent of cornea involved, condition of iris, redness, etc. Since live animals, not test tubes are used, scientists can also observe the healing process after exposure to an irritant.

For the skin irritancy test, the fur of as few as three animals is clipped and the test substance is placed in contact with the skin. The site is covered for a day and then the patch is removed to evaluate any irritation for up to three more days.

While the basic Draize procedures are essentially the same as those used since the 1940s, modifications to reduce or eliminate any pain or distress in test animals are now employed. These modifications include using ophthalmic anesthetics whenever possible, using diluted solutions or lower doses of test substance and eliminating the use of stocks for restraint so animals can move freely and have continual access to food and water.[8]

Is the Draize test really necessary?

As part of its overall effort to reduce or avoid unnecessary testing methods, the FDA has been carefully evaluating the use of the Draize test. It is the FDA's position that "the Draize eye irritancy test is currently the most meaningful and reliable method for evaluating the hazard or safety of a substance introduced into or around the eye."[9]

While in vitro studies can be useful as screening tools to indicate the relative toxicity or safety of a substance, the FDA states that "the responses and results of alternatives alone cannot, at the present time, be the basis for determining the safety of a substance. It is more likely that these efforts will result in reduction or refinement as opposed to the replacement of the Draize eye irritation test."[10]

What are the nonanimal alternatives for product safety testing?

As indicated above, in recent years the scientific community has been successful in reducing the number of animals used in product safety tests as well as in refining test methods to reduce any pain or distress these animals may experience.

Mathematical models can be useful in helping to predict an organism's response to varying levels of exposure to a particular substance and in improving the design of scientific experiments.

Computer data banks allow for the reduction of test duplication. Computers also are useful in the initial evaluation of chemicals slated for further study. Unsuitable chemicals can be eliminated from consideration prior to the institution of animal testing.

Cells, tissues and even whole organs obtained from animals and humans can be kept "alive" in the laboratory and used in preliminary screening of chemical compounds. However, testing on tissue culture will not reveal the effects of a substance on a complex living organism composed of many different physiological systems.

As useful as nonanimal methods have proven to be, they have only limited utility and cannot totally replace the use of animals in safety testing. Most researchers generally hold that nonanimal methods are adjuncts rather than alternatives to animals.[11]

For economic and ethical reasons, industry is actively committed to the search for alternatives to animal testing. Internal and external efforts to develop and evaluate promising nonanimal procedures are receiving significant industry support.

What is the truth about so-called "cruelty-free" products?

The public may be confused by announcements that some companies do not test their products in laboratory animals, thus providing "cruelty-free" products. This notion can be misleading because most of these products or their ingredients have been tested previously in animals, or the safety of the products has been well-established based on a long history of human exposure to them; therefore additional evaluation is not required. In some instances, animal testing may be conducted by raw material suppliers other than the manufacturer of the final product. If the safety of a consumer product has not been substantiated, federal agencies would require a prominent label declaration of this fact.

Because a given product may not require animal testing is not evidence that testing can be abandoned for all products. New chemicals, new uses of old chemicals and new mixtures of chemicals must be subjected to toxicity testing so that unsafe products will not be marketed inadvertently.

[1]James O. Mason, M.D., Assistant Secretary for Health, U.S. Department of Health and Human Services, Statement before the U.S. Senate Committee on Commerce, Science and Transportation, Consumer Subcommittee. November 8, 1989.

[2]U.S. Food and Drug Administration, Statement to the Maryland Governor's Task Force to Study Animal Testing. April 17, 1989.

[3]Society of Toxicology, Comments on the LD-50 and Acute Eye and Skin Irritation Tests. July 24, 1989.

[4]Environmental Protection Agency News. August 29, 1984.

[5]Gerald B. Guest, D.V.M. Director, U.S. Food and Drug Administration Center for Veterinary Medicine, *Animal Testing—The Present, The Future, Food & Drug Law Institute*. Sept. 7, 1989.

[6]Environmental Protection Agency, Revised Policy for Acute Toxicology Testing. September 22, 1988.

[7]Dr. David Rall, Director, National Institute of Environmental Health Sciences, National Institutes of Health, U.S. Department of Health and Human Services. Statement before the U.S. House of Representatives Energy and Commerce Committee, Subcommittee on Health and the Environment. May 16, 1983.

[8]Federal Register, Vol. 49 No. 105, pp. 22523. May 30, 1984.

[9]David A. Kessler, M.D., Commissioner, U.S. Food and Drug Administration. Letter to the Honorable Pete Wilson, Governor of California. September 3, 1991.

[10]Mason, see 1.

[11]Councils of the National Academy of Science and Institute of Medicine, Committee on the Use of Animals in Research. *Science, Medicine and Animals*. National Academy Press, 1991.

nabr *ISSUE UPDATE*

National Association for Biomedical Research
818 Connecticut Avenue NW, Suite 303,
Washington, DC 20006
Phone 202.857.0540 FAX 202.659.1902

Regulation of Biomedical Research Using Animals

Summary

Public polls reveal most people are unaware of the laws and regulations that govern the use of laboratory animals in biomedical research. Lacking this awareness, the average person is more likely to believe the consistently untrue charges of abuse made by those who totally oppose animal research. In fact, scientists who use animals in their work must comply with a comprehensive system of federal, state and local laws and regulations.

Even if such legal requirements did not exist, researchers know that laboratory animals must be cared for humanely for both ethical and scientific reasons. In order to understand the needs of these animals, the veterinary specialty of laboratory animal medicine was established. A host of organizations and programs exist within the scientific community dedicated solely to promoting excellence in laboratory animal care. In addition, academic and professional societies have longstanding policies, procedures and standards defining the ethical treatment of animals. To understand more about the scientific community's dedication to the highest-quality animal care, policies and procedures, please refer to the NABR *Issue Update* entitled *"Humane Care and Treatment of Laboratory Animals."*

The purpose of this Issue Update is to outline the major laws, regulations and guidelines that must be followed when laboratory animals are used. They include:

- U.S. Government Principles for the Utilization and Care of Vertebrate Animals Used in Testing, Research and Training
- Animal Welfare Act and the U.S. Department of Agriculture Animal Welfare Regulations and Standards
- Public Health Service Act and the U.S. Public Health Service Policy on Humane Care and Use of Laboratory Animals
- Guide for the Care and Use of Laboratory Animals prepared by the National Academy of Science Institute for Laboratory Animal Resources
- Good Laboratory Practice Standards of the Food & Drug and Environmental Protection Administrations
- Endangered Species Act
- Freedom of Information Act

Public accountability is essential to the future of biomedical research.

A majority of Americans, 77% according to an American Medical Association-sponsored Gallup Poll, support biomedical research using laboratory animals, provided the research is conducted in a humane and responsible manner. As a result of the general public's interest in animal welfare, a comprehensive system of federal, state and local laws and regulations governing the use of laboratory animals has evolved. This report outlines the mandatory legal requirements with which scientists must comply when using animals in research, testing and education programs.

U.S. Government Principles

U.S. Government Principles for the Utilization and Care of Vertebrate Animals Used in Testing, Research and Training express the tenets that underlie our current system of federal regulation. These principles succinctly describe the framework within which all activities involving laboratory animals must be conducted.

Federal Animal Welfare Act

The federal Animal Welfare Act[1] (AWA) was enacted in 1966 and has been amended by Congress four times since then. The Act applies to all research facilities—public or private, academic or industry based, whether or not they receive federal funds—that use animal species designated by the U.S. Secretary of Agriculture. Currently, the species so designated are guinea pigs, hamsters, gerbils, rabbits, dogs, cats, nonhuman primates, marine mammals, farm animal species when used in biomedical research and warm-blooded wild animals. Rats, mice and birds have traditionally not been covered in order to concentrate limited enforcement resources on other species.

All covered research facilities must register with the U.S. Department of Agriculture (USDA) and comply with USDA animal welfare regulations and standards (see below). Each facility must report to the USDA annually verifiying compliance and indicating the number and species of animals used by type of procedure (painless, pain relief/anesthesia given or not given because of scientific necessity). The USDA is required to inspect each research facility at least annually. These inspections are unannounced. More frequent unscheduled inspections are made if significant deficiencies are identified.

All registered research facilities are required to have an Institutional Animal Care and Use Committee (IACUC) that reviews and approves procedures involving animals before they take place and inspects facilities biannually for compliance with the AWA. At least one member of the Committee must be a veterinarian. At least one member must be a "public" member, not affiliated in any way with the institution, who represents general community interests in the care and treatment of animals.

USDA Animal Welfare Regulations and Standards[2]

As required by the AWA as amended, USDA's Animal and Plant Health Inspection Service (APHIS) has published and implemented detailed regulations defining the responsibilities of research facilities, the duties of IACUC's and specific standards for animal care. In experimental procedures, research facilities must ensure that any animal pain and distress are minimized, including adequate veterinary care with the appropriate use of anesthetic, analgesic, tranquilizing drug or euthanasia. Principal investigators must consider alternatives to any procedure using animals, especially those procedures likely to cause pain to a laboratory animal and provide written assurance of this consideration. They must also assure that activities do not unnecessarily duplicate previous experiments. Research facility personnel working with animals must have appropriate qualifications and training.

Beyond annual reporting to USDA, research facilities are required to keep extensive records on IACUC activities (minutes of meetings, reports of biannual inspections, etc.) and documentation of the source of animals. These records are available to USDA staff during inspection.

Animal welfare standards by species include requirements for handling, cage size, feeding, watering, sanitation, ventilation, temperature, humidity and adequate veterinary care. Research facilities also must have individualized, written plans for exercise programs for dogs and for addressing the psychological well-being of nonhuman primates.

Public Health Service Act

In 1985 Congress passed the Health Research Extension Act (P.L. 99-158), which created a mandate in federal law for the longstanding policies governing the use of animals supported by U.S. Public Health Service (PHS) funds. The PHS includes the Centers for Disease Control; the Food and Drug Administration, the Health Resources and Services Administration; the National Institutes of Health; the Substance Abuse and Mental Health Services Administration and other programs of the Office of the Assistant Secretary for Health, Department of Health and Human Services. These legal requirements are similar to and consistent with the federal Animal Welfare Act, and they apply to all PHS conducted or supported research, research training and biological testing activities involving the use of all vertebrate animals.

Public Health Service Policy on Humane Care and Use of Laboratory Animals

The *PHS Policy*[3] implements and supplements the laboratory animal-related provisions of the Public Health Service Act and the general U.S. Government Principles. Compliance with the *Policy* is required for activities conducted by PHS units and by awardee institutions as a condition of receiving PHS funds. PHS grants or contracts can be suspended or revoked for noncompliance.

A major provision of the *Policy* is the filing and annual updating of an Animal Welfare Assurance. The assurance document must fully describe the institution's animal care and use program. That program must comply with the Animal Welfare Act and other applicable federal laws and must adhere to the *Guide for the Care and Use of Laboratory Animals (Guide)*. Like the AWA, the *Policy* requires each institution to establish an IACUC with at least one outside member representing the public. As described in the *Policy*, the duties and responsibilities of the IACUC are comparable to AWA requirements. Also, each application for a PHS award includes the number and species of animals to be used, rationale for the use of animals, description of the proposed use, procedures to minimize pain and discomfort and method of euthanasia.

The NIH Office for Protection from Research Risks (OPRR) administers the *Policy* on behalf of the PHS. OPRR is responsible for reviewing and approving institutional assurances, advising research facilities about compliance, evaluating allegations of non-compliance with the *Policy* and conducting site visits as needed.

Other federal agencies, such as the National Science Foundation, rely upon the PHS Policy and assurance system. If a potential awardee is not a PHS-assured institution, special arrangements must be made.

Guide for the Care and Use of Laboratory Animals

The *Guide*[4] is widely accepted by scientific institutions as a primary reference on animal care and use. First published in 1963, it has been revised five times since then, most recently in 1985. The purpose of the 83-page *Guide* is to serve as a source of "information in common laboratory animals housed under a variety of circumstances" that will "assist institutions in caring for and using laboratory animals in ways judged to be professionally and humanely appropriate." The *Guide* was compiled by a panel of veterinary and other scientific experts brought together by the Institute of Laboratory Animal Resources (ILAR), Commission on Life Sciences of the National Research Council.

Sections of the *Guide* cover recommended institutional policies, laboratory animal husbandry (housing, cage size, social environment, food, water, bedding, sanitation and other issues); veterinary care including preventive medicine; control of disease; anesthesia and analgesia; surgery and post-surgical care and euthanasia; physical plant; special considerations such as control of hazardous agents; references and a bibliography.

Good Laboratory Practice Standards

Both the Food and Drug Administration (FDA) and the Environmental Protection Agency (EPA) enforce "Good Laboratory Practice" (GLP) rules.[5] The FDA regulations apply to all projects yielding data to support new drug, biologics and medical device applications. EPA regulations apply to all studies related to new pesticides or toxic substance approvals. The GLPs for both agencies address all areas of laboratory operations. Provisions relating to care and housing of test animals are identical in both GLP rules. Each has a full section on animal care, specifying Standard Operating Procedures for housing, feeding, handling and care, with additional standards on separation, disease control and treatment, identification; sanitation, feed and water inspection; waste and pest control.

Inspections are conducted by federal agency investigators, who visit each facility and are given access to all parts of the premises, all pertinent personnel and documentation. A final report and more detailed facility inspection reports are prepared after an audit is concluded. Noncompliance with GLPs can result in the federal agency's refusal to consider a study in support of an application; disqualification of the testing facility; or, in cases of alleged fraud, recommendation for criminal prosecution.

Endangered Species Act

A variety of federal and international laws and agreements exist to protect animals. Statutes such as the Endangered Species Act[6] prohibit or control acquisition of wild or captive-bred, domestic and non-domestic animals classified as "endangered" or "threatened." At a minimum, a permit or authorization from one or more federal agencies is required, if the animal can be obtained at all. Wild caught chimpanzees, for example, are classified as "endangered" and have not been imported into the U.S. since 1976. Only chimpanzees bred in this country are available for research purposes, and their use is strictly controlled by a national committee which reserves the captive-bred chimps for high priority research for which there is no alternative, such as AIDS vaccine studies.

Freedom of Information Act

The federal Freedom of Information Act (FOIA)[7] provides for public access to government information. A wide range of information about animal research and testing as well as its federal oversight is therefore available to the public. Requestors can and do obtain details about federally conducted or supported research projects, copies of PHS animal welfare assurance documents from awardee institutions, USDA annual reports filed by research facilities, USDA, EPA and FDA inspection reports.

State or local laws and regulations

Many states have statutes and regulations in place relevant to laboratory animals. Typical state requirements fall into these broad categories:

Regulation of Research Facilities—Twenty states and the District of Columbia have laws concerning licensing of research facilities. For licensing purposes, state officials may set rules, regulations and standards of animal care and treatment. A number of states also have facility inspection programs.

Availability of Pound Animals for Research—Currently thirteen states prohibit the use of animals obtained from in-state pounds; in one of these states the use of pound animals from inside and outside the state is prohibited. Five states and the District of Columbia require the release of pound animals for research purposes. In nine states, law permits research use of pound animals. The disposition of abandoned pound animals is not specifically prohibited, required or permitted in twenty-three states. In these twenty-three states, as well as the nine states that permit release, county and local governments often exercise jurisdiction over the question of whether unclaimed pound animals may be obtained for research.

Animal Cruelty Prevention—Longstanding laws against cruelty to animals exist in every state. In twenty-six states and the District of Columbia, properly conducted research is exempted.

For full details and specific citations for these laws, please refer to the NABR publication, *State Laws Concerning the Use of Animals in Research.*[8]

[1]Animal Welfare Act, 7 U.S.C. 2131 et seq.

[2]U.S. Department of Agriculture, Animal Welfare Regulations and Standards, 9 C.F.R., Ch. 1, Parts 1, 2 and 3.

[3]U.S. Department of Health and Human Services, Public Health Service, National Institutes of Health, Office of Protection from Research

Risks, *Public Health Service Policy in Humane Care and Use of Laboratory Animals*, revised September 1986, pursuant to Health Research Extension Act of 1985 (P.L. 99-158, Sect. 495, Nov. 20, 1985.).

[4]U.S. Department of Health and Human Services, Public Health Service, National Institutes of Health, *Guide for the Care and Use of Laboratory Animals*, NIH No. 86-23, 1985.

[5]Food and Drug Administration, Good Laboratory Practices, 21 C.F.R. 58 and Environmental Protection Agency, Good Laboratory Practices for Pesticide Program, 40 C.F.R. 160, and for Toxic Substances, 40 C.F.R. 792.

[6]Endangered Species Act, 16 U.S.C. 1531.

[7]Freedom of Information Act, 5 U.S.C. 552.

[8]*State Laws Concerning the Use of Animals in Research*, compiled by NABR, Third Ed., Sept., 1991.

nabr *ISSUE UPDATE*

National Association for Biomedical Research
818 Connecticut Avenue NW, Suite 303, Washington, DC 20006
Phone 202.857.0540 FAX 202.659.1902

The Use of Dogs & Cats in Research & Education

Summary

Although dogs and cats comprise less than 1% of all animals used in biomedical research and education programs in the U.S., they are critically important for both scientific and economic reasons. They are used in a wide variety of research, such as heart and kidney disease, brain injury, stroke, blindness and deafness, and for the education of future veterinarians and physicians. Dogs and cats used as research models are either purpose-bred (bred specifically for research) or random-source, animals whose genetic makeup is unknown and have not been bred for research. Based on a NABR survey of the biomedical research community, approximately 40% of all dogs and cats used in research are purpose-bred. The remaining 60% are random-source animals that are acquired from animal dealers or pounds.

Far less than 1% of the 10 to 16 million unwanted animals that otherwise would be put to death in pounds and shelters each year are released for research. Contrary to what many people believe, the small number of abandoned pound dogs and cats used for research and education are not people's pets. Rather, they are unwanted animals, often abandoned, whose owners have not claimed them or for which adoptive homes cannot be found. Research and educational institutions acquire them either directly from pounds or through dealers licensed and regulated by the U.S. Department of Agriculture.

Banning the use of random-source animals in biomedical research will not solve the problem of reducing the number of animals killed in pounds and shelters. However, it will create significant problems for biomedical researchers and veterinary and medical schools. Random-source dogs and cats are excellent models for many types of biomedical research, particularly those that require animals from a diverse genetic pool.

In addition, the cost of acquiring dogs and cats specifically bred for research can be at least two and as much as ten times as high as the cost of acquiring random-source animals. The high cost of acquiring purpose-bred animals to replace random-source dogs and cats may price many research projects out of existence. Some areas of professional education also may be curtailed. The result would be to slow the pace of research and the medical breakthroughs that come from that research.

If members of the public understand why and how random-source animals are used it is likely that they will support the continuation of responsible research and educational programs that involve random-source animals. The price of a ban is enormous: dogs and cats will not benefit and human and animal victims of disease will suffer.

Areas of Research Using Dogs and Cats

Laboratory animals are used for research into the cause, treatment and prevention of diseases in humans and animals. Dogs have been instrumental in our current understanding of the functions and diseases of the heart and lungs because their cardiovascular and respiratory systems resemble those of humans. Heart surgery techniques, such as coronary bypass surgery, artificial heart valve insertion and pacemaker implantation were tested and studied in dogs. Dogs were the first animals used to conquer rejection during organ transplantation. In fact, the 1990 Nobel Prize for Medicine was awarded to researchers who studied the immunologic basis of organ rejection by working with dogs. Through the use of dogs, scientists discovered that diabetics lack the hormone insulin. To this day, dogs, continue to play a vital role in diabetes research. Some orthopedic research, including the development of the artificial hip, relies on dogs, as does the training of emergency room physicians and nurses in lifesaving techniques to be used on trauma patients.

Cats are frequently the best animal models for research on the brain and on visual and auditory function and disease. Nobel-prize winning research on cats revealed the pathways that send information from the eye to the brain. Cats have contributed to a better understanding of so-called "lazy eye" and "cross-eye" and to research on glaucoma and cataract surgery. Cats are essential to promising research into improved treatment for stroke victims. They played a critical role in research leading to a vaccine for feline leukemia, and studies in this disease also made an important contribution to the identification of the virus believed to cause Acquired Immune Deficiency Syndrome (AIDS) in humans.

Number of Dogs & Cats Used in Research

Dogs and cats account for less than 1% of the approximately 20 million research animals used in the U.S.[1] According to the U.S. Department of Agriculture, a total (from all sources) of 106,191 dogs and 33,991 cats were used in research facilities in FY 1993.[2] These dogs and cats may be purpose-bred animals or random-source animals. Purpose-bred animals are specifically bred for research and have a known genetic background. According to a survey conducted by the National Association for Biomedical Research (NABR), approximately 40% of all dogs and cats used in research and education are purpose-bred animals. Animals whose genetic makeup is unknown and that are not bred for research are known as random-source. Based on the NABR survey, 60% of all dogs and cats used in research and education are random-source animals.[3] In 1993, this would have amounted to some 63,700 dogs and 20,400 cats, or less than 1% of the 10 to 16 million animals abandoned in pounds each year. Had they not been used by research and teaching institutions, these animals would likely have died in pounds or been euthanized by owners.

Number of Animals That Die Each Year in Pounds/Shelters

According to the most recent statistics available from the American Humane Association (AHA), between 10.3 and 17.2 million dogs and 5.9 and 9.8 million cats enter pounds/shelters each year. Of these, an estimated 59% of the dogs and 54% of the cats are lost or stray animals. The rest are "owner-relinquished." Once in the pound/shelter, the AHA national average indicates about 15% of the lost dogs and 2% of the lost cats are claimed by their owners, and approximately 19% of the animals are placed in new homes. The remaining 5.8 to 9.6 million dogs (about 66%) and 4.3 to 7.1 million cats (about 79%), not relaimed by owners or adopted by others, must be euthanized.[4] These 10 to 16 million animals are killed because pounds cannot afford to care for them indefinitely.

How Dogs & Cats are Acquired for Research and Education

Universities and other research institutions may acquire dogs and cats from United States Department of Agriculture (USDA) licensed, regulated and inspected Class A or Class B dealers; an inhouse breeding program; or pounds before they are killed.

Class A Dealers—Purpose-bred animals are acquired from USDA Class A dealers, commonly referred to as "breeders." Class A dealers supply only "animals that are bred and raised on the premises in a closed or stable colony and those animals acquired for the sole purpose of maintaining or enhancing the breeding colony."[5] Of the total number of laboratory dogs and cats, some 35% (43,400 dogs and 13,475 cats) are acquired from Class A dealers.[6]

Bred at the Research Institution—The second avenue used to acquire purpose-bred animals is to breed the animals at the research facility. Yet, without the correct environment, in-house breeding of animals can be a very costly undertaking. So, few institutions choose to breed their own animals. Only about 5% of all dogs (6,225 dogs) and cats (1,925 cats) used in research are bred at research facilities.[7]

Class B Dealers—The USDA term "Class B" dealer "includes the purchase and/or resale of any animal. This term includes brokers, and operators of an auction sale, as such individuals negotiate or arrange for the purchase, sale or transport of animals in commerce."[8] Class B dealers sell only random-source animals. They may acquire animals only from: 1) individuals who have bred and raised the animal on their own premises; 2) pounds and shelters; and 3) other Class B dealers. Approximately 45% of the dogs (55,800) and cats (17,325) purchased for research are bought from Class B dealers.[9]

Pound-Acquired—Approximately, 12% (13,640 dogs and 4,620 cats) of all lab dogs and cats are abandoned animals, purchased directly from pounds by research institutions.[10] The total number of some 18,000 animals is far less than 1% of the 10 to 16 million dogs and cats that are killed in pounds each year. Class B dealers are also permitted to obtain unwanted pound animals, but the number purchased is unknown.

By law in 13 states (Connecticut, Delaware, Hawaii, Maryland, Maine, Massachusetts, New Jersey, New Hampshire, New York, Pennsylvania, Rhode Island, Vermont and West Virginia) pounds/shelters cannot make animals available to research facilities or Class B dealers. In Massachusetts, the only state with a total ban, research facilities may not use pound animals from inside or outside the state. In the remaining 12 states, research and education programs must rely on USDA Class B dealers to provide dogs and cats obtained from out-of-state pounds/shelters and individuals who breed and raise animals, or obtain animals specifically bred for research from USDA Class A dealers. In other states, many city and county pounds and shelters do not make animals available for research as a matter of choice or by local ordinance. In these areas, like those with prohibitive state laws, research facilities are dependent upon USDA Class B dealers.

Why Use Random-Source Animals?

Random-source animals fulfill the scientific requirements of some research projects, as well as or better than, purpose-bred animals. The genetic makeup of random-source animals makes them excellent models for certain types of research because they come from a random genetic pool; that is, they have not come from controlled in-breeding. Some research, such as organ and cell transplantation, requires the use of "randomly outbred" animals. This diverse genetic makeup is analogous to the variations in genetic backgrounds among humans.

The cost of a research animal may also be one of the determining factors in the choice of a research subject. Funding for biomedical research is limited. Therefore, scientists must obtain animals in the most cost-effective way possible. If researchers can no longer use random-source animals, they must purchase animals from a Class A dealer. The resulting cost increase—at least twice and as much as ten times more—could retard or halt the progress of research in some vital areas such as heart disease, simply by pricing it beyond the reach of many research instituations.

	Dogs[11]	Cats[12]
Pound Animal	$ 0–100	$ 0–80
USDA Class B	$ 65–560	$ 15–350
Purpose-Bred	$260–995	$100–500

(Specially bred animals can cost as much as $1700)

Federal Regulation

One of the principal reasons for original passage of the Animal Welfare Act was to oversee the use of dogs and cats in research facilities. While pounds/shelters are not subject to the Act or any other federal law or regulation, all animal dealers and research facilities that use dogs and cats are subject to the provisions of the Act which is enforced by the USDA. The Act sets specific standards for the humane handling, care, treatment and transportation of research animals. USDA is responsible for inspection of dealers and facilities to ensure that federal requirements are met. If dealers or facilities fail to comply with these standards, USDA will impose financial penalties and temporarily suspend or permanently revoke their license/registration needed to conduct business.

Under the Act, it is a federal offense for any person to "buy, sell, exhibit, use for research, transport or offer for transportation, any stolen animal" or to obtain live random-source dogs and cats by using false pretenses, misrepresentation or deception. USDA will impose significant fines for noncompliance. In addition, USDA can and has revoked Class B dealers' licenses for repeated failure to comply with the law.

Before an unwanted pound dog or cat can be acquired by a Class B dealer, it must be held at the pound/shelter for not less than 5 full days, including a Saturday. If the Class B dealer obtains the animal from a public pound, it will then be held an additional 5 days on the dealer's premises before being made available to a research facility. If the animal is from a private or contract pound (ie. local humane society), it must be held for 10 days at the dealer's premises. Hence, USDA-mandated holding requirements for Class B dealers are 10 to 15 days. These holding periods ensure that if an owner is searching for a lost pet, he/she will have ample time to find the animal. The USDA requires all dealers to certify that holding periods are observed and to maintain a thorough and very specific description of and a record on each animal.

Consequences of Banning the Use of Random-Source Animals

A prohibition on the use of random-source animals for research and education would have numerous and far-reaching consequences.

Two animals would die instead of one. First, most of the 63,700 random-source dogs and 20,400 random-source cats currently being used in research annually would be killed in pounds. That number—about 84,000—would be replaced with purpose-bred animals that ultimately would be euthanized. This would result in an increase of 84,000 animal deaths. At a time when 10 to 16 million unwanted dogs and cats are killed annually in pounds and shelters, such waste is unjustifiable on ethical, scientific and economic grounds.

Without random-source animals, some research may not be performed, especially if older and/or larger animals are required. Class A dealers do not provide older/larger dogs because it is not cost-effective to breed and raise a dog for nine or ten years prior to sale. Class B dealers and pounds are the sole source for these animals.

Banning random-source animals and/or Class B dealers also could jeopardize the quality of research. It takes years to develop scientifically reliable animal models of individual diseases or to accumulate enough information on a particular species to design studies likely to yield data in the search for the cause, cure, treatment or prevention of diseases in humans and animals. If scientists are forced to abandon research with dogs and cats because random-source animals are not attainable and because of the prohibitive cost of purpose-bred animals, they will have to turn to other species. The time it will take to acquire the same amount of knowledge about a new animal model may slow the rate of research and subsequent medical breakthroughs. In cases where there is no other scientifically reliable animal model, the amount of research will be reduced significantly.

A prohibition against the use of random-source animals could severely impede or even price out of existence many research and educational programs currently dependent of such animals. Few scientific investigators or academic departments could support such cost increases.

Eliminating the use of random-source dogs and cats will not only have a devastating effect on research, but also has as a high ethical and economic cost. More dogs and cats will die, research may not be performed and research time and funds will be lost. In any event, the projected cost to human and animal life is beyond calculation.

[1]U.S. Congress, Office of Technology Assessment, *Alternatives to the Use of Animals in Research, Testing and Education*, (Washington DC: U.S. Government Printing Office, OTA-BA-273, February 1986), p. 5.

[2]United States Department of Agriculture, Animal and Plant Health Inspection Service, Regulatory Enforcement and Animal Care, *Animal Welfare Enforcement, Fiscal Year 1993: Report of the Secretary of Agriculture to the President of the Senate and the Speaker of the House of Representatives*, Appendix, Table 2.

[3]National Association for Biomedical Research (NABR), *Laboratory Dogs & Cats: An Overview of the 1994 NABR Survey*, (Washington DC, July 1994).

[4]Roger Nassar, Ph.D., and John Fluke, *American Humane Animal Shelter Reporting Study: 1988* (Denver: The American Humane Association, 1990), p. 2.

[5]United States Department of Agriculture, Animal Welfare Regulations and Standards, 9 C.F.R., Ch. 1, Part 1, Secton 1.1.

[6]NABR.

[7]NABR.

[8]United States Department of Agriculture, Animal Welfare Regulations and Standards, 9 C.F.R., Ch. 1, Part 1, Secton 1.1.

[9]NABR.

[10]NABR.

[11]NABR.

[12]NABR.

nabr *ISSUE UPDATE*

National Association for Biomedical Research
818 Connecticut Avenue NW, Suite 303, Washington, DC 20006
Phone 202.857.0540 FAX 202.659.1902

Animal Rights Extremists: Impact on Public Health

Summary

Over the past ten years, many organizations originally concerned with the humane care and treatment of animals have shifted their goals and now focus on the complete abolition of all animal use by humans. The term "animal rights" is based on the premise that animals are entitled to the same rights as human beings, and that animals are not to be utilized by humans for biomedical research, education, product safety testing, food, clothing, companionship or sport. According to Australian philosopher Peter Singer, author of the 1977 book *Animal Liberation*, any of the aforementioned uses of animals by humans is considered "specieism," the moral equivalent of racism.

Animal extremists believe that the end—finding cures and treatments for AIDS, cancer, heart disease and other grave afflictions, does not justify the means—humane animal research. Employing a wide range of tactics to attract media attention and public sympathy, the animal rights movement has entered the mainstream of public thinking and is regarded by many as the most important social issue of the decade, and the greatest threat to medical progress.

Today, there are well over 150 organizations in the United States dedicated to ending the use of animals in biomedical research, education, consumer product testing, farming and many other areas. These organizations are well funded—a conservative estimate of combined operating budgets in 1990 for ten of the largest groups opposed to animal research was $60,929,844.[1]

If animal rights extremists continue to stall and eventually halt necessary animal research, the impact on human health will be devastating. Animal studies have been an essential part of every major medical advance within the last century. These breakthroughs include the polio vaccine, insulin treatments for diabetics, organ transplant surgery, chemotherapy treatments for cancer, medication to control high blood pressure and countless other medical victories. Animal research has also contributed to advances in veterinary medicine including treatments for rabies, heartworms and feline leukemia. We can be sure that without animal research, cures and treatments for the many ailments that afflict all animals, both human and nonhuman, would be far beyond our reach.

At a June 1990 press conference in Washington, DC, former Health and Human Services Secretary Louis Sullivan delivered this strong statement regarding the animal rights movement: "The questions involved in this debate—questions about personhood, rights, responsibilities and ethics—are open questions. . . . One of the greatest threats of the animal activists lies in their bold claim to have correctly, definitely decided these questions . . . So I am saddened and a bit angry myself that we have had to put up with major disruptions to science by so-called 'animal rights' activists—who are, in fact, nothing more than animal rights terrorists. . . . They have tried to put us on the defensive through intimidation and even violence. . . . They will not succeed, because they are on the wrong side of morality."

Radical Animal Activism

Some animal rights groups feel that any action necessary to advance their cause is justified, even if that means engaging in radical, violent activities. The Animal Liberation Front (ALF), perhaps the most well-known group, is included on the FBI's list of 10 most dangerous terrorist organizations.[2] An offshoot of animal extremists in Great Britain called the Hunt Saboteurs Association and the Band of Mercy, the ALF has claimed responsibility for hundreds of illegal break-ins, animal thefts, fires and bomb threats at research laboratories, fur retailers, farms and meat production facilities throughout the United States, Canada and Great Britain.

Illegal activities to end animal research have steadily increased. Over the past ten years, there have been more than 90 reported arsons, break-ins and bomb-threats "in the name of animal rights," and the perpetrators of these crimes are almost never apprehended. In fact, there have been only three criminal convictions in the U.S. in connection with animal rights crimes, one of which was for attempted murder in a failed bombing incident in Connecticut. Other examples of illegal incidents include:

- A fire set at a lab under construction at the University of California-Davis in 1987. Damages set at $4.5 million.
- A 1989 fire at the University of Arizona-Tucson. Damages estimated at $100,000.
- A 1989 break-in at Texas Tech University. Property damages estimated at $70,000.
- A Utah State University research facility and faculty member's office firebombed in 1992. Damages estimated at over $100,000.

Recognizing the severity of animal rights terrorism, the U.S. Congress passed the "Animal Enterprise Protection Act of 1992." This law makes certain crimes against biomedical research facilities, farms, ranches, and other entities that use animals, federal offenses. Since 1988, at least 30 state legislatures have approved similar measures to protect animal facilities.

Because of criminal activities perpetrated by animal rights extremists, already limited research dollars are being diverted to pay for expensive security procedures, to restore damaged or destroyed facilities and equipment, and to replace stolen research animals. A 1990 survey of the 126 U.S. medical schools conducted by the Association of American Medical Colleges to assess the cost of extremists' illegal activities concluded the following conservative estimates over a five-year period:

- 76 schools reported damages or other losses totaling in excess of $2.2 million as a direct result of demonstrations, break-ins, vandalism and other disruptive incidents.
- Construction delays and installation of security systems have cost schools approximately $6.8 million.
- A collective estimate of $17.3 million will be needed to meet annual ongoing expenses to pay for security needs, increased costs to purchase animals and to adopt stricter research regulations.

Not only is the cost of physical damages astronomical, but other repercussions resulting from terrorist attacks on research that are almost impossible to compute into dollar figures include:

- The immeasure loss of thousands of hours of actual research that is destroyed or harmed each time a break-in or fire occurs.
- The large number of scientists who have abandoned or will not pursue biomedical research using animals for fear of personal attacks. Animal extremists engage in bomb and death threats, picketing of homes, vandalism and other forms of harassment against many researchers.

The highest price resulting from animal research sabotage will ultimately be paid by the millions of people suffering from diseases and ailments for which there are no cures or treatments.

"Moderate" Animal Rights

While many animal rights groups claim that they do not condone violence as a way to end animal research, their lack of condemnation of such crimes is viewed by some as passive sanctioning of criminal activities. While claiming no involvement in these actions, members of People for the Ethical Treatment of Animals (PETA) have acted as spokespersons for the ALF. PETA has provided the press with "evidence of animal abuses" including videotapes of actual laboratory raids after illegal incidents have occurred. Many times these press conferences are held the next day and thousands of miles from the scene of the crime.

PETA, with an annual budget of $10 million and a membership of approximately 300,000, is seen as the largest, richest animal rights group in the country. The group was co-founded in 1981 by Alex Pacheco, who obtained employment as a research assistant in a Silver Spring, Maryland laboratory and charged that 17 monkeys being used in neurological research were being abused. A raid by police followed, the primary researcher was convicted and later cleared of animal cruelty charges, the animals were confiscated and the 10-year legal custody battle of the infamous "Silver Spring Monkeys" was launched. Mr. Pacheco stated in a January 13, 1989 *New Yorks Times* interview " . . . animals have the same rights as a retarded human child, because they are equal mentally in terms of dependance."

PETA co-director Ingrid Newkirk is quoted in a September 3, 1991 USA Today article stating, "I don't agree with physical harm, but I . . . won't condemn it." In an August, 1986 *Washingtonian Magazine* article entitled "Who Will Live, Who Will Die," Ms. Newkirk explains her animal rights position by stating "Animal liberationists do not separate out the human animal . . . so there is no rational basis for saying that a human being has special rights. A rat is a pig is a dog is a boy. They're all mammals."

Other groups claim that animal research is no longer necessary, does not benefit human health, and believe that prevention is the key to eliminating disease. Such rhetoric is accepted as truth by many because it is being delivered by so-called "medical professionals." One such group, Physicians Committee for Responsible Medicine (PCRM) believes that a "growing number" of physicians question the necessity of animal research. PCRM has been denounced by the American Medical Association " . . . for implying that physicians who support the use of animals in biomedical research are irresponsible, for misrepresenting the critical role animals play in research and teaching, and for obscuring the overwhelming support for such research which exists among practicing physicians in the United States." The AMA has reported that although PCRM ". . . professes to speak for organized medicine . . . only 2,000 of the group's 30,000 members are allegedly physicians."

When PCRM suggested that meat and dairy products be eliminated from the traditional four basic food groups because they ". . . are simply not necessary in the human diet," the AMA charged that those recommendations were ". . . irresponsible and potentially dangerous to the health and welfare of Americans." The AMA further stated that PCRM is "blatantly misleading Americans on a health matter and concealing its true purpose as an 'animal rights' organization."

Still other groups that are believed to espouse a "moderate" stand on the animal research issue—the elimination of cruelty to laboratory animals—actually engage in activities that promote radical human rights philosophies. For example, the Humane Society of the United States claims in its *Statement of Principles and Beliefs*, "It is wrong to use animals for medical, educational, or commercial experimentation or research unless the following criteria are met: absolute necessity; no available alternative methods; and no pain or torment caused the animals." However, in a 1991 fund-raising appeal, HSUS described " . . . the needless and repetitive experimentation on animals in the 'research' laboratory," as one of several " . . . well-known examples of animal cruelty and neglect . . . ."

Protests and Letters to End Research

Not all extremist groups resort to violence to voice their opposition to animal research. Demonstrations and letter-writing campaigns to Members of Congress are common tactics used to gain media attention to the cause of animal rights and to end animal research projects. In a December 1987 bipartisan poll of the House and

Senate, a Washington, DC grassroots communication service firm reported that the top three mail issues received on Capital Hill were opposition to Medicare and Social Security cutbacks, concern over the budget deficit and animal rights.

The U.S. military has long been a target of animal rights activists despite the fact that military-related research has contributed to the development of modern surgical techniques including aseptic procedures, skin grafting, burn care, cardiovascular operations and treatments for vision problems. A deluge of letters to congressional offices against research conducted by the military caused Congress in 1989 to suspend funding for two Department of Defense projects pending investigation into animal abuse charges. One of the projects, which involved cats to study head injury, was reviewed by the General Accounting Office and several research experts. Although the project received favorable reports and recommendations that the research continue, funding was never reinstated.

Another example of how effective animal rights activists are at ending research projects occurred in New York at Cornell University in 1988. After an intense picketing and letter-writing campaign by the group Trans-Species, a researcher at the university returned a $600,000 research grant to the funding agency, the National Institute on Drug Abuse (NIDA). The grant was awarded to a researcher who, for 14 years, had been studying the effects of drug addiction using animals. NIDA officials said that, except for a case in which a scientist died, it was the first time a research grant had been returned after a project had already been funded. It was also the first time an animal study had been stopped based on charges that the information derived from the project did not justify the use of animals, instead of the more frequently used charges of animal abuse.

Animal Rights and the Law

The animal rights movement is beginning to push its agenda through the U.S. court system. Animal rights-initiated lawsuits against federal agencies, research facilities, universities and individuals are on the rise and vary from simple to complex issues. The increased use of the legal system to promote animal rights has led to the establishment of The Animal Legal Defense Fund (ALDF), a group "comprised of attorneys and law students dedicated to the legal protection of animals' rights, providing services to individuals in the movement and to enhancing the status of animals through litigation and legal action."

Animal rights groups have filed several suits in federal court alleging that the U.S. Department of Agriculture (USDA) failed to enforce the Animal Welfare Act (AWA). Because the AWA does not provide a cause of action for private individuals to bring such suits, courts have rejected these cases for lack of standing. That led to the introduction of federal legislation to grant any person the right to sue for enforcement or failure to comply with the AWA. Although "standing" legislation was introduced in three congressional sessions, it was never approved.

The federal Administrative Procedure Act (APA) has also been used by activists to obtain desired changes in policy. Activist groups including the ALDF and the Humane Society of the U.S. have won two separate suits filed against the USDA, the Department of Health and Human Services (HHS) and the Office of Management and Budget (OMB) The same federal judge has ruled twice in favor of the activists—in 1992 to have rats, mice and birds be covered species under the AWA and, in 1993, to have certain sections of new animal welfare standards declared unlawful and be rewritten.

Other common lawsuits filed by animal rights activists include suits to gain access to state university animal care and use committee meetings, records and proceedings, and student suits against schools and universities for alternatives to projects which involve animals.

Legal actions brought on by animal rights activists have stalled research projects and diverted funding earmarked for research to pay for expensive, time-consuming litigation.

Fund Raising

The animal rights movement has been very successful in the area of fund raising. Under the guise of "helping the animals," and with the use of slick, glossy photographs of "tortured" animals, extremist groups appeal to the sentiments of unsuspecting donors and receive millions of dollars annually in support. Large portions of these funds are many times channeled into programs to make more money for the organizations, or to pay legal fees for activists who find themselves in trouble with the law.

Target campaigns against specific research projects generate huge profits, as illustrated by the already mentioned "Silver Spring Monkeys" case. Initially, the appeal by PETA was to donate money to have the monkeys sent to an animal sanctuary. Because of the many years the animals had spent in "legal limbo," their health had begun to deteriorate. It was the opinion of several veterinarians that the humane course of action for the remaining animals in poor health should be euthanasia, to avoid prolonging their suffering. Ironically, each time the animals were scheduled to be put to sleep, PETA filed legal motions in court to stop euthanasia procedures, and appeals to "Save the Silver Spring Monkeys" continued.

The July/August 1990 issue of *Who's Mailing What!*, a monthly newsletter of the Direct Marketing Archive reported on how animal rights groups use direct mail services to solicit funds. According to the article, "Animal Activism: The New Pornography, ". . . animal activist organizations are raising more than enough money—well over $300 million a year—to save all the wild animals in the world, take care of all the abandoned dogs and cats in this country, with plenty left over for the environment. Yet . . . a ton of this precious money is being squandered on fund raising, agencies, salaries, administration, as well as redundant, useless, destructive programs."

A 1990 *Wall Street Journal* article entitled "Imperfect Pitch: Organized Charities Pass Off Mailing Costs As 'Public Education'" said "Doris Day's Animal League enriches its direct mailer, retains little for its work." As so often happens with solicitations for charities, "a great deal of the money raised is spent to raise the money. . . . Last year, about 90 cents of every dollar spent by the animal league went to sending out more mail, most of it asking for more money."

Grand juries have convened in at least five states to investigate criminal activities against animal facilities. PETA directors Newkirk and Pacheco, along with several other individuals have been subpoenaed to appear in court. These investigations have prompted the latest round of fund raising pleas, including telephone calls to donors, for funds to help pay legal fees.

Conclusion

Unfortunately, the animal rights message often based on distorted and misleading information, is being heard and widely accepted as truth, and valuable research that could one day solve medical mysteries is being stopped. The animal rights movement poses an ominous threat to the future of public health. Animal activists have been successful in impeding the research process, raising the cost of conducting research, and are even making it dangerous for those scientists who have dedicated their lives to relieving human suffering. Former HHS Secretary Sullivan has described statements that animals are not necessary for biomedical research as "ludicrous."

1. The Animals' Agenda, April 1992, p. 23. Figures obtained from 1990 Form 990 filings which all nonprofit groups are required to file with I.R.S.
2. U.S. Congress, Office of Technology Assessment, *Technology Against Terrorism: Structuring Security*, OTA-ISC-511 (Washington, DC: U.S. Government Printing Office, January 1992)

Understanding (and Misunderstanding) the Animal Rights Movement in the United States

Adrian R. Morrison, D.V.M., Ph.D.
Program for Animal Research Issues
National Institute of Mental Health
National Institutes of Health
Room 17C-26, Parklawn Building
5600 Fishers Lane
Rockville, MD 20857 USA
Tel: 301-443-1639
FAX: 301-443-3225

and

Professor of Anatomy
School of Veterinary Medicine
Department of Animal Biology
Laboratories of Anatomy
University of Pennsylvania
3800 Spruce Street
Philadelphia, PA 19104-6045 USA
Tel: 215-898-8891
FAX: 215-573-2004

To be published in Animal and Human Experimentation, Proceedings of a symposium on ethical consideration regarding biomedical experimental methods and techniques, Antwerp, Belgium, September, 10–11, 1993, John Libbey Co. Ltd., London.

Humans have battled for good health for as long as one can see back into history. Their enemy, Disease, was cloaked in disguises unwittingly provided by humans themselves through their ignorance: the will of the gods or miasmas. The recognition at last in the 19th century that many health problems stemmed from invasion of the body by microbes provided a point of attack leading eventually to the development of preventive health measures and antitoxins, vaccines and antibiotics. As Lewis Thomas, the noted medical scientist and essayist, reminds us:

> "Everyone forgets how long and hard the work must be before the really important applications become applicable. The great contemporary achievement of modern medicine is the technology for controlling and preventing bacterial infection, but this did not fall into our laps with the appearance of penicillin and the sulfonamides. It had its beginnings in the final quarter of the last century, and decades of the most painstaking and demanding research were required before the etiology of pneumonia, scarlet fever, meningitis and the rest could be worked out. Generations of energetic and imaginative investigators exhausted their whole lives on the problem. It overlooks a staggering amount of basic research to say that modern medicine began with the era of antibiotics."[39]

Basic biological and behavioral research, much of it depending on studies using animals, has done much more, of course: anatomists, physiologists, biochemists, psychologists, parasitologists, geneticists, zoologists (the list could go on) have all contributed the bits of information underlying modern medical and surgical practice that have resulted in the healthiest generation in history. Just how healthy we are is brought home by Paton's description of medical practice little more than a half century ago:

"One no longer sees infants with ears streaming pus, schoolboys with facial impetigo, beards growing from heavily infected skin, faces pocked by smallpox or eroded by lupus, or heads and necks scarred from boils or suppurating glands. Drugs and a better diet have transformed haggard patients with peptic ulcers. The languid, characteristically brown-skinned case of Addison's disease of the adrenals; the pale, listless patients of chronic iron deficiency or pernicious anaemia; and the cretin or, conversely, the young woman with 'pop eyes' and overactive emotional behaviour—due respectively to thyroid deficiency or excess—are all being treated. The soggy hulk of a patient in the oedematous stage of chronic kidney disease is relieved by diuretics. As a result of polio vaccine and control of tuberculosis, we see few crippled children; as one walks behind a group of youngsters today, varied as ever in shape and size, the marvel is how straight their limbs and backs are. The chronic arthritics with their sticks are being replaced by septuagenarians swinging along on their plastic hips. The patients now are rare that once one saw dying from an infected mastoid, struggling for breath in the last stages of heart failure, or dying from appendicitis, leukaemia, pneumonia, or bacterial endocarditis."[22]

Unfortunately, too many of the general public do not understand how advances from these conditions and how practical applications in medicine in general depend upon basic research. Because of their training and experience, biomedical professionals can readily accept and understand Comroe's and Dripps' findings[3] that 41% of the papers reporting work judged to be fundamentally important to the 10 most important advances in cardiology were concerned with studies that sought knowledge for the sake of knowledge itself. Non-scientists, however, have no way of evaluating this important study. And this lack of understanding provides fertile ground for the animal rights movement to sow their seeds of discontent about the use of animals in basic biomedical research.

Furthermore, fewer and fewer people have contact with the animal world, at least as active users of animals other than keeping them as pets. The United States is probably an extreme example. Only 2% of the citizenery live on a farm compared with 50% in 1880.[35] Having little understanding of how the living and dying of various animals contribute to their own existence in so many ways, too many are susceptible to the fanciful view of animal life that animal rightists promote.

The leaders of this movement have taken advantage of another aspect of the public's character: the willingness of most people to accept that others are at least trying to act honorably most of the time, however imperfectly. Even though we are, at the same time, skeptics to varying degrees, confidence ("con") men still do remarkably well in our society. And so, through clever misrepresentation of biomedical research and its value to society, animal rights leaders (con men really) have been able to attract quite a few intelligent people to their cause.

The animal rights movement in the United States has had continuing success in spite of the many well-publicized atrocities committed by some of its adherents. The movement has thrived on the phenomena I have already noted: relative ignorance of the nature of science amongst the public (science illiteracy in other words), naivete concerning the natural world, and the seemingly limitless capacity for distorting the truth revealed by the leadership of the animal rights movement. Other factors have also contributed to animal rightists' success: clever selection of targets in the form of laboratories doing very invasive studies on animals for purposes not well-understood by people—head injury and behavioral research rather than cancer research for example—to maximize contributions from a compassionate public not wishing to see animals harmed; timidity and ineptitude of institutional administrators in responding to attacks; and indifference and fear among scientists.

These factors together have translated into a considerable amount of money coming into the hands of activists from contributors. The 1992 budgets displayed on Internal Revenue Service forms of the 26 most visible and active animal rights, protection and welfare groups totalled $577 million.[36] Because of takeovers of traditional welfare organizations by radical followers of the animal rights philosophy[37] and a melding of the two concepts into "animal protection" [see below], contributors may be fooled into supporting something in which they really do not believe. Individuals willing to engage in terrorist acts under the aegis of the Animal Liberation Front have also created a negative balance of millions of dollars by destroying research facilities, and stealing animals and data, which stimulates others to spend more on added security.[44]

Even as oversight of animal use in the laboratory has resulted in a level of housing and care that outdistances that afforded any other animals in the United States, the terrorist attacks and attempts at introducing more burdensome legislation continue. Because of the tremendous human good done by medical researchers, the continuing relentless attacks on various researchers investigating serious problems, such as head injury, are clear evidence of the misanthropy that drives the most forceful and radical segment of the animal rights movement. Also, as Nicoll[18] has noted, activists continue to press for local and state legislation blocking the release of animals from pounds for research even though those animals will eventually be killed anyway. This drives up costs as the supply diminishes, forces purchase of costly purpose-bred animals and, of course, impedes research.

Although misanthropy may explain part of the drive for more legislation, at least on the part of radical activitists, a genuine concern for animals and ignorance of the nature of scientific inquiry among many of our citizens are significant elements as well. A variety of organizations supported by such individuals are currently grouped under the banner of the "animal protection community." However, those organizations with conventional animal welfare concerns are finding even the most radical animal rightists crowding under that honorable banner, and some are finding that their organization and its assets have been taken over by radical animal rightists.[37] It would seem that reference to animal protection rather than animal welfare had to emerge in the literature of organizations uncomfortable with research on animals and of groups overtly hostile to it once biomedical researchers (and other animal user groups for that matter) demonstrated that they, too, had animal <u>welfare</u> concerns, contrary to what the animal rights movement had been saying about them. Of course, emphasizing a common cause, "animal protection," leads to a usefully confusing blending of a belief shared by most of us that animals have a "right" to good care when in our hands with the extreme view that rats are our moral equals.

Do not interpret this analysis to mean that I am belittling those who believe that one has no right to use animals in any fashion, however impractical I think that essentially religious philosophy to be. Herzog, a behavioral researcher who works with animals, performed a very interesting qualitative survey of 23 rank-and-file animal rights activists; and it is clear that they have strongly held beliefs and struggle to live their convictions consistently.[9] He points out that they are essentially "religious" fundamentalists believing firmly in their cause. Herzog reported an antagonism toward science, not surprisingly; but there was no discussion of whether these individuals would refuse medical treatment for the rest of their lives as do some religious sects nor whether they had concerns that their activities might lead to unnecessary suffering of their fellow humans if medical research were impeded by activist activity. Because those who lead the movement continue to misrepresent the benefits that have come to us and how that research is conducted, I think those who follow have an obligation to investigate the matter and, at the very least, reject lies as a foundation for sincere concerns.

Having read the results of Herzog's survey,[9] I am convinced that some <u>followers</u> in the animal rights movement could be persuaded that they should not prevent others from benefitting from medical research. If educated in the <u>process</u> of science—I shall return to this theme at the end of this essay—how could they, for example, deny children with cystic fibrosis the chance to enjoy a longer, more comfortable life as a result of the gene therapy becoming available thanks to animal research conducted during the past few years in particular? Could the philosopher Tom Reagan?[24]

Coming from the animal protection community, undoubtedly due to a variety of motives and insights with differing levels of scientific (and political) sophistication, are calls for support of only that research that will obviously benefit human health. Some have "the illusion that anyone is qualified to make scientific judgements on the value of research, thereby successfully moving the issue from the scientific realm to an arena where moral judgements and personal opinions (however uneducated) determine public policy."[16] For example, Rollin[25]—a professor of philosophy, not a scientist—sees the need for permitting only "patently beneficial research."[26] He continues: "Obviously, this criterion would exclude a great deal of current invasive animal research, ranging from much, if not most, product testing to military research, to a great deal of "knowledge-for-its-own sake research."[27] Rollin would require a cost (to animals)-benefit (to humans) analysis leading to funding decisions that would be made, not by scientific experts but by "panels—grand juries as it were—of intelligent interested citizens who would look at research proposals and decide if the benefits exceeded the costs, or if the question being asked was the sort that truly needed to be answered."[28] He does say he would provide them with expert technical advice. Rollin claims that this scheme would "integrate science and concern for science far more closely into the fabric of society and help accelerate scientific literacy."[29] This incredibly naive proposal indicates to me that a priority concern of scientists should be raising the level of science literacy of our society, but not as Rollin would do it. Practical mechanisms for educating people about biomedical research will be proposed later.

Productive discussions can be held with those who genuinely care about animals and agree to their use but have a limited understanding of science and a concern (fueled by the animal rights movement) for what <u>might</u> be going on in laboratories. On the other hand, Horton[11] has noted calls for reaching agreement between animal rights activists and scientists imply that each group is at an extreme when, in fact, scientists are doing society's work. As Goodwin and Morrison have put it:

> "No consensus can be reached between those who accept the use of animals for important human needs and those who categorically reject animal use based on a philosophy that views all forms of sentient life as morally equivalent. If the question fundamentally is <u>whether</u> to use animals, no middle ground exists. Use cannot be partly ethical, it is either ethical or unethical.
>
> People who genuinely want to discuss the matter with researchers must first disavow the animal rights position. To engage in middle-ground discussions with true believers in animal rights is, at best, fruitless; at worst, it provides animal rights activists with a tactic to shift the uninformed a little closer to the ultimate goal of total abolition of animal use. Only if one starts from the principle that the use of animals is acceptable does the issue of <u>how</u> animals are used address the spectrum of legitimate issues.
>
> One end of this spectrum holds that we should spare no expense to provide absolute comfort and procedural safeguards for experimental animals; a corollary of this extreme animal welfare position is that research goals have to be secondary to the absolute commitment to animal comfort. The other end of the spectrum holds that humane treatment should not be a consideration if it involves any extra expense or inconvenience—that animals can be treated essentially like lab equipment. Between these two extremes, the biomedical research community occupies the middle ground. We are engaged in legitimate, important debates about regulations vs. costs and about animal welfare vs. human needs. . . ."[6]

Without question, though, a disdain for humanity coupled with an elevation of animals to equal status motivates the most radical and, unfortunately, the most successful animal rights organizations in the United States (and elsewhere for that matter). Repeating a couple of the statements made by a couple of these organizations' leaders will make my point: "Even if animal research resulted in a cure for AIDS, we'd be against it,"[17] is a famous utterance of Ingrid Newkirk, National Director of People for the Ethical Treatment of Animals (PETA), the most virulent and dominant of the many radical, "above-ground" US groups. Chris DeRose of Last Chance for Animals admits: "A life is a life. If the death of one rat cured all diseases, it wouldn't make any difference to me."[4]

Presently, the movement is attacking all uses of animals: biomedical research, toxicity testing, education, food, fur coats, hunting.[13] The extension of attacks to include all uses of animals is a relatively recent development, though; for the most publicized attacks were against researchers and research and testing establishments beginning in the early 1980's. Nicoll and Russell[19] analyzed the animal rights literature and found that, relative to the actual numbers of animals used, an inordinate amount of space was devoted to concerns about use of animals in research and testing. There is a likely reason for this: Few people engage in or understand the value of biomedical research; whereas many enjoy eating meat. If one wishes to obtain funds to fuel a social movement, it would not make sense to attack something most do—eat meat for example—and expect those consumers to contribute to the demise of a favorite activity. Singer revealed that, in fact, there were political considerations:

> "American animal researchers are a smaller and politically less powerful group than American farmers, and they are based in regions where animal liberationists live. They therefore made a more accessible, and slightly less formidable opponent. . . ."[33]

Now that the movement is well-funded, though, we are seeing attacks on every conceivable use of animals in the United States.

Goodwin[5] has noted that the attack on animal use in biomedical and behavioral research focused on several secondary "stop research" arguments when the true agenda was something else: promotion of an ideology that is really anti-societal in that it effectively equates animals with humans:

1. Research is inherently cruel—For those who value humans above animals, the idea that working to alleviate human suffering is a cruel activity has no merit. It is an idea based on a misconception, that scientists are cold and calculating, caring little about the animals on which they work. Granted, even the most caring scientist can make mistakes or lack veterinary knowledge—no human activity is perfect. These failings are countered, though, by the presence, by law, of institutional animal care and use committees (IACUC) and laboratory animal veterinarians that oversee animal experiments.[16] Finally, the United States Department of Agriculture (USDA) reported in 1990[42] that more than half (58 percent) of the experiments actually involved no pain or distress at all, 36 percent required analgesics or anesthetics and only 6 per cent involved study of pain mechanisms so that chemical pain relief was not possible (USDA). Experiments on pain are very carefully scrutinized by institutional committees, granting agencies and journals; and they can often be designed to permit the animal to escape intolerable pain.

2. Research is wasteful and duplicative—This argument reveals a lack of common sense and understanding of the process of science. What highly trained individual seeking recognition from his peers would pursue trivial or redundant research? At the same time, replication of findings by other scientists establishes the generality of one's own. With funding levels for approved grants from the National Institutes of Health now well below 20 per cent, uselessly redundant research has little chance of survival.

3. Research diverts funds that could go into prevention or treatment—For those who receive complaints from the public about animal

research, this is a frequently heard refrain. Actually, for every $100 spent on health care in the United States, only 40 cents is spent on animal research.[5] Furthermore, more than half of the grants awarded by the National Institutes of Health are given for studies on people or for those using methodologies not requiring animals (Personal communication, L Sibal, Office of Laboratory Animal Research, National Institutes of Health).

4. Alternative methods are now available, making animal research unnecessary—Another suggestion reflecting a profound lack of understanding of science—perhaps leaders of the movement are just being deceitful—is that modern science has passed animal research by, because other technologies have rendered animal use unnecessary. It is often stated that use of tissue cultures, computers and human clinical cases are sufficient to unravel the mysteries of disease. This claim, of course, does not take into account the complexities of the functioning organism. Also, one should not forget that scientists are the individuals who have devised the various alternative methods used in order to address scientific questions in the most appropriate way; they are not wedded to the idea of using animals.

The movement has also argued that biomedical research endangers pets. Of course, 90 per cent of the approximately 20 million animals[41] used in the United States are purpose-bred rats and mice and cannot possibly be stolen pets. The USDA licenses dealers who buy dogs and cats from those who raise them, or from pounds and shelters, where, in fact, several million abandoned or feral animals are killed in the latter each year. Although it is possible that some wanted but unclaimed pets could slip through this "safety net," the number would have to be small. There is also the possibility of false charges from the animal rights community: In the spring of 1993, a known activist charged a major university with having her pet dog in their animal quarters. In the presence of a USDA veterinarian, she pointed out the dog; however, it did not respond to her loving embrace. After comparing the dog and her pet's medical papers, the veterinarian pointed out that the dog in the facility was an intact male and that the medical records of her pet indicated it had been castrated! (Personal communication, anonymous source).

The over-riding principle guiding animal rights activists is that use or "exploitation" of animals for human benefit is nothing less than "speciesism," an idea generally identified with the Australian philosopher, Peter Singer, because of the popularity of his book Animal Liberation. Speciesism is "a prejudice or attitude of bias in favor of the interests of members of one's own species and against those of members of other species."[34] This notion makes no sense to a biologist.

I look at the problem in a practical way: Animals use other animal (and plant) species in the over-all scheme of things, and we are animals. Thus, in my opinion, we are also permitted to use other species. This doing-what-comes-naturally argument might appear to be weakened because we recognize right from wrong, and other animals do not operate in this framework. The argument seems to be that consequently we are obligated to act more nobly. In fact, we do. We are the ones who pass anti-cruelty laws. Yet, through our intelligence we can project our lives well into the future so that taking many rats' or dogs' lives because we can foresee a benefit to ourselves or our young seems no different from or less supportable than a cat's killing several animals to give herself energy to feed her young. Both activities meet the needs of survival. Viewed this way, our intelligence makes us no more culpable. One must ignore the realities of Nature and the power of natural impulses to conclude otherwise.

Cohen turned the concept of "speciesism" on its head in a way with which clearly agree:

> "I am a speciesist. Speciesism is not merely plausible; it is essential for right conduct, because those who will not make the morally relevant distinctions among species are almost certain, in consequence, to misapprehend their true obligations. The

> analogy between speciesism and racism is insidious. Every sensitive moral judgement requires that the differing natures of the beings to whom obligations are owed be considered. If all forms of animate life—or vertebrate animal life?—must be treated equally, and if therefore in evaluating a research program the pains of a rodent count equally with the pains of a human, we are forced to conclude (1) that neither humans nor rodents possess rights, or (2) that rodents possess all the rights that humans possess. Both alternatives are absurd. Yet one or the other must be swallowed if the moral equality of all species is to be defended.[2]

Vance[46] cautions that Singer and the other major philosopher of the animal rights movement, Tom Regan, cannot be brushed aside as promoting emotional arguments in favor of animals. Both present the rational arguments typical of analytical ethics although their arguments differ radically. Singer is a utilitarian arguing that animals should be accorded equal consideration because they equally feel pain and that one should act to bring about the best balance of good and bad consequences; while Regan holds that animals have inherent value and ultimately reaches the conclusion that they have moral rights, which affords them protection from experimentation. In addition to presenting quite different philosophical reasons for objecting to harming animals for human benefit, Regan and Singer presented their theses in quite different books. Except for a few sentences explicitly stating the usual distorted descriptions of medical research obtained from the animal rights literature, Regan's book, The Case of Animal Rights,[24] is a purely philosophical treatise. Singer, however, devotes more than half of Animal Liberation to proselytizing for animal liberation (not animal rights) and describing the evils of research, toxicity testing and agriculture—often out of context, even inaccurately, or ignoring new developments in animal care, as Katz,[12] Nicoll et al.,[21] Russell and Nicoll[31] and Verhetsel[42] have reported.

Vance[46] notes that additional work is under way in the philosophical community to shore up weaknesses in Regan's and Singer's thinking and that we should prepare ourselves for more carefully reasoned arguments. I am certain, however, that the most scholarly defense of the proposition that humans should, in general, not be afforded any more consideration than any other animal will not sway the vast majority of people. I am confident that the common man has more common sense.

We need to focus our efforts on three things: (1) assuring the public that we are caring for animals as well as we can while doing well-conceived research; (2) revealing the distortions of our literature by the more radical in the movement; and (3) educating the public, particularly young students, in the process of science. The remainder of this essay will address these items.

(1) The care of laboratory animals in the United States has improved over the years as researchers (and society in general) became more sensitive to animal needs and the specialty of laboratory animal medicine (and veterinary medicine itself) matured. Conditions were uneven, though, until a set of standards was mandated by law in 1985. Now, institutions receiving federal funds must have an attending veterinarian and a committee, the IACUC mentioned above, which approves all proposals for experiments using animals before they can begin, and oversees that use via semi-annual inspections. Public Health Service policy[43] states that the IACUC must have five members including a veterinarian experienced in laboratory animal medicine, a practicing scientist experienced in animal research, a person whose primary interest is non-scientific, and an individual with no ties to the institution. The USDA has similar requirements, and using their own inspectorate periodically inspects any facility using animals (excluding rats and mice), not just those receiving federal funds.[16]

Animal activists have made much of the fact that the USDA does not inspect rodent facilities even though rats and mice comprise 90 per cent of the 20 million animals used in research, but these animals are covered under Public Health Service policy.[45] Both the Office of Protection from Research Risks of the Public Health Service and the Regulatory Enforcement and Animal Care division of the Animal and Plant Health Service of the USDA have the power to investigate complaints against laboratories, demand corrective actions and invoke sanctions—and they use it. The system is working well, and researchers should tell the public so while striving to handle their animals as humanely as possible.

Researchers should also reveal their own ambivalence about working with animals. Herzog[7,8,10] has written a series of thoughtful articles that address this issue and the paradoxes that inevitably arise when putting animals to any human use. All the researchers I know are uncomfortable with having to harm animals, particularly when one has to kill an animal that one has studied over an extended period or when large numbers must be killed for a biochemical analysis. I have described my own feelings, which have been printed in newsletters of organizations concerned with protecting biomedical research and researchers: the incurably ill For Animal Research (iiFAR), National Animal Interest Alliance and Putting People First. Many scientists probably share my thinking:

> I believe animal use by humans is natural and no less appropriate in the scheme of things than animal use by other animals. Therefore, I reject as nonsense the notion of speciesism that the animal rights movement promotes. It is a perversion of biology, not a principle. Vivisection fits into the category of appropriate uses; for it is a function of a legitimate aspect of our nature, which is to explore and increase knowledge. Indeed, increasing knowledge in all spheres, even if it requires the deaths of some animals, is our obligation; for we are the only species capable of undoing "harm" to the planet caused by us. This is not to say that the invasiveness of vivisection is something I prefer or participate in without reflection. Because I do experimental surgery, I go though a soul-searching every few months, asking myself whether I really want to continue working on cats—my family has had up to four of them at a time as pets over the years—or other animals for that matter. The answer is always "yes" because I know that there is no other way for medicine to progress but through animal experimentation and that basic research leads ultimately to unforeseen benefits.
>
> My passion is to alleviate human misery; it takes precedence over everything. The animal rights movement, on the other hand, focuses on pain/suffering in "sentient beings" in the immediate present. Certainly pain should not be inflicted wantonly or left unalleviated if at all possible. Yet, we humans suffer the same and much more; there lies the difference. What animal can match in suffering the heartbreak of parents, who lose a child to illness or who have given birth to a child with severe birth defects, or the despair of a teenager who learns that life in the future will be incomplete as a result of the car accident that severed his spinal cord? Even chimpanzees cannot participate in the grief of others.[1] We can, even when learning of a tragedy in the newspaper. This makes us "special."

(2) The weakest point the animal rights movement has from an ethical standpoint is their willingness to lie and to misquote the scientific literature. The worst offenders in my eyes are medical professionals, who appear to have no compunction about engaging in this practice, fudging the data as it were.

Students gasp when I show a slide from a book[23] by Brandon Reines, a veterinarian, claiming that a scientist who developed a heart-lung machine for open-heart surgery did so using only human patients and then follow it with a slide revealing the full quotation[48] in which the scientists describe their use of dogs in a paragraph omitted from the center of Reines's quotation and replaced by ellipses. Thus, Reines quotes the researcher as follows:

"Lillihei himself says, 'We were feeling our way along as far as flow was concerned. . . . Then we got more experience with our heart-lung machine and it worked for all sizes of people.'[23]

Reines then continues: "Thus, Dr. Lillihei and his associates made the heart-lung machine safe and effective during the course of practice on actual patients—-not from studies on healthy laboratory animals."[23]

Audacious examples like this abound. Verhetsel[47] has reported many more misquotations from Reines's books and others often used to support animal rightists' claims, such as Ruesch's Slaughter of the Innocent[30] and Ryder's Victims of Science: The Use of Animals in Research[32] that try to present a negative image of medical research. Articles by Nicoll and Russell[20] have exposed the questionable practices of Neal Barnard, a psychiatrist who is President of the Physicians Committee for Responsible Medicine, an affiliate of PETA, as well as those of Singer,[21] noted above. Miller[15] has revealed additional misrepresentations of the literature by a psychologist[40] who has turned away from animal research, and by members of the Medical Research Modernization Committee, another group containing health professionals who reject the idea that animal research has been necessary for medical progress and twist the facts to promote their cause.

A useful exercise for students would be to have them read the original writings of scientists and then compare their conclusions with the quotations and interpretations found in the animal rights literature.

(3) Even though most of the public accepts that continued animal research and testing is necessary to ensure the medical progress from which they want to benefit, they are naturally uncomfortable with the idea that harm does come to other creatures, as are we scientists. They want that harm minimized. Again, so do we. Lacking a sophisticated understanding of the process of science, however, the public is prone to accept the idea that tighter and tighter control on animal use can only be a good thing. Rollin[25] predicts this as the wave of the future.

We need to help people learn how science is really done. Any experienced scientist knows that new knowledge comes with difficulty and that luck and insights obtained from unexpected sources play important roles. Of course, a project should start with an hypothesis formulated by a well-trained, well-read individual. It should be pursued, though, with a mind open to the very real possibility that the hypothesis will have to be modified or even abandoned along the way. Observations from a myriad of studies conducted in this manner eventually merge in unexpected ways that lead to a real medical advance. Thomas[39] sketched out this process in describing the development of antibiotics, and Comroe and Dripps[3] proved the value of basic research in their anlysis of the cardiology literature. One wonders what advances Rollin's "citizen panels" would have blocked by stopping "knowledge-for-its-own-sake research."

More of us should undertake exercises like that of Comroe and Dripps, but in a way that would make them accessible to students. Miller[14] has called for this and has written as a prologue to a textbook a review based on the excellent book by science historian Judith Swazey[38] on the development of chlorpromazine as a relief for schizophrenia. He compares the research to the assembling of a complex puzzle from many parts, involving study of the chemical structure of dyes and investigation of the fungus, ergot, and one of its components, histamine, and then antihistamines. Further study of a blue dye effective against malaria led to the development of new antihistamines, all producing drowsiness. A medical treatment for schizophrenia eventually resulted—but not without the additional help of developments in animal learning theory and behavioral tests useful for examining drug effects.

Those of us who have spent years in research have a story to tell even if we are not professional historians like Swazey. Thinking back on a career in sleep research, I marvel at how inspiration from a book outside my field, discussion of a reading with an undergraduate student, an accidential observation by another student, and even the experience of being attacked by the Animal Liberation Front eventually led me to investigate the pathophysiology of post-traumatic stress disorder. It is an interesting story and one I pledge to tell. I hope some of the readers of this chapter will consider doing the same.

1. Cheney, D.L. and Seyfarth, R.M. (1990): How monkeys see the world. Chicago: University of Chicago Press.
2. Cohen, C. (1986): The case for the use of animals in biomedical research. New Eng. J. Med. 315, 865–870.
3. Comroe, J.H., Jr. and Dripps, R.D. (1976): Scientific basis for the support of biomedical science. Science 192, 105–111.
4. DeRose, C. (1989): Los Angeles Times, April 12: E12.
5. Goodwin, F.K. (1992): Animal rights: medical research and product testing. Is this a "hang together or together we hang" issue? Contemp. Topics in Lab. Animal Sci. 31, 6–11.
6. Goodwin, F.K. and Morrison, A. R. (1993): In animal rights debate, the only valid moderates are researchers. The Scientist 17 (Sept. 6), 12.
7. Herzog, H.A., Jr. (1988): The moral status of mice. Amer. Psychol. 43, 473–474.
8. Herzog, H.A., Jr. (1991): Conflicts of interest: kittens and boa constrictors, pets and research. Amer. Psychol. 46, 246–248.
9. Herzog, H.A., Jr. (1993): "The movement is my life": the psychology of animal rights activism. J. Social Issues 9, 103–119.
10. Herzog, H. (1993): Human morality and animal research: confessions and quandries. Amer. Scholar. 62, 337–349.
11. Horton, L. (1989): The enduring animal issue. J. Nat. Cancer Inst. 81, 736–743.
12. Katz, L.S. (1993): How to attack animal agriculture with a pinch of fact and a pound of hyperbole. Coalition for Animals and Animal Research (CFAAR) Newsletter, P.O. Box 8060, Berkeley, CA 94707-8060 USA, 5/1, 30–31.
13. Marquardt, K., Levine, H. M. and LaRochelle, M. (1993): Animalscam: the beastley abuse of human rights. Washington, D.C.: Regnery Gateway.
14. Miller, N.E. (1985): Prologue. In Psychology and life, 11th edn, ed P. G. Zimbardo, pp. 1–vii, Glenview, Illinois: Scott Foresman.
15. Miller, N.E. (1991): Commentary on Ulrich: need to check truthfulness of statements by opponents of animal research. Psychol. Sci. 2, 422–424.
16. Morrison, A.R. (1993): Biomedical research and the animal rights movement: a contrast in values. Am. Biol. Teacher 55, 204–208.
17. Newkirk, I. (1989): Vogue, Sept.: 542.
18. Nicoll, C.S. (1991): A physiologist's views on the animal rights/liberation movement. The Physiologist 34, 303–315. (5 pp).
19. Nicoll, C.S. and Russell, S.M. (1990): Editorial: Analysis of animal rights literature reveals the underlying motives of the movement: ammunition for counter offensive by scientists. Endocrinol. 127, 985–989.
20. Nicoll, C.S. and Russell, S.M. (1992): Commentary: animal rights, animal research, and human obligations. Molec. Cell Neurosci. 3, 271–277.
21. Nicoll, C.S. and Russell, S.M. and Lau, A. (1992): Letter to the editor. The New York Review of Books, Nov. 5, 59–60.
22. Paton, W. (1993): Man and mouse: animals in medical research, 2nd edn, p. 93, Oxford: Oxford Univ. Press.
23. Reines, B. (1985): Heart research on animals: a critique of animal models of cardiovascular disease. Jenkintown, Pennsylvania: The American Anti-Vivisection Society, p. 52.
24. Regan, T. (1983): The case for animal rights. Berkeley: University of California Press.
25. Rollin, B.E. (1991): Some ethical concerns in animal research: where do we go next? In Animal experimentation: the moral issues. ed R.M. Band and S.E. Rosenbaum, pp. 151–158, Buffalo New York: Prometheus Books.
26. Ibid p. 154.
27. Ibid p. 154–155.
28. Ibid p. 156.
29. Ibid p. 156.
30. Ruesch, H. (1985): Slaughter of the innocent. Klosters, Switzerland: Civitas Publ.
31. Russell, S.M. and Nicoll, C.S. (1993): A dissection of Peter Singer. Manuscript to be published.

32. Ryder, R.D. (1983): Victims of science: the use of animals in research. London: National Anti-Vivisection Society.

33. Singer, P. (1990): Animal liberation, 2nd edn, p. 6, New York: Random House.

34. Singer, P. (1989): Unkind to animals. The New York Review of Books, Feb. 2, p. 37.

35. Strand, R. and Strand, P. (1993): The hijacking of the humane movement, p. 2, Doral Publ.: Wilsonville, Oregon.

36. Ibid, p. 55.

37. Ibid, p. 71.

38. Swazey, J.P. (1974): Chlorpromazine in psychiatry: a study of therapeutic innovation. Cambridge, Massachusetts: MIT Press.

39. Thomas, L. (1974): The lives of a cell, p. 117, New York: Viking Press.

40. Ulrich, R.E. (1991); Commentary: animal rights, animal wrongs and the question of balance. Psychol. Sci. 2, 197–201.

41. United States Congress Office of Technology Assessment (1986): Alternatives to animal use in research, testing and education. Washington, D.C.: US Government Printing Office, OTA-BA-73, Feb, p. 5.

42. United States Department of Agriculture (1990): Animal welfare enforcement, Fiscal Year 1990. Washington, D.C.: US Department of Agriculture.

43. United States Department of Health and Human Services. (1986): Public health service policy on humane care and use of laboratory animals. Bethesda, Maryland: National Institutes of Health, Office of Protection from Research Risks.

44. United States Departments of Justice and Agriculture (1993): Report to Congress on the extent and effect of domestic and international terrorism on animal enterprises.

45. United Sates Public Health Service (1992): Protecting laboratory animals (USPHS policy statement). Obtainable from author.

46. Vance, R.P. (1992): An introduction to the philosophical presuppositions of the animal liberation/rights movement. J. Amer. Med. Assn. 268, 1715–1719.

47. Verhetsel, E. (1986): They threaten your health: a critique of the antivivisection/animal rights (AV-AR) movement. Tucson, Arizona: Nutrition Information Center.

48. Wertenbaker, L. (1980): To mend the heart, p. 155, New York: Viking Press.

The World Medical Association, Inc.
September 1989

World Medical Association Statement on Animal Use In Biomedical Research

Adopted by the 41st World Medical Assembly

Preamble

Biomedical research is essential to the health and well-being of every person in our society. Advances in biomedical research have dramatically improved the quality and prolonged the duration of life thoughout the world. However, the ability of the scientific community to continue its efforts to improve personal and public health is being threatened by a movement to eliminate the use of animals in biomedical research. This movement is spearheaded by groups of radical animal rights activists whose views are far outside mainstream public attitudes and whose tactics range from sophisticated lobbying, fund raising, propaganda and misinformation campaigns to violent attacks on biomedical research facilities and individual scientists.

The magnitude of violent animal rights activities is staggering. In the United States alone, since 1980, animal rights groups have staged more than 29 raids on U.S. research facilities, stealing over 2,000 animals, causing more than 7 million dollars in physical damages and ruining years of scientific research in the process. Animal activist groups have engaged in similar activities in Great Britain, Western Europe, Canada and Australia. Various groups in these countries have claimed responsibility for the bombing of cars, institutions, stores, and the private homes of researchers.

Animal rights violence has had a chilling effect on the scientific community internationally. Scientists, research organizations, and universities have been intimidated into altering or even terminating important research efforts that depend on the use of animals. Laboratories have been forced to divert thousands of research dollars for the purchase of sophisticated security equipment. Young people who might otherwise pursue a career in biomedical research are turning their sights to alternative professions.

Despite the efforts of many groups striving to protect biomedical research from animal activism, the response to the animal rights movement has been fragmented, underfunded, and primarily defensive. Many groups within the biomedical community are hesitant to take a public stand about animal activism because of fear of reprisal. As a result, the research establishment has been backed into a defensive posture. Its motivations are questioned, and the need for using animals in research is repeatedly challenged.

While research involving animals is necessary to enhance the medical care of all persons, we recognize also that humane treatment of research animals must be ensured. Appropriate training for all research personnel should be prescribed and adequate veterinary care should be available. Experiments must comply with any rules or regulations promulgated to govern human handling, housing, care, treatment and transportation of animals.

International medical and scientific organizations must develop a stronger and more cohesive campaign to counter the growing threat to public health posed by animal activists. Leadership and coordination must be provided.

The World Medical Association therefore affirms the following principles:

1. Animal use in biomedical research is essential for continued medical progress.

2. The WMA Declaration of Helsinki requires that biomedical research involving human subjects should be based on animal experimentation, but also requires that the welfare of animals used for research be respected.

3. Humane treatment of animals used in biomedical research is essential.

4. All research facilities should be required to comply with all guiding principles for humane treatment of animals.

5. Medical Societies should resist any attempt to deny the appropriate use of animals in biomedical research because such denial would compromise patient care.

6. Although rights to free speech should not be compromised, the anarchistic element among animal right activists should be condemned.

7. The use of threats, intimidation, violence, and personal harassment of scientists and their families should be condemned internationally.

8. A maximum coordinated effort from international law enforcement agencies should be sought to protect researchers and research facilities from activities of a terrorist nature.

Public Veterinary Medicine: Food Safety and Handling

Euthanasia and slaughter of livestock

Temple Grandin, Ph.D.

From the Department of Animal Sciences, College of Veterinary Medicine, Colorado State University, Fort Collins, CO 80523. J. Amer. Vet. Med. Assoc. JAVMA, Vol 204, No. 9, May 1, 1994.

Euthanasia is defined as a humane death that occurs without pain and distress.[1] Physical methods of euthanasia are used in slaughter plants and in clinical practice. These methods are unpleasant to watch, but humane when used properly. In some cases, the most humane euthanasia method for an injured cow or horse is by gunshot, because fear and anxiety are minimized.

During the past 20 years, the author has worked in more than 100 slaughter plants as a consultant for improving handling techniques and equipment used for slaughter in an effort to improve the humaneness of slaughter. When slaughter is performed properly, it is euthanasia; however, when it is performed improperly, animal suffering may result. There is a tremendous need for veterinarians to increase their knowledge of physical methods of euthanasia and humane slaughter. Observations by the author at slaughter plants indicate that some veterinarians lack such practical and scientific knowledge. It is the responsibility of veterinarians to enforce the Humane Methods of Slaughter Act of 1978.[2] The purpose of this report is to provide veterinarians with practical and scientific information on the use of captive bolt guns, electrical stunning, CO_2 anesthesia, and ritual slaughter.

Captive Bolt and Gunshot

Gunshot and penetrating captive bolt are acceptable methods of euthanasia, according to AVMA guidelines.[1] The captive-bolt gun or firearm must be aimed at the correct location on the animal's forehead. It should not be aimed between the eyes. The hollow behind the poll also should be avoided, because it is less effective than the forehead position.[3] In sheep, the shot must be aimed at the top of the head because the front of the skull is thick. A captive-bolt gun must be placed firmly against the skull and a firearm must be held 5 to 20 cm away from the skull. Penetrating captive-bolt guns that are actuated by a blank cartridge can be obtained.[a] Practical experience has shown that a 22-caliber firearm is sufficient for cattle and horses. A larger caliber should be used on large bulls, boars, or buffalo.

A captive-bolt gun kills the animal by concussive force and penetration of the bolt.[4,5] A nonpenetrating, mushroom-head captive bolt only stuns the animal, thus, it cannot be used as the sole method of euthanasia.[1] A nonpenetrating captive bolt must be followed by an adjunctive method, such as exsanguination.[1] The gun must be carefully cleaned and maintained to achieve maximal hitting power. Observations by the author in slaughter plants indicate that poor gun maintenance is a major cause of poor stunning, in which more than one shot is required. In large, high-speed plants, maintenance personnel must be dedicated to servicing captive-bolt guns.[6] Many large slaughter plants use pneumatic powered captive-bolt guns. To obtain maximal hitting power, the gun must be supplied with sufficient air pressure and air volume per the manufacturer's recommendations.

When a standing animal is shot, it should instantly drop to the floor. In cattle, the neck contracts in a spasm for 5 to 10 seconds. Hogs have violent convulsions. Observations by the author indicate that this can occur even when the brain and part of the spinal cord have been destroyed. These are normal reactions. Rhythmic breathing must be absent and the animal must not moan, bellow, or squeal.[7] All eye reflexes should be absent.[7] Gasping or gagging reflexes are permissible because they are signs of a dying brain.[7] Within 10 seconds, the neck and head should be completely relaxed. In a clinical situation and in a slaughter plant, the animal's limbs may make uncoordinated movements for several minutes.

Electric Stunning and Electrocution

Electric stunning is not recommended for use in research facilities or on farms. Stunning and euthanasia by electrocution is acceptable only conditionally, because special skills and equipment are required.[1] Restraint equipment is required to hold the animal so that the electrodes can be placed in the correct position. Use of an electrical cord plugged into 115 V house current to kill piglets or other livestock is not acceptable. Piglets less than three weeks old should be euthanized by appropriate administration of pharmaceutical agents or by application of blunt trauma to the forehead.[8] Captive bolt or gunshot can be used on older pigs.

Properly applied electrical stunning in a slaughter plant induces instantaneous unconsciousness, and is approved under the Humane Methods of Slaughter Act of 1978.[2] Sufficient amperage must be applied through the brain to induce a grand-mal epileptic seizure.[9,10] The animal's brain must be in the current path between the two electrodes.

Hogs in small locker plants or meat science laboratories often are stunned with head-only reversible electric stunning. To prevent return to sensibility, the animal must be exsanguinated within 30 seconds,[10] and some researchers recommend 10 to 15 seconds.[11] The author has observed that delays between reversible head-only stunning and exsanguination are common in small locker plants because the hoist is too slow. Inadequate equipment causes severe welfare problems.

In large pork and sheep slaughter plants, animals are held in a conveyor restrainer, and stunning that induces cardiac arrest (cardiac-arrest stunning) is used. The interval between stunning to exsanguination is less critical. One electrode is placed on the forehead or in the hollow behind the ears, and the second electrode is placed on the back, side of the body, or forelimb. In pigs and sheep, this method simultaneously induces instantaneous insensibility and cardiac arrest.[10,12] The head electrode must not be placed on the neck, because failure to induce an epileptic seizure causes suffering.[12] Properly and poorly stunned animals look similar, because cardiac arrest masks the clinical signs of the grand-mal seizure.[12] The animal may be conscious, but appears to be properly stunned.

When cardiac-arrest stunning is used, electrode positions and electrical settings must be verified by measures of electrical or neurotransmitter activity in the brain.[9,13] Only scientifically verified electrode positions and electrical settings should be used. Visual assessment must not be used to verify new settings for cardiac-arrest stunning. New settings must be verified by scientific research to ensure that instantaneous insensibility occurs.

For large (108 kg) market-weight hogs, a minimum of 1.25 A at 300 V for one second should be used.[10] For slightly smaller pigs, the voltage can be dropped to 250 V.[7] For sheep, a minimum of 1 A at 375 V for three seconds should be used.[14] In New Zealand, electrical stunning is being successfully used in cattle.[15-17] Electrical stunning of cattle requires the use of a restraint device to hold the head.[17] Unlike pigs or sheep, cattle must have a stunning current (2.5 A) passed through the brain before the head-to-body cardiac-arrest current is applied.[17] In one study,[15] a single, 400 V 1.5 A current passed from the neck to the brisket failed to induce epileptic form changes in the electroencephalographic recordings. Equipment manufacturers have found that a minimum of 400 to 450 V is required to achieve insensibility. Practical experience indicates that greater amperages and voltages are required to achieve insensibility than to achieve cardiac arrest.

Amperage (current) is the most important factor in inducing unconsciousness.[9] Modern stunning circuits use a constant amperage power source, and voltage (electrical potential) varies with animal resistance. Some slaughter plants attempt to reduce meat quality defects by lowering the amperage or using high frequencies. This must not be permitted. Petechial hemorrhages can be minimized by the use of a constant amperage power supply.[18] Electrical frequencies of over 200 Hz must not be used unless they are scientifically verified.[9] In one report, a frequency of 500 Hz failed to induce unconsciousness. Most electrical stunning devices operate at 50 to 60 Hz.

Assessment of Stunning Efficacy in Slaughter Plants

Head-only reversible electric stunning of pigs and sheep causes an initial spasm (tonic phase), which lasts approximately 10 seconds. After the tonic phase, kicking begins (clonic phase).[9] Animals stunned with head-only reversible stunning kick more vigorously than animals in cardiac arrest. Animals are unconscious during the tonic and clonic phases. After a stunned animal is hung upside-down prior to exsanguination, the field methods for verifying insensibility are similar for captive bolt and electric stunning. Eye reflexes and blinking must be absent. In electrically stunned animals, eye reflexes should be checked 20 to 30 seconds after stunning. Prior to this time, eye reflexes are masked by the epileptic seizure. Veterinarians also need to check amperages, voltages, and electrode positioning to ensure that the stunner is being operated correctly.

In animals that have been shot with a captive bolt or stunned electrically, the limbs may move. Random limb movement should be ignored, but a limb that responds vigorously in response to a stimulus is a possible sign of return to sensibility. After the animal is hung on the overhead rail, the head must hang straight down and the neck should be limp. The tongue should hang out and the ears should droop down. Gasping and gagging reflexes are permissible,[7] but rhythmic breathing and vocalization must be absent. The animals must not have an arched-back righting reflex. Fully conscious animals suspended upside-down arch their backs in an attempt to lift their heads.

Carbon Dioxide Stunning

Carbon dioxide stunning is an approved method for inducing insensibility under the Humane Methods of Slaughter Act of 1978.[2] The AVMA states that CO_2 is an acceptable method of euthanasia, but other methods are preferable because large animals, such as swine, appear to be more distressed than small laboratory animals. In both pigs and human beings, there is great variability in reactions to CO_2.[19-23] Swedish Yorkshire pigs react well to CO_2, and the motoric excitation phase begins after the electroencephalograph, indicating second-stage anesthesia.[24] Hoenderken[10] found that the excitation phase started before the pig was unconscious.[10] Visual assessment by other investigators has shown that halothane-positive pigs have a greater amount of excitation than halothane-negative pigs.[21] Administration of halothane is used as a test to detect pigs that have porcine stress syndrome. Such pigs react to halothane by becoming rigid. There is concern that some of the pigs sensitive to halothane may be conscious during an initial excitation phase.[21] Observations by the author revealed that some pigs quietly lost consciousness when exposed to CO_2, whereas other pigs violently struggled when they first sniffed the gas. The encephalographic measurements that have been performed on the Swedish Yorkshire breed should be performed on pigs of various breeds that are sensitive and nonsensitive to halothane. At this time, available research data suggest that CO_2 is a good euthanasia method for certain genetic types of pigs, but may possibly cause discomfort in others.

To reduce excitation during anesthetic induction, pigs should be rapidly exposed to 80 to 90% CO_2.[21] Veterinarians at slaughter plants should monitor CO_2 concentrations because plant management may be tempted to lower concentrations to save money. Observations in the field have indicated that pigs that walk quietly into the CO_2 chamber have a milder excitation phase than agitated, excited pigs. A new CO_2 stunning system in Denmark appeared to greatly reduce excitement and squealing during handling because groups of five pigs were moved into the chamber at one time. The author observed that these pigs had little reaction when they first contacted the gas, and the motorific excitation phase appeared to occur after they became unconscious. One possible explanation for this observation is that Denmark has a low prevalence of pigs sensitive to halothane.

Preslaughter Stress

Properly performed slaughter induces cortisol concentrations equal to or less than that induced by on-farm handling and restraint when a captive bolt is used.[25-29] When preslaughter handling is performed properly, cattle should move through the chute at a slow walk and calmly enter the stunning area without balking. To reduce stress, cattle should be stunned immediately after they enter the stunning box or restrainer. Cattle should not have signs of visible agitation, such as bellowing or rearing. In a well-designed handling system, the author has been able to move 8 of 10 cattle into the stunning pen or restrainer without use of an electric prod.

Findings in studies[28-30] have indicated that cortisol concentrations can double or triple when cattle slip on slick floors, are restrained in poorly designed equipment, or are over-prodded. When this occurs, cortisol concentrations may greatly exceed on-farm handling concentrations. Epinephrine and norepinephrine are of limited value for evaluating preslaughter stress, because electric and captive-bolt stunning trigger massive release of these substances.[31,32] When stunning is performed correctly, the animal does not feel any discomfort because it is unconscious when the hormones are released.

In large (1,000 head/h) pork-slaughter plants, it is likely that hogs experience more stress than from on-farm handling because they squeal and jam together as they move through the single-file chute. To improve conditions for the hog's welfare and for pork quality, two restrainer systems may be required in high-speed plants. Providing confinement hogs with rubber hoses to chew on and additional contact with people during finishing results in calmer hogs that are easier to handle indiscriminate genetic selection for leanness and rapid growth tends to produce nervous excitable hogs.[33]

Behavioral Principles

People who handle animals must be trained to use behavioral principles. They need to understand the animal's flight zone and point of balance.[34,35] To make an animal move forward, the handler must stand behind the point of balance. Handlers also should work on the edge of the animal's flight zone because deep penetration of the flight zone may cause panic or vigorous escape reactions. Tame animals have a smaller flight zone than wild animals.

In new and old facilities, distractions that cause animals to balk must be eliminated. The author has observed that distractions, such as shadows, puddles, light reflections on the floor, and visible people ahead, can ruin the performance of a well-designed system. Moving an overhead lamp to eliminate a water reflection or installing a shield to prevent approaching animals from seeing a person ahead facilitates animal movement. Veterinarians need to look up chutes and see what the animals see. Lamps can be used to attract animals into dark chutes and restrainers. The light must not glare in the eyes of approaching animals or cause reflections off the floor or chute walls. Equipment should be designed to minimize noise. High pitched motor sounds, air hissing, and metal clanging and banging are more likely to cause excitement than the low rumble of a conveyor. Ventilation systems must not blow slaughter or rendering smells into the faces of approaching animals. The novelty of a smell or shadow causes animals to balk. Novelty is highly stressful when livestock are being handled in an unfamiliar environment. The author has observed that in a familiar environment, such as a feedlot pen, animals initially fear a novel stimulus, such as a loader used for pen cleaning. After they learn that it will not hurt them, it becomes environmental enrichment and the animals may approach and lick a parked machine. A spot of blood on the floor of the chute sometimes impedes animal movement. This appears to be attributable to visual contrast. The author has observed that throwing a piece of paper in a chute has the same effect.

Effect of Blood

Observations by the author during new restraint equipment start-ups in many plants indicate that blood from relatively calm cattle does not appear to frighten the next animal that enters a restrainer. The animal usually voluntarily enters a restrainer that is covered with blood. Some cattle may lick the blood. Blood or saliva from a highly stressed animal, however, appears to upset other cattle. If an animal becomes frenzied for several minutes, the cattle next in line often balk and refuse to enter the restrainer. After the equipment is washed, however, the cattle will enter. In one plant, a steer refused to walk over the spot where he had flipped over backward, and then refused to walk over dribbles of saliva that were smeared on the floor where he had flipped over. He voluntarily reentered the chute three times, but when he reached his saliva on the floor, he backed up through the chute for over 15 m. There is some evidence that there may be a "smell-of-fear" substance. In one study,[36,37] blood from stressed rats was avoided by other rats, but human or guinea pig blood had no effect. According to animal behaviorist, Eible-Eihesfeldt, if a rat is killed instantly by a trap, the trap can be used again, but if the traps fails to kill instantly, it will be avoided by the other rats.[38]

Possibly, the substance that the cattle are smelling is cortisol or some other substance that is secreted in conjunction with cortisol. Cortisol is present in the blood and saliva of cattle.[39] Cortisol is a time-dependent measure, up to 20 minutes is required to reach peak values.[40] The time course of cortisol secretion fits the author's observations. If an animal is stressed for only a few seconds by an electric prod, the next animal usually remains calm and walks into the restrainer. The most serious balking and refusals to enter occur after an animal has become seriously stressed by becoming jammed in a piece of equipment. The other cattle often balk and refuse to enter for several hours.

Design and Operation of Restraint Devices

Observations by the author in more than 100 slaughter plants indicate that the attitude of management is the single most important factor that determines how animals are treated.[41] Plants with good animals welfare practices have a manager who acts as their conscience. He or she is involved enough to care but not so involved that he or she becomes numb and desensitized to animal suffering. The author's observations also indicate that abuses, such as excessive prodding, dragging downed crippled animals, or running animals over the top of a downed animal, often occur when management is lax. The author has observed that in a few poorly managed plans, up to 10% of the cattle must be shot more than once with a captive bolt to render them insensible. It is the responsibility of the manager to enforce high standards of animal welfare. Good managers take the time to incrementally improve livestock handling. Perfecting handling techniques can take several months of sustained effort. There have been great improvements in equipment to handle and euthanitize livestock in slaughter plants. Unfortunately in the United States, advances in equipment have not been paralleled by similar advances in management. In many plants, management attitude toward animal treatment has improved, but in some plants animal handling has become rougher. This is attributable to an overemphasis on speed or management personnel who do not care.

Large slaughter plants use a variety of restraint devices for holding animals during stunning and slaughter.[35,42,43] Proper operation is essential for good animal welfare. The best equipment causes stress and suffering if it is operated roughly and animals are poked repeatedly with electric prods.

It is beyond the scope of this article to discuss equipment design in detail, but some basic principles should be mentioned. Solid sides should be installed on chutes, crowd pens, and restraint devices.[34] Solid sides keep animals calmer because they block outside distractions and prevent animals from seeing people deep inside their flight zone. Solid sides also make an animal feel more secure because there is a solid barrier between it and a threatening person. A basic principle is that an animal remains calmer in a restraint device if its vision is blocked until it has the feeling of restraint.[43,44] On conveyor restrainers, the solid hold-down over the entrance must be long enough to block the animal's vision until its rear feet are off the entrance ramp and it is completely settled down on the conveyor.[42]

A second basic principle is that sudden jerky motions of equipment or people excite animals, and slow steady movements have a calming effect. A third principle is the concept of optimal pressure. A restraint device should apply enough pressure to provide a feeling of restraint, but excessive pressure, which would cause pain, must be avoided. If an animal struggles because of excessive pressure, the pressure should be slowly released. When head restraint equipment is used, the animal should be stunned or ritually slaughtered immediately after the head is restrained. On restraint devices with moving parts that press against the animal, pressure limiting valves must be installed to prevent discomfort. A pressure limiting valve automatically prevents a careless operator from applying excessive pressure.

Ritual Slaughter

Ritual slaughter is slaughter performed according to the dietary codes of Jews or Muslims. Cattle, sheep, or goats are exsanguinated by a throat cut without first being rendered unconscious by preslaughter stunning. Ritual slaughter is exempt from the Humane Methods of Slaughter Act of 1978 to protect religious freedom.[2]

Because ritual slaughter is exempt, some plants use cruel methods of restraint, such as suspending a conscious animal by a chain wrapped around one hind limb. In other plants, the animal is held in a restrainer that holds it in an upright position.[43,45,46] Whether or not ritual slaughter conforms to the requirements of euthanasia is a controversial question. When ritual slaughter is being evaluated, the variable of restraint method must be separated from the act of throat cutting without prior stunning. Distressful restraint methods mask the animals' reactions to the cut.

The author designed and operated four state-of-the-art restraint devices that hold cattle and calves in a comfortable upright position during kosher (Jewish) slaughter.[35,42,45,46] To determine whether cattle feel the throat cut, at one plant the author deliberately applied the head restrainer so lightly that the animals could pull their heads out. None of the 10 cattle moved or attempted to pull their heads out. Observations of hundreds of cattle and calves during kosher slaughter indicated that there was a slight quiver when the knife first contacted the throat. Invasion of the cattle's flight zone by touching its head caused a bigger reaction. In another informal experiment, mature bulls and Holstein cows were gently restrained in a head holder with no body restraint. All of them stood still during the cut and did not appear to feel it. Disturbing the edges of the incision or bumping it against the equipment, however, is likely to cause pain. Observations by the author also indicated that the head must be restrained in such a manner that the incision does not close back over the knife. Cattle and sheep struggle violently if the edges of the incision touch during the cut.

The design of the knife and the cutting technique appeared to be critical in preventing the animal from reacting to the cut. In kosher slaughter, a straight, razor-sharp knife that is twice the width of the throat is required, and the cut must be made in a single continuous motion. For halal (Muslim) slaughter, there are no knife-design requirements. Halal slaughter performed with short knives and multiple hacking cuts resulted in a vigorous reaction from cattle. Fortunately, many Muslim religious authorities accept preslaughter stunning. Muslims should be encouraged to stun the cattle or use long, straight, razor-sharp knives that are similar to the ones used for kosher slaughter.

Investigators agree that kosher slaughter does not induce instantaneous unconsciousness.[47,48,b]In some cattle, consciousness is prolonged for over

60 seconds. Observations by the author indicated that near immediate collapse can be induced in over 95% of cattle if the ritual slaughterer makes a rapid, deep cut close to the jawbone.[45] Further observations indicated that calm cows and bulls lose sensibility and collapse more quickly than cattle with visible signs of agitation. The author has observed that cattle that fight restraint are more likely to have prolonged sensibility. Gentle operation of restraint devices facilitates rapid loss of sensibility.

Cattle do not appear distressed even when the onset of unconsciousness is delayed. Pain and distress cannot be determined by measurements such as an electroencephalogram. Behavioral observations, however, are valid measures for assessing pain.[49] The author has observed that cattle appear unaware that their throat is cut. Investigators in New Zealand have made similar observations.[50] Immediately after the cut, the head holder should be loosened slightly to allow the animal to relax. The author also has observed that ofter the head restraint is released, the animal collapses almost immediately or stands and looks around like a normal, alert animal. Within 5 to 60 seconds, cattle go into a hypoxic spasm and sensibility appears to be lost. The spasms are similar to those that occur when cattle become unconscious in a headgate that is used for restraint in feedlots. Practical experience has shown that pressure on the carotid arteries and surrounding areas of the neck from a V-shaped headgate stanchion can kill cattle within 30 seconds.

Even though exsanguination is not an approved method of euthanasia by the AMVA,[1] the author has observed that kosher slaughter performed with the long, straight, razor-sharp knife does not appear to be painful. This is an area that needs further research. One can conclude that it is probably less distressful than poorly performed captive-bolt or electrical stunning methods, which release large amounts of epinephrine.[31,32]

Welfare can be greatly improved by use of a device that restrains the animal in a comfortable upright position. For cattle and calves, a conveyor restrainer or an upright restraint pen can be used.[42,43,46] In small plants, sheep or goats can be held by a person. If an upright pen is used, vertical travel of the lift under the animal's belly should be restricted to 71 cm to prevent the animal from being lifted off the floor. A pressure limiting valve must be installed on the head holder and rear pusher gate.[43,45] Many existing upright restraint boxes apply excessive pressure. To prevent excessive bending of the neck, the head holder should position the animal's forehead parallel to the floor. Equipping the head holder with a 15-cm wide, rubber covered forehead bracket will make the head holder more comfortable. The animal should stand in the box with its back level. An arched back is a sign of excessive pusher-gate pressure. In some plants, animals are removed from the restrainer before they become unconscious. Discomfort to the animal can be minimized by allowing it to lapse into unconsciousness before it is removed from the restrainer.

During the past five years, many large kosher slaughter plants for cattle have replaced shackling and hoisting with upright restraint. Large numbers of veal calves and sheep, however, are still shackled and hoisted. Progressive plant owners have installed upright restraint equipment, but unfortunately there are some plant owners who still refuse to install humane restraint equipment because they are not legally required to do so. Animal handling guidelines published by the American Meat Institute recommend the use of upright restraint.[6]

Conclusions

The technology exists that allows slaughter and euthanasia to be one. Although some slaughter plants maintain high animal welfare standards, there are others in which management allows abuses to occur. After adequate equipment has been installed, the single most important determinant of good animal welfare is the attitude of management. Good equipment provides the tools that make humane slaughter and handling possible, but it is useless unless it has good management to go with it.

References

[a]Koch Supplies, Kansas City, Mo.

[b]Nangeroni LL, Kennett PD. Department of Physiology, Cornell University, Ithaca, NY: Unpublished data, 1963.

1. Andrews EJ, Bennett TB, Clark JD, et al. Report of the AVMA panel on euthanasia. *J Am Vet Med Assoc* 1993;202.229–249.
2. Federal meat inspection publication 95–445, part 313. Humane slaughter of livestock 3.3.2. In: *Humane Methods of Salughter Act of 1978.*
3. Daly CC. *Proceedings* symposium on humane slaughter of animals. Herts, UK: Universities Federation for Animal Welfare, 1987;15.
4. Blackmore DK. Energy requirements for penetration of heads of domestic stock and developments of a multiple projectile. *Vet Rec* 1985;116:36–40.
5. Daly CC, Whittington PF. Investigation into the principle determinants of effective captive bolt stunning of sheep. *Res Vet Sci* 1989; 46;406–408.
6. Grandin T. *Recommended animal handling guidelines for meat packers*. Washington, DC: American Meat Institute, 1991.
7. Gregory NG. Humane slaughter, in *Proceedings*. 34th Int Cong Meat Sci Technol, Workshop on Stunning Livestock 1988.
8. Blackburn PW. *The casualty pig*. Malmesbury, Wilts, UK: Pig Veterinary Society Grove Center, 1993.
9. Croft PS. Problem of electrical stunning. *Vet Rec* 1952;64:255–258.
10. Hoenderken R. Electrical and carbon dioxide stunning of pigs for slaughter. In: Eikelenboom G, ed. *Stunning of animals for slaughter*. Boston: Martinus Nijhoff Publishers, 1982;59–63.
11. Blackmore DK, Newhook JC. Insensibility during slaughter of pigs in comparison to other domestic stock. *N Z Vet J* 1981;29:219–222.
12. Gilbert KV, Cook CJ, Devine CE, et al. Electrical stunning in cattle and sheep: electrode placement and effectiveness, in *Proceedings*. 37th Int Congr Meat Sci Technol 1991;245–248.
13. Devine CE, Cook CJ, Maasland SA, et al. The humane slaughter of animals, in *Proceedings*. 39th Int Congr Meat Sci Technol 1993;223–228.
14. Gregory NG, Wotton SB. Sheep slaughtering procedures. III. Head-to-back electrical stunning. *Br Vet J* 1984;140:570–575.
15. Cook CJ, Devine CE, Gilbert KV, et al. Electroencephalograms and electrocardiograms in young bulls following upper cervical vertebrae to brisket stunning. *N Z Vet J* 1991;39:121–125.
16. Cook CJ. Stunning Science—A guide to better electrical stunning. *Meat Focus* 1993; 2(3):128–131.
17. Gregory NG. Slaughter technology, electrical stunning of large cattle. *Meat Focus* 1993; 2(1):32–36.
18. Grandin T. Cardiac arrest stunning of livestock and poultry. In: Fox MW, Mickley LD, eds. *Advances in animal welfare science*. Boston: Martinus Nijhoff Publishers, 1985/1986;1–30.
19. Grandin T. Possible genetic effect in pig's reaction to CO_2, stunning, in *Proceedings*. 34th Int Congr Meat Sci Technol 1988;23–24.
20. Dodman NH. Observations on the use of the Wernberg dip-lift carbon dioxide apparatus for pre-slaughter anesthesia of pigs. *Br Vet J* 1977; 133:71–80.
21. Troeger K, Waltersdorf W. Gas anesthesia of slaughter pigs. *Fleischwirtsch Int* 1991;4:43–49.
22. Griez E, Zandbergen J, Pols H, et al. Response to 35% CO_2 as a marker of panic and severe anxiety. *Am J Psychiatry* 1990;147:796–797.
23. Clark DH. Carbon dioxide therapy for neuroses. *J Ment Sci* 1954;100:722–726.
24. Forslid A. Transient neocortical, hippocampal and amygdaloid EEG silence by one minute inhalation of high concentrations of CO_2 in swine. *Acta Physiol Scand* 1987;130:1–10.
25. Grandin T. Farm animal welfare during handling, transport, and slaughter. *J Am Vet Med Assoc* 1994;204:372–377.
26. Mitchell G. Hattingh J. Ganhao M. Stress in cattle assessed after handling, transport and slaughter. *Vet Rec* 1988;123:201–205.
27. Zavy MT, Juniewicz PE, Phillips WA, et al. Effect of initial restraint, weaning, and transport stress on baseline and ACTH-stimulated cortisol responses in beef calves of different genotypes. *Am J Vet Res* 1992;53:551–557.
28. Ewbank R, Parker MJ, Mason CW. Reactions of cattle to head restraint at stunning: a practical dilemma. *Anim Welf* 1992;1:55–63.
29. Tume RK, Shaw FD. Beta-endorphin and cortisol concentrations in plasma of blood samples collected during exsanguination of cattle. *Meat Sci* 1992;31:211–217.

30. Cockram MS, Corley KTT. Effect of preslaughter handling on the behavior and blood composition of beef cattle. *Br Vet J* 1991;147: 444–454.
31. Althen TGK, Ono GK, Topel DG. Effect of stress susceptibility or stunning method of catecholamine levels in swine. *J Anim Sci* 1977;44: 985–989.
32. Pearson AM, Kilgour R, de Langen H. Hormonal responses of lambs to trucking, handling and electric stunning, in *Proceedings*. N Z Soc Anim Prod 1977;37,243–248.
33. Grandin T. Environmental and genetic factors which contribute to handling problems at slaughter plants. In: Eldridge C, Boon C, eds. *Livestock environment IV*. St Joseph, Mich: American Society of Agriculture Engineers, 1993;64–68.
34. Grandin T. Animal handling. *Vet Clin North Am Food Anim Pract* 1987;3:323–338.
35. Grandin T. Welfare of livestock in slaughter plants. In: Grandin T, ed. *Livestock handling and transport*. Wallingford, Oxon, UK: CAB International, 1993;289–311.
36. Hornbuckle PA, Beall T. Escape reactions to the blood of selected mammals by rats. *Behav Biol* 1974;12:573–576.
37. Stevens DA, Gerzog-Thomas DA. Fright reactions in rats to conspecific tissue. *Physiol Behav* 1977;18:47–51.
38. Stevens DA, Saplikoski NJ. Rats' reactions to conspecific muscle and blood evidence for alarm substances. *Behav Biol* 1973;8:75–82.
39. Fell LR, Shutt DA. Adrenocortical response of calves to transport stress as measured by salivary cortisol. *Can J Anim Sci* 1986;66:637–641.
40. Lay DC, Friend TH, Randel RD, et al. Behavioral and physiological effects of freeze and hot iron branding on cross-bred cattle. *J Anim Sci* 1991;70:330–336.
41. Grandin T. Behavior of slaughter plant and auction employees towards animals. *Anthrozoo* 1988;1:205–213.
42. Grandin T. *Double rail restrainer for handling beef cattle*. Paper No. 91–5004. St. Joseph, Mich: American Society of Agricultural Engineers, 1991.
43. Grandin T. Observations of cattle restraint devices for stunning and slaughtering. *Anim Welf* 1992;1:85–91.
44. Grandin T. The effect of previous experience on livestock behavior during handling. *Agri-Pract* 1993;14:15–20.
45. Regenstein JM, Grandin T. Religious slaughter and animal welfare, in *Proceedings*. 45th Annu Reciprocal Meat Conf 1992;155–160.
46. Grandin T. Double rail restrainer conveyor for livestock handling. *J Agric Eng Res* 1988;41: 327–338.
47. Daly CC, Kallweit E, Ellendorf F. Conventional captive bolt stunning followed by exsanguination compared to shechitah slaughter. *Vet Rec* 1988;122:325–329.
48. Blackmore DK. Differences between sheep and cattle during slaughter. *Vet Sci* 1984; 37:223–226.
49. Fraser AF, Broom DM. *Farm animal welfare*. London: Bailliere Tindall, 1990.
50. Bager F. Braggins TJ, Devine CF, et al. Onset of insensibility in calves: effects of electropletic seizure and exsanguination on the spontaneous electrocortical activity and indices of cerebral metabolism. *Res Vet Sci* 1984;37:223–226.

From the Corral

by Temple Grandin

Shackling, hoisting live animals is cruel

In Western Europe and Canada, shackling and hoisting of fully conscious live animals for ritual slaughter is forbidden. Plants that conduct ritual slaughter in these countries are required to hold the animal in a restraining device while the throat is cut. Hoisting by the hind leg is not permitted until after the throat has been cut. But the practice of shackling and hoisting fully conscious livestock for ritual slaughter is permitted in the U.S. This violates the humane religious principles of ritual slaughter and greatly endangers the safety of plant employees, many of whom are injured by kicking, thrashing animals.

Safety in packing plants has become a major issue in the news media. Safety alone is a sufficient reason for requiring the use of a restraining device that holds the animal in an upright position for religious slaughter. There is also increasing public pressure concerning animal rights: Just before I gave a speech at the recent Livestock Industry Institute in Toronto, Canada, an animal rights activist disrupted the meeting. Shackling and hoisting of conscious animals for ritual slaughter is an area of our profession in need of major housecleaning.

I have been in hundreds of slaughter plants, but I had nightmares after visiting one plant in which five big steers were hung up in a row to await slaughter. They were hitting the walls, and their bellowing could be heard out in the parking lot. To get the shackles on the live cattle, the operation was equipped with a pen with a false bottom that tripped the animal to make it fall down.

In some plants the suspended animal's head is restrained by a nose tong connected to an air cylinder. Stretching of the neck by pulling on the nose is painful. Suspension upside down also causes great discomfort because the rumen presses down on the diaphragm.

Hoisting live cattle can also be very painful for plant employees. Replacement of a shackle hoist with a restrainer can often reduce accident insurance premiums. In some plants accidents were greatly reduced when the shackling of live animals was stopped. Plants that continue to hoist live animals may also be in a precarious legal position. Owners could be sued by former injured employees on the grounds that an old and dangerous technology had been used when safer, newer technologies were available.

Many plants conducting religious slaughter use proper restraining devices and they are to be commended. Companies still using the shackle hoist need to change their ways. Many systems are available for humane ritual slaughter of both large and small animals. Below I will review the equipment available from U.S. manufacturers. Many excellent systems are also available from Canadian, English and French suppliers.

For high-speed ritual steer slaughter the Boss V conveyor restrainer with a headholder mounted at the discharge end works well. Speeds up to 200 per hour can be attained. (For more information see *Meat Industry*, August 1980, page 54.) Cincinnati Butchers Supply also sells the ASPCA pen, which can handling 50 to 60 bulls or steers per hour. Both of these systems also work well for conventional slaughter.

Koch Supplies, Inc., has an economical self-contained holding pen for large cattle that is suitable for small plants. It can be installed in one weekend, and major modifications of the building are not required. Plants that slaughter only

10–15 head per hour could install a headholder in the front wall of their existing stunning pen. The headholder could be fabricated in the plant maintenance shop.

For calves and sheep, the double-rail conveyor restrainer system is now commercially available from Clayton H. Landis in Souderton, Pa. (For more information on this system see *Meat & Poultry*, December 1986, and February 1987.) For small plants that ritually slaughter calves and sheep, an inexpensive manually operated restrainer has been developed by the University of Connecticut. The restrainer can be built in the plant shop for under $1,000.

Both Jewish and Moslem teachings emphasize the importance of humane treatment of animals. Causing pain to the animal is a violation of religious teachings. A. Shoshan, in his book *Animals in Jewish Literature*, states: "It is not permissible to tie the legs of any animal or bird in a way apt to cause pain." A chain attached to the rear leg of a suspended animal will cause a bruise. A bruise blemishes the animal. According to religious law the animal must be unblemished. Leviticus 22:24 states: "Ye shall not offer unto the Lord that which is bruised, or crushed or broken, or cut." The elimination of hoisting of livestock prior to cutting the throat would have the added advantage of bringing the plant into full compliance with religious law.

Temple Grandin is a livestock handling consultant with offices in Urbana, Ill.

Humanitarian Aspects of *Shehitah* in the United States

Temple Grandin

TEMPLE GRANDIN *is Assistant Professor of Animal Science at Colorado State University. She is also president of Grandin Livestock Handling Systems, Inc., Fort Collins, Colorado.*

"Preventing pain to an animal is a command of the Torah."[1] There are numerous passages in the Torah concerning humane animal treatment: "Thou shalt not muzzle the ox when he treadeth the corn." "Six days thou shalt labor and do thy work, but on the seventh day thou shalt rest in order that thine ox and thine ass may rest."[2] "The teaching of kindness to the living animal is reflected in the Jewish laws governing the slaughter of animals for food: they are designed to secure humane treatment and freedom from pain."[3] The Talmud also contains references to the kind treatment of animals: "A calf was being taken to slaughter, when it broke away, hid his head under Rabbi's skirts (referring to Rabbi Judah the Prince) and lowed (in terror). 'Go,' said he, 'for this wast thou created.' Thereupon they said (in heaven), "Since he has no pity, let us bring suffering upon him.'"[4] "Great importance is attached to the humane treatment of animals, so much so that it is declared to be as fundamental as human righteousness."[5]

Livestock slaughter by *shehitah* (ritual slaughter) are killed by severing the four major blood vessels in the throat. The esophagus (food pipe) and trachea (wind pipe) must also be severed. *Shehitah* is performed by a *shohet* (ritual slaughterer) with a very sharp knife,[6] whose blade must be perfectly free of nicks and imperfections,[7] since nicks in the blade are likely to cause the animal pain. If the knife becomes nicked or damaged, the animal is declared *treif* or not kosher. The leading rabbinic authorities have declared "the inadmissibility of any method of stunning prior to *shehitah*."[8] Thus, the animal must be fully conscious prior to slaughter.

Restraint for Shehitah

Shehitah is exempt from the 1958 Humane Slaughter Act,[9] which requires rendering an animal unconscious prior to cutting the throat or hoisting. The procedure of shackling and hoisting consists of placing a chain, called a shackle, around one back leg. After shackling, the animal is hoisted into the air until it is fully suspended by that leg. Traditionally, cattle and other livestock were cast into a lying position on the floor (casting) for *shehitah*. Many years ago, the United States Department of Agriculture forbade casting due to sanitary concerns. If *shehitah* is performed while the animal is lying on the floor the throat cut may have become contaminated with dirt or manure. When the Humane Slaughter Act was passed, restraining chutes to hold the animals were not readily available in the United States, and that is one reason why the Humane Slaughter Act does not contain legislation on livestock restraint for *shehitah*.

Today, many good restraint devices are available, but some kosher slaughter plants continue to restrain livestock by old fashioned, cruel, dangerous shackling and hoisting. The United States is very backward in this respect as compared to other countries. Shackling and hoisting for *shehitah* is prohibited in Canada, Australia, New Zealand, England and many other European countries. Rabbi Unterman, former Chief Rabbi of Israel, stated that: "In the land of Israel, we do not slaughter a beast hoisted in the air."[10] Rabbi B. Berkovits, in London, also stated that shackling and hoisting "are not an intrinsic part of *shehitah* and are not, in fact, allowed in England."[11]

The Union of Orthodox Jewish Congregations of America (OU) and the Joint Advisory Committee of the Synagogue Council both endorse the use of a restraining pen for cattle.[12] All OU-supervised slaughter plants are required to use a restraining pen for cattle, but some non-OU slaughter plants continue to engage in shackling and hoisting.

As of 1989, approximately 80 to 90% of kosher cattle were held in a restraining chute and only 10 to 12% were shackled and hoisted. Two very large plants and several smaller ones continue to use the shackle hoist for cattle, but fewer than 10% of the kosher veal and almost none of the kosher lambs and sheep are held in a restraining chute. Modern restraint equipment is used in only one kosher veal plant.

Shackling, Hoisting, Stress, and Injury

During my work as a livestock equipment designer, I have been in hundreds of slaughter plants and for days after visiting a high speed kosher shackle-hoist operation I had nightmares. The frantic bellowing of cattle could be heard throughout the office and the parking lot. Bellowing of this magnitude is never heard from a kosher plant that has a well-designed and operated restraining chute.

In some plants, steers and bulls are tripped by tilting a false floor in a narrow pen. A chain is then wrapped around one back leg and the animal is jerked into the air. When sheep or calves are shackled and hoisted, the chain is wrapped around a hind leg and an employee drags the animal over to the shackle hoist. Animals often struggle violently when they are suspended. In some plants five or six animals may be struggling on the chains while waiting for the *shohet*. In many shackle and hoist cattle plants, a clamp is placed in the nostrils and the neck is stretched by a powerful air cylinder, which may apply as much as 400 lbs. of pull. Suspending a 1300 lb. steer by one back leg is also very stressful. Cattle have more than one stomach of which the heavy first stomach, called the rumen, may weigh up to 130 pounds.[13] When the animal is suspended upside down, the rumen presses against the diaphragm and lungs, and breathing becomes difficult. Research at the University of Connecticut has indicated that shackling and hoisting is more stressful for sheep and calves than restraint in an upright position in a device that supported the animal under the belly and brisket. The stress is greatest for the heavier calves. Suspended calves and sheep have higher heart and breathing rates compared to animals restrained in an upright position.[14] Hanging by one back leg causes forced breathing in both species.

Shackling and hoisting of large cattle sometimes breaks the animal's leg. About half the legs are bruised prior to *shehitah,* whereas a well-designed restraining chute will not injure or bruise.[15] Jerking and injury to the back leg violates the humane intent of *shehitah.* Some American rabbinical authorities have stated that damage to the rear leg does not render an animal *treif,* but many people would agree that it violates the "spirit" of the law. A British rabbi was deeply disturbed when I told him about shackling and hoisting in this country. A. Shoshan, in his book, *Animals in Jewish Literature,* states that ". . . it is not permissible to tie up the legs of any animal or bird in a way apt to cause pain."[16] The Torah contains numerous passages which maintain that the animal must be unblemished: "Ye shall not offer unto the Lord that which is broken, or crushed or cut."[17] There are also prohibitions about eating limbs torn from animals: "And ye shall be holy men unto me, neither shall ye eat any flesh that is torn of beasts in the field, ye shall cast it to dogs."[18] Shackling and hoisting does not tear off an animal's limb, but it often jerks it hard and causes tearing of the tissues. Some *shohetim* mistakenly believe that hanging an animal upside down improves "bleedout," the process by which the blood is removed from the slaughtered animal, but University of Connecticut research indicated that bleedout in calves and sheep was slightly better while the animals remained upright in the restraint chute.[19]

Available Restraining Equipment

Over forty years ago, the first restraining device

was developed in Europe. The Weinberg cattle casting pen consists of a narrow stall which slowly inverts the animal until it is lying on its back. This system is less stressful than shackling and hoisting, but is more stressful to the animal than the newer upright restraint equipment. Recent research by C.S. Dunn indicates that cattle restrained in the Weinberg casting pen had higher cortisol (stress hormone) levels and more vocalizations compared to cattle restrained in the ASPCA pen which is described below.[20] The Weinberg method is slow, with a top speed of only 30 or 40 per hour.

In 1963, Cross Brothers Packing Co. patented an apparatus for holding cattle in a standing upright position for *shehitah.*[21] The patent was purchased by the American Society for the Prevention of Cruelty to Animals in 1964. This enabled any plant to use the device royalty free, and it became known as the "ASPCA pen." It is now in use in 50 to 60% of kosher cattle plants.

The pen is a narrow stall with an opening in the front for the animal's head. A lift supports the animal under the belly to prevent it from collapsing. A chin lift raises the animal's head and holds it still for the *shohet.* After *shehitah,* a shackle chain is placed around one back leg and the animal is hoisted out through the side of the pen. The ASPCA pen can handle 50 to 75 cattle per hour, and, with modifications, up to 10 cattle per hour. Some *shohetim* have stated that the ASPCA pen is more stressful than shackling and hoisting, but these complaints are usually due to a poorly designed "homemade" pen or a pen which has improper modifications added by the plant. For example, I have seen a sharp angle iron, instead of round pipe, used on parts of the pen that press against the animal. The pen, itself, must be adjusted to fit the type of cattle being slaughtered. The belly lift should not lift the animal off the floor, but simply support it.[22] Rabbi Menachem Genack of the Union of Orthodox Jewish Congregations of America states, "We have found that the pen not only mitigates pain to the animal, but also has advantages in terms of securing the safety of the shochet and carefully insures proper shechitah in that it immobilizes the head of the animal."[23] To minimize stress and maximize blood loss, the animal should not be fully restrained by the head holder and the belly lift until the *shohet* is ready to perform *shehitah.*[24] This general principle applies to all methods of restraint. The interval between complete restraint of the head and *shehitah* should be under 15 seconds.

The next development in kosher restraint equipment for large cattle was the high speed V conveyor restrainer with a head holder, which has a capacity of 200 large cattle per hour and in which cattle ride along between two conveyors which form a V.[25] The animal is held in a comfortable upright position, and its body is supported by the angled conveyors along each side, with its feet protroding through an opening between the bottom of the conveyors. The conveyor is stopped for *shehitah,* and a hyrdraulically operated head holder lifts the animal's head. Development of the headholder was done by Spencer Foods, Spencer, Iowa, with some assistance from the author.[26]

In 1986, the first upright restraining system for calves and sheep was installed at Utica Veal in Marcy, New York. A wooden laboratory prototype had been built in the early seventies by the University of Connecticut research team,[27] but many additional components needed to be invented to make the system work in a commercial plant. The system at Utica Veal was designed and constructed by Grandin Livestock Handling Systems of Fort Collins, Colorado[28] and Clayton H. Landis Equipment Company of Souderton, Pennsylvania. Funding of all phases of this project was provided by the Council for Livestock Protection, a consortium of national humane organizations, and Utica Veal. The restrainer can handle 150 kosher calves per hour. The animals ride astride a moving conveyor and are supported under the billy and brisket. This method of restraint is very comfortable and sheep and calves can be restrained in the apparatus for several hours with no signs of discomfort. They sit quietly and do not vocalize or struggle. The conveyor is stopped for each animal, a yoke holds the back of the neck, and a person holds the animal's head. A fully mechanized head holder could have been constructed, but Utica Veal found that the system worked adequately without it. Since the funding from the Council for Livestock Protection had run out, Utica Veal decided not to construct a fully mechanized head holder. *Shehitah* for Glatt Kosher meat has been performed successfully without the mechanized head holders for over three years. The University of Connecticut has also developed an economical, manually operated restrainer for sheep and calves which can be used in smaller plants that cannot afford the larger conveyor system.

Economic Factors

Decisions to purchase more humane restraining equipment for kosher slaughter are usually made by the plant owners and not by the supervising rabbi or the *shohtim*. A standard non-kosher beef plant can be converted to a kosher shackle-hoist operation at a minimum expense.

The majority of kosher plants are owned by non-Jews, but, unfortunately, there are some extremely bad shackle-hoist plants which are Jewish-owned. It is disturbing that two large shackle-hoist operators spent large amounts of money and time fighting the government about safety and animal rights groups. At one plant, the money already spent on politics would have paid for a restraining pen. Though some kosher plants are not willing to spend money for the sake of humaneness, safety for plant employees is sufficient justification for replacing a shackle-hoist with modern restraint equipment. For that very reason, two large kosher plants recently replaced their shackle-hoists and, at Utica Veal, the installation of the restrainer resulted in a drastic reduction in accidents.[29] Records reveal that for an 18 month period prior to the restrainer installation there were 126 working days lost due to five accidents. Three of these were very serious and one man had to have knee surgery. For an 18 month period after the restrainer was installed there was only one bruised hand, requiring two days off. The restrainer has been in place for over three years and there have been no additional accidents.

The shackling and hoisting of large cattle is very dangerous. Employees have been kicked in the head and have to wear football helmets to protect them from the flailing front feet of the cattle. At one plant, a man almost died from loss of blood because the *shohet* cut him while he was attempting to restrain the head of a struggling steer.[30] Shackling and hoisting of sheep is also dangerous. Employees have had teeth knocked out, and heavy shackle trolleys often fly off the rail. Modern restraint equipment is costly, but a reduction in Workmen's Compensation claims will often pay for it. Since injured employees are more inclined to sue today, avoidance of a single lawsuit would pay for the most expensive system. Plant managements could be sued on the basis that they use dangerous, obsolete equipment when modern safer equipment is available.

The large conveyor restrainer systems for kosher slaughter that can handle over 100 animals per hour require a major investment of money. The equipment and the building renovations usually cost about $250,000. An ASPCA pen, however, is much less expensive and can usually be installed in one weekend without disruption of plant operations or structural modifications to the building. Used ASPCA pens can often be purchased for $5,000 to $15,000, while a new one costs about $35,000. For calf and sheep plants which slaughter 100 animals or less per hour, a scaled down ASPCA type pen could be built for about $3,000 in materials. Small locker plants could build a simple pipe rack for holding sheep and calves for only a few hundred dollars.

Necessity is the mother of invention. If plants would have to get rid of their shackle hoists, many new restraining devices would be invented over night. During the last two years, the Occupational Safety and Health Administration (OSHA) has levied huge fines against several large packers for safety violations and, as a result, the industry has devised many innovative machines which improve safety and are making money for the plants because labor requirements are reduced.

Shehitah Controversy

One of the reasons why rabbinical authorities

have been reluctant to advocate a ban on shackling and hoisting is fear that *shehitah* itself will be attacked. In Sweden, Holland, Norway, Switzerland, and Iceland, slaughtering without stunning is forbidden,[31] and some animal welfare groups in England have also recently advocated a similar law.

Scientific researchers agree that sheep and goats lose consciousness within 2 to 15 seconds after the throat is cut.[32] Sheep, in particular, die very quickly after the throat is cut because blood flow to the brain travels through blood vessels that are severed during *shehitah*. In calves and cattle, some studies have shown that unconsciousness is sometimes delayed for over 60 seconds,[33] while other studies have shown that unconsciousness occurs very rapidly within 4 to 17 seconds.[34] The anatomy of cattle blood vessels is different from that of sheep.[35]

Observations made by the author indicate that the cutting technique of the *shohet* may explain the great variation in the results of scientific studies concerning the onset of unconsciousness in calves and cattle. At Utica Veal it was possible to observe accurately the reactions of hundreds of 375 lb. veal calves to *shehitah*. The most skillful *shohet* was able to cause over 95 percent of the calves to collapse immediately.[36] When a less skilled *shohet* performed *shehitah*, up to 30 percent of the calves righted themselves on the table and some animals even walked on the moving table like a treadmill.[37] Both cartoid arteries were severed in all animals. All *shohtim* cut the throat in the same location and depth, but the most skillful *shohet* used a rapid stroke and there were no sawing motions while the knife was passing through the carotid arteries.[38] The less skilled *shohet* used a slower knife stroke and observations indicate that the arteries were more likely to seal off that way.[39] To insure the humaneness of *shehitah, shohtim* could be tested periodically to make sure that their technique induces immediate collapse.

Good cutting technique will also prevent poor bleed-out and bloodspots in the meat. The latter are a common problem in cattle killed by *shehitah*[40] and are more likely to occur if the arteries seal off. They are also caused by tearing small capillaries in the muscles while the animal is being restrained. Shackling and hoisting or a poorly designed restraint system may increase bloodspots, but a well designed restrainer will help reduce them.[41] Many *shohtim* and plant managers do not realize that fluctuating temperatures and rapid weather changes also greatly sensitize animals to bloodspots.[42]

The Need to Eliminate Shackling-Hoisting

The elimination of shackling and hoisting as a method of restraint would help strengthen the position of *shehitah* in the United States. Many meat industry people think that shackling and hoisting is part of the slaughter ritual and, unfortunately, there are some *shohtim* who have defended it. Either they do not realize that much better restraint systems are available or they have had a bad previous experience with a "homemade," poorly designed restraining pen or a system that had been modified by either unknowledgeable or uncaring plant employees. People working in slaughter houses often become numbed and callous.[43] They sometimes deliberately abuse animals, though I have never observed a *shohet* who engaged in deliberate animal abuse.[44] However, some *shohtim* may also have become "numbed" and a few of them in the United States have declared that shackling and hoisting was not stressful, although, in England, where shackling and hoisting is not permitted, shackle hoist pictures horrified the viewers.

Consumers of kosher beef, veal, and lamb need to put pressure on the meat industry to replace the remaining shackle hoists. One way to do so is to buy only from plants that use proper restraining equipment. Every slaughter plant that ships across state lines has a USDA (United States Department of Agriculture) "Establishment Number" which is stamped on the carcasses, or is printed on the wholesale boxes. Unfortunately, the middlemen who sell kosher meat to the local butcher sometimes remove these numbers. One should insist on seeing the USDA inspected purple stamp on the carcasses or the printed label from the wholesale meat boxes. Plant names and addresses can be obtained through the USDA in Washington, D.C.

The Jewish community should also consider assisting in funding the installation of humane restraint equipment. In return, a plant would guarantee to continue its kosher operations and would be provided with a reliable market for its meat. Kosher beef plants need the financial support of the Jewish community. Some of these are dilapidated, old installations that got into the kosher business to keep a marginal business going. It is, however, becoming increasingly difficult to compete against huge, super-efficient, non-kosher plants which slaughter over 4000 cattle each day. Kosher plants have to compete aginst these mega-plants to sell the hind quarters and *treif* beef, though the competition problem does not exist in the veal industry where there are several highly profitable plants that have made no effort to replace their shackle hoists.

Conclusions

Elimination of the cruel and dangerous practice of shackling and hoisting as a method of restraint would be in accordance with this statement by Samuel Dresner:

Kashrut is a systematic means of educating and refining the conscience of those who observe it from early age to death . . . [T]he observance of Kashrut by the people of Israel has helped to do precisely this for them over the centuries, for to teach reverence for animal life is, all the more, to teach reverence for human life. The Jews are called *rahmanim b'nei rahmanim*, merciful ones and the children of merciful ones.[45]

To this end, therefore, the cruelty of shackling and hoisting should be abolished, and it can be easily and economically replaced.

1. A. Cohen, *Everyman's Talmud* (New York: E.P. Dutton Co., 1949), p. 236.
2. Deuteronomy XXV:4 and Exodus XXIII:12.
3. Bernard Homa, *Shehita* (London: The Board of Deputies of British Jews, Soncino Press, Ltd., 1967), p. 2.
4. The Babylonian Talmud, Seder *Nezikin*, Vol. I. *Baba Mezia* (London: The Soncino Press Ltd., 1948), p. 486.
5. Ibid., p. 486; see also the footnote on p. 382.
6. Rabbi Yacov Lipschutz, *Kashruth* (Brooklyn, New York: Mesorah Publications Ltd., 1988), pp. 17–21.
7. Ibid., p. 20.
8. Michael L. Munk and Eli Munk, *Shechita, Religious and Historical Research on the Jewish Method of Slaughter* (Jerusalem: Feldheim Press), p. 23.
9. P.L. 85–765, 85th Cong., H.R. 8308, August 27, 1958.
10. Letter of Isser Yehuda Unterman, former Chief Rabbi of Israel, to Dr. E.D. Ralbag, Chairman of the Federation of the Societies for Protection of Animals (Tel Aviv, n.d.; approx. 1961). (This and other letters referred to in the footnotes are available in the author's files.)
11. Letter of Rabbi B. Berkovits, Registrar, Court of the Chief Rabbi, Adler House, Tavistock Square, London, to Temple Grandin, August 6, 1986.
12. Morris Laub, *Why the Fuss Over Slaughter Legislation* (New York: Joint Advisory Council of the Synagogue Council of America, 1969); letter of Rabbi Menachem Genack, Rabbinic Coordinator, Union of Orthodox Jewish Congregations of America, New York, N.Y. to Edward David, San Francisco, CA, Oct. 20, 1982.
13. A.T. Phillipson, "Ruminant Digestion" in M.J. Swenson, ed., *Duke's Physiology of Domestic Animals*, 8th Edition (Ithaca, N.Y.: Cornell University Press, 1970), pp. 424–483.
14. Rudy G. Westervelt, Don M. Kinsman, Ralph P. Prince, and Walter Giger, Jr., "Physiological Stress Measurement During Slaughter in Calves and Lambs," *Journal of Animal Science*, Vol. 42, 1976:831–832.
15. Walter Giger, Ralph P. Prince, Rudy G. Westervelt, and Don M. Kinsman, "Equipment for Low-Stress, Small Animal Slaughter," *Transactions American Society of Agricultural Engineers*, Vol. 20, 1977:571–578.
16. A. Shoshan, *Animals in Jewish Literature: The Jew and his Animals* (Rechovot, Israel Shoshanim, 1971).

17. Leviticus, XXII:24.
18. Exodus, XXIII:31.
19. Westervelt, *et al., Op. cit.*, p. 836.
20. C.S. Dunn, "Stress Reactions of Cattle Undergoing Ritual Slaughter Using Two Methods of Restraint," *Veterinary Record*, 1990, Vol. 126:522–5; J. Rushen, "Aversion of Sheep for Handling Treatments: Paired Choice Studies," *Applied Behavior Science*, 1986, Vol. 16:363–370.
21. See diagrams of the ASPCA pen and a description of correct pen operation in Milton Marshall, Elwood E. Milbury, and Eugene W. Shultz, "Apparatus for Holding Cattle in Position for Humane Slaughtering," U.S. Patent Number 3,092,871, June 11, 1963. Sold commercially by the Cincinnati Butcher's Supply Co. in Cincinnati.
22. Ibid., Col. 3, line 24.
23. Rabbi Menachem Genack, *Op. cit.*
24. Temple Grandin, "How to Reduce Bloodsplash," *Meat Industry* (current magazine title is *Meat and Poultry*), August 1985;49–50.
25. See diagrams and descriptions of the V conveyor restrainer system for cattle in Oscar Schmidt, "Cattle Handling Apparatus," U.S. Patent Number 3,657,767, April 25, 1972, and Temple Grandin, "System for Handling Cattle in Large Slaughter Plants," *American Society of Agricultural Engineers* (collection of meeting papers on file with the Society in St. Joseph, Michigan), Technical Paper Number 83-4506, December 13–16, 1983.
26. Photographs and descriptions of a mechanized headholder for *shehitah* in the V conveyor restrainer. Temple Grandin, "Spencer Smooths Kosher Kill with 200 Head per Hour System," *Meat Industry*, August, 1980:54, and Temple Grandin, "Problems with Kosher Slaughter," *International Journal for the Study of Animal Problems*, 1980 Vol. 1 Number 6:375–390.
27. Westervelt *et al., Op. cit.*, pp. 831–837 and Giger *et al., Op. cit.*, pp. 571–578.
28. See description, diagrams and photographs of the double rail restrainer system for upright restraint of calves and sheep in Temple Grandin, "High Speed Double Rail Restrainer for Stunning or Ritual Slaughter," *33rd International Congress of Meat Science and Technology, Proceedings*, Paper Number 3:1, August 2-7, 1987, pp. 101–104 and Temple Grandin, "Double Rail Restrainer Conveyor for Livestock Handling," Journal of Agricultural Engineering Resource," 1988, pp. 327–338.
29. Letter of Victor Broccoli, Plant Manager of Utica Veal, Marcy, New York, to Temple Grandin, March 10, 1988, and Grandin, "Double Rail Restrainer...," p. 335.
30. Temple Grandin
31. Rabbi B. Berkovits, "European Schechitah: What Can Be Done," *Kashrus Magazine*, Summer 1989:26–31.
32. Scientific studies indicate that sheep lose consciousness rapidly after both carotid arteries are severed. David K. Blackmore, "Differences in Behavior Between Sheep and Cattle During Slaughter," *Research in Veterinary Science*, 1984, Vol. 37:223–226; Neville Gregory and S.B. Wotton, "Sheep Slaughtering Procedures," Time to Loss of Brain Responsiveness after Exsanguination or Cardiac Arrest," *British Veterinary Journal*, 1981. Vol. 110:354–360. Louis L. Nangeroni and Paul D. Kennett, "An Electroencephalographic Study of the Effect of Shehita Slaughter on Cortical Function of Ruminants," Unpublished Report, Dept. of Physiology, New York State Veterinary College, Cornell University, Ithaca, New York, September 1963; and W. Van Schulze, H. Schultz, A.S. Hazen, and R. Gross (no title available), *Deutsche Tierarzliche Wochenschrift*, 1978, Vol. 85:41–76.
33. Scientific studies indicate that the onset of unconsciousness in cattle and calves is sometimes delayed for more than 60 second after both carotid arteries are severed. David K. Blackmore, "Differences in Behavior Between Sheep and Cattle During Slaughter, *Research in Veterinary Science*, 1984, Vol. 37:223–226; and Clyde C. Daly and E. Kallweit, "Cortical Function in Cattle During Slaughter: Comparison of Conventional Captive Bolt Stunning Followed by Exsanguination With Schechita Slaughter," unpublished manuscript, Institute of Food Research, Bristol Laboratory, Langford, Bristol, England, 1986.
34. Scientific studies indicate that unconsciousness occurs rapidly in cattle and calves. Nangeroni and Kennett, *Op. cit.*, and I.M. Levinger, "Jewish Method of Slaughtering Animals for Food and its Influence on Blood Supply to the Brain and on Normal Functioning of the Nervous System," *Ani-*

mal Regulation Studies, 1979, Vol. 2:111–126. Neville G. Gregory and S.B. Wotton, "Time to Loss of Brain Responsiveness Following Exsanguination in Calves," Research in Veterinary Science, 1984. Vol. 37:141–143.
35. B.A. Baldwin and E.R. Bell, *Journal of Physiology,* 1963, Vol. 167:448-462, and B.A. Baldwin, "Anatomical and Physiological Factors Involved in Slaughter by Carotid Section," *Humane Killing and Slaughterhouse Techniques* (London: Universities Federation for Animal Welfare, 1971), pp. 34–43.
36. Temple Grandin, "Double Rail Restrainer for Stunning or Ritual Slaughter," International Congress of Meat Science and Technology, 1988, p. 102.
37. Temple Grandin, published discussion with 34th International Congress of Meat Science and Technology, Workshop on Stunning of Livestock, August 29 to Sept. 2, 1989, Brisbane, Australia, p. 27.
38. Ibid., p. 27 and Temple Grandin, "Double Rail Restrainer for Stunning or Ritual Slaughter," International Congress of Meat Science and Technology, 1988, p. 102.
39. Neville Gregory provides a possible physiological explanation for the difference between the efficacy of the more skilled *shohet*'s cutting technique. Published discussion, 34th International Congress of Meat Science and Technology, Workshop on Stunning of Livestock, August 29 to Sept. 2, 1989, Brisbane, Australia, p. 43.
40. Temple Grandin, "How to Reduce Bloodsplash," *Meat Industry*, 1985, p. 50.
41. Ibid., p. 50.
42.
43. Temple Grandin, "Behavior of Slaughter Plant and Auction Employees towards Animals," *Anthrozoos*, 1988, Vol. 4:205–213.
44.
45. Samuel H. Dresner, *The Jewish Dietary Laws, Their Meaning for Our Time* (New York: The Burning Bush Press, 1959, 1966), p. 38.

From Progress in Agricultural Physics and Engineering *Edited by* John Matthews CBE,BSc,CEng,CPhys, FIAgrE, FInstP, FErgS, FRAgrS, MemASAE, *Former Director, Silsoe Research Institute, Wrest Park, Silsoe,Bedford MK45 4HS, UK, CAB International*

Chapter 11

Principles of Abattoir Design to Improve Animal Welfare

Temple Grandin

Introduction

Restrainers, races, holding pens and unloading ramps in abattoirs must be properly designed to facilitate animal movement, prevent bruises and minimize stress. The livestock industry loses millions of dollars annually due to bruises and death losses (Livestock Conservation Institute, 1988; Marshall, 1977).

Stress-related meat quality problems such as dark cutting (DFD) and pale soft exudative (PSE) meat also cause huge losses (Canadian Meat Council, 1980). Well-designed abattoir facilities will also greatly improve animal welfare. Engineers must always remember that good facilities provide the tools to make humane handling possible, but management is essential to enforce good handling practices. Engineering is not a substitute for management.

Different countries will have specific requirements for facilities; for example, truck size will affect the size of the holding pens. Space and facilities must also be designed for specialized functions, such as weighing, animal identification, and washing. The engineer must be familiar with the specific requirements of the country in which the facility is being constructed.

Livestock behaviour facility design

Vision

Livestock have wide-angle vision. The visual field of cattle and pigs is over 300° (Prince, 1977). In sheep the visual field varies from 191–306°, depending on fleece thickness on the head. Races, crowd pens, and unloading ramps should have solid fences to prevent animals from seeing distractions outside the fence with their wide-angle vision (Grandin, 1980a, 1982; Rider *et al.*, 1974). Moving objects outside the fence will frighten livestock. The use of solid fences in races and crowd pens is especially important if animals are not completely tame, but solid fences in races and crowd pens will also help keep tame animals calmer.

Animals will often balk at a sudden change in fence construction or floor texture (Lynch and Alexander, 1973). Puddles, shadows, drains and bright spots will also impede animal movement. Poor depth perception may explain why livestock balk at many things. Livestock can perceive depth when they are still and have their heads down (Lemman and Patterson, 1964), but their ability to perceive depth while they are moving with their heads up may be poor (Hutson, 1985). To see depth accurately the animal has to stop and put its head down.

Indoor handling facilities should have even, diffuse lighting that minimizes shadows. Cattle, pigs and sheep have a tendency to move more easily from a dimly illuminated area to a more brightly illuminated area (Grandin, 1982; Hitchcock and Hutson, 1979; Kilgour, 1971; VanPutten and Elshof, 1978). At night or in enclosed facilities, lamps can be used to attract animals into races (Grandin, 1982). The lights should illuminate the floor and must not shine into the eyes of approaching animals. Livestock are more likely to balk if they are forced to move towards blinding sunlight. Pigs reared indoors under artificial illumination preferred to walk up a ramp illuminated at 80 lux (Phillips *et al.*, 1987); this was similar to the illumination of their living quarters. A dimly illuminated ramp with less than 5 lux was avoided and there was also a tendency to avoid an excessively bright ramp illuminated at 1200 lux (Phillips *et al.*, 1987).

Noise

Livestock have sensitive hearing and they are stressed by excessive noise (Kilgour and deLangen, 1970; Kilgour, 1983). They are specially sensitive to high-frequency sound around 7000–8000 Hz (Ames, 1974). Humans are most sensitive at 1000–3000 Hz (Ames, 1974). In steel facilities, gate strike posts should have rubber stops to reduce noise and in the stunning area, the shackle return should be designed to prevent clanging and banging. Air exhausts on pneumatically powered gates should be piped outside (Grandin, 1983b). If hydraulics are used to power gates or conveyors, the motor and pump should be located away from the animals and all hydraulic motors and plumbing should be designed to minimize noise; high-pitched sound from a hydraulic system is very disturbing to cattle. Cattle held overnight in a noisy yard close to the unloading ramp were more active and showed more bruising compared to cattle in a quieter pen (Eldridge, 1988).

Experience and genetic effects

An animal's previous experience at the farm or ranch of origin will affect its behaviour at the abattoir. Cattle which have been handled roughly at the feedlot of origin become more agitated and are more difficult to handle than cattle which have received gentle treatment.

At one abattoir, playing music throughout the stockyard and race area reduced excitement and agitation. The cattle became accustomed to the music in the holding pens and it provided a familiar sound which helped mask the sounds of abattoir equipment.

Pigs reared in an extremely barren environment with a lack of stimulation will be more excitable than pigs reared with additional stimulation, such as rubber hose toys or people walking in the pens (Grandin, 1989a). Pigs reared in confinement with a lack of stimulation were more difficult to load than pigs reared outside (Warriss *et al.*, 1983). Playing a radio in the fattening pens will reduce the animals' reactions to sudden noises such as a door slamming, and possibly stress could be reduced if animals were exposed to abattoir sounds on the farm—animals will readily adapt to reasonable sound levels. Ames (1974) found that continuous exposure to either instrumental music or miscellaneous sounds during fattening, improved weight gain of sheep at 75 dB and reduced weight gain when played at 100 dB. Exposure to reasonable sound levels during fattening will help reduce fear reactions to sudden unexpected noises.

Genetics is also an important determinant of how animals will behave at the abattoir. Brahman and Brahman-cross cattle are more excitable than Hereford or Angus (Tulloh, 1961). Since the mid-1980s the author has observed increasing problems with very excitable nervous pigs which are very difficult to handle at the abattoir. They have extreme shelter-seeking behaviour (flocking together), and the animals will not separate from the group and move up the race (Grandin, 1989b). This problem has increased and is often most evident in hybrid pigs which are selected for leanness and high productivity. Some of these animals are so 'crazy' that it is almost impossible to handle them humanely in a conventional race and crowd pen system. When these pigs become excited, handlers may over-use electric prods and meat quality will be reduced. Designing facilities to handle 'crazy' pigs will be very expensive. This problem should be corrected at its source by selective breeding for lean pigs with a calm temperament. Engineers should avoid the temptation to try to treat the symptoms of a problem with engineering, rather than correcting the problem at its source.

Layout of lairages and stockyards

Long, narrow pens are recommended in stockyards and lairages in abattoirs (Grandin, 1980a, b; Kilgour, 1971). One advantage of long, narrow pens is more efficient animal movement: animals enter through one end and leave through the other. To eliminate 90° corners, the pens can be constructed on a 60–80° angle. Each pen gate should be longer than the width of the alley, so that it opens on an angle to eliminate the sharp corner.

Long, narrow pens maximize lineal fence length in relation to floor area and this may help reduce stress (Kilgour, 1978; Grandin, 1980a, b). Cattle and pigs prefer to lie along the fenceline (Grandin, 1980b; Stricklin *et al.*, 1979). Observations indicate that long, narrow pens may help reduce fighting (Kilgour, 1976). Government regulations in some countries may require walkways in between the pens for observation of animals prior to slaughter. The layout remains the same except a 1 m wide walkway is placed between every other pen.

The size of the holding pens required for an abattoir is going to be at least partially dictated by size of the trucks. When small groups of animals are handled, block gates can be used in a long, narrow pen to keep different groups separated. Minimum space requirements for holding fattened, feedlot steers for less than 24 hours are $1.6m^2$ for hornless cattle and $1.85m^2$ for horned cattle (Grandin, 1979; Midwest Plan Service, 1980), and $0.5m^2$ for slaughter-weight pigs and lambs. During warm weather pigs require more space. Wild, extensively raised cattle may also require additional space. However, providing too much space may increase stress because wild cattle tend to pace in a large pen. Enough space must be provided to allow all animals to lie down at the same time. In countries with large trucks, larger pens and wider alleys will be required. To avoid bunching and trampling 25 m is the maximum recommended length of each holding pen, unless block gates are installed to keep groups separated, shorter pens are usually recommended.

Recommended alley and pen width will vary depending on the number of animals that will be moving through the facility. In pig and sheep abattoirs where the line speed is under 400 animals per hour or where trucks holding less than 80 animals are used, pen width should be 2 m and the alley width should be 1.3 m. In abattoirs served by trucks which hold over 150 pigs or sheep per load, pen width should be increased to 3.0–4.2 m and alley width will vary from 2.5–3 m. In large cattle abattoirs slaughtering 60 or more cattle per hour, 3 m alleys are recommended. Smaller abattoirs can reduce alley width to save space.

Avoid mixing strange animals

To reduce stress, prevent fighting, and preserve meat quality, strange animals should not be mixed shortly before slaughter (Barton-Gade; 1985; Grandin, 1983a; Tennessen and Price, 1980). Solid pen walls between holding pens prevent fighting through the fences. Solid fences are especially important in lairage and stockyard pens if wildlife such as deer, elk, or buffalo are handled.

To keep pigs separated presents some practical problems. In the USA, pigs are transported in trucks with a capacity of over 200 animals; however, they are fattened in much smaller groups. Observations at UK abattoirs indicate that mixing 200 pigs from three or four farms resulted in less fighting than mixing 6–40 pigs. One advantage of the larger group is that an attacked pig has an opportunity to escape. Price and Tennessen (1981) found a tendency towards more DFD carcasses and hence more stress when small groups of 7 bulls were mixed compared to larger groups of 21 bulls.

In Denmark, the design of the pig lairage at the abattoir is very specialized. Pigs are held in long, narrow pens equipped with manual push gates and a powered push gate moves pigs up the alley to the stunner. This system was invented by T. Wichmann of the Danish Meat Research Institute. The Danes have also developed automated block gates within the long, 2m wide pens to keep small groups of 15 pigs in separate groups (Barton-Gade, 1989). A powered gate moves along each pen and brings groups of 15 pigs up to the main drive alley. After a group of pigs is brought to the alley the powered push gate is reversed and moved back down the holding pen to bring up the next group of 15 pigs. It is raised so that it can pass over the animals and then lowered to bring them out of the pen. Block gates spaced at intervals along each pen keep three or four groups of 15 pigs separated. Since the biparting block gates swing back into the animals, overloading of each 15-head compartment is prevented (if the compartments are overloaded, the system will not work). This system can deliver a steady flow of 400–500 pigs per hour to the main drive alley and it greatly reduced damage and bruises caused by fighting. Each group of 15 pigs were pen mates on the farm. A disadvantage of this system is very high cost, and it may not work with some of the new genetic lines of pigs which are highly excitable and difficult to handle. The Danish abattoir that had this sytem had calm placid pigs which moved easily.

When strange bulls are mixed, physical activity during fighting increases DFD meat. The installation of either steel bars or an electric grid over the holding pens at the abattoir, prevented dark cutting in bulls (Kenny and Tarrant, 1987). These devices prevent mounting. The electric grid should only be used with animals that have been fattened in pens equipped with an electric grid. In Sweden and other countries where small numbers of bulls are fattened; individual pens are recommended at the abattoir (Puolanne and Aalto, 1981). In some European abattoirs, the holding area consists of a series of single file races which lead to the stunner. Bulls are unloaded directly into the races and each bull is kept separated by guillotine gates.

Flooring and fence design to reduce injuries

Floors must have a non-slip surface (Grandin, 1983a; Stevens and Lyons, 1997). For cattle, concrete floors should have deep 2.5 cm V-grooves in a 20 cm square or diamond pattern. Do not use the deep groove pattern for cattle living quarters such as cubicle housing or tied housing for milking cows. Prolonged standing for several weeks on the rough floor will damage the animals' feet. For pig and sheep abattoirs imprint the wet concrete with a stamp constructed from expanded steel mesh which has a 3.8 cm long opening (Grandin, 1982).

Concrete slats may be used in holding pens, but the drive alleys should have a solid concrete floor. Precast slats for cattle or swine confinement buildings will work well, and the slats should be specially ordered with a grooved surface. Slats or gratings used in pig and sheep facilities should face in the proper direction as sheep move more easily when they walk across the slats instead of parallel with them (Hutson, 1981; Kilgour, 1971), and the floor appears more solid when the animals walk across the slats. To facilitate animal movement, the animals must not be able to see light or reflection off water under the slats.

Animals will balk at sudden changes in floor texture or colour, so flooring surfaces should be uniform in appearance and free from puddles (Lynch and Alexander, 1973). In facilities that are washed, there should be concrete kerbs installed between the pens to prevent water in one pen from flowing into another. Drains should be located outside the areas where animals walk as livestock will balk at drains or metal plates across an alley (Grandin, 1987). Flooring should not move or jiggle when animals walk on it. Flooring that moves causes swine to balk (Kilgour, 1988).

Cattle and sheep can have bruises on the inside even though the hide is undamaged. Pigs are slightly less susceptible to bruising, but their meat quality is decreased when they become excited or hot. Edges with a small diameter will cause severe bruises. Never use steel angles or I-beams to construct livestock facilities: animals bumping the edges will bruise. Round pipe posts and fence rails are recommended. Surfaces which come into contact with animals should be smooth and rounded (Grandin, 1980c; Stevens and Lyons, 1977) and any exposed sharp ends of pipes should be bevelled or covered to prevent gouging. Areas which have completely solid fences should have all posts and structural parts on the outside away from the animals. An animal rubbing against a smooth flat metal surface will not bruise. A practical way to determine if a fence will cause bruising is to rub your shoulder against it; if it snags your shirt or jacket, it will cause a bruise.

All gates should be equipped with tie-backs to prevent them from swinging out into the alley and guillotine gates should be counterweighted and padded on the bottom with conveyor belting or large diamter hose (Grandin, 1983a).

Design of races, crowd pens and unloading ramps

Races

All races should have solid outer fences, and a curved single file race is especially recommended for moving cattle (Grandin, 1980a; Rider *et al.*, 1974). An inside radius of 5 m is ideal for cattle: if a shorter radius is used there must be a section of straight race at the junction between the single file race and the crowd pen. If the race is bent too sharply at the junction between the single file race and the crowd pen, it will cause animals to balk (Grandin, 1987). The race entrance must not appear to be a dead end.

Cattle races which are built correctly: walkways for the handler should run alongside the race; the walking platform should be 1 m from the top of the fence; the use of overhead walkways should be avoided. In abattoirs with restricted space a serpentine race system (Grandin, 1984) can be used, and if smooth continuous bends are used, the radius can be reduced to 1.5 m. The system must be laid out so that the animals standing at the race entrance can see a minimum of three body lengths up the race.

Curved races provide little or no advantages for pigs, because they have a strong urge to move forward in a race. A straight race will work efficiently. A race system at an abattoir must be long enough to ensure a continuous flow of animals to the stunner, but not be so long that animals become stressed waiting in line.

A common mistake is to build races too wide. Table 11.1 shows the correct lengths for cattle, pig, and sheep races. Cattle and sheep races should have straight sides. Jamming, and pigs rooting under each other, can be reduced by narrowing the bottom of the race.

In pig abattoirs, two races are sometimes built side by side, because the animals will enter more easily (Grandin, 1982). The outer walls of the race are solid, but the inner fence in between the two races is constructed from bars. This enables the pigs to see each other and promotes following behaviour. However, this system still causes stress at the stunner because pigs on one side have to wait. Stress could be greatly reduced by installing two stunners. This would enable two lines of pigs to continuously move forward, and is economically viable for large plants slaughtering over 500 pigs per hour.

A future possibility is to eliminate traditional single-file races. An idea [is] for multiple stunners with several parallel races. Funnelling the pigs down into single file is eliminated. This system would be very expensive, but may be required if pig breeders fail to select pigs for ease of handling. Swedish and Danish engineers have proposed putting pigs in individual boxes on a conveyor. This idea is so expensive that it is not practical. In Denmark there has been experimentation with group containers which hold a pen of pigs.

Crowd pens

Round crowd pens are very efficient for all species. The recommended radius for round crowd pens is 3.5 m for cattle, 1.83 m to 2 m for pigs, and 2.4 m for sheep. All crowd pens must have solid fences. For smooth animal movement

Table 11.1
Recommended single file race length for abattoirs.

Line speed (per hour)	Cattle (m)	Pigs and sheep (m)
20–75	14	6
80–150	24	11
150–300	30	11
300–500	n/a	11
>500	n/a	

never construct a building wall at the junction between the single file race and the crowd pen. The race must extend from the wall a minimum of two body lengths.

Cattle and sheep crowd pens should have a funnel design with one straight fence and the other fence on a 30° angle (Meat and Livestock Commission, no date). The funnel design should not be used with pigs as they will jam at the race entrance and jamming is very stressful for pigs (VanPutten and Elshoff, 1978). A crowd pen for pigs must be designed with an abrupt entrance to the race. A round crowd pen with an abrupt entrance to the single-file race is being successfully used in several US pig abattoirs, two crowd gates continually revolve. Another design is a single offset step equal to the width of one pig. It prevents jamming at the entrance of a single-file race (Grandin, 1982, 1987). Jamming can be further prevented by installing an entrance restricter at single-file race entrances. The entrance of the single-file race should provide only 1–2 cm on each side of each pig. A double race should also have a single offset step to prevent jamming. In Denmark an excellent double race system has been built with an offset step on both sides of the double race (Barton-Gade, 1989). A power push gate moves along the alley to urge the pigs into the races. Power push gates must never be used to forcefully push animals, and they must be designed to prevent animals from getting feet jammed under them.

For all species, solid sides are recommended on both the race and the crowd pen which leads to the race (Brockway, 1977; Grandin, 1980a, b; Grandin, 1982; Rider *et al.*, 1974; Vowles *et al.*, 1984). For operator safety, mangates must be constructed so that people can escape charging cattle. The crowd gate should also be solid to prevent animals from turning back. Wild animals tend to be calmer in facilities with solid sides. In holding pens, solid pen gates along the main drive alley facilitate animal movement (Grandin, 1980b).

When young pigs were given a choice of ramps, they preferred a ramp with either solid or woven wire sides (Phillips *et al.*, 1987). Ramps with vertical or horizontal barred sides were avoided. The overhead lighting used in the indoor experiment may have made the wire mesh appear solid.

A crowd pen must never be built on a ramp. The animals will pile up against the crowd gate. A small drainage slope will not affect animal movement. To facilitate entry into the single-file race, construct two to three body lengths of level single-file ramp before the animals reach the ramp.

Unloading

In abattoirs more than one unloading ramp is usually required to facilitate prompt unloading. During warm weather, prompt unloading is essential because heat rapidly builds up in stationary vehicles. In some facilities, unloading pens will be required. These pens enable animals to be unloaded promptly prior to sorting, weighing or identification checking. After one or more procedures are performed, the animals move to a holding pen.

Loading dock height is going to vary depending on the types of vehicles used. If vehicle heights vary a few centimetres, it is recommended to construct non-adjustable ramps level with the lowest vehicles used: this will enable the cross-over bridge that is attached to the higher vehicles to be used more effectively. Facilities used for unloading only should be 2.5 m to 3 m wide to provide the animals with a clear exit into the alley (Grandin, 1980d).

Ramps and slopes

Ideally an abattoir stockyard should be built at truck deck level to eliminate ramps for both unloading and movement to the stunner. Sheep move more easily on a level surface (Hitchcock and Hutson, 1979), and many animals are injured on excessively steep ramps. For fattened slaughter-weight pigs, a 15° angle is recommended (VanPutten, 1981), but the maximum angle for non-adjustable livestock unloading ramps is 20° (Grandin, 1979). If possible the ramp to the stunner should not exceed 10° for pigs, 15° for cattle, and 20° for sheep. A pig's heart rate increases as the angle of the ramp increases (VanPutten and Elshof, 1978), and the heart rate of a pig is faster when it is climbing a ramp compared to descending (Mayes and Jesse, 1980). Excessively steep ramps were avoided by pigs in a preference test; an angle of 20–24° was preferred compared to 28–32° (Fraser *et al.*, 1986; Phillips *et al.*, 1988). To reduce the possibility of falls, unloading ramps should have a flat dock at the top. This provides a level surface for animals to walk on when they first step off the truck (Agriculture Canada, 1984, Grandin, 1979; Stevens and Lyons, 1977), and this same principle also applies to ramps to the stunner. A level portion facilitates animal entry into the restrainer or stunning box.

Stairsteps are recommended on concrete ramps; they are easier to walk on after the ramp becomes worn or dirty. However, in new clean facilities small pigs expressed no preference between stairsteps or closely spaced cleat (Phillips *et al.*, 1987). The movements in this experiment were voluntary.

Recommended dimensions for stairsteps on unloading ramps are a 30 cm minimum tread width and a 10 cm rise for cattle, and a 25 cm tread width and a 5 cm rise for slaughter-weight pigs (Grandin, 1980d, 1982; United States Department of Agriculture, 1967). On a ramp to the stunner the tread width should be increased to 45 cm. The steps should be grooved to provide a non-slip surface. When cleats are used, space them 20 cm apart for large cattle and slaughter-weight pigs (Mayes, 1978). The 20 cm is measured from the beginning of one cleat to the beginning of the next cleat.

Design of restraint systems

Restraint devices to hold animals during stunning have improved since the days of large stunning boxes which hold more than one animal. One of the first major innovations was the V-conveyor restrainer for pigs which was patented in the USA by Regensburger in 1940. This device consists of two obliquely angled conveyors that together form a V-shape, and pigs ride with their legs protruding through the space at the bottom of the V. In the late 1970s the Nijhuis company in Holland developed an automatic stunner incorporated into two V-restrainers: one restrainer runs faster than the other to index the pigs for the stunner. The V-restrainer is a humane system for most types of pigs and sheep (Grandin, 1980d); the pressure exerted against the sides of a pig will cause it to relax (Grandin *et al.*, 1989). However, the V-restrainer is not suitable for restraining extremely heavy muscled pigs with large overdeveloped hams; the V pinches the large hams and the slender forequarters are not supported.

For smaller pig abattoirs the squeeze box restrainer works well. It was patented in the USA by Hlavacek *et al.* in 1963 and consists of two padded panels which squeeze the pig. After stunning, the animal is ejected. In the 1980s a modified version of this device was developed for small European abattoirs. In the late 1960s the V-restrainer was enlarged for use on cattle by Oscar Schmidt of the Cincinnati Butcher's Supply Company and Don Williams of Armour and Company in the USA. A complete description and dimensions can be found in Grandin (1980d, 1983b).

The development of the V-restrainer for adult cattle was a major innovation because it replaced dangerous multiple cattle stunning boxes in high-speed slaughter plants. After stunning, the shackle chain was attached to one rear leg while the animal was still held in the restrainer.

V-restrainer systems are suitable for fat cattle, but there are problems with small calves and thin animals. Small calves tend to cross their legs and fall through (Giger *et al.*, 1977). Lambooy (1986) also reported that 200 kg veal calves had difficulty entering the restrainer.

Researchers at the University of Connecticut developed a laboratory prototype for a new type of restrainer system (Giger *et al.*, 1977; Westervelt *et al.*, 1976). Calves and sheep are supported under the belly and the brisket by two moving rails. This research demonstrated that animals restrained in this manner were under minimal stress. Sheep and calves rode quietly on the restrainer and seldom struggled. The space between the rails provides a space for the animal's brisket. The prototype was a major step forward in humane restrainer design, but many components still had to be developed to create a system which would operate under commercial conditions. Some of the items that needed to be developed were an entrance device which would reliably position the animal's legs on each side of the double rail, a rapid adjustment system and compatibility with existing abattoir equipment.

In 1986 the first double-rail restrainer was installed in a large commercial calf and sheep slaughter plant. Dimensions and details of the system can be found in Grandin, 1988. Development of the commercial system was accomplished by Grandin Livestock Handling Systems, Inc. and Clayton H. Landis Company.

In 1989 a V-restrainer conveyor was replaced by a double-rail restrainer in a large cattle slaughter plant by Grandin Livestock Handling Systems and Swilley Equipment, Logan, Iowa (Grandin, 1991). There are now eight large cattle systems operating in commercial slaughter plants. The double-rail restrainer has many advantages compared to the V-restrainer. Stunning is easier and more accurate because the operator can stand 28 cm closer to the animal. Cattle enter more easily because they can walk in with their legs in a natural position. Shackling is facilitated because the legs are spread apart and cattle ride more quietly in the double-rail restrainer.

Proper design is essential for smooth humane operation. Incoming cattle must not be able to see light coming up from under the restrainer. It should have a false floor to provide incoming cattle with the appearance of a solid floor to walk in. The false floor is located about 20 cm below the animal's feet. The cattle restrainer also has a longer hold-down rack than the calf version. To keep cattle calm they must be fully restrained and settled down on the conveyor before they emerge from under the hold-down rack. Having a long enough hold-down rack is very important. If the hold-down is too short the cattle are more likely to become agitated. On both the calf and cattle versions of the double-rail restrainer there should be about 5 cm of clearance between an entering animal's back and the hold-down rack.

Conventional stunning boxes

The use of a device to hold the head during stunning is not required in either a V-restrainer or a double-rail restrainer. In a conventional stunning box stunning accuracy can be greatly improved by the use of a yoke to hold the head. Yokes and automatic head restraints for cattle have been developed in Australia, New Zealand and the UK. A common mistake is to build stunning boxes too wide. A 76 cm wide stunning box will hold all cattle with the exception of some of the largest bulls. Stunning boxes must have non-slip floors; humane stunning and handling is impossible if an animal is skidding and slipping in the stunning box.

Ritual slaughter

Ritual slaughter is increasing in many countries due to increased Moslem demand for Halal meat. In some countries, such as the USA, it is legal to suspend live animals by one back leg for ritual slaughter. This cruel practice is also very dangerous. Replacement of shackling and hoisting by a humane restraint device will greatly reduce accidents (Grandin, 1988). In Europe and Australia the use of humane restraining devices is required.

The first restraining device for ritual slaughter was developed in Europe over 40 years ago. The Weinberg casting pen consists of a narrow stall which slowly inverts the animal until it is lying on its back. It is less stressful than shackling and hoisting, but it is much more stressful than more modern upright restraint devices (Dunn, 1990). Animals restrained in the Weinberg had much higher levels of vocalizations and cortisol (stress hormone) compared to cattle restrained in the upright ASPCA pen.

A major innovation in ritual restraint equipment was the ASPCA pen. It was developed in 1963 at Cross Packing Company in Philadelphia (Marshall, *et al.*, 1963). It consists of a narrow stall with an opening in the front for the animal's head. A lift under the belly prevents the animal from collapsing after the throat cut. Proper design and operation is essential. The belly lift should not lift the animal off the floor. Air pressure which operates the rear pusher gate should be reduced to prevent excessive pressure on the rear, and the head holder must have a stop or a bracket to prevent excessive bending of the neck.

Further, developments in ritual slaughter equipment are the use of a mechanical head holder on a V-restrainer (Grandin, 1980d) and ritual slaughter of calves on the double rail (Grandin, 1988). The research team at the University of Connecticut has also developed a small inexpensive restrainer to hold calves and sheep during ritual slaughter (Giger *et al.*, 1977). For larger calves a miniature ASPCA pen could be built.

Conclusions

The use of more humane equipment and handling practices will improve animal welfare. It will also have the added advantage of improving both meat quality and employee safety.

References

Agriculture Canada (1984) Recommended code of practice for care and handling of pigs. Publication 1771/E, Agriculture Canada, Ottawa.

Ames, D. R. (1974) Sound stress and meat animals. *Proceedings*, International Livestock Environment Symposium ASAE. SP-0174, 324.

Barton-Gade, P. (1985) Developments in the pre-slaughter handling of slaughter animals. *Proceedings*, European Meeting of Meat Research Workers. Albena, Bulgaria. Paper 1: 1,1–6.

Barton-Gade, P. (1989) Pre-slaughter treatment and transportation in Denmark. *Proceedings*, International Congress of Meat Science and Technology, Copenhagen, Denmark.

Brockway, B. (1977) *Planning a Sheep Handling Unit.* Farm Buildings Centre, Kenilworth, Warwickshire, England.

Canadian Meat Council (1980) *Guide to PSE Pork.* Canadian Meat Council, Islington, Ontario.

Dunn, C. S. (1990) Stress reactions of cattle undergoing ritual slaughter using two methods of restraint. *Veterinary Record* 126, 522–5.

Eldridge, G. A. (1988) The influence of abattoir lairage conditions on the behavior and bruising of cattle. *Proceedings*, 34th International Congress of Meat Science and Technology, Brisbane, Australia.

Fraser, D., Phillips, P. A. and Thompson, R. K. (1986) A test of a free-access two level pen for fattening pigs. *Animal Production* 42, 269–74.

Giger, W., Prince, R. P., Westervelt, R. G. and Kinsman, D. M. (1977) Equipment for low stress animal slaughter. *Transactions of the American Society of Agricultural Engineers* 20, 571–8.

Grandin, T. (1979) Designing meat packing plant handling facilities for cattle and hogs. *Transactions of the American Society of Agricultural Engineers* 22, 912–17.

Grandin, T. (1980a) Observations of cattle behavior applied to design of cattle handing facilities. *Applied Animal Ethology* 6, 19–31.

Grandin, T. (1980b) Livestock behavior as related to handling facility design. *International Journal of the Study of Animal Problems* 1, 33–52.

Grandin, T. (1980c) Bruises and carcass damage. *International Journal of the Study of Animal Problems* 1, 121–37.

Grandin, T. (1980d) Designs and specifications for livestock handling equipment in slaughter plants. *International Journal of the Study of Animal Problems* 1, 178–200.

Grandin, T. (1982) Pig behavior studies applied to slaughter plant design. *Applied Animal Ethology* 9, 141–51.

Grandin, T. (1983a) Welfare requirements of handling facilities. In: Baxter, S. H., Baxter, M. R. and MacCormack, J. A. D. (eds), *Farm Animal Housing and Welfare.* Martinus Nijhoff, Boston/The Hague/Dordrecht/Lancaster.

Grandin, T. (1983b) System for handling cattle in large slaughter plants. American Society of Agricultural Engineers, Paper No. 83-406, ASAE, St Joseph, MI, USA.

Grandin, T. (1984) Race system for slaughter plants with 1.5 m radius curves. *Applied Animal Behavior Science* 13, 295–9.

Grandin, T. (1987) Animal handling. In: Price, E. O. (ed.), *Farm Animal Behavior*, vol. 3. Veterinary. Clinics of North America, W. B. Saunders, Philadelphia, 323–38.

Grandin, T. (1988) Double rail restrainer conveyor for livestock handling. *Journal of Agricultural Engineering Research* 41, 327–38.

Grandin, T. (1989a) 'Effect of rearing environment and environmental enrichment on behaviour and neural development in young pigs.' Ph.D. dissertation, University of Illinois, Champaign/Urbana.

Grandin, T. (1989b) Behavioral principles of livestock handling. *The Professional Animal Scientist*, American Registry of Professional Animal Scientists, Champaign, Illinois, 5, 1–11.

Grandin, T. (1991) Double rail restrainer for handling beef cattle. American Society of Agricultural Engineers, Paper No. 91-5004, ASAE, St Joseph, MI, USA.

Grandin, T., Dodman, N. and Shuster, L. (1989) Effect of Naltrexone on relaxation induced by flank pressure in pigs. *Pharmacology, Biochemistry and Beahvior* 33, 839–42.

Hitchcock, D. K. and Hutson, G. D. (1979) The movement of sheep on inclines. *Australian Journal of Experimental Agriculture and Animal Husbandry* 19, 176–82.

Hlavacek, R. J. (1963) Method for restraining animals, US Patent Number 3,115,670.

Hutson, G. D. (1981) Sheep movement on slotted floors. *Australian Journal of Experimental Agriculture and Animal Husbandry* 21, 474–9.

Hutson, G. D. (1985) Sheep and cattle handling facilities. In: Moore, B. L. and Chenoweth, P. J. (eds), *Grazing Animal Welfare*. Australian Veterinary Association, Queensland, 124–36.

Kenny, F. J. and Tarrant, P. V. (1987) The behavior of young Friesian bulls during social regrouping at an abattoir. Influence of an overhead electrified wire grid, *Applied Animal Behavior Science* 18, 233–46.

Kilgour, R. (1971) Animal handling in works; pertinent behavior studies. In: *Proceedings of the 13th Meat Industry Research Conference*, Hamilton, New Zealand, 9–12.

Kilgour, R. (1976) The behavior of farmed beef bulls. *New Zealand Journal of Agriculture* 13, 31–3.

Kilgour, R. (1978) The application of animal behavior and the humane care of farm animals. *Journal of Animal Science* 46, 1479–86.

Kilgour, R. (1983) Using operant test results for decisions on cattle welfare. *Proceedings*, Conference on the Human-Animal Bond, Minneapolis, Minnesota.

Kilgour, R. (1988) Behavior in the pre-slaughter and slaughter environments. *Proceedings*, International Congress of Meat Science and Technology, Part A, Brisbane, Australia, 130–8.

Kilgour, R. and deLangen, H. (1970) Stress in sheep from management practices. *Proceedings*, New Zealand Society of Animal Production 30, 65-76.

Lambooy, E. (1986) Automatic electrical stunning of veal calves in a V-type restrainer. *Proceedings*, 32nd European Meeting of Meat Research Workers, Ghent, Belgium, Paper 2:2, 77–80.

Lemman, W. B. and Patterson, G. H. (1964) Depth perception in sheep: Effects of interrupting the mother neonate bond. *Science* 145, 835–6.

Livestock Conservation Institute (1988) *Livestock Trucking Guide*, 6414 Copps Avenue, Madison, Wisconsin.

Lynch, J. J. and Alexander, G. (1973) *The Pastoral Industries of Australia*, University Press, Sydney, 371–400.

Marshall, B. L. (1977) Bruising in cattle presented for slaughter. *New Zealand Veterinary Journal* 25, 83–6.

Marshall M. Milburg, E. E. and Shultz, E. W. (1963) Apparatus for holding cattle in position for humane slaughtering. US Patent 3,092,871.

Mayes, H. F. (1978) Design criteria for livestock loading chutes. ASAE paper 78-6014.

Mayes, H. F. and Jesse, G. W. (1980) Heart rate data of feeder pigs. ASAE paper 80-4023.

Meat and Livestock Commission (no date) Cattle handling. Livestock Buildings Consultancy; Meat and Livestock Commission, Queensway, Bletchley, Milton Keynes, England.

Midwest Plan Service (1980) *Structures and Environment Handbook*, 10th edn. Iowa State University, Ames, 319.

Phillips, P. A., Thompson, B. K. and Fraser, D. (1987) Ramp designs for young pigs. American Society of Agricultural Engineers, Paper No. 87-4511, St Joseph, MI, USA.

Phillips, P. A., Thompson, B. K. and Fraser, D.

(1988) Preference tests of ramp designs for young pigs. *Canadian Journal of Animal Science* 68, 41-8.

Phillips, P. A., Thompson, B. K. and Fraser, D. (1989) The importance of cleat spacing in ramp design for young pigs. *Canadian Journal of Animal Science* 69, 483–6.

Puolanne, E. and Aalto, H. (1981) The incidence of dark cutting beef in young bulls in Finland. In: Hood, D. E. and Tarrant, P. V. (eds), *The Problem of Dark Cutting Beef*. Martinus Nijhoff, The Hague, 462–75.

Price, M. A. and Tennessen, T. (1981) Pre-slaughter management and dark cutting carcasses in young bills. *Canadian Journal of Animal Science* 61, 205–8.

Prince, J. H. (1977) The eye and vision. In: Swenson, M. H. (ed.), *Dukes Physiology of Domestic Animals*. Cornell University Press, New York, 696–712.

Regensburger, R. W. (1940) Hog stunning pen. US Patent 2,185,949.

Rider, A., Butchbaker, A. F. and Harp, S. (1974) Beef working, sorting and loading facilities. American Society of Agricultural Engineers, Paper No. 74–4523, St Joseph, MI, USA.

Stevens, R. A. and Lyons, D. J. (1977) Livestock bruising project: Stockyard and crate design. National Materials Handling Bureau, Department of Productivity Australia.

Stricklin, W. R., Graves, H. B. and Wilson, L. L. (1979) Some theoretical and observed relationships of fixed and portable spacing behavior in animals. *Applied Animal Ethology* 5, 201–14.

Tennessen, T. and Price, M. A. (1980) Mixing unacquainted bulls: The primary cause of dark cutting beef. *The 59th Annual Feeder's Day Report*. Agriculture and Forestry Bulletin, University of Alberta, Alberta, Canada, 34–5.

Tulloh, N. M. (1961) Behavior of cattle in yards. II: A study of temperament. *Animal Behavior* 9, 25–30.

United States Department of Agriculture (1967) Improving services and facilities at public stockyards. *Agriculture Handbook* 337, Packers and Stockyards Administration, United States Department of Agriculture, Washington, DC.

VanPutten, G. (1981) Handling slaughter pigs prior to loading and unloading the lorry. Paper presented at the Seminar on Transport, CEC, Brussels, 7-8 July.

VanPutten, G. and Elshoff, W. J. (1978) Observations on the effect of transport on the well being and lean quality of slaughter pigs. *Animal Regulation Studies* 1, 247–71.

Vowles, W. J., Eldridge, G. A. and Hollier, T. J. (1984) The behavior and movement of cattle through forcing yards. *Proceedings*, Australian Society for Animal Production 15, 766.

Warriss, P. D., Kestin, S. C. and Robinson, J. M. (1983) A note on the influence of rearing environment on meat quality in pigs. *Meat Science* 9, 271–9.

Westervelt, R. G., Kinsman, D. M., Prince, R. P. and Giger, W. (1976) Physiological stress measurement during slaughter of calves and lambs. *Journal of Animal Science* 42, 831-4.

Weyman, G. (1987) 'Unloading and loading facilities at livestock markets.' M.S. Thesis, Hatfield Polytechnic, UK.

Overviews
Research File
Meat Focus International—March 1994

Slaughter

Religious slaughter and animal welfare: a discussion for meat scientists

Temple Grandin[1] and Joe M. Regenstein[2]

The opinions expressed in this article are those of the authors and may not reflect the opinions of organisations with which they are affiliated.

Both the Muslim and Jewish faiths have specific requirements for the slaughter of religiously acceptable animals. The major difference from the general practices in most countries is that the animals are not stunned prior to slaughter. It is important that meat scientists understand the implications of these differences. They need to critically consider the scientific information available about the effects of different slaughter practices on animals before reaching any judgements about the appropriateness of a particular form of slaughter. It is also important that they understand the importance of these practices to the people who follow these religious codes. We hope to discuss some information that may be useful in evaluating religious slaughter.

The Jewish dietary code is described in the original five books of the Holy Scriptures. The Muslim code is found in the Quran. Both codes represented major advancements in the respect for animals and their proper handling in ancient times. For example, the Jewish code specifically forbid the use of limbs torn from live animals and the slaughter of both a mother animal and her children on the same day.

One way to view the rather comprehensive legal system of the Jewish faith is spelled out in the paragraphs below. We feel this explanation may help others understand the degree of significance of these religious practices to those of the Jewish faith (Grunfeld, 1972).

"'And ye shall be men of holy calling unto Me, and ye shall not eat any meat that is torn in the field' (Exodus XXII:30). Holiness or self-sanctification is a moral term; it is identical with . . .moral freedom or moral autonomy. Its aim is the complete self-mastery of man.

"To the superficial observer it seems that men who do not obey the law are freer than law-abiding men, because they can follow their own inclinations. In reality, however, such men are subject to the most cruel bondage; they are slaves of their own instincts, impulses and desires. The first step towards emancipation from the tyranny of animal inclinations in man is, therefore, a voluntary submission to the moral law. The constraint of law is the beginning of human freedom. . . .Thus the fundamental idea of Jewish ethics, holiness, is inseparably connected with the idea of Law; and the dietary laws occupy a central position in that system of moral discipline which is the basis of all Jewish laws.

"The three strongest natural instincts in man are the impulses of food, sex, and acquisition. Judaism does not aim at the destruction of these impulses, but at their control and indeed their sanctification. It is the law which spiritualises these instincts and transfigures them into legitimate joys of life."

We hope that the above quote suggests the importance of the kosher dietary laws to people of the Jewish faith. Similar religious philosophies underpin the Muslim requirements. Thus, the ability to carry out ritual slaughter is extremely important to people of these two faiths. The banning of such slaughter would certainly be viewed as a hostile act.

The actual reference to slaughter in the Jewish

Holy scriptures is quite cryptic: ". . . thou shall kill of thy herd and of thy flocks, which the Lord hath given thee, as I have commanded thee . . ." (Deuteronomy XII:21). Clearly it was assumed that people were familiar with the rules for kosher slaughter. These were a part of the "oral code". Eventually these rules were written down in the series of volumes referred to as the Talmud as well as in other religious texts. The Talmud contains an entire section on slaughter and the subsequent inspection of animals to ensure that they are religiously "clean". The text includes detailed anatomical information in order to teach the religious Jew exactly what was to be done during slaughter and the subsequent post-mortem inspection.

The Muslim rules with respect to animals and slaughter are contained in the Quran. Blood, pork, animals dying due to beating, strangulation, falls, goring or other damage from animals and animals dedicated to other religions are all forbidden. Any Muslim may slaughter an animal while invoking the name of Allah. In cases where Muslims cannot kill their own animals, they may eat meat killed by a "person of the book", i.e., a Christian or a Jew. Again stunning prior to slaughter is generally not the practice. However, a non-penetrating concussion stunning prior to slaughter has received approval from some Muslim authorities. Work in the 80's in New Zealand led to the development of a very sophisticated electrical stunning apparatus that met a Muslim standard where an animal must be able to regain consciousness in less than a minute and must be able to eat within five minutes. Head-only electric stunning prior to Muslim slaughter is used in almost all sheep slaughter plants in New Zealand and Australia. Electric stunning of cattle is used in many New Zealand Muslim cattle slaughter plants and the practice is spreading to Australia. Meat from electrically stunned cattle and sheep is exported to middle eastern countries with stringent religious requirements. "Halal" slaughter in New Zealand and Australia may be carried out by regular plant workers while Muslim religious leaders are present and reciting the appropriate prayers. However, the larger Halal slaughter plants in Australia, New Zealand, and Ireland do employ Muslim slaughtermen. Muslim slaughter without stunning is forbiddin in New Zealand. With Muslim slaughter in countries not using stunning, we are also concerned about the training given to the slaughtermen. More work is needed on training programs to teach proper sharpening of knives and to improve the actual slaughter techniques.

The Jewish religious codes require that allowed animals be slaughtered by a specially trained Jewish male, while the Muslims prefer that allowed animals be slaughtered by a person of that faith. In the case of the Jewish dietary laws, a specially trained person of known religiosity carries out the slaughter. This person, the "shochet", is specifically trained for this purpose. He is trained to use a special knife, called the "chalef", to rapidly cut in a single stroke the jugular vein and the carotid artery without burrowing, tearing or ripping the animal. The knife is checked regularly for any imperfections which would invalidate the slaughter. This process when done properly leads to a rapid death of the animal. A sharp cut is also known to be less painful.

Need for objective evaluation

Given the importance of religious slaughter to people of these two major faiths, it is important that scientists must be absolutely objective when evaluating these practices from an animal welfare standpoint.

Evaluation of religious slaughter is an area where many people have lost scientific objectivity. This has resulted in biased and selective reviewing of the literature. Politics have interfered with good science. There are three basic issues. They are stressfulness of restraint methods, pain perception during the incision and latency of onset of complete insensibility.

Restraint

A key intellectual consideration is separation of the variable of restraint stress from the animal's reaction to the slaughter procedure. Stressful or painful methods of restraint mask the animal's reactions to the throat cut. In North America some kosher slaughter plants use very stressful methods of restraint such as shackling and hoisting fully conscious cattle by one rear leg. Observations of the first author indicate that cattle restrained in this manner often struggle and bellow and the rear leg is bruised. Bruises or injuries caused by the restraint methods (or from any other cause) would be objectionable to observant Jews. In Europe, the use of casting pens which invert cattle onto their backs completely mask reactions to the throat cut. Cattle resist inversion and twist their necks in an attempt to right their heads. Earlier versions of the Weinberg casting pen are more stressful than an upright restraint device (Dunn 1992). An improved casting pen, called the Facomia pen, is probably less stressful than older Weinberg's pens but a well designed upright restrained system would be more comfortable for cattle. Another problem with all types of casting pens is that both cattle and calves will aspirate blood after the incision. This does not occur when the animal is held in an upright position.

Unfortunately some poorly designed upright American Society for the Prevention of Cruelty to Animals (ASPCA) restraint boxes apply excessive pressure to the thoracic and neck areas of cattle. In the interest of animal welfare the use of any stressful method of restraint should be eliminated. A properly designed and operated upright restraint system will cause minimum stress. Poorly designed systems can cause great stress. Many stress problems are also caused by rough handling and excessive use of electric prods. The very best mechanical systems will cause distress if operated by abusive, uncaring people.

In Europe there has been much concern about the stressfulness of restraint devices used for both conventional slaughter (where the bovine is stunned) and ritual slaughter. Ewbank *et al.* (1992) found that cattle restrained in a poorly designed head holder, i.e., where over 30 seconds was required to drive the animal into the holder, had higher cortisol levels than cattle stunned with their heads free. Cattle will voluntarily place their heads in a well designed head restraint device that is properly operated by a trained operator (Grandin 1992). Tume and Shaw (1992) reported very low cortisol levels of only 15 ng/ml in cattle during stunning and slaughter. Their measurements were made in cattle held in a head restraint (Shaw, personal communication, 1993). Cortisol levels during on-farm restraint of extensively reared cattle range from 25 to 63 ng/ml (Mitchell *et al.*, 1988; Zavy *et al.*, 1992).

Head stanchions used for electrical stunning of cattle in New Zealand work quite well. The first author observed these systems in two plants. Most cattle entered the stunning box voluntarily and quietly placed their heads in the stanchion. The animal was immediately stunned after its head was clamped. Immediate electrical stunning is essential in order to prevent the animal from fighting the stanchion. When this sytem was operated correctly the cattle were quiet and calm. The electric stun stanchion did not restrain the body. For ritual slaughter or captive bolt stunning devices to restrain the body are strongly recommended. Animals remain calmer in head restraint devices when the body is also restrained. Stunning or slaughter must occur within 10 seconds after the head is restrained.

Reactions to the throat cut

The variable of reactions to the incision must be separated from the variable of the time required for the animal to become completely insensible. Recordings of EEG or evoked potentials measure the time required for the animal to lose consciousness. They are not measures of pain. Careful observations of the animal's behavioural reactions to the cut are one of the best ways to determine if cutting the throat without prior stunning is painful. The time required for the animals to become unconscious will be discussed later.

Observations of over 3000 cattle and formula-fed veal calves were made by the first author in three different U.S. kosher slaughter plants. The plants had state-of-the-art upright restraint systems. The systems are described in detail in Grandin (1988, 1991, 1992, 1993a). The cattle were held in either a modified ASPCA pen or a double rail (centre track) conveyor restrainer. This equipment was operated by the first author or a person under her direct supervision. Very little pressure was applied to the animals by the rear pusher gate in the ASPCA pen. Head holders were equipped with pressure limiting devices. The animals were handled gently and calmly. It is impossible to observe reactions to the incision in an agitated or excited animal. Blood on the equipment did not appear to upset the cattle. They voluntarily entered the box when the rear gate was opened. Some cattle licked the blood.

In all three restraint systems, the animals had little or no reaction to the throat cut. There was a slight flinch when the blade first touched the throat. This flinch was much less vigorous that an animal's reaction to an eartag punch. There was no further reaction as the cut proceeded. Both carotids were severed in all animals. Some animals in the modified ASPCA pen were held so loosely by the head holder and rear pusher gate that they could have easily pulled away from the knife. These animals made no attempt to pull away. In all three slaughter plants, there was almost no visible reaction of the animal's body or legs during the throat cut.

Body and leg movements can be easily observed in the double rail restrainer because it lacks a pusher gate and very little pressure is applied to the body. Body reactions during the throat cut were much fewer than the body reactions and squirming that occurred during testing of various chin lifts and forehead hold-down brackets. Testing of a new chin lift required deep, prolonged invasion of the animal's flight zone by a person. Penetration of the flight zone of an extensively raised animal by people will cause the animal to attempt to move away (Grandin, 1993a). The throat cut caused a much smaller reaction than penetration of the flight zone. It appears that the animal is not aware that its throat has been cut. Bager *et al.* (1992) reported a similar observation with calves. Further observations of 20 Holstein, Angus and Charolais bulls indicated that they did not react to the cut. The bulls were held in a comfortable head restraint with all body restraints released. They stood still during the cut and did not resist head restraint. After the cut the chin lift was lowered, the animal either immediately collapsed or it looked around like a normal alert animal. Within 5 to 60 seconds, the animals went into a hypoxic spasm and sensibility appeared to be lost. Calm animals had almost no spasms and excited cattle had very vigorous spasms. Calm cattle collapsed more quickly and appeared to have a more rapid onset of insensibility. Munk *et al.* (1976) reported similar observations with respect to the onset of spasms. The spasms were similar to the hypoxic spasms which occur when cattle become unconscious in a V-shaped stanchion due to pressure on the lower neck. Observations in feedyards by the first author during handling for routine husbandry procedures indicated that pressure on the carotid arteries and surrounding areas of the neck can kill cattle within 30 seconds.

The details spelled out in Jewish law concerning the design of the knife and the cutting method appear to be important in preventing the animal from reacting to the cut. The knife must be razor sharp and free of nicks. It is shaped like a straight razor and it must be twice the width of the animal's neck. The cut must be made without hesitation or delay. It is also prohibited for the incision to close back over the knife during the cut. This is called "halagramah" (digging) (Epstein, 1948). The prohibition against digging appears to be important in reducing the animal's reaction to the cut. Ritual slaughtermen must be trained in knife sharpening. Shochets have been observed using a dull knife. They carefully obeyed the religious requirements of having a smooth nick-free knife, but they failed to keep it sharp. Observations of Halal cattle slaughter without stunning done by a Muslim slaughterman with a large, curved skinning knife resulted in multiple hacking cuts. Sometimes there was a vigorous reaction from the animal.

Further observations of kosher slaugher conducted in a poorly designed holder, i.e., one which allowed the incision to close back over the knife during the cut, resulted in vigorous reactions from the cattle during the cut. The animals kicked violently, twisted sideways, and shook the restraining device. Cattle which entered the poorly designed head holder in an already excited, agitated state had a more vigorous reaction to the throat cut than calm animals. These observations indicated that head holding devices must be designed so that the incision is held open during and immediately after the cut. Occasionally, a very wild, agitated animal went into a spasm which resembled an epileptic seizure immediately after the cut. This almost never occurred in calm cattle.

Time to loss of consciousness

Scientific researchers agree that sheep lose consciousness within 2 to 15 seconds after both carotid arteries are cut (Nangeroni and Kennett, 1963; Gregory and Wotton, 1984; Blackmore, 1984). However, studies with cattle and calves indicate that most animals lose consciousness rapidly, however, some animals may have a period of prolonged sensibility (Blackmore, 1984; Daly *et al.*, 1988) that lasts for over a minute. Other studies with bovines also indicate that the time required for them to become unconscious is more variable than for sheep and goats (Munk *et al.*, 1976; Gregory and Wotten, 1984). The differences between cattle and sheep can be explained by differences in the anatomy of their blood vessels.

Observations by the first author of both calf and cattle slaughter indicate that problems with prolonged consciousness can be corrected. When a shochet uses a rapid cutting stroke, 95% of the calves collapse almost immediately (Grandin, 1987). When a slower, less-decisive stroke was used, there was an increased incidence of prolonged sensibility. Approximately 30% of the calves cut with a slow knife stroke had a righting reflex and retained the ability to walk for up to 30 seconds.

Gregory (1988) provided a possible explanation for the delayed onset of unconsciousness. A slow knife stroke may be more likely to stretch the arteries and induce occlusion. Rapid loss of consciousness will occur more readily if the cut is made as close to the jaw bone as religious law will permit, and the head holder is loosened immediately after the cut. The chin lift should remain up. Excessive pressure applied to the chest by the rear pusher gate will slow bleed out. Gentle operation of the restrainer is essential. Observations indicate that calm cattle lose consciousness more rapidly and they are less likely to have contracted occluded blood vessels. Calm cattle will usually collapse within 10 to 15 seconds.

Upright restraint equipment design

Good upright restraint equipment is available for low stress, comfortable restraint of sheep, calves and cattle (Giger *et al.*, 1977; Westervelt *et al.*, 1976; Grandin, 1988, 1991, 1992, 1993). To maintain a high standard of animal welfare, the equipment must be operated by a trained operator who is closely supervised by plant management. Handlers in the lairage and race areas must handle animals gently and induce each animal to calmly enter the restrainer. Unfortunately, some very poorly designed restraint systems have recently been installed in Europe. The designers had little regard for animal comfort. Below is a list of specific recommendations.

All restraint devices should use the concept of optimal pressure. The device must hold the animal firmly enough to provide a "feeling of restraint" but excessive pressure that would cause discomfort should be avoided. Many people operating pens make the mistake of squeezing an animal harder if it struggles. Struggling is often a sign of excessive pressure.

1. To prevent excessive bending of the neck, the bovine's forehead should be parallel to the floor. This positions the throat properly for ritual slaughter and stretches the neck skin minimising discomfort. There is an optimal tightness for the neck skin. If it is too loose, cutting is more difficult. If it is too tight, the Jewish rule which prohibits tearing may be violated as the incision would have a tendency to tear before being cut by the knife. This also would be likely to cause pain. Some head restraints cause great distress to the cattle due to excessive bending of the neck in an attempt to obtain extreme throat skin tightness. This is not necessary for compliance

with religious law. One must remember that 4000 years ago hydraulic devices which could achieve such extremes of throat tightness were not available.

All head holders must be equipped with pressure limiting devices. Pressure limiting valves will automatically prevent a careless operator from applying excessive pressure. A 15 cm wide forehead bracket covered with rubber belting will distribute pressure uniformly and the animal will be less likely to resist head restraint. The forehead bracket should also be equipped with an 8 cm diameter pipe that fits behind the poll. This device makes it possible to hold the head securely with very little pressure.

2. The rear pusher gate of the ASPCA pen must be equipped with a pressure limiting device. The animal must not be pushed too far forward in the head holder. The pressure must be regulated so that the animal stands on the floor with its back level. Arching of the back is a sign of excessive pressure. A calm relaxed animal will stand quietly in the pen and will not attempt to move its head. If the animal struggles, this is due to excessive pressure or being thrown off balance by the pusher gate.

3. The animal must not be lifted off the floor by the belly lift of an ASPCA pen. The list is for restraint not lifting. Lift travel should be restricted to 71 cm from the floor to the top of the lift. Other restrainers such as the double rail system are designed to give full support under the belly. The conveyor slats must be shaped to fit the contours of the animal's sternum.

4. All parts of the equipment should always move with a slow steady motion. Jerky motions or sudden bumping of the animal with the apparatus excites and agitates them. Jerky motion can be eliminated by installing flow control valves. These valves automatically provide a smooth steady motion even if the operator jerks the controls. All restraint devices should use the concept of optimal (NOT maximum) pressure. Sufficient pressure must be applied to give the animal a feeling of being held, but excessive pressure that causes struggling must be avoided. Animals will often stop struggling when excessive pressure is slowly reduced.

5. All equipment must be engineered to reduce noise. Air hissing and clanging metal noises cause visible agitation in cattle. Air exhausts must be muffled or piped outside. Plastic guides in the sliding doortracks will further reduce noise.

6. A solid barrier should be installed around the animal's head to prevent it from seeing people and other distractions in its flight zone. This is especially important for extensively reared cattle, particularly when they are not completely tame. On conveyor systems the barrier is often not required because the animals feel more secure because they are touching each other.

7. Restraint equipment must be illuminated to encourage animals to enter. Lighting mistakes or air blowing back at the animals will cause cattle to balk (Grandin 1993b). Distractions that cause balking must be eliminated.

For plants which slaughter small numbers of sheep and goats a simple upright restrainer can be constructed from pipe (Giger *et al.*, 1977). For veal calf plants a small ASPCA pen can be used. For large high speed plants a double rail restrainer can be equipped with a head holding device.

Some rabbinical authorities prefer inverted restraint and cutting downward because they are concerned that an upward cut may violate the Jewish rule which forbids excessive pressure on the knife. There is concern that the animal may tend to push downward on the knife during an upward cut. Observations indicate that just the opposite happens. When large 800 to 950 kg bulls are held in a pneumatically powered head restraint which they can easily move, the animals pull their heads upwards away from the knife during a miscut. This would reduce pressure on the blade. When the cut is done correctly, the bulls stood still and did not move the head restraint. Equal amounts of pressure were applied by the forehead bracket and the chin lift.

Upright restraint may provide the additional advantage of improved bleed out because the animal remains calmer and more relaxed. Observations indicate that a relaxed, calm animal has improved bleedout and a rapid onset of unconsciousness. Excited animals are more likely to have a slower bleedout. The use of a comfortable upright restraint device would be advantageous from a religious standpoint because rapid bleedout and maximum loss of blood obeys the biblical principle of "Only be sure that thou eat not the blood: for the blood is life" (Deuteronomy 12:23).

Rapid bleedout and a reduction in convulsions provide the added advantage of reducing petechial haemmorrhages and improving safety. Convulsing animals are more likely to injure plant employees. A calm, quiet animal held in a comfortable restraint device will meet a higher animal welfare standard and will have a lower incidence of petechial haemorrhages.

Welfare aspects of slaughter

Many welfare concerns are centered on restraint. In Europe and the U.S. highly stressful restraint devices are still being used. Many of these systems apply excessive pressure or hold the animal in a position that causes distress. The recent 1992 decision by the Swedish Board of Agriculture to uphold its ban on slaughter without stunning was largely driven by their concerns about forceful immobilisation and clamping of cattle (Andersson *et al.*, 1992). Proper design and operation of restraint devices can alleviate most of these concerns with cattle and sheep. Restraint devices will perform poorly from an animal welfare standpoint if the animals balk and refuse to enter due to distractions such as shadows, air hissing or poor illumination. These easily correctable problems will ruin the performance of the best restraint system. Abusive workers will cause suffering in a well designed system. For more information about properly operating pens, see Grandin, 1993.

Restraint devices are used for holding animals both for ritual slaughter and for conventional slaughter where animals are stunned. The use of a head restraint will improve the accuracy of captive bolt stunning. In large beef slaughter plants without head restraint captive bolt stunning has a failure rate of 3 to 5% i.e., a second shot is required.

Captive bolt and electric stunning will induce instantaneous insensibility when they are properly applied. However, improper application can result in significant stress. All stunning methods trigger a massive secretion of epinephrine (Van de Wal 1978; Warrington 1974). This outpouring of epinephrine is greater than the secretion which would be triggered by an environmental stressor or a restraint method. Since the animal is expected to be unconscious, it does not feel the stress. One can definitely conclude that improperly applied stunning methods would be much more stressful than kosher slaughter with the long straight razor sharp knife. Kilgour (1978), one of the pioneers in animal welfare research, came to a similar conclusion on stunning and slaughter.

Halal (Muslim) slaughter performed with a knife that is too short causes definite distress and struggling in cattle. We recommend to those Muslim religious authorities who require slaughter without stunning that they require that the knife must be razor sharp with a straight blade that is at least twice the width of the neck. Unless the animals are stunned, the use of curved skinning knives is not acceptable. Due to the fact that Muslim slaughtermen do not usually receive as extensive special training in slaughter techniques as Jewish Shochtim, preslaughter stunning is strongly recommended. As stated earlier, reversible head-only electrical stunning is accepted by most Muslim religious authorites. Preslaughter stunning allows plants to run at higher line speeds and maintain high standards of animal welfare.

In some ritual slaughter plants animal welfare is compromised when animals are pulled out of the restraint box before they have lost sensibility. Observations clearly indicated that disturbance of the incision or allowing the cut edges to touch caused the animal to react strongly. Dragging the cut incision of a sensible animal against the bottom of the head opening device is likely to cause pain. Animals must remain in the restraint device with the head holder and body restraint loosened until they collapse. The belly lift should remain up during bleedout to prevent bumping of the incision against the head opening when the animal collapses.

Since animals cannot communicate, it is impossible to completely rule out the possibility that a correctly made incision may cause some unpleasant sensation. However, one can definitely conclude that poor cutting methods and stressful restraint methods are not acceptable. Poor cutting technique often causes vigorous struggling. When the cut is done correctly, behavioural reactions to the cut are much less than reactions to air hissing, metal clanging noises, inversion or excessive pressure applied to the body. Discomfort during a properly done shechitah cut is probably minimal because cattle will stand still and do not resist a comfortable head restraint device. Observations in many plants indicate that slaughter without stunning requires greater management attention to the details of the procedures than stunning in order to maintain good welfare.

Ritual slaughter is a procedure which can be greatly improved by the use of a total quality management (TQM) approach to continual incremental improvements in the process. In plants with existing upright restraint equipment significant improvements in animal welfare and reductions in petechial haemorrhages can be made by making the following changes: training of employees in gentle calm cattle handling, modifying the restrainer per the specifications in this article, eliminating distractions which make animals balk, and careful attention to the exact cutting method. There needs to be continual monitoring and improvements in technique to achieve rapid onset of insensibility. A high incidence of prolonged sensibility is caused by poor cutting technique, rough handling, excessive pressure applied by the restraint device, or agitated excited animals.

The meat industry and other animal industries need to constantly strive to improve their methods and to use the best available technology. The industry must be the leader in bringing about legitimate animal welfare goals. The veterinarian, the animal scientist, and the meat scientist can often be an important and positive contributor to this process. With your knowledge of animal biology and behaviour, you should be speaking up in a positive way for the best possible processes to slaughter animals while respecting the religious needs of others. The responsibility of all those involved in animal agriculture is to assure that animals are properly handled at all times.

References

[1]*Department of Animal Science, Colorado State University, Fort Collins, CO 80523.*

[2]*Cornell Kosher Food Initiative, Department of Food Science, Cornell University, Ithaca, NY 14853-7201, USA*

Andersson, B.; Forslid, A.; Olsson, K.; Ronnegard, J.O. *(1992) Slaughter of Unstunned Animals, Swedish Board of Agriculture, Report 1992,37.*

Bager, F.; Braggins, T.J.; Devine, C.E.; Graafhus, A.E.; Mellor, D.J.; Taener, A.; Upsdell, M.P. *(1992) Onset of insensibility in calves: Effects of electropletic seizure and exsanguination on the spontaneous electrocortical activity and indices of cerebral metabolism. Res. Vet. Sci.* ***52,*** *162–173.*

Blackmore, D.K. *(1984) Differences in the behaviour of sheep and calves during slaughter. Res. Vet. Sci.* ***37,*** *223–226.*

Daly, C.C.; Kallweit, E.; Ellendorf, F. *(1988) Cortical function in cattle during slaughter: Conventional captive bolt stunning followed by exsanguination compared to shechita slaughter. Vet Rec.* ***122,*** *325–329.*

Dunn, C.S. *(1990) Stress reactions of cattle undergoing ritual slaughter using two methods of restraint. Vet. Rec.* ***126,*** *522–525.*

Epstein, I. *(Editor) (1948) The Babylonian Talmud. Socino Press, London.*

Giger, W.; Prince, R.P.; Westervelt, R.G.; Kinsman, D.M. *(1977) Equipment for low stress animal slaughter. Trans. Americ. Sco. Agric. Eng.* ***20,*** *571–578.*

Grandin, T. *(1987) High speed double rail restrainer for stunning or ritual slaughter, Int. Congress.*

Meat Sci. & Tech. pp. 102–104.

Grandin, T. *(1988) Double rail restrainer for livestock handling. J. Agric. Eng. Res.* ***41,*** *327–338.*

Grandin, T. *(1991) Double rail restrainer for handling beef cattle. Technical Paper 915004, Amer. Soc. Agric. Eng., St. Joseph, MI.*

Grandin, T. *(1992) Observations of cattle restraint devices for stunning and slaughtering, Animal Welfare* ***1,*** *85–91.*

Grandin, T. *(1993) Management commitment to incremental improvements greatly improves livestock handling. Meat Focus, Oct. pp. 450–453.*

Gregory, N. *(1988) Published Discussion, 34th International Congress of Meat Science and Technology, Workshop on Stunning of Livestock. CSIRO Meat Research Laboratory, Brisbane, Australia, p. 27.*

Gregory, G.; Wotton, S.D. *(1984) Time of loss of brain responsiveness following exsanguination in calves. Res. Vet. Sci.* ***37,*** *141–143.*

Grunfeld *(1972) The Jewish Dietary Laws. Soncino Press, London.*

Kilgour, R. *(1978) J. Anim. Sci.* ***46,*** *1478.*

Nangeroni, L.L.; Kennett, P.D. *(1963) An Electroencephalographic Study of the Effect of Shechita Slaughter on Cortical Function of Ruminants, Unpublished report. Dept. of Physiology, New York State Veterinary College, Cornell University, Ithaca, NY.*

Mitchell, G.; Hahingh, J.; Ganhao, M. *(1988) Stress in cattle assessed after handling, transport and slaughter. Vet. Rec.* ***123,*** *201–205.*

Munk, M.L.; Munk, E.; Levinger, I.M. *(1976) Shechita: Religious and Historical Research on the Jewish Method of Slaughter and Medical Aspects of Shechita. Feldheim Distributors, Jerusalem.*

Tume, R.K.; Shaw, F.D. *(1992) Beta endorphin and cortisol concentration in plasma of blood samples collected during exsanguination of cattle. Meat Sci.* ***31,*** *211–217.*

van der Wal, P.G. *(1978) Chemical and physiological aspects of pig stunning in relation to meat quality. A review. Meat Sci.* ***2,*** *19–30.*

Warrington, R. *(1974) Electrical stunning: A review of the literature. Vet. Bull.* ***44,*** *617–633.*

Zavy, M.T.; Juniewicz, P.E.; Phillips, W.A.; Von Tungeln, D.L. *(1992) Effect of initial restraint, weaning and transport stress on baseline ACTH stimulated cortisol response in beef calves of different genotypes. Amer. J. Vet. Res.* ***53,*** *551–557.*

Recommended Animal Handling Guidelines for Meat Packers

Published by
American Meat Institute—1991

Written by
Temple Grandin, Ph.D.
Assistant Professor
Colorado State University

Table of Contents

Credits

Many people worked with the American Meat Institute (AMI) to develop this book. AMI extends special thanks to the AMI Animal Welfare Committee for initiating the idea, and to Dr. Temple Grandin, assistant professor in the Animal Science Department at Colorado State University, for writing the guidelines. AMI also thanks the many industry executives, USDA inspection officials and animal handling experts who reviewed the text.

Introduction

Proper livestock handling is extremely important to meat packers for obvious ethical reasons. Once livestock arrive at packing plants, proper handling procedures are not only imporant for the animals' well-being, but can also mean the difference between profits and losses due to meat quality or worker safety. The Humane Slaughter Act of 1978 dictates strict animal handling and slaughtering standards for packing plants. Those standards are monitored by some 7,000 federal meat inspectors nation-wide. The meat packing industry takes these standards very seriously.

For the best results in animal handling, management must make proper handling and stunning a high priority. Top management must play an active role. Plants with the best handling and stunning practices have managers who closely monitor stunning and handling practices. This manual provides employees and managers with information which will help them improve both handling and stunning. Employees handling hundreds of animals day after day sometimes need reminders from management that animals must always be handled carefully.

Healthy animals, properly handled, keep the meat industry running safely, efficiently and profitably. This handbook is designed to help ensure that proper handling guidelines and recommended practices are widely understood.

Guidelines for Livestock Holding Facilities and Trucking

Preventing Injuries and Bruises—Non-slip flooring is essential to prevent falls and crippling injuries. Humane, efficient handling is impossible on slick floors. All areas where livestock walk should have a non-slip surface. Existing floors can be roughened with a light jack hammer or a grooving machine. On scales, crowd pens and other high traffic areas, a grid of one-inch steel bars will provide secure footing. Construct a 12-inch by 12-inch grid and weld each intersection. Use heavy rod to prevent the grid from bending.

New concrete floors for cattle should have an 8-inch diamond or square pattern with 1 1/2 inch X 1 1/2 inch V grooves. For hogs and sheep, stamp the pattern of 1 1/2 inch raised expanded metal into the wet concrete. A rough broom finish will become worn smooth. Floors should be grooved. It is also essential to use the right concrete mix for maximum resistance to wear.

Gates, fences and chutes should have smooth surfaces to prevent bruises. Sharp edges with a small diameter, such as angle irons, exposed pipe ends, and channels, will cause bruises. Round pipe posts with a diameter larger than 3 inches are less likely to bruise. Vertical slide gates in chutes should be counter-weighted to prevent back bruises. The bottom of these gates should be padded with cut tires or conveyor belting. The gate track should be recessed into the chute wall to eliminate a sharp edge that will bruise. Gates in drive alleys should be equipped with tie backs to prevent them from swinging out into the alley. Livestock are easily bruised if they become caught between the end of the gate and the fence. This is a common cause of bruises in the valuable loin area.

Pressing up against a smooth flat surface such as a concrete chute fence will not cause bruises. However, a protruding bolt or piece of metal will damage hides and bruise the meat. Bruise points can be detected by tufts of hair or a shiny surface. Contrary to popular belief, livestock can be bruised moments before slaughter, and stunned cattle can be bruised until they are bled. The entrance to the restrainer should be inspected often for broken parts with sharp edges.

Surveys show that groups of horned cattle will have twice as many bruises as polled (hornless) cattle. A few horned animals can do a lot of damage and tipping horns does not reduce bruises.

Improving Animal Movement—Cattle, hogs and sheep have wide-angle vision and they can see behind themselves without turning their heads. This explains why they will often balk at shadows or puddles of water on the ground. Balking slows production and can be prevented. Drains should be located outside of the areas where animals walk. A drain or a metal plate running across an alley will cause balking. Flapping objects such as a coat hung over a fence will also make livestock balk. When wetting hogs in the chute, be sure not to spray the animals' faces with water, because they will back up.

Animals tend to move from a darker area to a more brightly lighted area. Lamps can be used to attract animals into chutes. The light should illuminate the chute up ahead and it should never glare directly into the eyes of approaching animals. Another approach is illuminating the entire chute area. This approach eliminates patches of light and dark which may confuse animals.

Solid sides which prevent the cattle from seeing outside the fence should be installed on the chutes which lead to the stunner and the crowd pen which leads up to the chute. Solid sides in these areas help prevent cattle from becoming agitated when they see activity outside the fence—such as people. Cattle tend to be calmer in a chute with solid sides. The crowd gate on the crowd pen should also be solid to prevent animals from attempting to turn back, towards the stockyard pens they just left.

In some hog plants, one solid side and one open side on fences is useful, allowing employees to view hogs and correct problems if necessary. When one fence side is open, plants should restrict employee traffic on that side to reduce hog agitation.

It is important to reduce noise in the stunning area. Animals are more sensitive to high pitched noise than people. Animals will be calmer and easier to handle if noise levels are reduced. Install mufflers on air valve exhausts or put them outside. Rubber stops on gates can be used to stop clanging. Braking devices on the shackle return improve safety and reduce noise. Use large diameter plumbing and replace noisy pumps with quieter ones. Some brands of pumps are quieter than others. Rubber hose connection between the power unit and metal plumbing will help prevent power unit noise from being transmitted throughout the facility. Any new equipment that is installed in animal handling or stunning areas should be engineered for quietness.

Improving Meat Quality and Animal Welfare—All livestock must have access to clean drinking water. During hot weather (over 70 degrees Fahrenheit), hogs should be sprinkled with water in the stockyard pens. For maximum cooling effect, the sprinklers should have a coarse enough spray to wet the animals. Sprinklers that make a fine mist should not be used because they tend to increase humidity rather than cool the hogs. Sprinkler water should flow intermittently to keep hogs wet at all times. Keeping hogs cool is very important because a hot hog will have more PSE (pale, soft, exudative, stressed pork.) Hogs become over heated easily because they are covered by a layer of fat and they do not sweat. All hog plants should have a heated staging area prior to the stunning chute where hogs can be showered prior to stunning during cold weather.

Using Prods and Persuaders Properly—Electric prods should be used sparingly to move livestock. They must never be wired directly to house current. A transformer must be used. Hogs require lower voltages than cattle. A doorbell transformer works well for hogs. Low prod voltages will help reduce both PSE and blood spots in the meat. Fifty volts is the maximum voltage for prods hooked to an overhead wire. Battery-operated prods are best from a livestock handling standpoint because they provide a localized directional stimulus between two prongs.

The use of electric prods can be greatly reduced by using other driving aids such as plastic streamers or strips cut from garbage bags attached to a stick. The plastic can be clamped in a mop handle. Cattle can be easily turned and moved in the crowd pen by shaking the streamers near their heads. Canvas slappers can be used for moving hogs. Rattles work well for moving sheep.

Truck drivers should be careful with prods. Rushing livestock during unloading is a major cause of bruises. Serious loin bruises are often caused by two cattle wedged in a truck door. Management should closely supervise truck unloading.

Providing Adequate Pen Space—Stockyards at packing plants should have sufficient capacity so that animals can be promptly unloaded from trucks. Heat builds up rapidly in a stationary vehicle. To reduce PSE, hogs should be rested two to four hours prior to stunning. In large plants, pens should be designed to hold one or two truckloads. A few smaller pens will also be required for small lots. Pen space allocations may vary depending upon weather conditions, animal sizes and varying holding times. As a guideline, 20 square feet should be alotted for each 1,200-pound steer or cow, and six square feet per hog. These stocking rates will provide adequate room for "working space" when animals are moved out of the pen. If the animals are stocked in the pen more tightly, it will be more difficult for the handler to empty the pen. The recommended stocking rates provide adequate space for all animals to lie down.

Recommended Handling Facility Layout—All the animal movement is one-way and there is no cross traffic. Each long narrow pen holds one truckload. The animals enter through one end and leave through the other. The round crowd pen and curved chute facilitate movement of cattle to the stunner.

A curved chute is more efficient for cattle because it takes advantage of their natural circling behavior. It also prevents them from seeing the other end while they are standing in the crowd pen. A curved chute should be laid out correctly. Too sharp a bend at the junction between the single file chute and the crowd pen will create the appearance of a dead end. All species of livestock will balk if a chute looks like a dead end.

Round crowd pens are efficient for moving all species into a single file chute.

Crowd gate lengths for hog operations may vary based on line speeds and plant sizes. As a guideline, the recommended radii (length of crowd gate) are: Cattle, 12 feet; hogs, seven feet; and sheep, eight feet. The basic layout principles are similar for all species, but there is one important difference. Cattle and sheep crowd pens should have a funnel entrance and hog crowd pens must have an abrupt entrance. Hogs will jam in a funnel. A crowd pen should never be installed on a ramp because animals will pile up in a crowd pen. If ramps have to be used, the sloped portion should be in the single file chutes. In hog facilities, level stockyards and chute systems with no ramp are most efficient.

Unloading Animals Properly—For all species, a plant should have sufficient unloading ramp capacity so trucks can be unloaded promptly. In large plants, at least two and preferably three ramps are required. Unloading ramps should have a level dock before the ramps go down so that animals have a level surface to walk on when they exit the truck. The slope of the ramp should not exceed 20 degrees. On concrete ramps, stairsteps are recommended because they provide better traction than cleats or grooves when ramps become dirty.

For cattle, the recommended stair step dimensions are 3 1/2 inch rise and a 12-inch long tread. If space permits, an 18-inch long tread will create a more gradual ramp. For hogs, a 2 1/2 inch rise and a 10-inch tread works well. On adjustable ramps, cleats with eight inches of space between them are recommended. All flooring and ramp surfaces should be non-slip. Many animals are injured on slippery unloading ramps.

Recommended Trucking Practices—Trailers should be kept in good repair. To comply with environmental regulations, truck floors should be leak-proof to prevent urine and manure from dripping onto the highway. With today's modern taller cattle, it is essential that semi-trailers have sufficient height between decks to prevent back injuries. Overloading of trucks will increase bruising. In one survey, overloading two extra head of cattle increased bruising. Overloading of hog trucks will increase death losses and PSE. To prevent skin blemishes, hog trucks should be cleaned after each load.

When the temperature is over 60 degrees, use wet sand or wet shavings to keep hogs cool. If the temperature is over 80 degrees, sprinkle hogs with water prior to loading at buying stations or on the farm. Never bed hogs with straw during hot weather. When the temperature is below 60 degrees, bed hogs with straw or deep, dry shavings to keep them warm. In the northern regions, approximately half of the air holes in aluminum trailers transporting hogs should be covered with plywood when the temperature drops to 10 degrees.

Veal calves can require special care in transport because they are so young. Take care in cooler temperatures (below 60 degrees) to provide straw bedding and plug some airholes in trucks so the calves do not become too cold.

People trucking and handling animals need to understand both wind chill factors and heat stress. Wind chill can kill livestock. Wetting the haircoat on cattle destroys its ability to insulate the animal's body. Death losses in cattle are often greatest when the temperatures are near freezing and either rain or freezing rain blows into a truck. Dry cold weather is less hazardous to cattle because the coat retains its ability to insulate. Wind chill can make the back of a trailer very cold. When a truck is moving 50 mph on a 20-degree day, the wind chill factor for hogs is minus 23 degrees.

The combination of high temperature and humidity is especially detrimental for hogs. When the Livestock Weather Safety Index is in the Danger and Emergency zone, try to schedule hog shipments for the early morning.

If this is not possible, trucks should be kept moving and drivers should not be allowed to stop at the coffee shop with a loaded trailer. When the trucks reach the plant, they must be unloaded promptly.

Livestock Facility and Trucking: Safety Tips for Workers

1. Prods must always be wired through a transformer. A light bulb wired in series is dangerous to both people and livestock.

2. Mangates and other devices must be installed so people can easily escape from agitated cattle. This is especially important in areas with solid fences. In concrete fences, toe holds can be formed in the walls.

3. Be alert around the unloading dock. A truck driver backing in may not be able to see you.

Recommended Basic Livestock Handling Principles

The principles of good livestock handling are similar for the different species. All livestock are herd animals and will become agitated when separated from the herd. If a lone animal becomes agitated, put some other animals in with it. Never get in the crowd pen or other confined space with one or two agitated, excited cattle. Animals will remain calmer when they are in a group.

Be gentle with them. Cattle, hogs and sheep become agitated and stressed when they are poked with electric prods or hit with a stick. Many people ask, "Do they know they are going to be slaughtered?" Animals behave in the same manner in a packing plant and in the veterinary chute at the farm. They have no concept of being slaughtered. They know they are being handled and they probably think they are getting on a truck. In one study, cattle actually had calmer behavior at the slaughter plant compared to a squeeze chute back at the feedlot.

Livestock may become agitated or balk at strange sounds or the smell of the rendering plant. They are reacting to the strangeness of the smell. They do not know what it is, because they have never smelled these smells before.

Understanding Flight Zone and Point of Balance—Handlers who understand the concepts of flight zone and point of balance will be able to move animals more easily. The flight zone is the animal's personal space, and the size of the flight zone is determined by the wildness or tameness of the animal. Completely tame animals have no flight zone and people can touch them. An animal will begin to move away when the handler penetrates the edge of the flight zone. If all the animals are facing the handler, the handlers is outside the flight zone.

Recommended Truck Loading Densities

(Livestock Conservation Institute)

Feedlot Fed Steers or Cows, Avg. Weight	Horned or Tipped or more than 10% Horned and Tipped	No Horns (polled)
800 lbs.	10.90 sq. ft.	10.40 sq. ft.
1000 lbs.	12.80 sq. ft.	12.00 sq. ft.
1200 lbs.	15.30 sq. ft.	14.50 sq. ft.
1400 lbs.	19.00 sq. ft.	18.00 sq. ft.
Market Weight Hogs	Winter	Summer (Temp. is 75+ degrees)
200 lbs.	3.50 sq. ft.	4.00 sq. ft.
250 lbs.	4.26 sq. ft	5.00 sq. ft.
Slaugher Weight Lambs and Sheep	Shorn	Full Fleece
60 lbs.	2.13 sq. ft.	2.24
80 lbs.	2.50	2.60
100 lbs.	2.80	2.95
120 lbs.	3.20	3.36

To keep animals calm and move them easily, the handler should work on the edge of the flight zone. He penetrates the flight zone to make the animals move and he backs up if he wants them to stop moving. The handler should avoid the blind spot behind the animal's rear. Deep penetration of the flight zone should be avoided. Animals become upset when a person is inside their personal space and they are unable to move away. If cattle turn back and run past the handler while they are being driven down a drive alley in the stockyard, overly deep penetration of the flight zone is a likely cause. The animals turn back in an attempt to get away from the handler. If the animals start to turn back, the handler should back up and increase the distance between himself and the animals. Backing up must be done at the first indication of a turn back. If a group of animals balk at a smell or shadow up ahead, be patient and wait for the leader to cross the shadow. The rest of the animals will follow. If cattle rear up in the single file chute, back away from them. Do not touch them or hit them. They are rearing in an attempt to increase the distance between themselves and the handler. They will usually settle down if you leave them alone.

The point of balance is at the animal's shoulder. All species of livestock will move forward if the handler stands behind the point of balance. They will back up if the handler stands in front of the point of balance. Many handlers make the mistake of standing in front of the point of balance while attempting to make an animal move forward in a chute. Groups of cattle or hogs in a chute will often move forward without prodding when the handler walks past the point of balance in the opposite direction of each animal in the chute. It is not necessary to prod every animal. If the animals are moving through the chute by themselves, leave them alone. Often they can be moved by tapping the side of the chute.

Using Animals' "Follow the Leader" Instinct—Livestock will follow the leader, and handlers need to take advantage of this natural behavior to move animals easily. Animals will move more easily into the single file chute if it is allowed to become partially empty before attempting to fill it. A partially empty chute provides room to take advantage of following behavior. Handlers are often reluctant to do this because they are afraid the line will run out and miss notches on the power chain. Once a handler learns to use this method, he will find that keeping up with the line will be easier. As animals enter the crowd pen they will head right up the chute.

One of the most common mistakes is overloading of the crowd pen. Crowd pen sizes and densities in hog operations may vary depending on the size and line speeds of plants. In cattle plants, 18 cattle is the maximum number which should be placed in the crowd pen.

In hog plants, 15 hogs is the recommended maximum for chain speeds under 300 an hour and 25 hogs for chain speeds over 800 per hour. Handlers must also be careful not to push the crowd gate up too tightly. Animals need room to turn. The crowd gate should be used to follow the animals and should never be used to forcibly push them. The handler should concentrate on moving the leaders into the chute instead of pushing animals at the rear of the group.

One-way or sliding gates at the entrance to the single file chute must be open when livestock are brought into the crowd pen. Cattle will balk at a closed gate.

One-way flapper gates can be equipped with a rope to open them by remote control from the crowd pen. Less prodding will be required if a stick with plastic streamers is used to turn cattle towards the chute entrance. To turn an animal, block the vision on one side of its head with the streamers. If the leader balks right at the chute entrance, a single poke with the prod may be required. Once the leader enters, the rest of the animals will follow.

Dealing with Excitable Hogs — During the last five years, there have been increasing problems with highly excitable hogs which are difficult to drive at the packing plant. These animals squeal, bunch and pile up. It is difficult to make these hogs separate and walk up the chute. They will constantly balk. Only five percent of the hogs in the Midwest have this problem, but about 10 to 20 percent of the hogs in the Southeast are overly excitable. It is caused by a combination of genetic selection for rapid weight gain and a lack of stimulation in confinement buildings.

Packers and producers need to work together to produce hogs which will have both rapid weight gain and good meat quality. Highly excitable hogs have severe PSE due to agitation during handling. PSE levels will be high even though these hogs are negative on the halothane test for PSS (Porcine Stress Syndrome). Quiet, calm handling is almost impossible due to their bad temperament.

Excitability problems can also be reduced by providing confinement hogs on the farm with rubber hoses to chew on, greater contact with people in their pens and a radio. Playing a radio in the finishing barn gets the animals accustomed to different kinds of sounds. Bad flooring in the finishing barn can also cause problems. Hogs finished on metal or plastic floors will have excessive hoof growth. These animals will balk constantly in the plant.

Recommended Basic Livestock Handling: Safety Tips for Workers

1. A single, lone, agitated steer is very dangerous. Many serious cattle handling injuries are caused by a single agitated steer or cow. One man received twenty-seven stitches after he got in the crowd pen with a lone animal and teased it.

2. Escaped cattle must never be chased. An animal which is loose on the plant grounds will return to the stockyard if it is left alone. If an animal gets loose inside the plant, employees should stay quiet while one designated person either stuns it or eases it out a door.

3. Stay out of the blind spot behind a steer's rear end. If he cannot see you, he is likely to kick you.

4. Install a safety fence consisting of upright posts around the cattle shackling area to prevent cattle from entering other parts of the plant.

Recommended Stunning Practices

Stunning an animal correctly will provide better meat quality. Improper electric stunning will cause bloodspots in the meat and bone fractures. Good stunning practices are also required so that a plant will be in compliance with the Humane Slaughter Act and for animal welfare. When stunning is done correctly, the animal feels no pain and it becomes instantly unconscious. An animal that is stunned properly will produce a still carcass that is safe for plant workers to work on.

Captive Bolt Stunning—To produce instantaneous unconsciousness, the bolt must penetrate the brain with a high concussive impact. For cattle, the stunner is placed on the middle of the forehead on an "X" formed between the eyes and the base of the horns. Some plants which save brains, place the stunner on the hollow behind the animal's poll on the back of the head. This position is less effective, therefore the frontal position on the forehead is recommended.

For sheep, a captive bolt is placed on the top of the head. This position is more effective for sheep because they have a very thick skull over the forehead. For hogs, the captive bolt is placed on the forehead.

A good stunner operator learns not to chase the animal's head. Take the time and aim for one good shot. The stunner must be placed squarely on the animal's head. All equipment manufacturers' recommendations and instructions must be followed. Pneumatic stunners must have an adequate air supply. Low air pressure is one cause of poor stunning.

Poor maintenance of captive bolt stunners is a major cause of bad stunning. Pneumatic captive bolt stunners require cleaning and seal replacement every night in large plants. For example, a cattle plant which is double shifted and has a chain speed of 250 head per hour will require two to three pneumatic stunners to be completely serviced every night. It is important to keep stunner cartridges dry and the correct cartridge strength must be used.

Eye reflexes should be checked often to insure that stunning is making the cattle unconscious. When the eyelid or cornea is touched there should be no response. An animal that blinks is not properly stunned. Breathing should have stopped and there should be no indication of a righting reflex when the animal is hanging on the rail. Reflexes may cause a stunned animal's legs to move, but the head should hang straight down and be limp.

Proper Cattle Restraint for Stunning—If a stunning box is used, it should be narrow enough to prevent the animal from turning around. The floor should be non-slip so the animal can stand without losing its footing. It is much easier to stun an animal that is standing quietly. Only one animal should be placed in each stunning box compartment to prevent animals from trampling on each other

Most large plants restrain cattle in a conveyor restrainer system. There are two types of conveyor restraints, the V restrainer and the new center track system. In a V restrainer system, the cattle are held between two, angled conveyors. In the center track system the cattle ride astride a moving conveyor. The center track system provides the advantages of easier stunning and improved ergonomics because the stunner operator can stand closer to the animal. Either type of restrainer system is much safer for workers than a stunning box. Restrainer conveyors are recommended for all plants which slaughter over 100 head per hour. Stunning boxes are difficult and dangerous to operate at higher speeds. In a plant which slaughtered 160 cattle per hour, replacement of multiple stunning boxes with a conveyor restrainer eliminated at least one serious accident each year.

Lighting in the restrainer room over the top of the conveyor will help induce cattle to raise their heads for the stunner. Cattle should not be able to see light coming up from under the restrainer because it may cause balking at the entrance. Restrainer systems should be equipped with a long, solid hold-down rack to prevent rearing. The hold-down should be long enough so that the animal is fully settled down onto the conveyor before it emerges from under it.

If an animal is walking into the restrainer by itself, do not poke it with an electric prod. Center rack systems require less prodding to induce cattle to enter it. Workers need to break the "automatic prod reflex" habit.

Recommended Captive Bolt Stunning Techniques: Safety Tips for Workers

1. Cartridge-fired stunners must ALWAYS be unlocked before they are set down.

2. NEVER, NEVER throw a cartridge-fired stunner to another person.

3. Inspect latches on stunning boxes to make sure they latch securely. Before the next animal is admitted to the box, check the latch.

4. All guards must be kept in place over exposed pinch points which could be easily touched by employees during normal operation of the restrainer system equipment.

5. If a worker has to get inside a restrainer conveyor system to unjam it, lock it out first to prevent somebody else from turning it on.

6. Cartridge-fired stunners must always be kept unloaded when they are carried away from the stunning area.

7. Good maintenance is essential on pneumatic stunners to prevent excessive recoil which can strain and injure the operator's hands, arm or back.

Electric Stunning of Hogs and Sheep—To produce instantaneous, painless unconsciousness, sufficient amperage (current) must pass through the animal's brain to induce an epileptic seizure. Insufficient amperage or a current path that fails to go through the brain will be painful for the animal. It will feel a large electric shock or heart attack symptoms, even though it may be paralyzed and unable to move. When electric stunning is done correctly, the animal will feel nothing.

There are two types of electric stunning, head only and cardiac arrest stunning which stops the heart. Most large plants use cardiac arrest head to back or head to side of body stunning. It produces a still carcass that is safer and easier to bleed. Cardiac arrest stunning requires the use of a restraining device to prevent the animal from falling away from the stunning wand before it receives the complete stun. Cardiac arrest stunning kills the animal by electrocution. Head only stunning is reversible. Hogs and sheep which are stunned with a head only stunner must be bled within a maximum interval of 30 seconds to prevent them from regaining consciousness. An interval of 10 to 17 seconds is recommended.

When cardiac arrest stunning is used, one electrode must be placed on either the forehead or in the hollow behind the ears, and the other electrode is placed on either the back or the side of the body. The head electrode should not be allowed to slide back onto the neck. When head only stunning is used, the electrodes may be either placed on the forehead or clamped over around the sides of the head like ear muffs. Hogs should be wetted prior to stunning. The stunning wand must be applied to the animal for two to three seconds to stun properly. Stunners should be equipped with a timer.

Meat packers should use amperage, voltage and frequency settings which will reliably induce unconsciousness. Both properly and improperly stunned cardiac arrested animals can look similar (Gilbert and Devine, 1991). Current flow through the spine masks the epileptic seizure. If there is any question, electrical parameters should be verified by scientifically valid measurements.

To prevent bloodspots in the meat and pain to the animal, the wand must be pressed against the animal before the button is pushed. The operator must be careful not to break and make the circuit during the stun. This causes the animal's muscles to tense up more than once and bloodspots may increase. Stunning wands and wiring should be checked often for electrical continuity. A worn switch may break the circuit enough to cause bloodspots. Electrodes must be kept clean to provide a good electrical contact. Operators must never double stun animals or use the stunning wand as a prod.

Modern stunning circuits use a constant amperage design. The amperage is set and the voltage varies with hog or sheep resistance. Older style circuits are voltage regulated. These circuits are inferior because they allow large amperage surges which can fracture bones and cause bloodsplash. The distance between the head electrode and the back electrode should not exceed 14 inches. Hog stunners should be equipped with blunt electrodes which do not stick into the animal. The most modern sheep stunners from New Zealand utilize water jets to conduct electricity down through the wool.

Preventing Bloodsplash (Bloodspots)— Handle animals gently. Gentle handling prevents damage to small blood vessels caused by excited animals jamming against each other or equipment. Electric prod usage should be kept at a minimum. Animals should never be left in the restrainer system during breaks and lunch. Bloodsplash may also be increased if one side of a V restrainer runs faster than the other. This causes stretching of the skin which damages blood vessels. Both sheep and hogs should be bled within 15 seconds after stunning to minimize meat damage. The slats on the V restrainer and hold-down rack should be insulated to prevent current leakage which can cause bloodsplash.

Rapid temperature fluctuations and periods of extremely hot weather can greatly increase the incidence of bloodsplash. In these circumstances, plants should take extra care in handling animals to minimize bloodsplash problems.

Electric Stunning of Sheep and Hogs: Safety Tips for Workers

1. The stunner operator's station must be kept dry. Stunning wands should be designed so that they can be operated by one hand. Avoid designs where the two electrodes are held separately in each hand. These increase a shock across the chest electrocution hazard.

2. The operator should wear rubber boots and stand on non-conductive plastic grating.

3. The restrainer frame and worker walkway structure should be grounded to a perfect ground. However, the side of the restrainer that the stunner operator can touch should be covered with heavy insulating materials such as a plastic meat cutting board.

Recommended Ritual Slaughter Practices— For both humane and safety reasons, plants which conduct ritual slaughter should install modern upright restraining equipment. The practice of hanging live cattle and calves upside down should be eliminated. There are many different types of humane restraint devices available.

ASPCA Pen—This device consists of a narrow stall with an opening in the front for the animal's head. After the animal enters the box, it is nudged forward with a pusher gate and a belly lift comes up under the brisket. The head is restrained by a chin lift for the rabbi to perform shehita. Vertical travel of the belly lift should be restricted to 28 inches so that it does not lift the animal off the floor. The rear pusher gate should be equipped with either a separate pressure regulator or special pilot-operated check valves to allow the operator to control the amount of pressure exerted on the animal. The pen should be operated from the rear toward the front. Restraining the head is the last step. The operator should avoid sudden jerking of the controls. Many cattle will stand still if the box is slowly closed up around them, and less pressure will be required to hold them.

An ASPCA pen can be easily installed in one weekend with minimum disruption of plant operations. It has a maximum capacity of 100 cattle per hour and it works best at 75 head per hour. A small version of this pen could be easily built for calf plants.

Conveyor Restrainer Systems—Either V restrainer or center track restrainer systems can be used for holding cattle, sheep or calves in an upright position during shehita. The restrainer is stopped for each animal and a head holder holds the head for the rabbi. Research in Holland indicates that the center track design provides the advantage of reducing bloodspots in the meat.

Small Restrainer Systems—For small locker plants which ritually slaughter a few calves of sheep per week, an inexpensive rack constructed from pipe can be used to hold the animal in a manner similar to the center track restrainer.

Recommended Ritual Slaughter Practices: Safety Tips for Workers

1. Shackling and hoisting large cattle and calves can be very dangerous. It has caused many serious accidents such as loss of an eye, permanent knee damage and head injuries from kicking and falling shackles. In one plant, replacement of the shackle hoist with a restrainer resulted in a 500 percent reduction in accidents. Shackling and hoisting of live sheep is also hazardous. There have been several incidents of teeth knocked out.

Recommended Handling of Disabled or Crippled Livestock

Although injured or incapacitated animals (sometimes called "downers") represent less than .1 percent of all livestock arriving at packing plants, they are significant because they require special attention in the areas of handling, transporting, holding pens and inspection.

As a general principle, all disabled animals arriving at packing plants should be dispatched to slaughter as promptly as possible to minimize the animal's suffering.

Offloading from Trucks—All trucks carrying livestock should be unloaded promptly. Trucks carrying disabled animals should unload ambulatory animals first, then promptly unload the animals unable to walk. Delayed unloading can cause death losses and downer animals due to extreme temperatures and stress.

To offload a non-ambulatory animal from a truck, plants should use the truck exit nearest to the animal and should place as little stress as possible on the animal. In some cases, a slide board or cripple cart may be helpful. The board can then be dragged off the truck and the animal loaded into a suitable mechanical device for transport to an inspection area.

The Humane Slaughter Act prohibits dragging of downed or crippled livestock in the stockyards, crowd pen or stunning chute. By using slideboards and cripple carts, animals can be transported humanely and efficiently to a pen or other area where they can be examined by an inspector, stunned and moved to slaughter.

In all cases, disabled livestock should be handled and moved as little as possible. Trucks carrying downers should park as close to the slaughter area as possible, and disabled animals should be inspected by a USDA veterinarian, stunned and moved to slaughter as quickly as possible.

Inspection and Slaughtering Considerations—USDA rules require that any "suspect" animal—an animal with signs of abnormalities or diseases—must be held separately and closely examined by a USDA inspection service veterinarian. For meatpackers, this means that downer animals must be held apart from other animals in a "suspect" pen for USDA inspection. "Suspect" animals must be slaughtered separately so inspectors can carefully examine the animals' carcasses and parts.

Plants should call for the USDA veterinarian as soon as a disabled animal arrives. Once the animal has been examined by the USDA inspector, plants should identify the earliest possible point in production when that animal may be slaughtered "separately." This separation point should be discussed with the USDA inspector. It should be noted that plants need not always wait until the end of a shift to slaughter a "suspect" animal. Waiting can prolong a disabled animal's suffering.

If a steer or cow goes down in the single file chute which leads to the stunner, it must be stunned prior to dragging. A cartridge-fired captive bolt on a long handle is recommended. If blood gets on the chute, wash it off to prevent balking. In hog plants, the stunning chute should be equipped with side doors so that stressed-out downer hogs can be easily removed.

Emergency Slaughter for Injured Animals — If a suffering, injured animal arrives at a plant and a USDA inspector veterinarian is not available to examine the animal, it may be necessary to perform an "Emergency Slaughter." USDA regulation 311.27 allows plants to perform such an emergency act for humane reasons. The regulations requires that the carcass and all parts must be held for inspection, with the head and all viscera except the stomach, bladder and intestines held by the natural attachments. If all parts are not kept in this manner for inspection, the entire carcass will be condemned. When an inspector is available, the held carcass and viscera will be inspected and may be passed or condemned, depending on the evidence.

Animals that are sick, dying or have recently been treated with drugs or chemicals and are presented for slaughter before the required withdrawal period are not covered by emergency slaughter.

Handling Downer Hogs—Many problems with splitter and downer hogs are genetic. Weak hindquarters in hogs is correctable by breeding. Hogs will sometimes collapse from PSS (Porcine Stress Syndrome). This is a genetic defect which make a hog prone to a heart attack. PSS downers will sometimes recover if they are left alone. Never throw cold water on a disabled hog, the shock to its system will kill it. Wet the floor around the hog and allow it to cool by evaporation.

Conclusion

By applying the principles you have learned in this manual, you can ensure that animal handling and stunning is more efficient and humane. These recommended practices will also help provide a safer work place.

Proper livestock handling is a critical element of good business management. It ensures that the animals, the workers and practices employed are treated with understanding and respect.

American Meat Institute
Serving the Meat Industry Since 1906
P.O. Box 3556, Washington, DC 20007
1700 North Moore St., Arlington, VA 22209
Telephone: 703/841-2400

Drug Information Journal, Vol. 26, pp. 85–94, 1992 0092-8615/92

Current Ethical Issues Surrounding Animal Research

Linda Compton
Medical Research Specialist, The Upjohn Company, Kalamazoo, Michigan

Cathy Taylor
Regulatory Affairs, GCP Compliance Specialist, The Upjohn Company, Kalamazoo, Michigan

Presented at the Drug Information Association 26th Annual Meeting, "The Future of Clinical Research," San Francisco, California, June 6, 1990.

Reprint address: Linda Compton, Dermatology Research, The Upjohn Company, 7000 Portage Road, Kalamazoo, Michigan 49007.

It is becoming increasingly important to focus on the research and safety testing that precedes human clinical trials. This research is conducted in animals, and a segment of today's society vehemently and sometimes violently opposes this use. Both researchers and animal welfare groups agree that the appropriate care and use of animals, the avoidance of unnecessary duplication, and the use of nonanimal models when possible are essential for an effective research program. The scientific community is proud of the drastic reduction in the total number of laboratory animals used today, thanks to new in vitro screening methods, and the refinement of traditional animal models. In spite of this encouraging reduction, animals are still needed in product safety testing, in education, and in basic research. While it is clear that the need for animals is present, research, along with other animal use, is under attack.

Key Words: Animal rights; Animal research; Regulations; Safety testing

WORLDWIDE REVULSION to reports of Nazis' experimentation on prisoners resulted in three International Codes and Conventions (Nuremburg in 1946; The World Medical Assembly in Helsinki, modified in 1989 but dating from 1958; and the Council for International Organizations of Medical Sciences in 1949 established jointly by the World Health Organization [WHO] and United Nations Education, Scientific, and Cultural Organization [UNESCO]) (1). Out of these conferences came a clear mandate that informed consent and preclinical safety and efficacy information is required before human testing or clinical trials begin. The Declaration of Helsinki states:

> Biomedical research involving human subjects must conform to generally accepted scientific principles and should be based on adequately performed laboratory and animal experimentation and on a thorough knowledge of the scientific literature.

Also, US Federal laws were enacted in response to the injuries and deaths caused by unsafe products. Before 1938 over 100 children died from untested products like elixir of sulfanilamide. Cosmetics also caused disfigurement and death. Today three U.S. agencies share responsibility for ensuring consumer safety. In addition, with the advent of world-wide corporations, non-US regulatory requirements have to be considered in marketing and development plans.

The Food and Drug Administration (FDA) requires information about a substance's biological effects before clinical trials in humans begin. FDA regulations do not specify *which* tests must be done, but refer to toxicological effects in animals and acute, subacute, and chronic toxicity testing. They set guidelines and will reject data considered inadequate or incomplete. Animal testing is integral to the acceptability of safety data. The safety information obtained through animal testing influences not only decisions on whether to expose human subjects to the substance, but also the manner in which human clinical drug trials will be performed. These requirements are contained in the Code of Federal Regulations, and listed in Table 1.

The Environmental Protection Agency, enforces several laws protecting human health and the environment. The Toxic Substances Control Act of 1976 applies to most chemicals in the US. Specific animal safety tests are ordered to determine if a substance poses an unreasonable risk of injury to health or the environment.

The Consumer Product Safety Commission has the authority to regulate all products that pose an unreasonable risk of injury or illness to consumers. The 1960 Federal Hazardous Substances Labeling Act details specific animal toxicity tests that are required to support decisions about appropriate labeling and continued marketing of products. Skin and eye irritancy tests are required to assure we have adequate information in the event of accidental exposure or ingestion.

Table 1
Federal Regulations Protecting Human Safety

Protection of Human Health Mandated by Federal Regulation:

The Food and Drug Administration:
— 1938: Congress passed the Federal Food, Drug and Cosmetics Act that mandates safety of products and ingredients which go into products. Coal-tar eyelash dye, that had caused blindness was the first product seized. Today the FDA has legal authority to move against cosmetics, because today there is a provision in the law prohibiting the sale of harmful cosmetics.
— 21 CFR Part 312: governs the investigational use of drugs in humans and 312.23(a)(8) requires submission of animal safety data.
— 21 CFR Part 314: governs the submission of a new drug for FDA approval and 314.50(c)(2) specifically states nonclinical acute, subacute, and chronic toxicity test results are required.

The Environmental Protection Agency:
— The Toxic Substances Control Act of 1976: mandates testing to determine potential for environmental risks.
— Fed. Reg. 40, CFR Part 798, Subpart E, Section 798.4500, September 27, 1985: codifies this mandate.

The Consumer Product Safety Commission:
— The CPSC, has the authority to regulate all products that pose an unreasonable risk of injury or illness to consumers.
— The 1960 Federal Hazardous Substances Labeling Act details specific animal toxicity tests that are required to support decisions about appropriate labeling and continued marketing of products. (16 CFR, 1500.41, 1979): A revision in primary irritancy testing requiring a shorter skin exposure time (Fed.Reg.37(244): 27635–27636, 1972) has not yet been codified into the CFR.
— In 1966 and 1977 additional laws were passed which mandated honest labelling. Skin and eye irritancy tests are required under these laws to assure we have adequate information, and appropriate handling precautions included on packaging, shipping and marketing labels.

Safe handling precautions must be included in shipping and marketing labels. All chemicals and chemical intermediates must be tested in order to properly label them for consumers, and to protect workers who may be exposed while handling or shipping potentially dangerous substances.

Contrary to claims by animal rights activists, pound animals are not used for product safety tests. The mixed genetic background of the pound animal introduces too much variability. Rabbit, mouse and rat tests are most common. Animal testing of personal care and household products is required only when the product contains a chemical ingredient for which no safety data exists, or when combined with new chemicals such that the potential toxicity is unknown.

Animal Rights: Opposition to Research

Animal welfare has been a concern for more than one hundred years. In contrast, the concept of Animal Rights first appeared in 1975 in a book called *Animal Liberation* by an Australian philosopher, Peter Singer (2). Singer makes a case that animals think, know fear, and feel pain to a greater degree than a human infant. Yet, he states, no one but a monster would dream of experimenting on a baby or a retarded person. He further dismisses the differences between animals and people as ethically irrelevant and assigns equal status to creatures based on their common ability to feel pain. Therefore, he concludes that all human use of animals is "speciesism"—morally equivalent to racism.

By the mid-1980's at least 400 animal-related organizations had been formed. Today they are backed by millions of dollars; they are lobbying and organizing boycotts, civil disobedience, harassment, and litigation. They have targeted the fur industry, cosmetics testing, and factory farming, as well as the use of animals for sport and research. The American Medical Association (AMA) estimates that more than $200 million is contributed annually to the 400 various animal welfare and rights organizations (3). This is significantly more money than the scientific establishment has to fight them.

The animal rights movement is able to attract legislators and recruit celebrities; it is politically sophisticated, media-wise and adept at raising money. An anti-suffering theme has found a following under a banner of anti-fur and cruelty-free cosmetics. Some short term goals of the animal rights movement include banning product safety testing; granting animals legal standing and humans the right to sue in their behalf; banning the use of primates as well as pound animals in research; and banning all psychological testing on animals. For every $1 spent on this issue by the pharmaceutical industry, animal rights groups spend $50 (4).

Activists have moved from targeting individual cases of animal abuse to targeting institutions; from demanding larger, cleaner cages to campaigning for the elimination of whole facilities. Frederick Goodwin, head of the US Alcohol, Drug Abuse and Mental Health Administration, explains:

> Animal Rights activists have gradually taken over highly respectable humane societies by using classical radical techniques—packing memberships and steering committees and electing directors. They have insidiously gained control of one group after another (5, personal communication).

Often, average supporters have no idea that many organizations which traditionally promoted better treatment for animals, taught pet care, built shelters, and cared for strays, are now dedicated to ending animal research and all human use of animals. The Humane Society of the United States (HSUS), for example, today calls bacon and eggs the "Breakfast of Cruelty" (6).

The policy statement of the HSUS strongly condemns the use of abandoned pound animals for research. "Humane care is simply sentimental, sympathetic patronage," according to Michael

W. Fox, environmental-studies director of the HSUS (7, p. 190).

People for the Ethical Treatment of Animals (PETA) is one of the richest and most well known Animal Rights organizations. PETA was founded by Alex Pacheco in 1981 with only 18 members. Today it claims 250,000 members and an eight million dollar annual budget. The evolution of Singer's premise into an even more radical focus is demonstrated by these quotes: from PETA's Ingrid Newkirk: "A rat is a pig is a dog is a boy," (8, p. 114) and HSU's Michael Fox: "There are no clear distinctions between us and animals" (7, p. 194). Newkirk has further made this sad commentary on human life: she has no reverence for life. She says "All I can do—all you can do—while you are alive is try to reduce the amount of damage you do by being alive" (9, B10). She has also drawn angry responses with what is now called her Auschwitz analogy: "Six million people died in concentration camps, but six billion broiler chickens will die this year in slaughterhouses" (10, p. 41).

Violence

The Animal Liberation Front (ALF) was founded in 1976 in Great Britain by Ronnie Lee. Its purpose is to inflict economic damage on "animal torturers." Damage to laboratories has cost six million British pounds annually since 1980. This is happening in the country with the *most restrictive laws governing research animal use.* ALF is on the FBI's list of terrorists organizations, although they believe there are not more than 100 ALF members in the US. It is well organized, well funded, and totally committed to "the cause."

In 1984, the raid on a University of Pennsylvania Head Injury Laboratory introduced ALF to the US. Videotapes were stolen from the laboratory and PETA edited 60 hours of tape to 28 minutes; it appeared technicians were eating, smoking, and making callous remarks in the laboratory. Allegations of cruelty were never proven, in fact never formally charged; but the National Institute of Health (NIH) funding was stopped, research ended, and a $4,000 fine was imposed by the US Department of Agriculture (USDA). Table 2 is only a partial listing of animal rights activity including the type of research that has been damaged. In addition, at the University of California at Berkeley, environmental laws have been invoked to halt construction of state-of-the-art animal facilities. Six court actions over four years have cost the University $500,000 in legal fees. Similar tactics have failed at Stanford and the University of California at San Francisco, but have cost the Universities millions of dollars in legal fees. In 1987, ALF set fire to an animal diagnostic facility under construction at the University of California at Davis, causing nearly $5 million damage. This facility was designed to *improve* animal welfare. By 1986, during an ALF raid at the University of Oregon, the intent of the raid was openly acknowledged to be destruction of property. Clearly, the focus has moved from improving conditions for animals; animal activists want to stop research.

Harassment—including slashed tires, threatening calls, bomb threats, and other terrorist activities—intimidates researchers. Labs close down, security costs go up, and death threats drive scientists out of the field. At least a dozen researchers at Columbia University have had their lives or the lives of their children threatened by animal-rights activists. PETA's medical adviser and head of the animal rights group Physicians Committee for Responsible Medicine, is psychiatrist Neal Barnard. He claims "We're demoralizing the people who think there's a buck to be made in animal research. And they're starting to get scared, and they're starting to get angry, and they're starting to give way" (7, p. 190).

Table 2
Summary of Animal Right's Activity: Partial Listing

Year	Facility	Type of Research Disrupted
1982	Howard University	Nerve transmission Sickle Cell Anemia
1982	University of California (Berkeley)	Eye Research
1983	Naval Medical Research Institute	Environmental stress Infectious diseases Adult respiratory diseases (new surgical procedures) Bone marrow transplantation Malaria vaccine Decompression sickness
1983	Harbor-UCLA Medical Center	Improved pacemakers Improve grafting in coronary bypass surgery Drugs to treat heart damage
1984	California State University (Sacramento)	Compulsive behaviors (life-threatening) Autistic children, education and treatment
1984	University of Pennsylvania	Head injury
1984	City of Hope Research Institute	Lung cancer detection Emphysema Herpes prevention
1985	University of California (Riverside)	Sonar as an aid to blind infants Causes of birth defects Malaria treatments Insecticide toxicity Birth control and Infertility Vision
1986	University of Oregon	Blindness Seasonal Affective Disorder (SAD)/Biological rhythms Hurler syndromes Neurotransmitter/Parkinson's disease
1986	SEMA, Inc.	AIDS Hepatitis vaccine
1987	University of California (Davis)	Veterinary diagnostics
1987	University of California (Davis)	Sublethal rodenticide effects on reproduction of wild birds (including the California Condor)
1987	USDA Animal Parasitology Institute	Toxoplasmosis in cattle and cats
1988	University of California (Irvine)	Smog effects during moderate exercise Cause and cure for sleep apnea
1988	Loma Linda University	Organ rejection in infant transplants
1989	University of Arizona	Drugs to treat skin disease (including cancer) Cryptosporidium parasitic infection in third world countries, lethal to AIDS patients Intestinal and vaginal parasitic infections (human and animal) Selenium (perhaps a nutritional requirement as well as carcinogen in large quantities) Swine dysentery vaccine
1989	Texas Tech. University	Sudden Infant Death Syndrome (SIDS)
1990	University of Pennsylvania	Sleep disorders (apnea, SIDS)
1990	State University of New York (Buffalo)	Transplantation immunology (rejection) Schistosomiasis parasite vaccine (plague of the third world)

Product Safety Testing

Product safety testing is under vigorous public attack. This includes toxicity testing of new drugs, testing for skin and eye irritancy of new chemicals and household products, and testing for environmental damage. The FDA, The Environmental Protection Agency, and The Consumer Product Safety Commission mandate specific test requirements which allow the proper marketing, labeling, and handling of products to insure consumer, worker, and environmental safety.

Today few of us remember when product safety, labeling, and advertising were not regulated. People were injured, and some even died before cosmetics were tested. In the 18th Century, face powder was made with lead. As a result, people seldom looked youthful past the age of 30; and some died of lead poisoning. In the 1930s American women used a coal tar based-eyelash dye. Some became allergic with dermatitis and itching around the eyes. In 1933, one 52 year old woman died; another woman became blind in both eyes. After weeks of intense pain, she discovered her corneas had eroded away. She testified before congress that year regarding the need for stronger food and drug laws. The result of these congressional hearings was the Food, Drug and Cosmetic Act of 1938 (11).

Before the 1938 FDA regulations requiring products to be tested, more than 100 children were killed by a sulfanilamide elixir made up in a toxic solvent—the product had never been tested in animals or humans (12). Now the FDA reviews all safety data before any product can be marketed. Over 7 million harmful human exposures to products have been reported to Poison Control Centers in the US since 1983 (13). All of the information given to physicians, victims, or their families initially comes from product safety testing on animals.

The dramatic drop in numbers of animals used by researchers in recent years (14), has not impressed all State and Federal lawmakers. They continuously introduce legislation that mandates use of non-animal alternatives to the Draize and LD-50 tests. This, in spite of the fact that the classical LD-50 is rarely done any more. It has been replaced by an acute toxicity test, the range-limit test, which requires many fewer animals, is scientifically validated, and is acknowledged as the test of choice. The scientific community accepts the fact the precision of the LD-50 is not needed to predict a range for potential human toxicity.

Table 3 lists just a few of the companies that claim to manufacture products that are not tested on animals (15). This claim further confuses the issue, because "cruelty-free" and "non-animal tested" labels do not present the complete picture. It is illegal to sell untested products without a specific warning on the label. Frequently, product ingredients are tested before purchase by "cruelty-free" companies; often, established ingredients are supplied in standard formulations, negating the need for re-testing. Companies engaged in the development of innovative formulations, that recognize they must conduct safety and irritancy tests on animals before new products can be marketed, now have a public image problem.

Table 3
Partial Listing from PETA's "Cruelty-Free Shopping Guide"

The following companies have joined the "cruelty-free" band wagon:
Avon
Benetton Cosmetics
Clinique
Estee Lauder
Germaine Monteil
Max Factor
Nexxus
Revlon

Source: PETA, 1990.

Education

Animal use in primary and secondary education is generally limited to observation or an introduction to functional anatomy. Dissection is occasionally encountered as an education issue; and opinions vary within the scientific and educational community regarding value of mandatory participation in dissection. There is strong agreement that standard college anatomy, biology, biochemistry, physiology, and ecology laboratories do require some animal use. Certainly, medical and veterinary education require animal use for appropriate training. Much of today's basic animal research is conducted at colleges and universities. In addition to requisites of anatomy, and physiology, the medical community relies on animals for advanced trauma/life support training. Continuing education for trauma physicians requires the use of pound animals for maintaining skill in trauma intervention techniques such as open heart massage and intubation. Surgeons also use pound animals to develop and refine new techniques. All of this animal use is under attack, in spite of the demonstrated critical human need.

Basic Research

Animal use in basic research is also controversial. Researchers use a variety of animals to understand basic physiological mechanisms, study complex interaction between body systems, and develop new drugs and surgical techniques. The information gained helps animals as well as humans. "Nearly every medical advance in the 20th century . . . has resulted directly from experiments involving animals" (16, p. 784). The Council on Scientific Affairs of the American Medical Society clearly attributes advances in treatment of diseases such as cancer, trauma and shock, cardiovascular disease, and diabetes to availability of animal models (17).

Regulations That Protect Animals

Care and use of Research Animals is regulated by the Federal Animal Welfare Act through the USDA Animal and Plant Health Inspection Service (Table 4). The 1985 amendments to the Animal Welfare act mandate Institutional animal-care and use committees, the animal equivalent of Institutional Review Boards. Procedures involving discomfort are subject to scrutiny; numbers and species of animals used *are* justified; veterinary care, anesthesia, cage size, feeding schedules and ventilation *are* regulated. In fact, the USDA's Animal and Plant Health Inspection Service has codified new standards for dogs, cats and nonhuman primates based on the NIH Guide for Care of Laboratory Animals already followed by most researchers.

Pound Animals

The use of pound animals continues to be a major ethical controversy, and has been the target of much new legislation. Restrictions in pound animal availability is a major concern for university and hospital researchers. Most industrial research programs can afford to buy purpose bred animals, or raise the size and type of dog needed. Extra costs are passed along to the consumers of their products.

About 20 million unclaimed dogs and cats are put to death each year in animal pounds; less than 1% of those are needed for research (18). Pound animals are used for teaching physiology, developing surgical techniques, and basic pharmacological research including cardiac, renal, transplant, and respiratory research. Efforts to end the use of pound dogs in research have resulted in 13 states prohibiting the sale of pound animals for research. Institutions in these states must purchase animals specifically bred for research at prices 4 to 5 times the cost of pound animals treated and conditioned for laboratory use. An inability to use pound animals means that twice as many animals die—the pound animal as well as the research animal.

Massachusetts has the tightest restrictions against animal use; most states with restrictions on pound animal use for research can import animals from other states; Massachusetts law also forbids importation. Critcal work on transplantation, heart disease, cardiovascular surgery,

Table 4
Federal Regulations Protecting Research Animals

Treatment of research animals in U.S. Laboratories is governed by two laws:

The Health Research Extension Act:
— Statutory basis for the National Institutes of Health Guide for the care and Use of Laboratory Animals
— Any institution receiving Public Health Service grants must provide written documentation of its animal care and use program, commitment to animal welfare, and details concerning veterinary care, animal husbandry, and animal facilities.
— Requires that grant-receiving institutions eatablish institutional animal care and use committees (IACUCs) which have a veterinarian, a scientist experienced in animal research, and a member of the public unaffiliated with the institution. The IACUC evaluates the institution's animal research program, inspects animal facilities every 6 months, and reviews research proposals for procedures, use of analgesics or anesthetics, method of euthanasia, and follow-up medical care.

The Animal Welfare Act (Originally passed in 1966, and amended in 1970, 1976, and 1985):
— Administered by the Department of Agriculture's Animal and Plant Health Inspection Service (APHIS). APHIS must inspect each licensed research facility at least once each year.
— The 1985 amendments require standards set to minimize animal pain and distress during experimental procedures, that alternatives to painful procedures are considered, and that no animal is used for more than one major operation from which it is allowed to recover.
— On March 18, 1991 regulations became effective governing the research use of dogs, cats, and nonhuman primates (Amended 9 CFR Part 3; Subpart A and D). These regulations were based in large part on NIH's "Guide for the Care and Use of Laboratory Animals."

and neurosurgery—all of which use dog models—has been slowed or halted. Dr. Harold Wilkinson, a neurosurgeon at the University of Massachusetts, did research on brain swelling after head trauma until three years ago when it became too expensive to obtain animals. Each year 105,000 people aged 15 to 25 suffer neurological damage as a result of head injuries. At Harvard University, Dr. Anthony Monaco has stopped work in organ transplantation. He is attempting to induce tolerance in organ grafts by pretreating the recipient's bone marrow. Although preliminary rejection problems have been worked out in rodents, they cannot translate their findings to human patients until testing in dogs is done. The NIH is no longer able to provide enough funding to compensate for quadrupled animal costs. In short, when money runs out, research stops (7).

Alternatives

Claims of "alternatives" to whole-animal safety testing have been promoted by the animal rights community. The success of these systems is dramatically overstated. The Draize score, for example, is a multi-dimensional measure of cell lysis, organelle dysfunction, breakdown of tight junctions, and the inability of cells to divide and repair the test lesion. While some cell and tissue culture methods, computer models, and bacterial test systems have been developed as adjuncts to animal tests, predicting the possibility an eye will not heal after a chemical insult requires more than a cell culture. Further, Dr. Alan Goldberg, director of the Center for Alternatives to Animal Testing at the Johns Hopkins School of Hygiene and Public Health, stresses that specific questions about some toxic substances can be answered in cell and tissue cultures, but the more complex questions about disease origins and body systems continue to require whole-animal testing (19). Alternatives have not yet been adequately tested or validated as replacement for animal tests. Today, the only real alternatives to using animals for testing, education, and research are human injury, disease, and death.

Conclusions

Animal activists are attempting to stop medical research. The general public, the media, and legislators need reassurance that laboratory animals receive proper care which is already mandated by legislation, and that medical experts agree that the testing methods are reasonable and necessary.

The historic ostrich with its head in the sand reaction of the scientific community has not worked well. Expressed views carry more weight than unexpressed views, and silent majorities are often run over. There are many ways you can help shape the future of important medical research. More information on current legislation is available from: NABR—National Association for Biomedical Research, 818 Connecticut Avenue NW, Suite 303, Washington, DC 20006, (202) 857–0540. A grass-roots organization that has been incredibly successful in defending the use of animals for research is iiFAR—incurably ill For Animal Research, P.O. Box 1873, Bridgeview, IL 60455, (708) 598–7787. Concerns associated with the possible misuse of animals can be expressed to the institution's veterinarian responsible for the animals.

References

1. Borwein B. Provincial board member of Canada's Partners in Research. *Keynote Address: Michigan Society for Medical Research (MiSMR) Annual Meeting, October 1990.*
2. Singer P. *Animal Liberation. A New Ethic for Our Treatment of Animals.* NY: Random House; 1975.
3. AMA White Paper. *Use of Animals in Biomedical Research: The Challenge and Response.* 1988.
4. Horton L. The enduring animal issue. *Journal of the National Cancer Institute.* 1989;81(10):36–42.
5. Goodwin F. US Alcohol, Drug Abuse and Mental Health Administration; 1990.
6. Rockwell L. Animal rights activists are going crackers. *American Trapper.* 1990; Mar/Apr.
7. McCabe K. Beyond cruelty. *Washingtonian.* Feb, 1990;73–77; 185–195.
8. McCabe K. Who will live, who will die? *Washingtonian,* Aug. 1986;112–118; 154–187.
9. Brown C. She's a portrait of zealotry in plastic shoes. *Washington Post.* November 13, 1983; pp. B1, B10.
10. Horton L. A look at the politics of research with animals: Regaining lost perspective. *The Physiologist.* 1988;31(3):41–44.
11. Whittman JH ed. *An Overview of Product Safety and Cosmetic Safety.* New York, NY: Marcel Dekker Inc; 1987;1–7.
12. Temple RJ. Access, science, and regulation. *Drug Inf J.* 1991;25:1–11.
13. 1989 Annual Report of the American Association of Poison Control Centers. 3800 Reservoir Road NW, Washington, DC 20007.
14. Loeb JM, Hendee WR, Smith SJ, and Schwarz MR. Human vs Animal Rights. *JAMA.* 1989;262(19):2716–2720.
15. PETA's Caring Consumer Campaign. *Caring Consumer Cruelty-Free Shopping Guide.* Products that are not tested on animals. Co-sponsor: The International Fund for Animal Welfare. Autumn, 1990. Distributed by: Kalamazoo Humane Society. Kalamazoo, Michigan. August 1991.
16. Bernstein SL. Animal rights activists distort issues. Viewpoint. *JAMA* 1989;261(5):784.
17. Council on Scientific Affairs. Animals in Research. *JAMA.* 1989;261(24):3602–3606.
18. Concannon PW ed. Animal Research, Animal Rights, Animal Legislation. 1990 Society for the Study of Reproduction. 309 West Clark Street, Champaign, IL 61820.
19. Goldberg AM and Frazier JM. Alternatives to animals in toxicity testing. *Scientific American.* 1989;261(2):24–30.

North American Equine Ranching Information Council
P.O. Box 43968 • Louisville, KY 40253-0968 • (502) 245-0425 • Fax: (502) 245-0438 •
http://www.naeric.org

Statement of Purpose

The North American Equine Ranching Information Council (NAERIC) is a not-for-profit association representing horse breeders and ranchers in North America engaged in the collection of pregnant mares' urine (PMU). NAERIC is an educational association whose mission is to provide the general public with accurate information relating to PMU, a key source of hormones used in estrogen replacement therapy. NAERIC works to promote this unique partnership between the profession of horse breeding and women's health care. NAERIC serves as a forum of professional equine ranchers to exchange and analyze information relevant to the horse industry. In order to foster greater awareness of the industry among the general public, NAERIC also works to present factual information and pertinent knowledge related to the practice of equine ranching.

A Supplement to The Horse
July 1995
AAEP Report
News and Notes from the American Association of Equine Practitioners

AAEP Officials Inspect PMU Farms

Recent concerns surrounding the management of mares on PMU (pregnant mare urine) farms prompted AAEP officials to accept an invitation to participate in a CANFACT (Canadian Farm Animal Care Trust) sponsored tour and inspection in Manitoba, Saskatchewan and Alberta in February and March.

AAEP Equine Welfare Committee Chairman Dr. Nat Messer IV, and Dr. Venaye Reece, a committee member, shared the responsibility with Dr. Messer inspecting sites in Saskatchewan and Alberta, and Dr. Reece conducting the same inspections at Manitoba facilities. The purpose of the tour was to determine if management practices employed by the farms ensured that the care and welfare of the mares being kept on the farms was satisfactory. Another focus centered around determining if the PMU farm industry was a responsible use of horses.

Pregnant mare urine or PMU is used in the production of Premarin™, an estrogen supplement which counteracts the effects of menopause, and is the most widely used prescription drug in the United States.

PMU farms are privately owned facilities which are contracted by Wyeth-Ayerst Labs, a subsidiary of Ayerst Organics, Ltd., to produce a designated quantity of urine from mares. Producers are paid by the company on the basis of the quality of urine collected, which is determined by the concentration of estrogen per volume of urine. Each farm is required to conform to a 1990 "Code of Practice," which was developed by a committee initiated by Manitoba Agriculture and Wyeth-Ayerst.

During the inspection tour Messer and Reece examined the facilities, including barns, corrals and harnesses; the types of and frequency of feeding; water availability, including how much and how often; bedding, including types and use; exercise, including how much and how often, and general health and health care. They were accompanied by two representatives of Wyeth-Ayerst who supervised the inspectors employed by the company to monitor compliance to the "Code of Practice."

Messer and Reece report that farms visited ranged in approximate size of 75-350 mares "on line" or confined to their stalls with a collection apparatus (harness) in place. Mares go "on line" in mid-October and are removed near the end of March. This results in a total of five to six months of the 11-month gestation period being spent in a stall by the pregnant mare. The remainder of the year, mares are turned out in either improved or native pastures to foal, be bred and to rear their foals. Most farms house other horses not involved in the collection in either pastures or in corrals and include replacement stock, pregnant mares not "on line," barren mares, stallions and riding horses used by the owners.

Inspectors are assigned a specific number of farms in a geographic area, with a total of six inspectors and two supervisors (one serving as a seventh inspector) conducting inspections on 65-70 farms each. Each farm is inspected approximately every four to six weeks and a report is filed with the company. Inspection reports are then used to determine if contracts will be renewed. Inspectors also work to advise producers regarding equipment, feed and other management problems.

During the course of their inspection, neither Dr. Messer or Dr. Reece noticed any strikingly inhumane or improper management practices, as charged by such groups as PETA (People for the Ethical Treatment of Animals) and other animal rights organizations. However, each did note that improvements in management practices could be made in a number of areas, many of which were not dissimilar to recommendations often made to other horse farm operations.

"The facilities, in general, were satisfactory," Dr. Messer reported. "Even though the older barns did not always comply with the 'Code of Practice,' they were frequently more safe and comfortable for the horses." Messer noted head to head configurations created a greater atmosphere of contentment for the mares than other configurations, however, lead ropes were not always long enough to allow mares to lie down comfortably and very few were designed for a "quick release" in the case of an emergency.

Messer also noted that the harness or collection systems were very well tolerated by the mares, with an occasional chafing in the perineal or flank areas. Lesions were very superficial and not likely to cause any noticeable discomfort. Types of harness varied from producer to producer, and cleaning and maintenance varied considerably between farms and were not always up to "code."

While the mares are fed very well (concentrate two to three times a day and roughage in the form of hay or straw free choice), water was a highly controlled commodity. "Controlled" watering, as used on PMU farms, provided the mares with approximately the same amount of water they would normally consume on a voluntary basis, but the effects of not having free access to water on their welfare has yet to be determined. There were no clinically apparent signs of dehydration in the mares that were examined. Made available either manually or automatically, mares received five to eight gallons of water within a 24-hour period, which is in line with published data regarding voluntary water consumption by horses at rest, Messer reported.

"Producers indicate to us that the amount of water intake had nothing to do with grade of urine," he said. "They utilize 'controlled' water intake in order to avoid having mares spill water into the stalls and mangers, and to avoid having to clean food out of the water buckets. There is little scientific data to indicate what effect unlimited water intake has on grade of urine, so it remains to be determined."

Messer also indicated that the mares were exercised intermittently, but noted that producers gave several reasons as to why mares are not exercised more regularly, including inclement weather, risk of injury (kicking, fighting and running) and that it simply wasn't necessary. However, they did acknowledge that exercise was routinely prescribed for mares which developed limb edema or swelling, joint stiffness or bad attitudes while "on line."

"Some have a system of rotating mares on and off line so that each stall is occupied at all times, but certain mares have several days off," he related. "Regardless of the exercise protocol used or not used, there does not appear to be any serious health problems occurring as a consequence of the restricted exercise."

In general, Messer reported, the mares appeared to be in good health, with body condition ranging from good to excellent. Only one mare was observed to be too thin in the Alberta province, while in some cases the mares would be considered overweight to most brood mares in the third trimester of pregnancy.

"There was a surprisingly low incidence of signs of respiratory disease," he remarked. "Occasional mares coughed during the visits, but there was no significant evidence of chronic obstructive pulmonary disease (heaves) or infectious respiratory disease (i.e. influenza)."

He did note, however, there were numerous lower limb abnormalities (old wounds, edema, abrasions, scar tissue and abnormal hoof growth), most of which, excluding abnormal hoof growth, occurred on the hind legs and were presumably due to kicking at other mares and striking stall dividers.

"Since we did not observe every mare out of her stall, we cannot say how many mares were actually lame," he explained. "Lameness is one of the primary reasons for culling mares. One of our recommendations is to conduct research to see what factors result in the development of lameness in PMU mares and how lameness can be minimized and prevented."

Veterinary care varied from farm to farm, and generally, veterinarians were used only in the case of emergency or pregnancy testing. Routine medical problems were handled by producers, and several did not demonstrate a satisfactory knowledge of routine veterinary problems. Routine preventative medicine consisting of immunization and de-worming varied considerably, and were often contrary to the "Code of Practice," he reported.

Drs. Messer and Reece included a number of recommendations for improvement in their reports which will be reviewed by AAEP's Animal Welfare Committee, as well as other officials at CAN-FACT and Wyeth-Ayerst.

North American Equine Ranching Information Council
P.O. Box 43968 • Louisville, KY 40253-0968 • (502) 245-0425 • Fax: (502) 245-0438 •
http://www.naeric.org

Checks and Balances in PMU Ranching Ensure High-Quality Care for Horses

Pregnant mares' urine (PMU) ranching has more checks and balances to ensure animal care and welfare than any other livestock sector, making it one of the most regulated and closely inspected equine-related activities in the world. The way in which this checks and balances system functions is described below:

Care and Handling of Horses

- PMU ranchers are contractually obligated to adhere to the *Recommended Code of Practice for the Care and Handling of Horses in PMU Operations*. The Code was developed in 1990 to codify previously adopted ranching practices, thus ensuring high standards of care for horses involved in PMU ranching. Company inspectors, agriculture/equine specialists and veterinarians all refer to the Code's guidelines when inspecting or observing PMU ranches.

Inspections

- PMU ranches have undergone many state and provincial reviews conducted by different animal welfare groups and individuals representing these groups. The Canadian Farm Animal Care Trust; the United States Department of Agriculture; the American Association of Equine Practitioners (AAEP); Nat T. Messer IV, D.V.M., chairman of the AAEP Equine Welfare Committee and Art King, D.V.M., president of the Ontario Equestrian Federation and past chairman of the Canadian Equestrian Federation's Ethics and Equine Welfare Committee found PMU horses to be healthy and well cared for. For example:
  - ***"The general standard of care and welfare of the PMU horses compared favorably with those in other segments of the horse industry. Many examples of much care and compassion were observed."***—The Canadian Farm Animal Care Trust, May 1995.
  - ***"The PMU farmers were genuinely concerned about the health and welfare of their horses, and it was obvious they worked very hard to comply with the 'Code of Practice'."***—Nat Messer IV, D.V.M., chairman of the AAEP Equine Welfare Committee, July 1995.
- An international team of equine experts, comprised of veterinary representatives from the AAEP, the Canadian Veterinary Medical Association and the International League for the Protection of Horses, will visit PMU ranches during the 1996-97 collection season to observe the health and welfare of the horses.
- Qualified company inspectors visit each ranch on a monthly basis, throughout the year, to check conformance with the guidelines in the Code.
- Each PMU ranch is reviewed by independent, practicing veterinarians or qualified company inspectors a minimum of 15 times per year. After each visit, veterinarians and inspectors submit formal reviews of their findings.
- Provincial and state veterinarians have the authority to inspect PMU ranches, throughout the year, at their own discretion.

Veterinary Care

- Contracts with the ranchers require three herd health reviews per collection season. Thus, veterinary care on PMU ranches exceeds the norm for the U.S. "household-owned" horse population, as reported by the American Veterinary Medical Association (AVMA), Center for Information Management in their 1991 report. In contrast to the fact that 100 percent of PMU ranches are reviewed by a veterinarian at least three times during the course of a year, the AVMA reported that more than 46 percent of U.S. "household-owned" horses did not receive a visit from a veterinarian.

- A herd health review program was developed in conjunction with, and is monitored by, a committee of 11 individuals; eight of those being equine veterinarians officially appointed by the Alberta, Manitoba and Saskatchewan veterinary medical associations and the North Dakota Board of Animal Health. These individuals include:

  Roxy Bell, D.V.M. and Darrel Dalton, D.V.M. — Alberta Veterinary Medical Assn.
  Sidney Griffin, D.V.M. and Wayne Dunnigan, D.V.M.—Saskatchewan Veterinary Medical Assn.
  Allan Preston, D.V.M. and Ron Mentz, D.V.M. —Manitoba Veterinary Medical Assn.
  Bill Rotenberger, D.V.M. and Judith Gibbens, D.V.M.—North Dakota Board of Animal Health

The remaining committee members include: Nadia Cymbaluk, M.S. , D.V.M., Managing Veterinarian, Linwood Equine Ranch; Ross Chambers, Field Supervisor, Ayerst Organics and Norman K. Luba, Executive Director, North American Equine Ranching Information Council (NAERIC).

- There are more than 90 veterinarians participating in the herd health review program. After reviewing the ranches their reports are filed and compared with company inspection results.

Contract Approval

- All PMU ranches contract independently with Wyeth-Ayerst Laboratories to provide pregnant mares' urine, used in the production of an estrogen replacement medication. Before receiving a contract, all ranching facilities must be examined and approved by company inspectors.

NAERIC

The North American Equine Ranching Information Council is a not-for-profit association representing ranchers in the United States and Canada who are involved in the collection of pregnant mares' urine (PMU), a key source of hormones used in the production of estrogen replacement therapy. NAERIC seeks to provide accurate information on PMU ranching to the general public.

North American Equine Ranching Information Council
P.O. Box 43968 • Louisville, KY 40253-0968 • (502) 245-0425 • Fax: (502) 245-0438 •
http://www.naeric.org

For Immediate Release:
Contact: Norm Luba (502) 245-0425
Barbara Biggar (204) 883-2699

North American Equine Ranching Information Council (NAERIC) Seeks to Inform, Educate Public About Mare, Foal Husbandry

(Louisville, Ky/Winnipeg, MB, Canada) — They are part of an industry that brings $85 million into the economy of Canada and North Dakota, yet their occupation is one that seldom leads them to venture outside the confines of their home and their livelihood is one that very few people know anything about.

They are men and women who live and work on some 500 prairie farms in Canada and North Dakota, breeding, raising and caring for horses. The focus of much of their daily labor is devoted to harvesting pregnant mares' urine (PMU), an unlikely but essential source of hormones necessary to produce an estrogen replacement therapy used to treat osteoporosis and relieve short-term symptoms of menopause.

The ranchers responsible for raising and handling the mares own and operate their farms. They contract independently to provide specific quantities of PMU that are then shipped to a leading pharmaceutical manufacturer to undergo a 125-step producton process to refine the raw materials into a leading estrogen replacement therapy. Largely because the prairie farmers work in such remote areas, they have formed their own association, the North American Equine Ranching Information Council (NAERIC) to serve as the representative voice for the PMU ranching industry.

NAERIC's Executive Director, Norm Luba, says that the ranchers hope the organization will serve to inform and educate the general public about the industry providing accurate information related to equine ranching and the collection of PMU.

According to Luba, "The prairie farmers working within this segment of the ranching industry are equine professionals who work seven days a week to earn a modest living breeding, raising and caring for horses. They form the heart and soul of a profession that has been a staple of the North American agricultural industry for more than 50 years".

Luba, who holds a master's degree in reproductive physiology and a bachelor's in animal science, has spent all of his life in the ranching industry. As former Executive Director of the non-profit Kentucky Horse Council and former Equine Industry Liaison at the University of Louisville, Luba is a past founder and coordinator of numerous educational, information and research programs in Kentucky, Maryland and New York. He has also spent much of his life owning and raising horses.

"The ranchers in North Dakota, Saskatchewan, Manitoba and Alberta are people whose entire livelihoods revolve around breeding and caring for horses," Luba said, "yet they remain largely unknown and misunderstood by the general public. That's why NAERIC was created — to help introduce the public to the people who are the very fiber of the agricultural fabric of North America."

Charlie Knockaert is a typical rancher within the NAERIC network. He owns a modest stable in the Pembina Hills of Canada's prairie country. Knockaert, a former dairy farmer, works with his wife and son in tending, feeding and administering care to 40 Belgian draft horses he acquired from some Amish farmers in Ohio.

"I feel like I know each one of my horses like a friend, and there's no doubt in my mind that they know me," Knockaert said.

He got started in the business at the urging of his 19-year-old son, Aaron, who shares his father's lifelong love of horses. Knockaert, his wife Lorraine and their son tend each day to the horses on the same land once farmed by his grandfather.

Kirk and Gail Bridgeman, a husband and wife team, have managed their own PMU ranch for the past 14 years in Brandon, Canada. The Bridgemans care for some 110 Appaloosa mares on their farm. It's a way of life to which the Bridgemans, who both grew up on farms, always aspired. Kirk Bridgeman is the son of a longtime PMU rancher, while Gail returned to the farming life after spending her early adult years in the insurance business.

"This is definitely a labor of love," Gail Bridgeman says. "I've always loved horses and wanted to work with them. These horses are first and foremost a part of the family. It's a special bond that develops between us that is hard to explain unless one has actually experienced all the hard work and caring that goes into raising these animals."

The Bridgemans take pride in the fact that many of the offspring of the mares involved in the PMU operation have been entered in and won prizes in regional Canadian horse shows. Two of the Bridgeman geldings, both foaled by mares involved in PMU production, will be entered this August in the Canadian Appaloosa Horse Show.

"Ranching is a way of life that turns into a livelihood," Gail Bridgeman says. "I wouldn't trade breeding and caring for these animals for anything."

Rod Hiatt and his father Howard, who operate a 200-head horse farm in Bottineau, N.D. are among a small breed of 30 mostly family-owned farms in North Dakota involved in ranching.

Rod Hiatt typically begins pitching grain and hay each morning at 6 a.m. in a temperature-controlled barn. "My father and I look after all the horses under our care with the kind of carefulness and attention that most families tend to reserve for their children. Some of these horses have been through many winters with us. They're part of the family," Hiatt said.

Luba noted that ranchers like the Knockaerts, the Bridgemans and the Hiatts are representative of the people within an industry that, up until now, has gone unappreciated and sometimes misunderstood by the general public. "PMU ranching is an important—but little known—part of the equine industry," Luba said. "NAERIC has a big job to do in terms of familiarizing the public with a remarkable industry that means so much to agriculture and to women's health care."

North American Equine Ranching Information Council
P.O. Box 43968 • Louisville, KY 40253-0968 • (502) 245-0425 • Fax: (502) 245-0438 •
http://www.naeric.org

Recommended Code of Practice for the Care and Handling of Horses in PMU Operations

What is the recommended Code of Practice and why does Pregnant Mares' Urine (PMU) ranching need one?

- The *Recommended Code of Practice for the Care and Handling of Horses in PMU Operations* is a set of guidelines developed to ensure the humane treatment and best possible care for horses involved in PMU ranching.
- The Code builds upon existing, scientifically-based management recommendations to provide ranchers with guidelines they can tailor to the needs of each individual horse.
- PMU ranchers are contractually obligated to follow the Code and ranches are reviewed by independent equine experts and company inspectors on a regular basis. (For more information on this subject, please refer to the fact sheet entitled *Checks and Balances in PMU Ranching Ensure High-Quality Care for Horses).*

Who developed the Recommended Code of Practice, and how?

- The Code was first published in 1990 and was drafted by a committee of independent veterinary experts, PMU ranchers and government veterinary specialists. The joint committee, initiated by the Manitoba government and PMU producers, studied PMU ranching and wrote its recommendations based on these findings.
- The Code is reviewed and updated annually by the independent committee.
- The Code has been endorsed by the Ministers of Agriculture in Alberta, Manitoba and Saskatchewan and the North Dakota Commissioner of Agriculture.

What aspects of horse care does the Recommended Code of Practice address?

- The Code sets high standards for all animal husbandry practices on PMU ranches, including, but not limited to, the provision of temperature-controlled barns; stalls large enough for the mares to move about and lie down comfortably; watering and a nutritionally balanced diet based on guidelines established by the National Research Council's *Nutrient Requirements of Horses*; transportation; veterinary care; exercise and restraint-free, lightweight collection systems that guarantee a full range of emotion.

Do other equine-related activities have a Recommended Code of Practice?

- The Code for PMU ranching is the only existing code for horses in North America. In Canada, "recommended codes" are available for the care and handling of several species of animals. However, "the PMU Code is the most visible, the most used, and the most compiled with, of all the Codes of Practice for farm animals."—Harry J. Enns, Manitoba Minister of Agriculture, May 1996.

Conclusion

By creating a Recommended Code of Practice, PMU ranching is promoting high standards of animal care and handling and building on existing laws and regulations. Moreover, the Code allows ranchers to be flexible in providing individual care to each of their horses, while following guidelines that guarantee the best care and attention for all of them.

NAERIC

The North American Equine Ranching Information Council (NAERIC) is a not-for-profit association representing ranchers in the United States and Canada who are involved in the collection of pregnant mares' urine (PMU), a key source of hormones used in the production of estrogen replacement therapy. NAERIC seeks to provide accurate information on PMU ranching to the general public.

North American Equine Ranching Information Council
P.O. Box 43968 • Louisville, KY 40253-0968 • (502) 245-0425 • Fax: (502) 245-0438 •
http://www.naeric.org

For Immediate Release:
Contact: Norm Luba Executive Director (502) 245-0425

Independent Equine Veterinary Practitioners To Conduct 1,450 Herd Health Reviews in 96-97 Season

(Louisville, Ky. — September 3, 1996)—With the 1996-1997 collection season right around the corner, the North American Equine Ranching Information Council (NAERIC) announced today that during the upcoming season there will be approximately 1,450 equine veterinary visits to pregnant mares' urine (PMU) ranches to review the health and welfare of the horses. This enhanced veterinary supervision is the result of a new herd health program implemented last year by Ayerst Organics, the company that contracts with ranchers for PMU. The program requires three veterinary reviews of each herd during a collection season.

"This program has increased veterinary involvement on PMU ranches. No other livestock industry that I am aware of has such a regulated requirement for routine care," said Ron Mentz, D.V.M., of Manitoba, a practicing veterinarian and veterinary review committee member. "I have been impressed with the serious approach this industry has taken to assure the continued outstanding care of horses on PMU ranches."

The program features 91 private-practice veterinarians — 48 in Manitoba, 19 in Saskatchewan, 16 in Alberta and 8 in North Dakota — that visit and complete full-herd health reviews on each PMU ranch three times per year. There are 483 PMU ranches in North America: 282 in Manitoba, 106 in Saskatchewan, 66 in Alberta and 29 in North Dakota.

A committee of eight equine veterinarians appointed by the Alberta, Manitoba and Saskatchewan veterinary medical associations and the North Dakota Board of Animal Health, along with two representatives of Wyeth-Ayerst—the purchaser of PMU—and one from NAERIC, originally developed a comprehensive herd health review form. They have also enhanced it for the coming collection season.

As with last year, this year's program will have veterinarians assessing elements of each PMU herd's overall health, including the physical condition of the mares, general condition of the facilities, watering and nutrition practices, and the condition of other horses kept on the ranch. Veterinarians use the herd health reviews form to evaluate the herd on each visit.

"Our ranchers welcome these herd health reviews. Veterinarians provide additional health and welfare expertise, which is beneficial to the management of horses," said Norm Luba, executive director of NAERIC. "Equine veterinarians are the best-qualified people to help the ranchers assess the ongoing health and welfare of the herds."

"On my ranch, we work together with the visiting veterinarians to keep detailed herd health records," said Kevin Frith, PMU rancher from Devil's Lake, N.D. "Keeping these records enables me to ensure continued good health of my horses."

In fact, PMU mares are much more likely to receive veterinary care than even a privately-owned horse. According to a recent (1992) American Veterinary Medical Association (AVMA) survey, only 53 percent of American pleasure horses in the survey were examined by a veterinarian during 1991. It also reported that only 21.9 percent of horses were visited by a veterinarian three or more times that year. PMU ranchers have 100 percent of their horses examined at least three times each year as a requirement of their production contract.

"The veterinary care standards of PMU ranches are among the highest of any equine industry," Luba said. "Government officials have indicated that none of the approximately 300 cases of reported animal abuse in Manitoba were made against a PMU rancher."

NAERIC is a not-for-profit association representing ranchers in Canada and the United States who are involved in the collection of pregnant mares' urine, a key source of hormones used in the production of a leading estrogen replacement therapy. NAERIC seeks to provide accurate information on PMU ranchers and the ranching practices they use in the management of their horses.

North American Equine Ranching Information Council
P.O. Box 43968 • Louisville, KY 40253-0968 • (502) 245-0425 • Fax: (502) 245-0438 •
http://www.naeric.org

For Immediate Release:
Contact: Norm Luba Executive Director (502) 245-0425

PMU Ranching's Veterinary Care Standards Surpass Those of Private Ownership

(Louisville, Ky.—July 23, 1996)—Which is more likely to get regular veterinary attention, a privately owned pleasure horse or a horse being raised by a pregnant mares' urine (PMU) rancher? According to the most recent American Veterinary Medical Association (AVMA) survey, it is often the latter.

Horses involved in PMU ranching, a horse breeding industry which supplies the PMU used in the manufacturing of a hormone replacement therapy, are examined by veterinarians at least three times each year, according to Norm Luba, executive director of the North American Equine Ranching Information Council (NAERIC).

"The AVMA survey shows that about 46 percent of American pleasure horses don't get examined by veterinarians each year. PMU ranchers have 100 percent of their horses examined at least three times each year as a requirement of their production contract," Luba said.

The most recent AVMA survey, completed in 1992, showed that only 53.6 percent of the American pleasure horses counted by the survey were examined by a veterinarian during 1991. It also reported that only 21.9 percent of horses saw a veterinarian three or more times that year. The next AVMA study will be conducted in January, and the Canadian Veterinary Medical Association does not track this information.

"Many people who own pleasure horses tend to only call veterinarians if their horses become ill, while professional horse breeders more often implement preventive herd health care programs," Luba said.

Approximately 84 private-practice veterinarians in Manitoba, Alberta, Saskatchewan, and North Dakota are involved in completion of full-herd health reviews on PMU ranches three times per year, according to Luba.

"Keep in mind, many of the veterinarians visit the PMU ranches they work with more often than the three reviews mandated by the ranchers' contracts with the pharmaceutical company," said Dr. Bill Rotenberger, North Dakota state veterinarian, Department of Agriculture and a member of the PMU Veterinary Review Committee.

"Our veterinarian is involved in almost every aspect of this ranch and our breeding process, and he plays an integral role during our foaling season," said Kevin Frith, who owns a PMU ranch in Devil's Lake, ND. "We work together to keep detailed health records on the horses, which enables us to identify potential problems before they arise," he said.

Frith said that his veterinarian visits his ranch approximately once a month throughout the year to provide services from deworming and vaccinations to pre- and post-natal care.

"Veterinary standards in our segment of the equine community are among the highest," Luba said. "When your livelihood depends on keeping your horses as healthy as possible, you take every opportunity to consult with veterinarians and make sure your horses are getting the best care available," he said.

NAERIC is a not-for-profit association representing ranchers in the United States and Canada who are involved in the collection of pregnant mares' urine (PMU), a key source of hormones used in the production of estrogen replacement therapy. NAERIC seeks to provide accurate information on PMU ranchers and the ranching practices they use in the management of their horses.

North American Equine Ranching Information Council
P.O. Box 43968 • Louisville, KY 40253-0968 • (502) 245-0425 • Fax: (502) 245-0438 •
http://www.naeric.org

For Immediate Release:
Contact: Norm Luba (502) 245-0425
Barbara Biggar (204) 883-2699

NAERIC Updates Equine Welfare Committee at Largest AAEP Convention Ever

(Louisville, Ky.—January 5, 1996)—Emphasizing that pregnant mares' urine (PMU) ranchers are taking a number of steps to confirm, validate and continually improve the management of horses in their care, the Executive Director of the North American Equine Ranching Information Council (NAERIC) addressed the Equine Welfare Committee during the 41st annual meeting of the American Association of Equine Practitioners (AAEP), held in December.

Norm Luba, NAERIC's Executive Director, informed the AAEP about the additional veterinary herd reviews, fire awareness information, research on continuous versus interval watering and other activities that PMU ranchers have implemented to address concerns regarding their industry.

"The ranchers involved in the PMU industry are the most regulated of any segment of the horse breeding industry," he said. "NAERIC also continues to examine the management practices utilized in the industry and appreciates the constructive assistance and recommendations of professionals in the horse industry who have provided review and input, such as the AAEP."

Charlie Knockaert, President of NAERIC and a PMU rancher from Bruxelles, Manitoba, was on hand throughout the convention to meet with the equine veterinarians. "The convention was worth the time away from the farm," he said. "Horse veterinarians told me our efforts of providing information based on facts, studying our management practices and making improvements where appropriate are the correct responses to the animal activist campaigns being waged against us."

Dr. Peter Fretz, a Saskatchewan veterinarian who works at the Western College of Veterinary Medicine and is a member of the Executive Committee of the Canadian Veterinary Medical Association, welcomed the information NAERIC provided to those in attendance. "An association like the AAEP enables different segments of the equine industry to share information regarding equine disease, management, research, and welfare through continuing education," he said. "We are very pleased to have the PMU industry providing equine professionals information on their initiatives. That is the way responsible industries react to challenges."

The AAEP is an international organization whose mandate is to meet with veterinarians and other opinion leaders about important issues regarding the health and welfare of the horses. The 1995 convention was the largest in AAEP history with a total of 3,673 in attendance.

NAERIC was established by PMU farmers in Canada and North Dakota to serve as a forum for professional equine ranchers to exchange and analyze information relevant to the horse industry. NAERIC also works to present factual information and pertinent knowledge related to the practice of equine ranching to the horse industry and the broader community.

North American Equine Ranching Information Council
P.O. Box 43968 • Louisville, KY 40253-0968 • (502) 245-0425 • Fax: (502) 245-0438 •
http://www.naeric.org

For Immediate Release
Contact: Norm Luba (502) 245-0425
Kevin Frith (701) 662-4386

PMU Horses Command Record Prices at South Dakota Livestock Exchange

(Louisville, Ky. — October 18, 1995) — A crowd of hundreds paid record sales prices for a grouping of nearly 100 PMU weanlings at the 2nd Annual Paint Breeders and Rush Valley Farms Production Sale in Douglas County, South Dakota.

The majority of horses sold, with the exception of eight Quarterhorse weanlings, were either Paint or Paint Breeding Stock.

A total of 235 buyers were on hand from 10 states, including Wisconsin, Minnesota, North Dakota, Iowa, South Dakota, Nebraska, Illinois, Missouri, Colorado and Kansas.

Some of the highlights of the sale included:

- A top price of $2,950, breaking the record for the highest dollar weanling ever sold at a Douglas County livestock auction.
- An average of $1,537 for the top 15 weanlings and $2,340 for the top five horses.
- An average sale price of $713 for 99 weanlings.

The sale was a tremendous success according to Kevin Frith, the President of the North Dakota Equine Ranching Association."The prices paid for the horses are indicative of the quality, health and strength of the breeds turned out by participating ranchers within the PMU industry", Frith said. "The horses that come out of the PMU industry have developed a high caliber reputation that is clearly reflected by the results of the auction."

Norm Luba, the Executive Director of the North American Equine Ranching Information Council (NAERIC) called the results "a barometer of sound breeding and handling practices. A horse that's been weaned on a PMU ranch is clearly recognized as a cut-above the rest by professionals within the horse industry."

Luba noted that ranchers are generally very selective both about the purposes for which they sell their horses as well as the auction facilities with which they will do business.

Seventy-six percent of ranchers recently surveyed by NAERIC indicated they deal selectively with certain categories of buyers while 50 percent participate only in private sales.

Other sales averages included:

- $1,003 each for 27 yearlings
- $1,361 each for 9 broodmares
- $1,538 each for 19 broke mares and geldings

North American Equine Ranching Information Council
P.O. Box 43968 • Louisville, KY 40253-0968 • (502) 245-0425 • Fax: (502) 245-0438 •
http://www.naeric.org

For Immediate Release
Contact: Norm Luba (502) 245-0425

PMU Ranchers To Benefit from Breeding Enhancement Program

(Louisville, Ky.—August 28, 1995)—Ranchers involved in the collection of pregnant mare's urine (PMU) for production of estrogen replacement therapy will soon have the opportunity to participate in an effort designed to introduce Thoroughbred stallions into their breeding program. The results of this effort will be improvements to the pedigree and value of foals produced in the PMU program, according to the North American Equine Ranching Information Council (NAERIC), the association representing over 450 PMU ranches.

Ranchers who participate will have their mares evaluated using a scoring criteria adapted from the Alberta (Canada) Horse Improvement Program. A certificate will be issued by NAERIC confirming the physical attributes of the mare, with descriptive data entered into a master database. Mares that meet the criteria will be paired with loaned or leased Thoroughbred stallions. The stallions will be housed with the participating rancher, or stand at stud in cooperative fashion to groups of ranchers.

Since the resulting foals will be Half-Thoroughbred, they will be eligible for registration in The Jockey Club's "Performance Horse Registry" (PHR), which was formed to record the Thoroughbred's influence in non-racing equine activities. Both NAERIC and the PHR will be promoting such registration to all PMU ranchers.

"The PMU industry maintains a fine record of breeding and supplying valuable foals for ranch work, recreational riding and horse show markets," said Norm Luba, executive director of NAERIC. "The addition of Thoroughbred stallions and a certification program for brood mares, coupled with the ability to register the resulting foals, will help assure a continued breeding improvement program."

"This documentation will help increase the sales potential for foals coming from PMU ranches," said Mr. Luba. "It will increase our viability in existing markets, and create new sales opportunities for NAERIC members."

The Performance Horse Registry was established in 1994 to register horses, that are at least half Thoroughbred, throughout North America. PHR leadership was enthusiastic about the introduction of Thoroughbreds into the PMU breeding program and the impact it will have on the quality of horses produced.

"I commend NAERIC for introducing Thoroughbred stallions into their breeding program," said Edward S. Bonnie, chairman of the Performance Horse Registry, which is based in Lexington, Kentucky. "Doing so will certainly increase the market value of the foals produced on PMU ranches and do so at a reasonable cost."

"We have two goals as a responsible industry," said Charles Knockaert, president of NAERIC and head of the Manitoba Equine Ranching Association. "First, as lifelong ranchers, we must produce quality horses that can lead active and productive lives. Second, as business people, we must develop the kinds of programs that can assure our industry members the highest rate of return."

According to Mr. Luba, pilot programs are expected to begin in each Canadian prairie province and North Dakota by spring of 1996.

Canada and U.S. join forces in the War on Animal Rights

By Terri Greer
PRCA Administrative Assistant

Canadian and U.S. stockmen, ranchers and honest animal welfare organizations are waging a war against an enemy whose tactics know no bounds and whose ethics do not exist.

We are talking about a movement involving individuals who have an inherent desire to force their misplaced values and thoughts onto the rest of society. It matters little to them which targets they seek out as long as the target has the potential to provide them easy media access and fundraising possibilities.

To those who study the activities of the leadership of the animal rights movement, it soon becomes astonishingly clear these purveyors of fiction prey on the uninformed public who have great empathy for suffering, especially if it involves an animal. The great crime is that this segment of society will open their hearts along with their wallets without first verifying what they are being told is based on documented fact. We see mounds of materials travelling through the mail from animal rights groups presenting themselves an animal welfarists who present abolutely no substantiation to accompany their allegations of abuse.

No animal user is immune from their attacks. Rodeo, circus, racing, horse and dog shows, aquariums, pets, livestock breeders; we all have a common thread, the animal rights movement. We must learn about what is happening in other animal industries and support the responsible use of animals in a humane and caring manner as a united force. Stockmen, Rodeo contestants, circus performers and pet owners must work together to take the truth about animal rights to the public.

A number of groups involved in the animal rights movement frequent both sides of the border. One such group becoming more visible in Canada is the People for the Ethical Treatment of Animals (PETA). Possibly due to receiving an increasing amount of bad press in the United States, PETA is looking for fresher ground.

Recent press coverage in Saskatoon, SK, would support this theory. The Western Producer, December 15 issue, appears to have a very savvy reporter by the name of Mary MacArthur. Apparently she has done her homework on PETA and has a clear understanding of their motives. Within Mary's article, she quotes Montreal author Alan Herscovici from his book, *Second Nature: The Animal Rights Controversy*; "They have to keep coming up with new sensational campaigns. Animal rights activists make a living attacking people, and their main objective is to cause as much trouble as they can."

Truer words were never spoken.

One of PETA's most recent targets is the PMU farms in North Dakota and Manitoba. PMU farms collect pregnant mare's urine for the production of estrogen hormones. PETA claims to have gone undercover to 18 farms and contends abuse to some 75,000 mares in the U.S. and Canada.

First, the very fact that PETA, as Rodeo and others have experienced, often prints statements that lack creditability should be cause for reasonable people to question the validity of their allegations on any subject. PETA statements come from staff members lacking qualification to make accurate judgements. Did they have a large animal veterinarian or a recognized livestock expert with them? Doubtful.

While there may be random instances of poor animal welfare practices, PETA spokesperson, Margaret Lawson's statement to the Western Producer of, "I saw neglect of horses on every single farm," leaves one with the impression this woman had a pre-conceived notion. It would be interesting to see if a list of all 18 farms exists at all.

Never forget, PETA and others move from one subject to the other. Rodeo has been and will be a target again, when it is convenient and when they feel it will be profitable.

Keep your ear to the ground and let us know about activities you hear about in your area. Please contact Terri at PRCA headquarters: 101 Pro Rodeo Drive, Colorado Springs, CO 80919–2339 U.S.A. or telephone (719) 593-8840.

Canadian Rodeo News
February 1995

Animal Welfare Issues

Rights stunt backfires

By **Terri Greer**
PRCA Animal Welfare Issues Coordinator

According to the Toronto *Globe and Mail*, December 28th issue, the Animal Rights Militia (ARM), pulled a stunt which not only backfired on the animal rights movement, but harmed the animals as well.

Prior to Christmas, Vancouver media outlets and two Vancouver supermarket chains received notices from ARM that some turkeys had been injected with rat poison. The tactics backfired because the stores immediately pulled the turkeys off their shelves, offered to replace any previously sold turkeys and replaced those pulled from the stores.

Since more than 20,000 birds were returned, it required the killing of more birds to replace those pulled. No contamination was found in any of the pulled birds.

This threat caused problems for a great many people and has furthered the continued loss of respect the animal rights movement is beginning to see from both the media and the public. This radical group gave no thought to the elderly who had no transportation and could not easily return the turkeys. Needy families receiving charity baskets were also greatly harmed.

The president of Save-On-Foods expressed his anger this way: "Arguably, everybody has the right to make their point and to promote their own particular cause, but at the risk of health and safety of other individuals ... that's just not acceptable in the general public's viewpoint."

This group definitely deserves the "dumb stunt" of the year award. However, remember, every incident such as this causes further harm to the animal rights' movement. In the end, they may be their own worst enemy.

PMU arns receive passing grade from the Canadian Farm Animal Care Trust

Last issue, we reported on the attacks on the PMU farms in Canada by the People for the Ethical Treatment for Animals (PETA). As usual, PETA failed to do any significant research and thus set themselves up for yet another black eye.

According to a report issued by a farm animal welfare organization, the level of animal care in the Pregnant Mare's Urine industry is "impressive." The statement continued, "the mares used for this purpose receive a standard of care which is probably better, on average, than the care provided for many privately owned riding or pleasure horses." Upon inspection, the animal welfare organization found the barns to have good air quality, water, feed, housing and stall space.

Many groups differ in philosophy from the radical organizations like PETA. However, as we see PETA on a steady decline in the United States, it appears to be trying to gain attention in Canada. A sharing of any information regarding this group will be useful in putting them where they should be ... bankrupt.

Readers are responding

A special thank you to the Canadian friends who have actively been sending articles and information for future animal welfare issues. Your input is vital in spreading the message to everyone. We must share what is occurring in other industries as well as the sport of Rodeo.

The animal rights movement is the issue which connects us together and by working as a united front, we have a voice that will be heard. Please send any information you come across or call 1-800-955-3816.

We can and are making a difference. We are beginning to educate more and more of the public and the media about the true and hidden agenda of animal rights. Bringing these groups to a halt must be a priority of horse owners. Rodeo contestants, fans, pet owners, honest animal organizations, and everyone who enjoys their relationships with animals.

We must promote the use of animals in our society, provided it is done in a humane and responsible manner. Take your message to the public and continue to educate . . . .

You may contact Terri at PRCA headquarters, 101 Pro Rodeo Drive, Colorado Springs, CO, 80919—2399 U.S.A.

Why the Study of Animal Behavior Is Associated with the Animal Welfare Issue[1]

Harold W. Gonyou

Prairie Swine Centre, Inc., Saskatoon, Saskatchewan, Canada S7H 5N9

Abstract: Of the various disciplines within the animal sciences, the issue of animal welfare has been most closely associated with ethology, the study of animal behavior. Prior to the modern welfare movement, applied ethology was primarily involved in studies on feeding and reproductive behavior. The emphasis on freedom of movement and mental experiences in animal welfare resulted in the field of applied ethology developing its current welfare interests. During the past 30 yr, applied ethology has been used to gather appropriate information to develop alternate management systems that accommodate normal behavior. The issue of behavioral needs has been addressed and research interest in motivation has developed. Preference tests have been used for their traditional role of improving comfort and have been modified to assess motivation as well. We have used abnormal behaviors as indicators of poor welfare and are shifting our emphasis to causative factors of these behaviors. The emotional states and cognitive abilities of animals have been studied but will become an increasingly important component of behavior research into animal welfare in the future.

Key Words: Animal Behavior, Animal Welfare

Introduction

Animal welfare is an issue that involves several scientific disciplines that are part of the "animal sciences." Perhaps the discipline that has been most closely associated with the welfare issue is the study of animal behavior, known as ethology. The term "applied ethology" is often used to designate the subdiscipline involved in studying the behavior of animals that are managed in some way by humans, whether they be on farms, in laboratories or zoos, or managed wildlife. Applied ethology involving agricultural species has become so closely associated with the scientific study of animal welfare that some use the terms *behavior, ethology* and *welfare* as virtual synonyms. Such was not always the case. The first major book on the behavior of domestic animals (Hafez, 1962), contained over 600 pages of information, indicating that the discipline was based on a large volume of knowledge. However, the terms "welfare" or "well-being" did not appear in either the first or second editions of the book (Hafez, 1969). The initial thrust of applied ethology was not related to animal welfare, but rather to production.

This paper examines why ethology and the animal welfare issue became and remain so closely associated. The approach will be to discuss three aspects of this association. These are 1) the identification of ethology as a discipline relevant to animal welfare concerns, 2) the rationale for ethological studies on animal welfare, and 3) the future role of ethology in resolving animal welfare concerns.

©J. Anim. Sci. (1994) 72:2171–2177.

Identification of Ethology as Relevant to Welfare

It is generally acknowledged that the publication of *Animal Machines* (Harrison, 1964) played a pivotal role in the beginning of the modern animal welfare movement. Harrison was critical of the intensive animal production practices that had become increasingly common after the second world war, particularly the use of battery cages for hens, crates for veal calves, and large-scale broiler production. Her concerns included not only the welfare of the animals but also the use of drugs in animal production, the quality of animal products, and the esthetics of modern farming. Relatively little discussion of animal behavior was included in the book. She did consult with Lorenz, who later was awarded a Nobel prize for his work in ethology, on two issues. The first point, which Lorenz answered in the negative, was whether chickens experienced an anticipation and fear of death in poultry processing plants. Lorenz did indicate, in response to the second question, that social behavior would affect the productivity of broilers in large groups. Other references to behavior in the book included vices such as feather pecking and cannibalism in poultry and excessive licking in calves, and the thwarting of instinct in terms of neonatal behavior, food selection, boredom, dunging patterns, and sleep cycles. Harrison aroused the public's concern for agricultural animals through her implication that animals are viewed as machines, in a similar way that anti-vivisection groups formed in response to Descarte's reference to animals as automata.

In response to Harrison's book, the British government appointed a technical committee composed of two veterinarians, four agriculturalists, a surgeon, and two zoologists to enquire into the welfare of intensively farmed animals. One of the zoologists was an ethologist, W.H. Thorpe, well known for his work on bird song, and the other, F.W.R. Brambell, chaired the group that is often referred to as the Brambell Committee (Command Paper 2836, 1965). The committee accepted that animals can experience pain, suffering, and stress and such emotions as rage, fear, apprehension, frustration, and pleasure. It is not surprising, then, that the report's widely cited statement on welfare refers to "both physical and mental well-being," and that evaluation of animal welfare must include "scientific evidence available concerning the feelings of the animals that can be derived from their structure and functions and also from their behaviour." In addition to an emphasis on animals' feelings, the report stressed freedom of movement, in what became known as the "Five Freedoms": An animal should at least have sufficient freedom of movement, to be able without difficulty, to turn around, groom itself, get up, lie down and stretch its limbs." The committee identified the study of animal behavior as critical to the animal welfare issue, and one that had not attracted the attention it deserved. The report of the Brambell Committee, more than any other document, identified ethology as relevant to the issues of the modern animal welfare movement.

The report of the Brambell Committee was frequently cited in terms of freedom of movement and the mental well-being of agricultural animals. Many other aspects of welfare discussed in the report were often overlooked. The government did act on one recommendation of the committee, and that was to appoint a standing advisory committee on animal welfare. The Farm Animal Welfare Committee (FAWC) of the United Kingdom continues to function and make recommendations to the government. The FAWC published what are known as the "New Five Freedoms," perhaps to correct an imbalance in the reporting of the Brambell Committee's suggestions. These freedoms, as recently revised (FAWC, 1993), are as follows:

1. Freedom from thirst, hunger and malnutrition by ready access to fresh water and a diet to maintain full health and vigor.

2. Freedom from discomfort by providing a suitable environment, including shelter and a comfortable resting area.

3. Freedom from pain, injury, and disease by prevention or rapid diagnosis and treatment.

4. Freedom to express normal behavior by providing sufficient space, proper facilities, and company of the animal's own kind.

5. Freedom from fear and distress by ensuring conditions that avoid mental suffering.

In commenting on these freedoms, Stookey (1992) points out that freedoms 1–3 have traditionally been accepted and addressed by agriculturalists, but that the latter two freedoms reflect the current concerns raised by society at large. Webster (1993) divides the freedoms into production traits (1–3) and ethological issues (4, 5). This latter division emphasizes that ethology is critical to addressing the concerns of the modern welfare movement.

Within months of the publication of the Brambell Committee's report, the Society for Veterinary Ethology, currently the International Society for Applied Ethology, was formed. The question of the relationship between ethology and animal welfare was raised by the first secretary of the society at the founding meeting (Fraser, 1980). The result was that one of the aims of the society was "to promote the exchange of information between veterinarians and between them and others concerned with the behaviour and well-being of animals" (Petherick and Duncan, 1991). Thus, a scientific society recognized that animal welfare should be addressed by applied ethologists.

Rationale for Past Ethological Studies

Having identified ethology as an important discipline related to animal welfare, agricultural research groups studying welfare began including measurements of behavior on their studies. However, a few of the researchers in the 1960s and 1970s were actually trained in ethology. Even today, most departments of animal science do not have applied ethologists. As a result, the rationale of some research has been poorly understood. Although it is not possible to discuss all the themes in ethological studies on animal welfare in this paper, I will examine five that have received significant attention. The studies mentioned were chosen because they illustrate a particular approach or philosophy, even though the results may not have been definitive or necessarily supportive of the original hypotheses.

Accommodating Normal Behavior

The FAWC's fourth freedom states that animals should be able to express their normal behavior patterns. One interpretation of this freedom is represented by Kilgour (1978), who suggested, using terminology of McBride (1969; cited by Kilgour, 1978), that we fit farms to animals, not animals to farms. By designing our farms to accommodate normal behavior, we should prevent or minimize distressful situations. Such an approach may seem obvious, but Kilgour suggests that farms have been designed to accommodate humans or economic considerations and not the behavior of the animals that have been forced to live on them. Ewbank (1988) suggests that providing for normal behavior is a means of ensuring that animals have a reasonable opportunity to maintain adequate welfare. This approach can be summarized as follows: in general, animals that can perform normal behavior are more likely to achieve better welfare than those that cannot.

If normal behavior should be accommodated in production settings, then we need to know what normal behavior is for each species. This is the basis of studies that develop ethograms of our livestock (Banks, 1982). These studies have usually been conducted in seminatural environments to ensure that animals have the opportunity to respond to a wide variety of environmental features and to aid in the interpretation of the behaviors observed. Three major efforts of this nature have been conducted with pigs. Feral pigs were studied on an island off the Georgia coast by Graves (1984). Stolba and Wood-Gush (1989) observed groups of pigs in a park near Edinburgh, and Jensen (1986) reported on pigs kept in a wooded area in Sweden. A common misperception of these studies has been that the purpose was to raise pigs commercially in seminatural conditions.

Jensen (1991) states that the studies are part of the information gathering step in the scientific method and form the basis for developing hypotheses, which must then be tested before they are applied to animal production in more conventional surroundings.

All these studies have provided information used in designing new housing and management systems for pigs. Graves (1984) reported that sows formed groups of three to six adults with their offspring. A group of similar size was selected for an alternative group housing system for sows (Morris and Hurnik, 1990). Stolba (1981) incorporated findings on social behavior into his "family pen" system that kept pigs with their dams until market. A new system developed in Sweden, which groups sows with their litters together approximately 10 d after farrowing (Hogsved, 1990), was based on the observations by Jensen (1986). Whether these systems will eventually replace current production methods remains to be seen, but the approach of designing management around behavior is moving forward.

Clearly successful, but less dramatic, examples of this approach are available. Taylor (1990) studied the movements of sows while they were eating and designed a feeder that provided the space envelope necessary for these movements. As a result, sow feeder design has changed dramatically throughout the industry in the past few years. Space envelopes have also been developed for free stalls for dairy cattle and gestation stalls for cows. The former have been adopted by the industry and the latter are currently being investigated.

Behavioral Needs

The first approach to accommodating normal behavior is very general and does not consider any specific behaviors as being essential. That is, the goal is to accommodate as many behaviors as possible, but none is considered more important than others. However, a second approach to accommodating behavior considers the possibility that certain behaviors are essential. The concept of a behavioral need is separate from that of a physical need, in that it is the performance of the behavior that is critical, not its physical consequences. Motivation is a critical part of this issue; thwarting a motivation is seen as a means of causing psychological suffering. Dawkins (1983, 1988) and Dellmeier (1989) have addressed the concept of behavioral needs and their importance to the welfare issue.

The concept of behavioral needs has been controversial, partly because ethology was developing new models of motivation at the time that welfare became an issue. Wood-Gush (1973) discussed the conflict of the traditional Lorenzian model, which emphasized internal sources of motivation, and newer models with greater emphasis on external sources. Hughes (1980) suggested that behaviors that are primarily internally motivated are of critical importance in animal welfare, whereas those that are externally motivated are less likely to be considered essential. Jensen (1993) has recently addressed the issue and believes that the source of motivation is less critical than the strength of the motivation. However, most research on motivation has emphasized internal sources, and the increase in motivation if the opportunity to perform the behavior is unavailable. Vetstergaard (1980) has studied dust bathing in hens using deprivation or control of early experience. Dellmeier et al. (1985) studied calves in veal crates as a model for deprivation of general movement. Both studies concluded that motivation for these behaviors increased during deprivation. Jensen (1993) examined the prefarrowing behaviors of sows and suggested that they could be divided into those that are triggered by the external environment and those that are primarily under internal control.

The issue of behavioral needs is largely a question of motivation and the consequences of thwarting a highly motivated behavior. As we have seen, part of the research in this area has emphasized the source of the motivation, either internal or external factors. Other research has addressed the response of animals if a particular behavior is prevented from occurring, in terms of increasing levels of motivation or performance of displaced behavior. Of critical importance is the measurement of motivation (Dawkins, 1983). This will be addressed in the next section of this paper.

Preference Tests

The Brambell Committee (Command Paper 2836, 1965) recommended that chicken wire, thin wire woven in a hexagonal pattern, not be used as flooring in cages for hens. They suggested that a thicker wire in a rectangular pattern would be more comfortable. This judgement was based on conjecture, rather than scientific evidence. This statement serves as an example of the problems of insufficient behavioral research. When a study was conducted, Hughes and Black (1973) concluded that chicken wire was probably more comfortable than the suggested alternative, because hens spent more time on the chicken wire than on other floors when given a choice among several. Hughes and Black (1973) used a preference test; this type of test has been widely used to study floors and lighting and thermal conditions to improve the comfort of animals.

The traditional preference test has allowed the animal to choose between two conditions, both of which are conducive to the same behavior. For example, animals have access to two floors, to determine which is preferred for lying on. Dawkins (1983) introduced two new concepts to the use of preference tests in welfare related research. The first is to examine motivation, as opposed to comfort. In such a test the animal is given access to two situations that are suitable for different behaviors. By giving hens access to food in one choice and dust in the other, the test determines which motivation, for feeding and dustbathing, is greater (Dawkins, 1983).

The second concept introduced by Dawkins is the measurement of the strength of a motivation (Dawkins, 1983). Using consumer demand theory, Dawkins argued that by increasing the cost of obtaining access to a condition in which a behavior was possible, we could measure the strength of the relevant motivation. Increasing the cost has often been accomplished by increasing the reinforcement ratio in operant conditioning studies. Matthews and Ladewig (1987) reported that pigs demonstrate a preference to be with other pigs and that they will work to obtain access to them. However, as the reinforcement ratio increased, pigs quickly ceased to work. The motivation to be with other pigs exists, but it is not very strong. Conversely, Hutson (1991) reported that sows fed typical gestation diets will continue to work for food at very high reinforcement ratios. The sows were highly motivated to eat, suggesting that their freedom from hunger was not being met.

Problem Behaviors and Indicators of Poor Welfare

Agriculture is a goal-oriented industry and we are encouraged to move our research quickly to a definitive stage. In the case of animal welfare, there is considerable interest in assessment, in determining whether welfare is good or poor. In terms of ethology, the interest has been in abnormal or deleterious behaviors that indicate that welfare is poor. Some examples of these behaviors are aggression, injurious behaviors such as feather pecking or tail biting, and stereotypies. However, assessment does not improve the welfare of animals unless it is accompanied by a determination and elimination of the causes. Unfortunately, many behaviors that are considered indicative of poor welfare have multiple causes, and the determination of causation has been the focus of much ethological research.

Aggression is one aspect of social behavior. By increasing our understanding of the social behavior of our domestic species, we might be able to reduce aggression through management. Studies such as those of McBride and James (1964), Meese and Ewbank (1973), and McGlone (1985) are examples of this approach involving pigs. Craig (1992) has studied aggression in hens in a similar manner. Other injurious behavior, such as feather pecking, tail biting, and the buller syndrome, have also been studied as social behaviors. All these behaviors may be indicative of other problems in the animals' environment. Frustration and discomfort may result in increased aggression. Lack of an enriching substrate, such as straw in the environment, may result in greater tailbiting (Fraser et al., 1991) or feather pecking (Blokhuis and Arkes, 1984). Applied ethologists working with companion animals recognize that aggression problems in dogs fall into several categories and must be diagnosed and treated appropriately (Borchelt and Voith, 1982). A similar diagnostic approach is needed in animal welfare research.

Stereotypies have been the subject of much applied ethological research. Cronin and Wiepkema (1984) suggested that stereotypies were the result of retraint and began as attempts to escape. Boredom or lack of environmental enrichment have been considered possible causes. Duncan (1970) demonstrated that frustration could lead to stereotypies in chickens. Appleby and Lawrence (1987) reported that low feeding levels, typical for gestating sows, are a cause of stereotypies. Most recently, Lawrence and Terlouw (1993) developed a model for the development of stereotypies in sows that includes hunger (internal feeding motivation), external feeding cues, general arousal, and a barren environment as contributing factors.

Emotional States and Cognitive Abilities

The question of animal cognition is controversial. Emotions and cognitive abilities are said by some to be outside the realm of scientific investigation and the result of anthropomorphism. The Brambell Committee (Command Paper 2836, 1965) stated that feelings (emotional states) must be considered when discussing animal welfare, but also indicated that the feelings of animals are likely to differ from those of humans. Nevertheless, the committee maintained that animals can experience such emotions as "rage, fear, apprehension, frustration and pleasure." It has recently been argued that not only should animal feelings be included in considerations about welfare, but that welfare is entirely a question of the animal's mental, psychological, and cognitive needs (Duncan and Petherick, 1991).

Studying the mental experiences of animals is not easy, but it can be accomplished by careful design. Bateson (1991) describes not only the importance, but also the methodology, of studying pain in animals. Pain associated with standard agricultural practices such as beak trimming (Duncan et al., 1989) and castration (McGlone et al., 1993) has been evaluated. We have also developed methods to study fear (Jones and Faure, 1981) and frustration (Duncan, 1970) in poultry. We have often attributed normal behavior to boredom in our animals, but the study of boredom itself is in its infancy (Wemelsfelder, 1991). In terms of congitive abilities, Harrison (1964) suspected that chickens were able to anticipate death. It is appropriate that anticipation of future events by chickens, even if only related to feeding, is now the subject of research (Petherick and Waddington, 1991).

The Future of Ethology in Animal Welfare

Although a considerable knowledge base on farm animal behavior existed before the modern welfare movement began, much of the information was obtained from studies on nutrition and reproduction (Hafez, 1962). Few agricultural institutions had researchers trained or specializing in ethology. The International Society for Applied Ethology now has over 350 members, most of whom are interested in the welfare of agricultural animals. Duncan (1993) identified approximately 65 applied ethologists working with agricultural animals at institutions in the United States, Canada, the United Kingdom, Denmark, and The Netherlands. Again, most of these would be involved in welfare-related research. On a global basis, the number of applied ethologists is now sufficient to make significant contributions to improving animal welfare. However, the effort directed at resolving welfare problems remains disproportionately low, even among the well-developed agricultural countries listed above (Duncan, 1993).

In addition to having adequate expertise and personnel, countries must also include ethology in their long-term research plans if their efforts are to be effective. The FAWC continues to recommend ethological studies to the British government (FAWC, 1993). In Canada, an expert committee of the Ministry of Agriculture recommends that welfare and ethological research be maintained or expanded (Expert Committee, 1993). The recent Food Animal Integrated Research conference in the United States recommended animal well-being and behavior as an important part of future agricultural research with animals (FAIR, 1993). Assuming these recommendations are accepted and appropriate funding is provided, applied ethology should be maintained or expand in these countries.

The direction of welfare-related research in applied ethology in the future will include many of the approaches used in the past 30 yr. However, some change in emphasis may occur. It is my opinion that too much emphasis has been placed on using behavior as a means of assessing welfare, and too little on the behavioral basis of welfare problems (Gonyou, 1993). I predict that research on the use of abnormal behaviors as indicators of welfare will receive less emphasis in the future, and that ethology will be seen as a means of answering questions about the animals' welfare requirements. Thus, preference testing to determine comfort and motivation will increase. Design and development of management systems will be increasingly based on ethological issues. The emphasis on animal cognition in recent theoretical papers (Duncan and Dawkins, 1983; Dawkins, 1990; Duncan and Petherick, 1991) will likely lead to greater research being directed toward determining the cognitive abilities of our animals and the degree to which they experience pleasant and unpleasant emotional states.

Implicatons

Applied ethology will continue to play a major role in animal welfare research. Countries in Europe, North America, and Australasia will lead in this field as their governments recognize and fund appropriate research. As research shifts to address the basis of animal welfare, as opposed to its measurement, a greater degree of expertise will be required to answer the difficult questions arising. To achieve and maintain competence in applied ethology, agricultural and veterinary colleges will need to expand teaching of applied ethology at both the undergraduate and graduate levels. As students who have taken such courses enter industry and academia, the general understanding of animal welfare issues and the role of ethology in addressing these issues will increase.

---

[1]Invited paper on animal well-being presented at the ASAS 85th Annu. Mtg., Spokane, WA.

Received September 27, 1983.

Accepted March 18, 1994.

Literature Cited

Appleby, M. C., and A. B. Lawrence. 1987. Food restriction as a cause of stereotypic behaviour in tethered gilts. Anim. Prod. 45: 103.

Banks, E. M. 1982. Behavioral research to answer questions about animal welfare. J. Anim. Sci. 54:434.

Bateson, P. 1991. Assessment of pain in animals. Anim. Behav. 42: 827.

Blokhuis, H. J., and J. G. Arkes, 1984. Some observations on the development of feather-pecking in poultry. Appl. Anim. Behav. Sci. 12:145.

Borchelt, P. L., and V. L. Voith. 1982. Classification of animal behavior problems. Vet. Clin. North Am. Small Anim. Prac. 12:571.

Command Paper 2836. 1965. Report of the Technical Committee to Enquire Into the Welfare of Animals Kept Under Intensive Livestock Husbandry Systems. Her Majesty's Stationery Office, London.

Craig, J. V. 1992. Measuring social behavior in poultry. Poult. Sci. 71:650.

Cronin, G. M., and P. R. Wiepkema. 1984. An analysis of stereotyped behaviour in tethered sows. Ann. Rech. Vet. 15:263.

Dawkins, M. 1983. Battery hens name their price: Consumer demand theory and the measurement of ethological 'needs'. Anim. Behav. 31:1195.
Dawkins, M. 1988. Behavioural deprivation: A central problem in animal welfare. Appl. Anim. Behav. Sci. 20:209.
Dawkins, M. S. 1990. From an animal's point of view: Motivation, fitness and animal welfare. Behav. Brain Sci. 13:1.
Dellmeier, G. R. 1989. Motivation in relation to the welfare of enclosed livestock. Appl. Anim. Behav. Sci. 22:129.
Dellmeier, G. R., T. H. Friend, and E. E. Gbur. 1985. Comparison of four methods of calf confinement. II. Behavior. J. Anim. Sci. 60:1102.
Duncan, I.J.H. 1970. Frustration in the fowl. In: B. M. Freeman and R. F. Gordon (Ed.) Aspects of Poultry Behaviour. pp 15–31. British Poultry Science, Edinburgh, U.K.
Duncan, I.J.H. 1993. The science of animal well-being. Anim. Welfare Info. Center Newsletter 4:1.
Duncan, I.J.H., and M. S. Dawkins. 1983. The problem of assessing "well-being" and "suffering" in farm animals. In: D. Smidt (Ed.) Indicators Relevant to Farm Animal Welfare. pp 13–24. Martinus Nijhoff, The Hague.
Duncan, I.J.H., S. S. Gillian, E. Seawright, and J. Breward. 1989. Behavioural consequences of partial beak amputation (beak trimming) in poultry. Br. Poul. Sci. 30:479.
Duncan, I.J.H., and J. C. Petherick, 1991. The implications of cognitive processes for animal welfare. J. Anim. Sci. 69:5017.
Ewbank, R. 1988. Animal welfare. In: Management and Welfare of Farm Animals: The UFAW Handbook (3rd Ed.). pp. 1–12. Ballière Tindall, London.
Expert Committee. 1993. Farm Animal Welfare and Behavior in Canada. Report of the Expert Committee on Farm Animal Welfare and Behaviour, Canada. Agriculture Canada, Agassiz, B.C.
FAIR. 1983. Enhance animal well-being throughout the life cycle of food-producing animals. Food Animal Integrated Research. Meeting in St. Louis, October, 1992. Association Headquarters, Champaign, IL.
FAWC. 1993. Second Report on Priorities for Research and Development in Farm Animal Welfare. Farm Animal Welfare Council. MAFF Tolworth, U.K.
Fraser, A. F. 1980. Ethology, welfare and preventive medicine for livestock. Appl. Anim. Ethol. 6:103.
Fraser, D., P. A. Phillips, B. K. Thompson and T. Tennessen, 1991. Effect of straw on the behaviour of growing pigs. Appl. Anim. Behav. Sci. 30:307.
Gonyou, H. W. 1993. Animal welfare: Definitions and assessment. J. Agric. Ethics 6(Spec. Ed.):37.
Graves, H. B. 1984. Behavior and ecology of wild and feral swine (Sus scrofa). J. Anim. Sci. 58:482.
Hafez, E.S.E. 1962. The Behaviour of Domestic Animals. Ballière Tindall, London.
Hafez, E.S.E. 1969. The Behaviour of Domestic Animals (2nd Ed.). Ballière Tindall, London.
Harrison, R. 1964. Animal Machines—The New Factory Farming Industry. Vincent Stuart Publishers Ltd., London.
Högsved, O. 1990. Current status and future prospects for group housing systems in Sweden. Proc. of a Seminar on Group Housing of Sows, November 7–8, Brussels. European Conference Group on the Protection of Farm Animals, Horsham, U.K.
Hughes, B. O. 1980. The assessment of behavioural needs. In: R. Moss (Ed.) The Laying Hen and Its Environment, pp 149–159. Martinus Nijhoff Publishers, Boston.
Hughes, B. O. and A. J. Black. 1973. The preference of domestic hens for different types of battery cage floor. Br. Poult. Sci. 14:615.
Hutson, G. D. 1991. A note on hunger in the pig: Sows on restricted rations will sustain an energy deficit to gain additional food. Anim. Prod. 52:233.
Jensen, P. 1986. Observations on the maternal behaviour of free-ranging domestic pigs. Appl. Anim. Behav. Sci. 16:131.
Jensen, P. 1991. Back to nature: The use of studying the ethology of free-ranging domestic animals. In: M. C. Appleby, R. I.

Horrell, J. C. Petherick, and S. M. Rutter (Ed.) Applied Ethology: Past, Present and Future, Proc. of the Int. Congress, Edinburgh. 1991. pp 62–64. Universities Federation for Animal Welfare, Hertfordshire, U.K.

Jensen, P. 1993. Nest building in domestic sows: The role of external stimuli. Anim. Behav. 45:351.

Jones, R. B. and J. M. Faure. 1981. The effects of regular handling on fear responses in the domestic chick. Behav. Processes 6:135.

Kilgour, R. 1978. The application of animal behavior and the humane care of farm animals. J. Anim. Sci. 46:1478.

Lawrence, A. B., and E.M.C. Terlouw, 1993. A review of behavioral factors involved in the development and continued performance of stereotypic behaviors in pigs. J. Anim. Sci. 71:2815.

Matthews, L. R., and J. Ladewig. 1987. Stimulus requirements of housed pigs assessed by behavioural demand functions. Appl. Anim. Behav. Sci. 17:369 (Abstr.)

McBride, G. 1969. Fitting farms to fowls, Proc. 16th Annu. Poult. Convention, Massey Univ., Palmerston North, p. 1. [Cited by Kilgour (1978).]

McBride, G., and J. W. James. 1964. Social behaviour of domestic animals IV. Growing pigs. Anim. Prod. 6:129.

McGlone, J. J. 1985. A quantitative ethogram of aggressive and submissive behaviors in recently regrouped pigs. J. Anim. Sci. 61:559.

McGlone, J. J., R. I. Nicholson, J. M. Hellman, and D. N. Herzog, 1993. The development of pain in young pigs associated with castration and attempts to prevent castration-induced behavioral changes. J. Anim. Sci. 71:1441.

Meese, G. B., and R. Ewbank. 1973. The establishment and nature of the dominance hierarchy in the domesticated pig. Anim. Behav. 21:326.

Morris, J. R. and J. F. Hurnik, 1990. An alternative housing system for sows. Can. J. Anim. Sci. 70:957.

Petherick, J. C., and I.J.H. Duncan. 1991. Society for Veterinary Ethology 1966–1991, 25th anniversary review. In: M. C. Appleby, R. I. Horrell, J. C. Petherick, and S. M. Rutter (Ed.) Applied Ethology: Past, Present and Future. Proc. of the Int. Congress, Edinburgh, 1991. pp 11–16. Universities Federation for Animal Welfare, Hertfordshire, U.K.

Petherick, J. C., and D. Waddington. 1991. Can domestic fowl (*Gallus gallus domesticus*) anticipate a period of food deprivation. Appl. Anim. Behav. Sci. 32:219.

Stolba, A. 1981. A family system in enriched pens as a novel method of pig housing. In: Alternatives to Intensive Husbandry Systems. Proc. Symp. at Wye College, July 13–15, University of London, Ashford, Kent. pp 52–67. The Universities of Federation for Animal Welfare, Hertfordshire, U.K.

Stolba, A., and D.G.M. Wood-Gush, 1989. The behaviour of pigs in a semi-natural environment. Anim. Prod. 48:419.

Stookey, J. M. 1992. Animal welfare in Canada-the next 20 years. Paper presented to the Alberta Feed Industry Conf., September 23–24, Lethbridge, AB. (In press).

Taylor, I. A. 1990. Design of the sow feeder: A systems approach. Ph.D. Dissertation. University of Illinois at Urbana-Champaign.

Vestergaard, K. 1980. The regulation of dustbathing and other behaviour patterns in the laying hen: A Lorenzian approach. In: R. Moss (Ed.) The Laying Hen and Its Environment. pp 101–113. Martinus Nijhoff Publishers, Boston, MA.

Webster, A.J.F. 1993. The challenge of animal welfare. Proc. VII World Conf. on Animal Production, June 28–July 2, 1993, Edmonton, AB, Canada.

Wemelsfelder, F. 1991. Animal boredom: Do animals miss being alert and active? In: M. C. Appleby, R. I. Horrell, J. C. Petherick, and S. M. Rutter (Ed.) Applied Animal Behavior and Animal Welfare Ethology: Past, Present and Future, Proc. of the Int. Congress, Edinburgh. 1991. pp 120–124. Universities Federation for Animal Welfare, Hertfordshire, U.K.

Wood-Gush, D.G.M. 1973. Animal welfare in modern agriculture. Br. Vet. J. 129:167.

The Use of Animals in Medical Research Are There Alternatives?

by
California Biomedical Research Association

According to the Director of the National Institutes of Health virtually every medical advance in this century can be attributed to the use of animals in research. But in the current wave of agitation opposing animal research, simplistic assertions are frequently made about "alternatives" that could replace animals if only researchers would take advantage of them. Computer and mathematical models and tissue and cell cultures are most commonly said to be the methods which could eliminate the need for animal studies.

To be sure, modern medicine owes its success to a wide variety of research methods and strategies. Scientists explore the complexities of biology through some of the most interesting and sometimes surprising avenues—both with animal and non-animal models.

The animals most commonly used by researchers are rats and mice bred specifically for laboratory purposes. But the National Science Foundation recently pointed out that researchers are making significant discoveries with an interesting variety of animals. One study of electric eels, for example, contributed to the understanding of a neuromuscular disease, myasthenia gravis, and to improvements in its treatments. The same panel noted the potential for treatment of nerve cell damage from strokes or injury through studies of certain types of song birds which forget their songs at the end of the mating season and relearn them the following spring through a process that involves a loss of nerve cells and the formation of new ones. These examples point out the complex and interrelated biology we share with other animals and the vast potential for understanding human illness they offer.

And there are biomedical scientists working with non-animal methods which are fascinating as well. Several computer programs in developmental stages can flag some possibly dangerous substances by comparing their characteristics to those of compounds known to be toxic. Such "pattern recognition" programs, however, stand on the shoulders of years of animal research which provide the data necessary to develop the programs in the first place. And devising computer programs to detect agents of cancer or birth defects is proving much more difficult. Furthermore, computer models promise only to reduce the number of animals used in testing—not eliminate them altogether. Other suggested "alternatives" such as cell and tissue cultures have been adjuncts of biomedical research for many years. They are neither shunned nor underestimated by researchers. They are natural outgrowths of the scientific process.

A celebrated and successful method for detecting the possible cancer-causing properties of compounds, the Ames test, is but one example of a useful cell culture technique. By observing a substance after it is introduced into a bacteria culture, it is possible to determine with accuracy whether or not the substance is a mutagen. From animal research, we know that mutagens frequently cause cancer, so the Ames test is a widely used first stage screening procedure for drug and chemical testing. It cannot, however, actually determine whether or not something will cause cancer—thus the need for tests on animals. Again, such cell culture techniques have reduced the number of animals necessary for research, not eliminated them.

Tissue cultures are also widely used by researchers, for example, to observe the effects of drugs and chemicals on tissue. Recent studies of the effect of radiolabelled antibodies on human breast tumor tissue samples indicate promising possibilities for a new therapy. No responsible researcher would suggest, however, that such tissue culture testing can substitute for trials on complete, functioning living animal systems—in this case to assure that the time it takes for the radioactive antibodies to reach cancerous tissue in a living system does not produce damage to other healthy tissue in the process.

Will research models which reduce the number of animals needed for experimentation continue to be developed and used by researchers? If recent history is any guide, it seems most likely. Between 1968 and 1978, the U.S. Department of Agriculture reported a 40% decrease in the number of animals used for scientific research in this country. That reduction is no doubt due in part to the development of adjuncts such as tissue and cell cultures and computer modelling which allow a more efficient use of animals, particularly in toxicity testing.

But, are these non-animal methodologies really "alternatives" to animal research? The complexity of entire living systems—the interrelations of the nervous systems, blood and brain chemistry, gland and organ secretions, immunologic responses—makes it impossible to either explore, explain or predict the course of many diseases or the effects of many treatments without observing and testing entire living systems. At some point, of course, human beings become test subjects. But unless we are prepared to volunteer for highly experimental surgery, to suffer a significant reduction in the safety and reliability of drugs or the thousands of new chemicals being developed each year, or to seriously limit the possibility of soon finding a cure or treatment for AIDS and Alzheimer's disease, to name but two current medical mysteries, the continued use of animals in research simply is essential for the foreseeable future.

The Importance of Animals in Biomedical and Behavioral Research

(U.S. Public Health Service Coordinating Committee on Animal Research)

Virtually every medical achievement of the last century has depended directly or indirectly on research with animals. The knowledge gained from animal research has extended human life and made it healthier through many significant achievements, as illustrated by the following examples: vaccines to prevent poliomyelitis and other communicable diseases; surgical procedures to replace disease heart valves; corneal transplants to restore normal vision; new medicines to control high blood pressure and reduce death from stroke; broad spectrum antibiotics to treat infections; and chemical agents to cure or slow childhood cancers. Of course, there are many other diseases and disorders, such as AIDS, many forms of cancer, common cold, Alzheimer's disease, schizophrenia, hepatitis, arthritis, systic fibroses, and brain and spinal cord injures—just to name a few—for which no effective prevention, treatment, or cure now exists.

The use of living animals remains an important way to solve a medical problem. Researchers continually seek other models to understand the human organism, study disease processes, and test new therapies. In seeking more rapid and less expensive ways to obtain basic biological information that can be applied to human disease, scientists often study simpler organisms, such as bacteria, yeasts, roundworm, fruit flies, squids, and fishes. Researchers have spent decades learning how to sustain cells, tissues, and organs from both animals and humans outside the body to understand biological processes and develop new medical treatments. Mathematical, computer, and physical models, complement animal experimentation as well. Although computers alone cannot produce new biological information, they enable scientists to analyze vast amounts of data and test ideas. In the end, the validity of the results obtained from these model systems must be verified in appropriate animal systems and, possibly as the final step, in clinical trials using human volunteers.

Like most people, scientists are concerned about animal well-being. Elaborate safeguards in the form of Federal laws have been implemented to ensure that institutions comply with the regulations and policies affecting the care and use of animals in research. Before beginning a project, all research proposals involving animals must be carefully reviewed and approved at each research facility by an Institutional Animal Care and Use Committee comprised of scientists, veterinarians, and private citizens. Veterinarians trained in laboratory animal medicine are responsible for observing and caring for animals, providing guidance to researchers, and overseeing institutional animal care programs. In addition, institutions conducting animal research are routinely inspected by the U.S. Department of Agriculture and monitored by the U.S. Public Health Service. Many institutions are further accredited by an independent evaluating body, the American Association for Accreditation of Laboratory Animal Care.

For more than a century, there have been organized groups and individuals who have objected to using animals in biomedical research. This opposition has increased markedly in the last two decades. Animal activists organizations spurred by a philosophy that there is no moral justification for the use of animals in research—even to save human lives—have attempted to slow or halt the work of scientists. Some disseminate misleading information, intimidate or harass individual scientists, conduct mass demonstrations, or even commit acts of vandalism or terrorism. The few health professionals who support the activist movement stand apart from the vast majority of the Nation's physicians, and most Americans readily accept the fact that animal research is necessary to achieve medical progress.

Institutions receiving support from the Public Health Service are obliged to adhere to the highest possible standards for the humane care and responsible use of laboratory animals. And scientists themselves have adopted the principle: "Good Animal Care and Good Science Go Hand in Hand."

March 1994

Protecting Laboratory Animals

A Statement from the United States Public Health Service

As a result of a recent lawsuit brought by two animal protectionist organizations, a Federal court ordered the U.S. Department of Agriculture (USDA) to reconsider its exclusion of rats, mice, and birds from coverage under the Animal Welfare Act. In the judge's opinion, "the USDA's decision not to regulate these species sent a message that researchers may subject these animals to cruel and inhumane conditions."

People who are familiar with the extensive system of U.S. laws, regulations, guidelines, and principles that protect the welfare of laboratory animals would not necessarily agree with the judge's comment. **The Public Health Service (PHS) wants to reassure the American people that other laws exist to safeguard the welfare of rats, mice, and birds, species that comprise about 90% of research animals.**

According to the **Health Research Extension Act**, over 1,000 institutions receiving funds from the PHS to conduct animal experiments are required to comply with the provisions of the Act and to follow the recommendations in the Guide for the Humane Care and Use of Laboratory Animals (Guide). The Guide was prepared to assist researchers in maintaining high quality care for all commonly-used laboratory animals. It includes the Government principles for animal care and use adopted by all agencies and institutions that conduct federally-supported animal research. This guide also applies under another Federal law, the **Good Laboratory Practices Act**. Research laboratories that conduct studies using rats and mice are regulated by the PHS's Food and Drug Administration and are subject to inspections.

In addition, most institutions that do not receive PHS funding follow the Guide. For example, laboratory animal breeders, pharmaceutical manufacturers, and commercial research laboratories that may not be subject to USDA and PHS regulations voluntarily participate in a national program of certification by the American Association for Accreditation of Laboratory Animal Care. This private organization monitors institutional animal care programs to be sure they maintain the standards set forth in the Guide.

Animal use is an integral component of biomedical and behavioral research and testing. The vast majority of scientists recognize that good science and good animal care go hand-in-hand and would not tolerate or condone cruelty to, or inhumane treatment of, any laboratory animal.

December 1992

Public Health and the Role of Animal Testing

A Statement from the United States Public Health Service

This statement has been prepared to inform the general public about the need for animal testing to ensure that medications, vaccines, environmental chemicals, and a wide variety of consumer products, including cosmetics, are safe for the public when used appropriately. The Public Health Service (PHS) is concerned that animal activist organizations are trying to convince the public incorrectly that product testing in animals is outdated and no longer necessary.

Consumers may be further confused by announcements that some companies have stopped testing their products in laboratory animals. For example, two ways in which a company can make such a claim are by using only ingredients that historically are known to be safe or that have been previously tested in animals and found to be non-toxic. When new ingredients need to meet testing and safety requirements, it is often necessary to test them in one or more animal species.

To protect the public from unexpected or unintended effects of toxic substances, some PHS agencies conduct and support toxicological testing to determine the harmful effects of commonly used products. To judge whether a product may be unhealthy, or even deadly, for humans and animals, scientists called toxicologists must know how the substance is absorbed, distributed, used, stored, and released by the body. For some products, it may be necessary to identify long-term, cumulative health effects, such as the potential to cause cancer, promote birth defects, affect reproduction, or harm the nervous system. Without laboratory animals, scientists would lose a fundamental method for obtaining the data needed to make wise decisions about potential health risks.

The PHS agencies support many initiatives to develop and validate systems to reduce dependency on animal testing. Scientists have become skilled in culturing a wide variety of tissue and organ cells outside the living body (in vitro) and in writing computer programs that simulate human and animal systems.

Human and animal cell cultures are being used increasingly to screen toxic substances before progressing to whole-animal testing. When in vitro studies show that a substance is toxic, testing it in animals may not be necessary. Computer models are also being used to help predict the properties of substances and their probable actions in living systems. Although computers can store and analyze enormous amounts of data, some information must come from experimental animals. These non-animal research tools have reduced our dependence on animals, but they cannot completely replace experimental animals for the foreseeable future.

Toxicologists have the responsibility to treat laboratory animals with great care and compassion. Today, all projects involving animal testing supported by funds from the PHS must comply with the regulations of the Animal Welfare Act, as amended, and the Health Research Extension Act. These laws were enacted to protect research animals. An institution that uses laboratory animals for any purpose must operate a sound animal care program. The PHS fosters quality control in animal care and has a high regard for the welfare of laboratory animals.

The American people want assurance that the products they use in recovery from illness and daily living are safe; the U.S. Congress has enacted laws that require the safety of products; and the scientific community endeavors to promote the public health through animal testing. Dr. James O. Mason, Assistant Secretary for Health, has put it this way: "Whole animals are essential in research and testing because they best reflect the dynamic interactions between the various cells, tissues, and organs comprising the human body."

The number of products used by society has increased greatly since animal testing began, but adverse health effects are relatively uncommon. This is, in itself, compelling evidence for the predictive value of animal testing of products for human use.

December 1992

iiFARsighted Report
lab animals save lives!
incurably ill For Animal Research®
November 1988 Vol. 2 No. 4

iiFAR
P.O. Box 1873
Bridgeview, IL 60455
(312) 598-7787

Animals in Diabetes Research

If you know 20 people, chances are one of them is among the 11 million Americans suffering from diabetes. In 1989, an additional 600,000 will develop the disease. Because diabetes is so prevalent, almost everyone is familiar with it. Or are they?

Most people are really unaware of just how serious a disease diabetes is. For instance, did you know diabetes is the third leading cause of death by disease? Or that it is the leading cause of new blindness for people under 65? Or that a person with diabetes has twice the chance of a heart attack or stroke? Or that kidney disease is 17 times more common among diabetics?

And the dollar cost of diabetes is enormous—**$20.4 billion** in 1987. It accounted for more than 11.5 million days of hospital care costing $6.9 billion. Another $1.7 billion was spent for doctor visits and medications, and $942 million for nursing-home care. $10.8 billion was attributed to lost productivity.

What is diabetes?

Diabetes is a disease where the body fails to produce or effectively use insulin. Diabetes in children, if not treated, can kill in a short time; in older patients, it may go undetected for years. Even with treatment, complications (such as blindness, kidney failure, loss of limbs, heart disease, or stroke) almost inevitably occurs.

What kinds of diabetes are there?

Diabetes occurs in two forms—Insulin-Dependent Diabetes Mellitus (IDDM) and Non-Insulin-Dependent Diabetes Mellitus (NIDDM). Insulin-dependent diabetics require regular injections of insulin to keep their blood sugar levels within acceptable limits. IDDM is usually diagnosed at an early age, and is often referred to as juvenile onset diabetes. IDDM's major characteristic is a loss of the beta cells that secrete insulin from the islet of Langerhans located in the pancreas. IDDM appears most often in people with a genetic tendency towards diabetes and who then acquire a virus infection. The infection sets in motion a chain of events resulting in a death of beta cells. Once the cells die, these people become insulin-dependent.

Nearly **one million** IDDM patients live in this country, but if insulin were not available, they would not be insulin-dependent, they would be dead.

NIIDM is the most prevalent form of diabetes, with approximately **ten million** NIIDM patients in America, many undiagnosed. A non-insulin-dependent diabetic is able to maintain proper blood sugar levels through special diets. Many non-insulin-dependent diabetics are able to live their entire lives with dietary control alone—others eventually become insulin-dependent.

Are humans the only animals to get diabetes?

In addition to humans, diabetes also naturally occurs in rats, cats, dogs (Michigan State University has a colony of diabetic dogs) and several primate species (the Oregon Regional

Primate Research Centers has several insulin-dependent *Macac nigra* monkeys). For this reason, these animals are used to study diabetes, and the new treatments developed through this research benefit those animals as well as man.

What has animal research done to help diabetics?

The major turning point for treatment of diabetes came in 1921 when Dr. Frederick Banting and Charles Best developed a means of extracting insulin from the pancreases of dogs (**pound dogs**, by the way). In a short time, insulin from cows and pigs was being produced in large quantities and saving the lives of thousands of diabetics.

Since then, many important developments have come about through animal research. Information about dietary management of diabetes has increased due to controlled animal experiments. Research and safety testing on dogs made possible external insulin pumps, now becoming available for diabetics having difficulty regulating their blood sugar levels through conventional injections. The first human trials of internal insulin pumps (PIMS), are underway at Johns Hopkins University, following several years of carefully evaluation and refinement of dogs.

What hope does animal research hold for the future?

Perhaps the most exciting phase of diabetics research is that which is currently underway into curing the disease. Researchers are pursuing two promising courses to reach this goal—gene therapy and islet cell transplantation—and animals are vital to both of these pursuits.

What about transplantation?

In recent years, surgeons have successfully performed several transplants of entire pancreases. However, all organ transplants carry with them the drawback of organ rejection. For most people with diabetes, the side effects of the required anti-rejection drugs are worse than the symptoms and complication of diabetes itself. However, if a diabetic has kidney failure and the only hope of survival is a kidney transplant, a simultaneous pancreas transplant is often considered. iiFAR has a member in California who was an insulin-dependent diabetic since childhood and suffered kidney failure. She had a combination kidney/pancreas transplant, and is now completely cured of her diabetes.

The most promising of transplant techniques is that of transplanting only the insulin producing islet cells. This has proven successful in inbred mice, and is now being perfected in random source dogs (pound dogs). Islet cell transplantation is not a major surgery, and the problem of rejection, while still the major problem left to overcome, is not nearly as great as with an entire organ.

What is gene therapy?

The really high-tech approach to curing diabetes is through gene therapy. Scientists have been able to isolate the human gene that produces insulin. When this gene is injected into a single-cell mouse embryo and the embryo re-implanted into the mouse, the resulting offspring produces human insulin and the insulin production is regulated in the same way as in humans. This type of gene therapy is called germ-line, and is the easiest to accomplish.

Gene therapy research for diabetes has now progressed to the more difficult technique, known as somatic cell therapy. Genes put into somatic cells only affect the animal—or person—treated. By contrast, genes put into germ-line cells (such as embryos) would be inherited by the offspring. For ethical reasons, most scientists believe germ-line gene therapy is inappropriate for humans.

Mice are currently being used in somatic cell gene therapy. Researchers have successed in injecting human insulin-producing genes into mice with artificially induced diabetes, and the human gene causes insulin production. Too much, in fact. Before this process can progress upwards to dogs then humans, more work with rodents is needed to understand the complex ways in which the insulin gene is normally regulated.

Animals are imporant to diabetes research!

Animal research is critical to the understanding of diabetes. If research on diabetes is stopped because laboratory animals are not available, the additional cost in human suffering and the financial burden to society would be enormous. Both will be dramatically reduced when diabetes can be prevented and/or cured, hopefully in the not-too-distant future.

From the Medical Community

It was a crisp evening in the fall of 1935. I was at the Municipal Opera House when an emergency call came for me from the Michael Reese Hospital. By the time I reached the entrance an ambulance was waiting for me, and in a few minutes I was at the hospital getting a Drinker respirator (the iron lung) ready for service. By the time the patient arrived the respiratory was ready—none too soon.

The patient was a very beautiful girl of about twelve years of age. Quite unconscious, her lips a deep blue with a bluish cast to her skin. Placed in the Drinker respirator the bluish color faded from her skin and she began a conversation with her father as though she had awakened from a deep sleep.

Poliomyelitis was all too prevalent in the 1930's. We didn't know much about the dreaded malady. We weren't sure about the cause, but there were those who thought that it was caused by a virus. Laboratories around the world were busy, and all types of non-primate animals were inoculated in order to attempt to find an animal in which the organism (if it was an organism) would grow so that it could be studied. Dogs, cats, sheep, goats, horses, cattle, rabbits, laboratory rats and mice, ferrets, and all sorts of wild animals were inoculated. For some time the results were uniformly negative, and those who were opposed to the use of animals in research had a field day. Finally a wild rodent from the swamps of Florida, *Sigmodon hispidus*, was found to be susceptible to the virus, giving undisputed proof that "polio" was caused by a virus, and enabling virologists to study the organism. A vaccine was finally made available.

Society owes a great debt to Endens, Sauk, and others who developed our ability to reduce the suffering brought on by polio, but we also owe a great debt to the Florida cotton rat, *Sigmodon hispidus*. Poliomyelitis is no longer the terrible menace it was in the 1930's when many parents became discomposed over the threat of "polio" to their children, when not even the president of our country was safe from its attack.

The story of the conquest of polio is one more appealing reason for the continued support for the use of animals in research.

N.R. Brewer, D.V.M., Ph.D., D.Sc. (Hon)
Department of Animal Care
University of Chicago

iiFARsighted Report
lab animals save lives!
incurably ill For Animal Research®
Winter 1990 Vol. 4 No. 1

iiFAR
P.O. Box 1873
Bridgeview, IL 60455
(312) 598-7787
FAX: (312) 598-7814

Cats and Medical Research

Cats are now the most popular pet in America. It is estimated that there are more than 34 million cats inhabiting 24% of the households in America, and another 15 million plus who are homeless.

In the past 30 years, the health of the house-dwelling cat has been improved to such a degree that their life expectancy has increased by 6 to 8 years, allowing many to reach 20 years of age. This more than 60% increase is comparable to the increase in human life expectancy from 45 in 1900 to more that 70 years of age today. This increase in longevity for both cats and humans, has been brought about by many of the same reasons: better nutrition, new and improved diagnostic procedures, surgical techniques, drug therapies, and vaccines. All of these are the result of medical research—research which often relied on animals, a few of which were cats.

While less than 1/2% of the animals used in medical research are cats (90% are rodents), they have made tremendous contributions to medical progress. Only a few of these advances are presented in this newsletter.

Cats, like humans, are susceptible to loss of hearing from exposure to high-volume sounds and have been used in testing the effectiveness of hearing protectors. Humans and cats also share in an unpleasant condition known as tinnitus, a constant ringing of the ear. Studies in cats have known this disorder to be related to changes in middle ear pressure and possible therapies are under investigation.

Cancer is also quite prevalent in cats, especially mammary cancers. These tumors more closely resemble human breast cancer than do those of either the mouse or the dog. Therefore, cats have been used as a model to develop and test new forms of therapy for this disease which affects 1 woman in 11.

Of the many diseases cats and humans share, probably one of the best known is leukemia. Feline leukemia is the most common fatal disease of domestic cats today. Through research on cats, it is now known that this disease is caused by a retrovirus and vaccine is now available to protect our pets from its wrath. Scientists have also developed ways to test cats for feline leukemia in its early stages, before symptoms are present. Researchers are building upon this work to develop early warning test for human leukemia. Also, the information about retroviruses learned from these cat studies gave researchers a head start on understanding AIDS, which is also caused by a member of the retrovirus family.

Diabetes occurs naturally in 1 out of every 800 cats, and is usually accompanied by deposits of amyloid in the pancreatic islets that secrete insulin. This is significant because it only happens with regularity in cats and humans. Therefore, cats are important to this aspect of diabetes research.

Lupus is another disease that cats and humans share. More than 16,000 people, primarily women of child-bearing age, develop lupus each year. Research with cats has shown that genetic factors and perhaps a virus are possible causes for Lupus.

The pathways between the brain and the sensory organs (eyes, ears, and nose) are remarkably similar in cats and humans. This has allowed basic researchers to learn much about how the brain receives and understands these signals. This knowledge, in turn, has allowed other researchers to develop treatments for several neurological disorders, some of which are covered in more detail later in this newsletter, including one Nobel Prize winning project!

The debate over the use of animals in medical research is growing more ferocious everyday. Recently this uproar has centered around the use of cats in neurological research, specifically that dealing with sleep and respiration. Research involving these creatures has been called ghastly, horrifying and unjustifiable by animal rights supporters. Most of us are familiar with the stolen or allegedly staged photographs presented by animal activists. These photos are used in an attempt to convince an uninformed public, through emotionalism, that research involving cats is unwarranted.

Every year in this country 8,000 infants die from Sudden Infant Death Syndrome (SIDS). This condition, which results from an unexplained loss of respiration during sleep, has lead to insurmountable grief for parents around the world. Combined with this staggering statistic are the untold numbers of individuals suffering from disorders such as sleep apnea, insomnia and narcolepsy. It is estimated that the most common clinical complaint reported to physicians is, in one way or another, sleep related.

Sleep is a naturally occurring phenomenon that for most is rejuvenating, refreshing, and for all of us, essential. While our understanding of non-wakefulness has increased over the last twenty years, thousands of questions still remain. How do we fall asleep? Is there a specific chemical which initiates sleep? Why do

The cat's contribution to the clinical treatment of sleep abnormalities is well documented. For example, every year individuals (mostly men) from around the country are treated for a sleep disorder which often results in severe physical injury. Normally, the body is completely paralyzed during Rapid Eye Movement (REM) sleep, the phase of sleep in which we dream. This keeps us, as well as <u>all</u> mammals, from acting out our dreams. However, some individuals lose this paralysis. When this happens, normal, docile men can become extremely violent. Some have severely injured themselves after leaping into nightstands or door frames during dream sleep. Others have been awakened by their bed partner after violent or even potentially murderous acts. Until recently, the causes of this disorder, now known as REM Sleep Behavior Disorder, were misunderstood and these men were believed to suffer from an unknown psychiatric illness.

Dr. Adrian Morrison, a world renowned sleep researcher at the University of Pennsylvania, discovered REM sleep without paralysis during basic research on "REM-less" sleep in cats. He learned that the paralysis of REM sleep could be eliminated by a small amount of damage in a particular part of the lower brain called the pons. It was this research that lead Drs. Schenk and Mahowald, physicians who treat patients at the Minnesota Regional Sleep Disorders Center, to the discovery of REM Sleep Behavior Disorder in humans. Furthermore, these animals provided a model to test possible drug treatments. Dr. Mahowald recently stated, "had it not been for research conducted on cats, we would have had no way of understanding this disorder." As a result, patients are now able to take medication which **completely** relieves them of this dangerous and annoying disorder.

There are many more examples of the tremendous contribution which cats have played in advancing our understanding of human and animal health. While animal rights and antivivisection forces use stolen, falsified, and misrepresentative information in an attempt to further their emotionalist cause, it is obvious that we all owe a great deal to research that was, and which continues to be, conducted with cats.

people with respiratory problems suffer more serious problems during sleep? What is the physiological basis for narcolepsy? As well as a multitude of yet unthought-of questions which will, within their answer, hold the secret to treating these more devastating disorders.

Not unlike the vast reaches of space, the mammalian brain is an incredibly uncharted and unknown segment of our universe. Consisting of literally billions of individual cells the size of those found in the blood, the brain is truly a testament to the complexity of life. Adding to this complexity is the fact that each cell can receive and then transmit information to hundreds or even thousands of other cells. Sleep, as with many other body functions, is directly controlled by this maze of circuitry, thus requiring extensive neurological examinations in order to understand both normal and abnormal sleep patterns.

If we are to understand the specific mechanics of this cell-to-cell communication which enables areas of the brain to assemble information, recordings of cell activity must be taken. Different cells within the brain do different things. For example, when an electrode, which is attached to amplification equipment, is placed next to the cells which are responsible for respiration, a noise similar to a machine gun can be heard and recorded every time a breath is taken. Because of this, researchers are able to insert the hair thin probes into the brain (which is incapable of sensing pain) and determine which cells are firing, thus giving them some insight into which cells are, and are not, functioning during sleep. This type of research is vital in understanding other body functions including respiration, vision, balance, and thousands of other unconscious acts. Insertion of these recording rods takes place under sterile conditions no different than those found in our local hospitals.

A device known as a head cap is often attached to the cat's skull with dental cement in order to insure that the recording electrodes remain in place. Researchers have been able to record electrical transmissions from a single cell for as long as 12 hours because of this cap. Equally important is the fact that this device protects against infection and trauma while at the same time allowing the animal to move freely about when not being used in a recording session. In fact, the cats lead a very normal and playful life, completely oblivious to the lightweight and painless head cap. Thanks to the techniques developed by researchers, the process of determining which cells are where and what simulus they respond to (brain mapping), several questions have been answered. However, this work must continue, because until we understand exactly what areas, cells and chemicals within the brain are involved in initiating acts such as sleep, we will simply have no way of finding a cure for these disorders.

Cats have long been used in our search for the answers to the mysteries of sleep. Along with humans, researchers have learned that cats experience both Rapid Eye Movement (REM) sleep and non-REM sleep, a very important factor in studying sleep and sleep related disorders. Since felines are similar in size across almost all breeds, and since the cat's known neurological make-up so closely parallels that of humans, scientists have been able to do extensive mapping and recording of brain activity on otherwise unwanted shelter animals who were doomed to euthanasia. Furthermore, cats are good sleepers. Their many "cat naps" during the day provide scientists with the opportunity to study the process of naturally falling asleep several times in just one day. A great deal of what we know today about the brains involvement in sleep has been obtained through intuitive animal researchers seeking answers to their questions.

As with any field of science, neurological examinations require very controlled situations. Cats used in sleep research must meet very rigid standards before they can help in answering a researcher's questions. Among these are that the animals be healthy. Many types of diseases or disorders directly interfere with the collection of accurate data. The animals must also be free from pain. While this is an issue which animal activists often misrepresent, the truth of sleep research is; if the animals are in pain, they don't undergo normal sleep and therefore scientists are unable to collect the information they require.

Basic research involving cats is vital in order to understand the yet unanswered questions about sleep and respiration. While activists argue otherwise, there is no question that humane treatment and accurate data go hand in hand when studying the areas of sleep and respiration. And while many animal rights supporters criticize this type of work, the fact remains that 8,000 infants will die this year in their sleep, the helpless victims of SIDS. In the words of David Hubel, the Nobel Prize winning neurobiologist, "There are people who think you can do the research on computers. That's just sheer nonsense. We understand quite well how computers work; it's the brain we don't understand."

As a young man, David Hubel was always involved in some sort of science. As the son of a chemist, he began to dabble in his father's trade at a very young age. By his college years, however, he had decided to pursue physics and chose it as his major while attending McGill University. After graduation, he became interested in medicine and enrolled in medical school at McGill, even though he had never even taken a biology course. In 1955, Hubel found himself part of the "doctor's draft" and was sent to Walter Reed Army hospital where he began to do research.

After three years at Walter Reed, Hubel moved to Johns Hopkins where he met Torsten Wiesel. Together, these two basic researchers would answer hundreds of questions regarding how the visual system within the brain functions. Their work was so revolutionary that they earned the Nobel Prize in medicine and physiology in 1981. Their work in understanding the mechanisms of vision has laid the groundwork for intuitive researchers around the world.

Thanks to the cats which Hubel and Wiesel used in their studies, these two pioneers in the field of visual science were able to discover the pathways that send information from the eye to the brain. In ensuing studies they uncovered that when vision was impaired at a young age in one eye, the ability to see out of that eye was lost because the pathways within the brain did not develop normally. Hubel and Wiesel's work lead subsequent researchers to ask some very important questions.

In the years following these Nobel Laureates' ground-breaking work, basic researchers began to study exactly how the phenomenon of brain-eye interconnection occurs and why the brain develops abnormally in some patients. Also important is the fact that this basic research continues to give physicians a better understanding of an eye disorder known as ambloypoia, or lazy eye, which affects some ten million Americans and which previously was not understood.

It is now known that patients with ambloyopia often suffer from some type of obstruction in their vision at a young age, a condition which can be simulated by closing one eye of young cats. As with the cat, this obstruction in humans causes malformation of the pathways leading to the brain, and thus, varying degrees of vision are permanently lost in the obstructed eye. This discovery has encouraged pediatric ophthalmologist to pay special attention to the vision of infants and to correct any problems early in their development in order to avoid irreversible vision loss.

Prior to Hubel and Wiesel's discoveries, disorders such as congenital cataracts, for example, were not corrected until an infant matured. Today, because of their work, cataract surgery on infants is routine. Due to this revolutionary change in ophthalmology, thousands of children are spared from a world of irreversible darkness.

Equally as important is the ongoing research, building upon the work of Hubel and Wiesel, which holds the answers to treating individuals who were not fortunate enough to have their vision disorder corrected as an infant. Their work is also critical in understanding strabismus or "cross eye" which is the most common cause of human blindness.

In response to an attack aimed at a fellow researcher who conducts work of this type, and which an animal rights group dubbed the "most cruel, worthless animal research of the year" in 1988, Dr. Hubel is quoted as saying: "As to the assertion that such work is 'useless', I should add that in 1981 the Nobel Committee in Stockholm awarded the Prize for Medicine and Physiology to Dr. Torsten Wiesel and me, for just this type of work. To attack work aimed at preventing blindness in children as 'useless' seems wildly irresponsible."

In a candid conversation with Doug Stewart of Omni Magazine, Dr. David Hubel shares his thoughts on his early basic research, his friend and co-worker Torsten Wiesel, and the impact of the animal rights movement. Portions of that interview, which appears in its entirety in the February 1990 edition of Omni, are presented here:

Omni: ***What first got you interested in this question [how do we see]?***

Hubel: My original aim was to study sleep using microelectrodes. ...I first observed it *[Rapid Eye Movement, or REM]* in cats, just when the first papers on REM sleep were published. I was astonished. My electrodes happened to be put in the part of the cat's cortex that deals with vision. I decided to use a visual stimulus to see if the responses differed in sleeping and waking animals. They did, but gradually I became more interested in these responses of the visual cortex and less and less in changes occurring during sleep.

Omni: ***How did your collaboration with Wiesel begin?***

Hubel: The lab was being remodeled and things were at a standstill. So Steve Kuffler suggested I come to his lab and work with Torsten. We decided to anesthetize our animals and to our delight we could still make the cells fire despite the anesthesia.

Omni: ***You could get the brain cells to fire by shining a light on an animal's eyes while it was out cold?***

Hubel: Yes, because the anesthetic turned out to have no effect on the first dozen or so stages in the visual pathway. We pushed the electrode forward into the *[cats]* brain tissue a fraction of millimeter and listened until we came upon a cell... Using the microelectrode to record from the cortex, we suddenly had a very powerful tool.

Omni: ***Have animal rights protesters caused problems for you?***

Hubel: It's a big problem now. Some of the best people in the field have been targeted. There are certain animal rights people you're not going to persuade any more than you could have persuaded the late Ayatollah he shouldn't try to kill Rushdie. But there are a huge number still uncommitted to whom arguments in defense of animal experiments must be made—or we won't be doing research. People no longer hesitate to send their kids to summer camp because they fear they'll get polio. That's because of medical research. ...But many people prefer to ignore what fantastic things medical research has accomplished. In our lab we set out to understand the visual system, but along the way it occurred to us that we might get at the cause of blindness that follows strabismus, or cross-eyedness... Our animal studies suggested that if nonparallel eyes were operated on early in a person's life, the blindness could usually be avoided. So our work had a real impact on ophthalmology.

If you want to get at problems like that, the only way to do it is animal research. There are people who think you can do research on computers. That's just sheer nonsense. We understand quite well how computers work; it's the brain we don't understand. You can't design a new heart valve or something like that without using animals.

Educators For Responsible Science
Connecticut Chapter
10 Bay Street • Suite 63 • Westport • CT 06880 • (203) 222-7933

Policy Statement

Dissection of Animals in the Classroom

The dissection of animals has a long and well established place in the teaching of life sciencies. Well constructed dissection activities, led by thoughtful instructors can illustrate important and enduring principles in biology [3]. Learning theory scholars recognize that active participation by students is preferable to passive reception of the content to be learned [1, 2]. Educators for Responsible Science (Connecticut Chapter) believes that dissection of animals is a valuable method of giving students a motivating, active biological experience in the study of anatomy and physiology of organisms. When dissection is used in the classroom:

- Teachers should thoroughly explain the learning objectives of the lesson and utilize additional worksheet/audio-visual materials to maximize the educational benefits of the experience
- All specimens should be treated with respect
- All students will be informed, prior to dissection, that they have the option of discussing individual concerns about dissection with the appropriate teacher/administrator. Decisions regarding these concerns will be made on an individual basis by the appropriate teacher/administrator
- Upon completion of dissection, the remains should be appropriately disposed of as recommended by the local department of public health

References

1. ***Code of Practice on Use of Animals in Schools.*** Washington, DC: National Science Teachers Association, 1985
2. Doll, R. ***Curriculum Improvement: Decision Making and Process.*** Boston: Allyn & Bacon, Inc., 1986
3. ***Policy Statement — Animals in Biology Classrooms.*** Reston, Virginia: National Association of Biology Teachers, 1989

Rev. June 1992

Policy on Humane Care and Use of Laboratory Animals

The American Association for Laboratory Animal Science (AALAS) endorses the United States Government **Principles for the Utilization and Care of Vertebrate Animals Used in Testing, Research, and Training**, and requires that all papers published in our journal *Contemporary Topics in Laboratory Animal Science* report research conducted in comformance with these principles. Research for papers submitted from outside the United States must be in conformance with the guidelines of that country's government. The Editor reserves the right to reject papers reporting results of research not adhering to these principles.

Principles for the Utilization and Care of Vertebrate Animals Used in Testing, Research, and Training

The development of knowledge necessary for the improvement of the health and well-being of humans as well as other animals requires *in vivo* experimentation with a wide variety of animal species. Whenever U.S. Government agencies develop requirements for testing, research, or training procedures involving the use of vertebrate animals, the following principles shall be considered; and whenever these agencies actually perform or sponsor such procedures, the responsible institutional official shall ensure that these principles are adhered to:

1. The transportation, care, and use of animals should be in accordance with the Animal Welfare Act (7 U.S.C. 2131 et seq.) and other applicable federal laws, guidelines and policies.
2. Procedures involving animals should be designed and performed with due consideration of their relevance to human or animal health, the advancement of knowledge, or the good of society.
3. The animals selected for a procedure should be of an appropriate species and quality and the minimum number required to obtain valid results. Methods such as mathematical models, computer simulation, and *in vitro* biological systems should be considered.
4. Proper use of animals, including the avoidance or minimization of discomfort, distress, and pain when consistent with sound scientific practices, is imperative. Unless the contrary is established, investigators should consider that procedures that cause pain or distress in human beings may cause pain or distress in other animals.
5. Procedures with animals that may cause more than momentary or slight pain or distress should be performed with appropriate sedation, analgesia, or anesthesia. Surgical or other painful procedures should not be performed on unanesthetized animals paralyzed by chemical agents.
6. Animals that would otherwise suffer severe or chronic pain or distress that cannot be relieved should be painlessly killed at the end of the procedure or, if appropriate, during the procedure.
7. The living conditions of animals should be appropriate for their species and contribute to their health and comfort. Normally, the

housing, feeding, and care of all animals used for biomedical purposes must be directed by a veterinarian or other scientist trained and experienced in the proper care, handling, and use of the species being maintained or studied. In any case, veterinary care shall be provided as indicated.

8. Investigators and other personnel shall be appropriately qualified and experienced for conducting procedures on living animals. Adequate arrangements shall be made for their in-service training, including the proper and humane care and use of laboratory animals.

9. Where exceptions are required in relation to the provisions of these principles, the decisions should not rest with the investigators directly concerned but should be made, with due regard to principle II, by an appropriate review group such as an institutional animal research committee. Such exceptions should not be made solely for the purposes of teaching or demonstration.

For guidance throughout these Principles please refer to the *Guide for the Care and Use of Laboratory Animals* prepared by the Institute of Laboratory Animal Resources, National Research Council.

Index